Anion-Binding Catalysis

Anion-Binding Catalysis

Edited by Olga García Mancheño

WILEY-VCH

Editor

Prof. Olga García Mancheño
Westfälische Wilhelms-Univ. Münster
Organic Chemistry Institute
Corrensstr. 36
49148 Münster
Germany

Cover Image: Image provided by Olga García Mancheño

Library of Congress Card No.: applied for

British Library Cataloguing-in-Publication Data
A catalogue record for this book is available from the British Library.

Bibliographic information published by the Deutsche Nationalbibliothek
The Deutsche Nationalbibliothek lists this publication in the Deutsche Nationalbibliografie; detailed bibliographic data are available on the Internet at <http://dnb.d-nb.de>.

Print ISBN: 978-3-527-34857-2
ePDF ISBN: 978-3-527-83064-0
ePub ISBN: 978-3-527-83065-7
oBook ISBN: 978-3-527-83066-4

Cover Design SCHULZ Grafik-Design, Fußgönheim, Germany
Typesetting Straive, Chennai, India
Printed and Binding CPI Group (UK) Ltd, Croydon, CR0 4YY

Printed on acid-free paper

C110167_171121

Contents

Preface

The purpose of this book is to introduce the reader to the field of anion-binding catalysis and provide an overview of how this activation approach can contribute to the advance of organic synthetic chemistry. Inspired by the crucial and ubiquitous anion-binding processes involving ionic species in nature, as well as further important chemical applications such as in sensing, analytical, or material sciences, anion-binding processes have recently found new directions in the research area of catalysis. As a result, this field has experienced a remarkable development in the past few years. Therefore, this textbook is aiming at assisting the reader in evaluating the power and potential of the field by covering the most important aspects and presenting the state of the art according to the type of anions and catalyst structures involved.

The field of anion-binding catalysis is covered from different perspectives, starting with an introductory Chapter 1 that illustrates the historical development, activation modus-operandi, and importance of anion-binding in chemical catalysis.

In Chapter 2, an introduction to the most commonly used analytic methods and some problematics on the determination of the anion-binding affinity of the catalysts is included, aiming at providing a solid platform to the new readers in the field for their practical implementation.

The next Chapters 3–5 describe the catalytic and synthetic applications, especially comprising asymmetric catalysis, of common NH-based hydrogen-donor catalysts such as (thio)ureas and squaramides involving different anions. While Chapter 3 deals with the more explored and broader use of halogens as counter-anions, Chapter 4 describes the use of acid-cocatalysis for achieving unprecedented activations and reactivities, and Chapter 5 covers other types of systems including cyanide, alkoxylates, and other oxygen-based anions.

In the following Chapters 6–8, the structure, properties, and catalytic applications of further NH- and OH-based H-donors (Chapter 6), C–H triazole/triazolium catalysts (Chapter 7), as well as onium salts (Chapter 8) are presented. This organization of the chapters by the type of catalyst instead of reaction type was made in order to provide the reader a better overview on the existing classes of effective catalysts in this area.

In Chapter 9, the combination of (thio)urea and squaramide-based anion binding with other types of organocatalysis is presented. In particular, amino-catalysis, involving enamine, dienamine, or iminium activation approaches, as well as nucleophilic cocatalysis, are discussed.

Anion-binding catalysis by halogen-, chalcogen, pnictogen, and tetrel bonding is then covered in Chapter 10. The idea behind this chapter is to present the application of secondary interactions based on the elements of the seventh, sixth, fifth, and fourth main group of the periodic table as a highly powerful alternative to hydrogen bonding for the binding and activation of ionic substrates.

Finally, Chapter 11 deals with the new tendencies in the field of anion-binding catalysis, which include the assistance and combination of anion binding with other types of activations such as photocatalysis and metal catalysis, as well as their extension to supramolecular catalytic systems and anion–π catalysis, hence allowing for new reactivities and broadening the scope of this field.

This book was not possible without the outstanding contributions of renowned experts in different areas of catalysis and anion-binding applications, for which I would like to truthfully thank all the authors for putting together their knowledge to present the key concepts, examples, and perspectives to this still young area of research. Moreover, I am grateful to the publishers of Wiley, especially to Dr. Elke Maase and Ms. Katherine Wong, for coordinating and assisting the project.

Münster, Germany *Olga García Mancheño*
06 June 2021

List of Abbreviations

A^3	three-component reaction
Ac	acetyl
ACDC	asymmetric-counterion-directed catalysis
Ar	aromatic
B^*	chiral Brønsted acid
$[(BArF_4)]^-$	tetrakis(3,5-bis(trifluoromethyl)phenyl)borate
BEMP	2-*tert*-butylimino-2-diethylamino-1,3-dimethylperhydro-1,3,2-diazaphosphorine
BIFOL	biphenyl-2,2′-bisfenchol
BINAM	1,1′-bi(2-naphthylamine)
BINOL	1,1′-bi-2-naphthol
Boc	*tert*-butyloxycarbonyl
Bu	butyl
Cat.	catalyst
Cbz	benzyloxycarbonyl
CD	circular dichroism
ChB	chalcogen bond; chalcogen bonding
conv.	conversion
CS	catalyst–substrate complex
CSA	camphorsulfonic acid
D	diffusion coefficient
DABCO	1,4-diazabicyclo[2.2.2]octane
DBN	1,8-diazabicyclo[5.4.0]undec-7-ene
D_T	dissociation energy for a given temperature
DCM	dichloromethane
de	*diastereomeric excess*
DED	dispersion energy donor
DFA	difluoroacetic acid
DFT	density functional theory
DHP	3,4-dihydo-2*H*-pyran
DIM	7,7′-diamido-2,2′-diindolylmethane
DIPEA	diisopropylethylamine (Hünig′s base)
DMAP	4-dimethylaminopyridine

DMSO	dimethyl sulfoxide
DOSY	diffusion ordered spectroscopy
DPAP	4-(*N*,*N*-dipropylamino)pyridine
DPZ	dicyanopyrazine
DTBP	di-*tert*-butyl phenol
d.r.	diastereomeric ratio
E	energy
ε	dielectric constant
E_{rot}	rotational energy
EDA	electron donor–acceptor complex
EDSA	ethane disulfonic acid
ee	enantiomeric excess
equiv	equivalent
e.r.	*enantiomeric ratio*
ESI	electrospray ionization
ESP	electrostatic potential
G	guest
H	host
ΔH	solvation enthalpy (kcal mol^{-1})
HB	hydrogen bond; hydrogen bonding
HBD	hydrogen bond donor
HG	host–guest complex
HOESY	heteronuclear Overhauser effect spectroscopy
HSAB	hard and soft acids and bases
HTS	high-throughput screening
i	iso
IBO	2,2-dimethyloxiran
IDP	imidodiphosphoric acid
IR	infrared
ITC	isothermal titration calorimetry
K_a	association or affinity constant
K_d	dissociation constant
KIE	kinetic isotope effect
k_{rel}	relative rate constant (relative to uncatalyzed reaction)
L^*	chiral ligand
LA	Lewis acid
LB	Lewis base
ln	natural logarithm
LUMO	lowest unoccupied molecular orbital
M	molarity ($mol{\cdot}L^{-1}$); metal
Me	methyl
MEP	molecular electrostatic potential
MD	molecular dynamics
MIM	mechanically interlocked molecule
MOP	2-methoxypropene; 2-methoxypropyl

MS	molecular sieve; mass spectrometry
MTBE	methyl *tert*-butyl ether
n	normal
NADPH	nicotinamide adenine dinucleotide phosphate
NBSA	nitrobenzenesulfonic acid
NCI	non-covalent interaction
NCS	*N*-chlorosuccinimide
NDI	naphthalenediimide
NHC	*N*-heterocyclic carbene
NMR	nuclear magnetic resonance
NOE	nuclear Overhauser effect
NOESY	nuclear Overhauser enhanced spectroscopy
NPSS	*N*-phenylselenyl succinimide
Nu	nucleophile
OMCOS	organometallic catalysis directed toward organic synthesis
OYE	old yellow enzyme
PC	photoredox catalyst
PDI	perylenediimide
PDPY	4-piperidinopyridine
Ph	phenyl
PnB	pnictogen-bond; pnictogen bonding
ppm	parts per million
PPY	4-(pyrrolidino)pyridine
Pr	propyl
PTC	phase-transfer catalysis; phase-transfer catalyst
pTSA	*p*-toluene sulfonic acid
Q_{ZZ}	quadrupole moment
quant	quantitative
R^*	chiral organic residue
rac	racemic
ROESY	rational frame nuclear Overhauser effect spectroscopy
ROP	ring-opening polymerization
rt	room temperature
s-factor	selectivity factor
t	tertiary
TBA	tetrabutylammonium
TBDMS	*tert*-butyldimethylsilyl
TBS	*tert*-butyldimethylsilyl
TeB	tetrel-bond; tetrel bonding
TES	triethylsilyl
Tf	triflate
TFAA	trifluoroacetic anhydride
TIPS	triisopropylsilyl
TMG	1,1,3,3-tetramethylguanidine
TMS	trimethylsilyl

THF	tetrahydrofuran
THP	2-tetrahydropyranyl
TOF	turnover frequency (h^{-1})
TON	turnover number
TPP^+	2,4,6-triphenylpyrylium
Troc	2,2,2-trichloroethoxycarbonyl
Ts	(para-toluenesulfonyl; tosyl)
TS	transition state
VANOL	3,3′-diphenyl-2,2′-bi-1-naphthol
X^*	chiral counterion
X^-	anion
XB	halogen bond; halogen bonding

1

From Anion Recognition to Organocatalytic Chemical Reactions

Friedemann Dressler and Peter R. Schreiner

Justus Liebig University, Institute of Organic Chemistry, Heinrich-Buff-Ring 17, 35392 Giessen, Germany

1.1 Introduction and Background

Non-covalent organocatalysis [1–14] utilizes small organic molecules for the activation of substrates through hydrogen bonding ("*partial protonation*") to neutral or negatively charged (*anion-binding*) electron lone pairs. This catalytic concept mimics nature where non-covalent interactions (NCIs) are ubiquitous and essential for many processes essential to life [15], e.g., through catalyzing many biochemical reactions [16–18]. NCIs empower enzymes with the high selectivity in substrate recognition, activation, and the stereocontrol in biotransformations [19]. The reason why nature uses these bonding types in catalytic systems such as ribonucleases, antibodies, and enzymes is that the binding (recognition) of the substrates is associated with small changes in Gibbs free energies [20]. Crystal structure analyses, spectroscopic investigations [21, 22], and computational studies [23, 24] reveal that enzymes typically do not contain strong Lewis acidic sites [25, 26]. Binding to electron-deficient metal ions, which would be the obvious alternative for catalyzing chemical reactions, would result in strong enthalpic binding [1]. Copper, iron, or zinc ions in enzymes are considered "soft" in the Pearson HSAB sense, and they are embedded in large and polarizable structures, e.g., in carbonic anhydrase [27]. The recognition processes and the details of enzyme catalysis are difficult to deconvolute because there are multiple interactions, and each of them are necessary for high selectivity and high reactivity [15, 20, 28–31]. The dominant parts of these interactions are hydrogen bonding [32–37] and anion binding [38, 39], but also, NCIs such as aromatic π–π stacking [40, 41], van der Waals [42, 43], and dipole–dipole interactions [44, 45] are included in the enzyme's active site. Scheme 1.1 depicts four examples of enzymes using either hydrogen bonding or anion binding for substrate activation and reaction initiation in their active sites. Epoxide hydrolase (**1**) activates an epoxide for a nucleophilic opening reaction *via* double hydrogen bonding [46–48] similar to haloalcohol dehalogenase (**2**) that activates the same species for reversible ring opening with chloride [49]. Formate dehydrogenase (**3**) promotes the oxidation of formate, which binds *via* the

Anion-Binding Catalysis, First Edition. Edited by Olga García Mancheño.

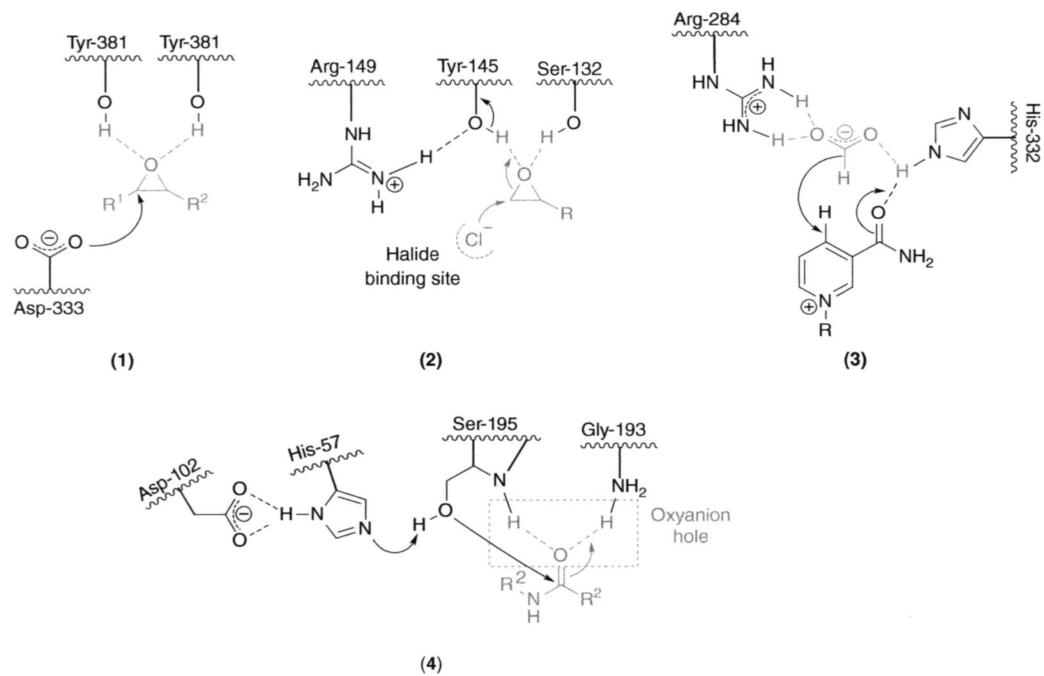

Scheme 1.1 Enzyme active sites utilizing either hydrogen bonding or anion binding for substrate recognition and activation: Epoxide hydrolase (**1**), halo alcohol dehydrogenase (**2**), format dehydrogenase (**3**), and serine protease (**4**).

guanidine function of an arginine residue [28, 50]. Serine protease (**4**) is an excellent example of multifunctional activation for the cleavage of amides. The protease binds the amide utilizing double hydrogen bonding, while at the same time, the imidazole moiety of a histidine moiety activates in concert with a carboxylic acid function of an aspartic acid motif, the hydroxy function of a serine, which cleaves the amide [51].

These examples show that (double) hydrogen bonding and anion binding in enzymes are very closely related. Catalysis by hydrogen-bonding enzymes shows no apparent product inhibition [52] but high turnover frequencies (TOFs), e.g., acetylcholinesterase with TOF = 25 000 s^{-1} [53], while operating in water and under aerobic conditions [54, 55]. These characteristics offer attractive perspectives for the development of metal-free artificial enzyme moieties [56], the so-called "*synzymes*" [57–61], which represent an alternative to traditional metal(-ion)-containing catalysts that are often toxic and sensitive to moisture and air [62]. *Synzymes* designed on the basis of Pauling's [63, 64] and related paradigms [26, 65] have the goal of imitating the complete enzyme architecture with its structurally complex protein backbone and active site(s). This design strategy leads to reaction mechanisms that follow general acid catalysis [44, 66–68] and Michaelis–Menten kinetics [69, 70] comparable to natural enzymes. The structural design of *small-molecule catalysts* such as (non)-covalent organocatalysts, however, builds on the idea that only the active site of an enzyme will be mimicked, resulting in small organic molecules with a minimum of complexity [28, 71] such as *explicit double hydrogen-bonding* and *anion-binding catalysts*. Scheme 1.2 depicts the reduction of nitro-olefins utilizing either enzyme catalysis (**5**) or organocatalysis (**6**). Enzymes from baker's yeast or the Old Yellow Enzyme (OYE) [72], which was termed a nitroalkene reductase, efficiently catalyze the NADPH-linked olefin reduction through activation of the nitroalkene by hydrogen bonding to His-191 and Asn-194. The hydride transfer to the β-position occurs from reduced flavin generated with NADPH, followed by proton transfer from the Tyr-196 hydroxyl to the α-position [73–75]. Zhang and Schreiner adapted this complex enzymatic arrangement to a simple organocatalytic system and utilized a thiourea moiety for nitro-olefin activation through hydrogen bonding and a Hantzsch ester as an NADPH analog.

The long success story of anion-binding chemistry using hydrogen bonding apparently started in 1940 when Bartlett and Dauben observed an increased acid

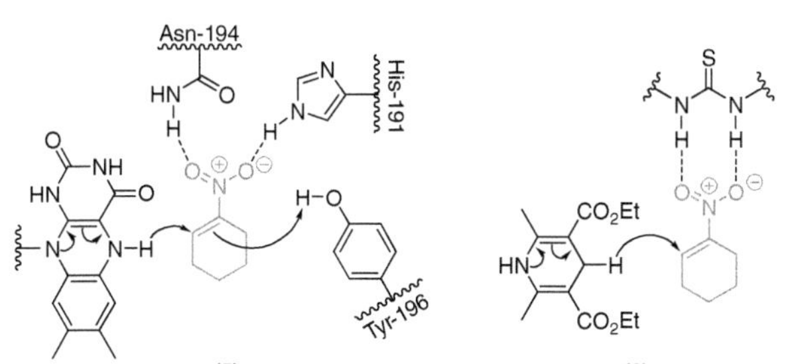

(**5**) (**6**)

Scheme 1.2 Proposed models for the reduction of nitroalkenes utilizing enzyme catalysis (**5**) and the adapted thiourea-catalyzed biomimetic reduction (**6**).

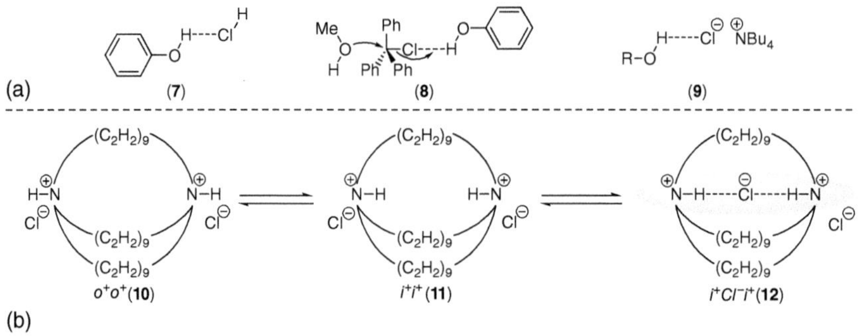

Scheme 1.3 (a) First examples of anion binding through hydrogen bonding and (b) isomerization of the first cage compound diazabicyclo[9.9.9]nonacosane dihydrochloride and the corresponding cage structure (**12**).

strength of hydrogen chloride in dioxane by the addition of alcohols [76]. They interpreted this by a hydrogen bond of the hydroxy function of the alcohols to the chlorine of hydrogen chloride (**7**) as depicted in Scheme 1.3 and found a correlation between the ability to form a hydrogen bond and the acid strength of phenol derivatives, except for ortho-substituted phenols that preferentially form intramolecular hydrogen bonds. In 1948, Swain investigated the kinetics of the S_N1 reaction of trityl chloride and alcohols in benzene with an excess of pyridine [77]. When methanol was used as the nucleophile, he found first-order kinetics for trityl chloride and second-order kinetics for methanol; the same result was obtained using phenol as the nucleophile, but the reaction was slower. When Swain used a 1 : 1 mixture of methanol and phenol as nucleophiles, the methyl ether formed seven times faster than the sum of the individual reaction rates, with first order in relation to trityl chloride, methanol, and phenol (**8**). Furthermore, only the methyl ether formed. He postulated a mechanism where one hydroxy group binds to the chloride *via* hydrogen bonding, thereby weakening the carbon–chlorine bond. Simultaneously, the second alcohol solvates the incipient carbocation. The reason why only methyl ether forms is the higher Brønsted acidity of phenol, causing a stronger hydrogen bond to chloride. These two examples derive from a strong intermolecular hydrogen bond to the *in situ* formed halide, even though the term "anion-binding" was not used. In 1958, Lund introduced the term "*hydrogen-bonding to an anion*" when he investigated the effect of quaternary ammonia halides on the O–H stretching frequency of various alcohols such as ethanol, phenol, and 2,2,2-trichloroethanol (**9**) [78]. He observed that Brønsted acidity and the strength of anion binding did not correlate and proposed to use Lewis acidity instead. Bufalini and Stern confirmed Lund's work and described the "*hydrogen bond formation to an anion*" as a solvation process [79]. In addition, they calculated the "*solvation enthalpy*" for the $n\mathrm{Bu_4Br}$–MeOH system as $\Delta H = -6.7\,\mathrm{kcal\,mol^{-1}}$ in good agreement with the accepted value of $\sim 5\,\mathrm{kcal\,mol^{-1}}$ for the formation of the hydrogen bond in the water dimer [80, 81]. The determined entropy change of $\Delta S = -14.0\,\mathrm{cal\,K^{-1}\,mol^{-1}}$ matches the reported value of $\Delta S = -13.6\,\mathrm{cal\,K^{-1}\,mol^{-1}}$ closely [81]. In 1963, Schleyer and Allerhand investigated covalent and ionic halides as proton acceptors in hydrogen-bonding complexes *via* IR spectroscopy [82]. They used butyl halides and $HgCl_2$ as covalent halides and observed small shifts of the O–H stretching frequencies of methanol and phenol as covalent halides are weak proton acceptors. However, quaternary ammonium halides are much better proton acceptors and lead to a larger hydrogen-bonding frequency shift. In addition, they observed that the magnitudes of these shifts significantly depend on extent on the halide anion, in the order $Cl^- > F^- > Br^- > I^-$; when ionic halides were used, for covalent halides, the order was $F^- < Cl^- < Br^- < I^-$, but in total much smaller. The variation of the cations made little difference for bromine and iodide but had a much bigger influence for chlorine. This effect was not investigated for fluoride because the synthesis of quaternary fluorides was laborious and only a few derivatives were described in the literature at that time. In 1968, Simmons and Park described a series of bridged macrobicyclic diamines **10** [83–85]. First, they investigated the *in–out*-isomerization of the bis-hydrochlorides and found that alkyl chains with even numbers of carbon atoms favor the *in–in-* and those with odd numbers the *out–out*-conformation. By treating diazabicyclo[9.9.9]nonacosane with HCl, the authors obtained crystals, which revealed the presence of only the *katapinate* (**12**) in the ^1H NMR (Scheme 1.3). A similar behavior was observed with HBr, which was less favorable than HCl; however, with iodide, no encapsulation in the cavity was observed. Simmons and Park also treated diazabicyclo[10.10.10]triacontane with NaCl in 50% aqueous trifluoroacetate and observed a high field shift of the α-CH_2 protons in the ^1H NMR, which indicated the formation of the in^+–Cl^-–in^+ conformer. Similar results were obtained using bromide and iodide salts. The authors concluded that the size of the cavity is an important factor in halide "*katapinosis*." A few years later, Marsh and coworkers confirmed the structure of the host–guest complex by X-ray single-crystal analysis [86].

The work of Simmons and Park arguably marks the beginning of the supramolecular chemistry of artificial organic host molecules with anions as guest molecules [87–90]. In the following years, the field of supramolecular anion recognition was applied to anion sensing [89, 91–94], extraction [95–97], anion transport [98–105], and self-assembling of molecular capsules [106–109]. Nevertheless, despite the large number of publications in the field of supramolecular chemistry, the selectivity of the host molecules for a given anion species was limited to a few examples. In the last years, especially "fluoride-only" receptors appeared, often containing boron because of its high affinity to fluoride [87, 110–112]. "Halide-only" hosts were also developed [113], but the differentiation between chloride and bromide is still a challenge [114]. The development of anion-binding (recognition) organocatalysts and their application in synthetic organic chemistry emerged from supramolecular chemistry and goes hand in hand with the introduction and improvement of explicit double hydrogen-bonding catalysts [1, 2]. Both catalysis types emerged at about the same time, and the mode of action is also similar: The catalyst coordinates and activates either a neutral substrate (hydrogen-bonding catalysis) or an anion substrate (anion-binding catalysis). This indicates that many hydrogen-bonding catalysts can also be used as anion-binding catalysts and vice versa, and so the anion-binding catalyst story cannot be told without the story of the double hydrogen-bonding catalyst

(Scheme 1.4). Hydrogen-bonding interactions have been known since the fundamental works of Errera and Mollet in 1936 [118], and, in particular, Wolf, Prahm, and Harms in 1937 [119]. Errera and Mollet measured the IR spectra of ethanol solutions in carbon tetrachloride (0.1 M) at different temperatures and observed a band at 3640 cm^{-1} (56% absorption) and a large band at 3350 cm^{-1} (94% absorption) at 0 °C [118]. At 70 °C, the authors observed the band at 3640 cm^{-1} with increased absorption (64%), whereas the band at 3350 cm^{-1} decreased to a low percentage. Errera and Mollet described the band at 3640 cm^{-1} as the O–H stretching vibration of "*isolated alcohol molecules*," and in contrast, the band at 3350 cm^{-1} "*to intermolecular actions, for example, the hydrogen atom of one molecule vibrating with the oxygen atom belonging to another one (hydrogen-bond, as interpreted by Sidgwick)* [120]." Wolf, Prahm, and Harms examined the mixing heat, density, dissociation, and solvation energy of various solvent mixtures of aliphatic alcohols (methanol, ethanol, and propanol), acetone, carbon tetrachloride, and hydrocarbons such as *n*-hexane and benzene. The authors described interactions between a ligand and a substrate and introduced the term supra- or supermolecule ("*Übermolekül*") [119].

Hine and coworkers investigated the double hydrogen bonding of 1,8-biphenylenediol **13** with hexamethylphosphoramide, 1,2,6-trimethyl-4-pyridone, and 2,6-dimethyl-γ-pyrone using cocrystals for X-ray crystal structure analysis (Scheme 1.4) [115]. They reported two simultaneous and identical strong hydrogen bonds of the almost coplanar adduct of 1,8-biphenylenediol **13** to the Lewis bases and reported the shortest intermolecular hydrogen bonds (2.548 Å) detected in an X-ray crystal structure analysis at that time. In 1985, the same group reported the aminolysis of phenyl glycidyl ether **24** with diethyl amine in butanone at 30 °C using phenol-based catalysts (Table 1.1) [116, 117]. In the absence of catalyst, they observed a small autocatalytic effect by the product and proposed that amino alcohol **25** catalyzed the reaction *via* hydrogen bonding to the epoxide oxygen. A Brønsted plot suggested a strong linear correlation between the catalysts pK_a values and the reaction rate, with two exceptions: first, catechol **30** catalyzed the reaction somewhat faster than expected from the Brønsted plot because hydrogen bonding can occur *via* two hydroxyl groups. This effect could easily be corrected by specifying the reaction rate per hydroxyl group of the catalyst. The second exception was 1,8-biphenylenediol, which was as efficient as would be expected for a phenol that was about 600 times as acidic, and catalyzed the reaction with a TOF of 0.1 h^{-1}. Catechol, whose hydroxyl groups did not form simultaneous double hydrogen activation to one substrate molecule, was less active, which indicates that both hydroxyl groups of **13** are involved in the epoxide activation. Further investigation of the ionization constants (pK_a values) using electron-rich and electron-poor 1,8-biphenylenediol derivatives indicated that the acidity of the hydroxyl functions correlates with the ability to double hydrogen bond formation. The most acidic 4,5-dinitro-1,8-biphenylenediol **14** (Scheme 1.4) was found to be the strongest double hydrogen bond donor [121–124].

In 1990, Kelly's group synthesized 3,6-dipropyl-4,5-dinitro-1,8-biphenylenediol **15**, which is more soluble in CD_2Cl_2, for the Diels–Alder reactions between various α,β-unsaturated aldehydes as well as ketones with cyclopentadiene at 55 °C and

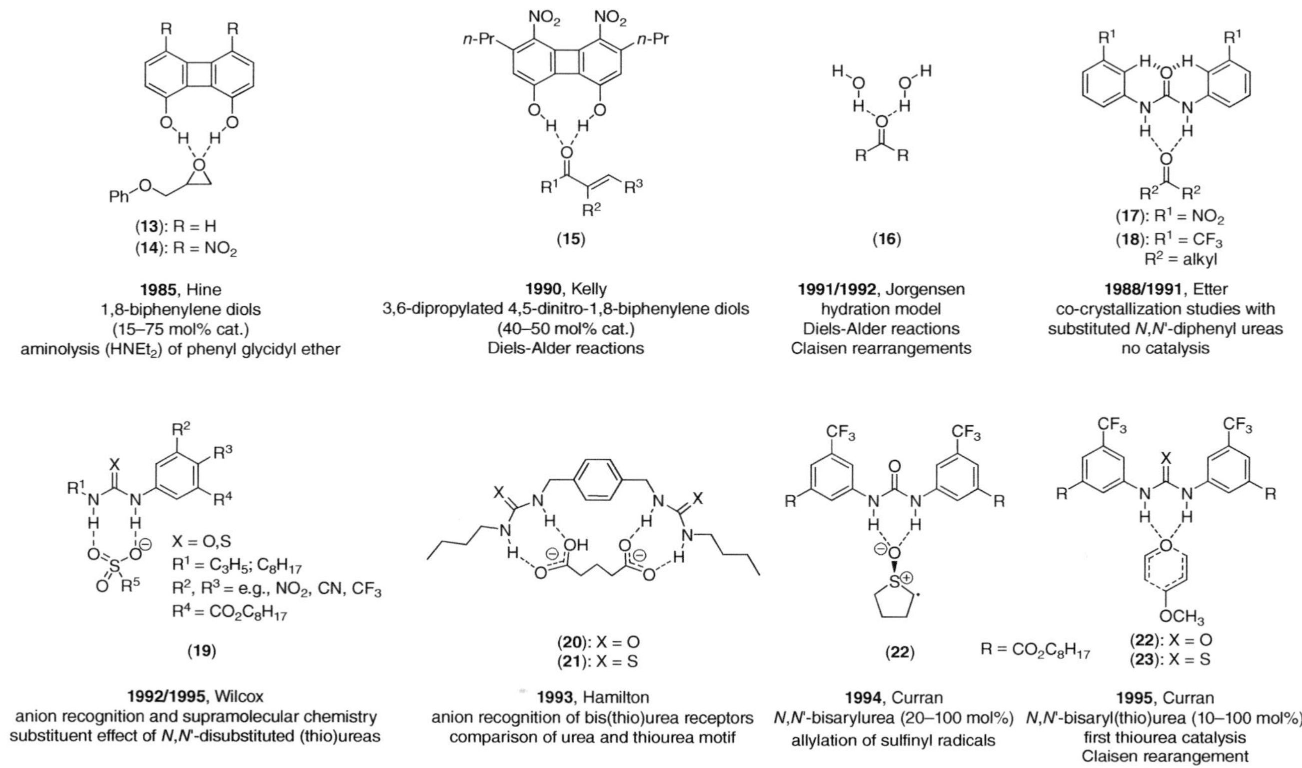

Scheme 1.4 Historic milestones on the way to hydrogen-bonding and anion-binding catalysts that display explicit double hydrogen bonding interactions for substrate recognition.

Table 1.1 Aminolysis of phenyl glycidyl ether **24** catalyzed by phenol-based derivatives. The reaction rates k_c are given relative to the uncatalyzed reaction.

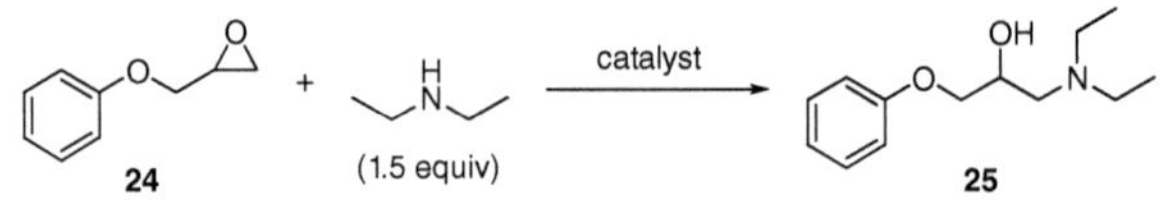

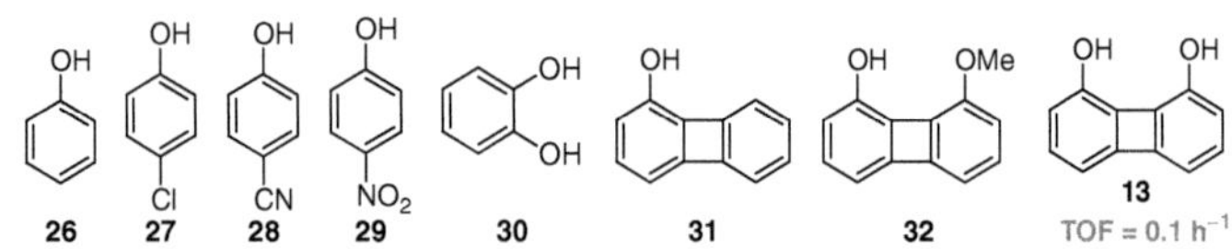

	26	27	28	29	30	31	32	13
$10^5 k_c/L^2\ mol^{-2}\ s^{-1}$	6.0	7.7	15.3	17.0	11.9	11.5	7.3	75
pK_a	9.99	9.14	7.97	7.15	9.36	8.64	9.15	8.01

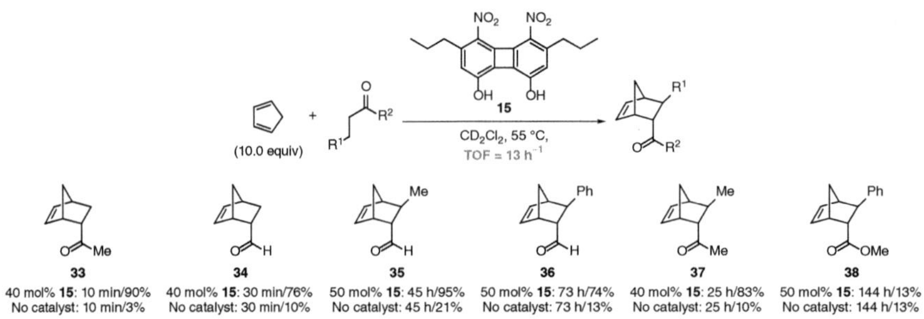

Scheme 1.5 Typical products of the 15-catalyzded Diels–Alder reaction of cyclopentadiene and α,β-unsaturated carbonyls.

40–50 mol% loading of **15** achieving a TOF of $13\,h^{-1}$ (Scheme 1.5) [125]. They did not observe significant rate acceleration in the case of ester dienophiles, e.g., product **39**, which they explained by inhibition of complex formation as a consequence of repulsive interactions. The double hydrogen bond activation of dienophiles resulting in a substrate–catalyst complex as depicted in Scheme 1.4 was interpreted as the reason for the observed up to 30-fold rate enhancements. Monte Carlo simulations confirmed the catalyst–substrate association motif and the accelerating solvent effect of water compared to aprotic solvents in Diels–Alder reactions [126, 127] and Claisen rearrangements [128], for which Jorgensen and coworkers computed a 100-fold rate enhancement. They explained these results not by cooperative hydrogen-bonding effects but rather with an aqueous solvent effect described through clamp-like hydrogen-bonding interactions (**16**) of two water molecules to the carbonyl group leading to transition state stabilization [128].

From 1988 to 1991, Etter *et al.* investigated the cocrystallization of imides and *N,N′*-diphenyl ureas with Lewis bases such as triphenylphosphine oxide [129], ethers, nitro aryl species, ketones, and sulfoxides by X-ray structural analysis [130, 131]. The N–H urea protons act as partial proton donors (Lewis acid) with the carbonyl oxygen as the proton acceptor (Lewis base), leading to intermolecular hydrogen-bonding interactions as observed in *N,N'*-bisphenylurea [132]. *N,N'*-bis(3-nitrophenyl) -urea **17** co-crystallizes with cyclohexanone in a 1 : 1 ratio through hydrogen bonding of the N–H protons and the carbonyl oxygen. Etter and coworkers found many receptors that are better hydrogen bond acceptors than the urea oxygen of **17** and termed this derivative an "*only intermolecular proton donor*" [130]. Urea derivatives with electron-withdrawing substituents in ortho- or para-positions or *meta*-alkyl substituents did not form cocrystals. In addition to **17**, *N,N'*-bis[(3-trifluoromethyl)phenyl]urea **18** cocrystallizes with carbonyl compounds, which was explained by the nearly planar urea conformation and short distances between the weakly acidic ortho protons and the carbonyl oxygen of 2.23–2.29 Å [130] that is below the limit for hydrogen-bonding interactions of about 2.4 Å [133, 134]. This distance is 2.49–2.66 Å in *N,N'*-bisphenylurea [132], therefore not leading to hydrogen-bonding interactions. Wilcox *et al.* examined the hydrogen-bonding recognition of sulfonates, phosphates, carboxylates, and zwitterionic 4-tributylammonium-1-butanesulfonate with *N*-allyl- [135] and *N*-octyl-*N'*-phenyl(thio)ureas **19** [136]. The host–guest complexes were investigated by ^{1}H NMR titration experiments *via* quantification of the downfield shifts of the N–H protons upon complexation. In addition, these studies confirmed the stronger double hydrogen bonding with anions, in particular, when electron-deficient substituents in meta- or para-position in the ureas were used, and that acidity can be considered a *useful parameter* of hydrogen-bonding strength [137]. In 1993, Hamilton's group examined the anion recognition of *N,N''*-[1,4-phenylenebis(methylene)]bis[*N'*-butyl]urea **20** and the thiourea analog **21** to glutarate. They found a 15-fold increase in the binding constants for the thiourea derivative compared with the corresponding urea in accord with Wilcox' results that thiourea binding is stronger than that of the corresponding urea derivatives [138]. In 1994, Curran and Kuo used 20–100 mol% of urea catalyst **22** as the first double hydrogen bonding urea organocatalyst (or additive in the case of equimolar amounts) for the allylation of α-sulfinyl radicals generated from **39** with allyltributylstannane (Scheme 1.6) [140]. Instead of the well-established NO_2 groups, electron-withdrawing CF_3 substituents were incorporated in meta position in combination with lipophilic ester substituents to improve the solubility in common organic solvents. In the presence of the catalyst, Curran and Kuo observed small rate accelerations (TOF = 0.1 h^{-1}) and increased *cis*/*trans* selectivity.

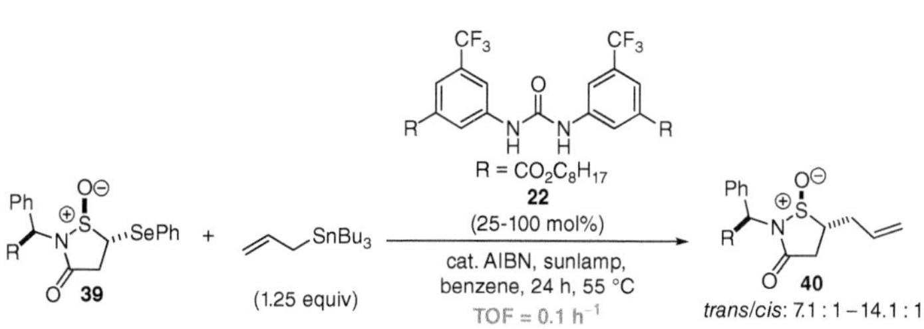

Scheme 1.6 Allylation of α-sulfinyl radicals catalyzed with **22**.

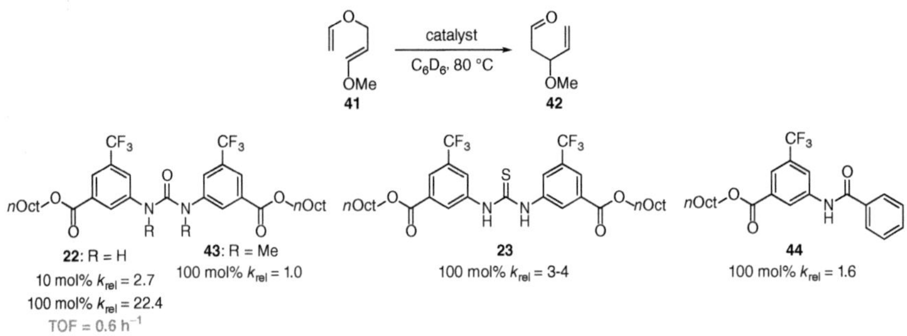

Scheme 1.7 Claisen rearrangement catalyzed by urea **22** and thiourea **23**.

The same group utilized urea **22** in the Claisen rearrangement of 1-methoxy-3-vinyloxy-propene **41**, where they observed with 10–50 mol% catalyst loading a rate acceleration enhancement by 1.7- to 5.0-fold; using 100 mol% gave 22-fold acceleration (Scheme 1.7) [139]. The authors probed the double hydrogen-bonding motif by using the *N,N′*-dimethyl derivative **43** (k_{rel} = 1.0), and a corresponding benzanilide **44** (k_{rel} = 1.6). These findings confirmed the work of Hine [115–117, 122] and Kelly [125] with **13** and **15** for epoxide and carbonyl activation, respectively, namely, that the double hydrogen-bonding motif is important for catalytic activity. Furthermore, Curran and Kuo examined thiourea analog **23** in the same reaction, and although slow decomposition was observed under the same reaction conditions, some rate acceleration was detected (at <10% decomposition: k_{rel} = 3–4) [139].

1.1.1 Evolution of Thiourea-Based Catalysts

Schreiner's group intensified the research on thiourea derivatives as organocatalysts starting in early 1997 (when this work was first presented, oddly at a conference geared toward organometallic chemistry ("OMCOS") in Göttingen where one of the authors was a local co-organizer) [141]. The idea was to bring hydrogen-bonding catalysis to the forefront of organic synthesis where it would be practical in the sense of ease of preparation of the catalysts, catalyst efficacy that was clearly rather limited at that time when one considers the high loadings (20–100 mol%), and the poor rate accelerations (k_{rel} < 5) and generality. One early consideration was that electron-deficient thiourea catalysts would be more practical than their urea analogs because of the following [142]:

(1) Thiourea derivatives are typically much more soluble in various organic solvents because of the lower tendency of the sulfur atom to serve as a hydrogen bond acceptor.
(2) Thiourea derivatives are easier to prepare (for instance, liquid thiophosgene is much easier to handle than gaseous phosgene).
(3) The acidity of thiourea (pK_a = 21.0) is about six orders of magnitude higher than that of urea (pK_a = 26.9) [138, 143], which should lead to stronger hydrogen-bonding and anion-binding interactions.

The strong hydrogen-bonding ability of thiourea derivatives became apparent in various areas such as supramolecular chemistry [144, 145], molecular anion recognition [146–148], crystal engineering [149, 150], herbicides [151, 152], and inclusion compounds [153–155]. Schreiner's group introduced the 3-(trifluoromethyl)phenyl (e.g., in **45**) and 3,5-bis(trifluoromethyl)phenyl (e.g., in **46**) motifs for thiourea catalysts with the idea that the ester function in Curran's **23** would obstruct the catalytic process by engaging in hydrogen-bonding interactions as well (Scheme 1.6) [156]. Early DFT computations (B3LYP/6-311+G(d,p)) suggested that **45** activates ketones through explicit double hydrogen-bonding interactions, which is conceptually similar to Jorgensen's realization of two water molecules acting as activators in aqueous Claisen rearrangements (Scheme 1.4) [156, 157]. Binding studies of catalyst **45** with the bidentate dienophiles *N*-acyloxazolidinone **52** revealed that **45** acts as a bidentate Lewis acid through explicit double hydrogen bonding (Scheme 1.8) [158].

This was confirmed experimentally by low-temperature IR measurements and comparison of the observed shifts of the carbonyl groups with computations on a reduced model system – assuming that **45** adopts a *syn*-conformation (*trans*/*trans* rotamer) as suggested by the crystal structure data of urea **17** [130, 134], other bisaryl-ureas [132, 159], and in particular structurally related thiourea **46** [1, 160]. Utilizing temperature-dependent ^{1}H NMR studies, it was found that **46** exhibits a large dimerization entropy ($\Delta S = -35.6\,\text{cal K}^{-1}\ \text{mol}^{-1}$), which leads to free available catalyst at room temperature and consequently favored complexation with 1,3-diketone substrates ($\Delta S = -9.6\,\text{cal K}^{-1}\ \text{mol}^{-1}$). The high diastereoselectivity of the Diels–Alder reaction between cyclopentadiene and *N*-acyloxazolidinone **52**

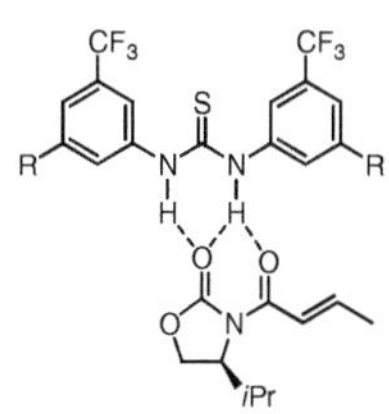

(**45**): R = H
(**46**): R = CF_3
1997/2000/2002, Schreiner
electron-deficient *N,N'*-bisaryl thioureas (1 mol% and 25 mol%)
Diels-Alder reactions

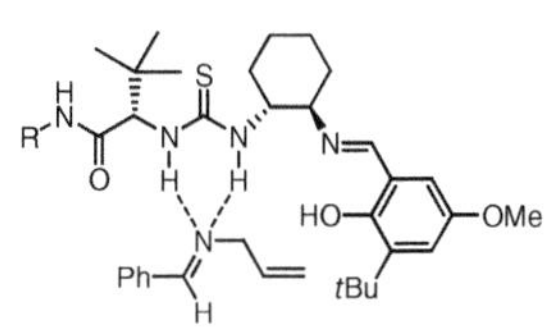

(**47**): R = Bn
(**48**): R = Polystyrene
1998/2002, Jacobsen
Schiff base thiourea catalyst (2 mol%)
first enantioselective thioureas
Strecker reaction

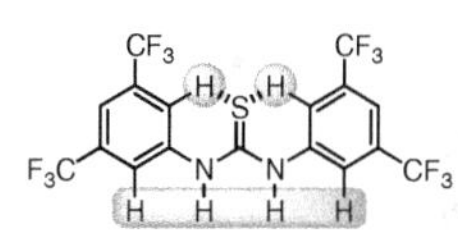

46
2003/2012, Schreiner
guidelines for (thio)urea catalyst structure design, binding studies,
N-H and *ortho* C–H act as hydrogen bond donors

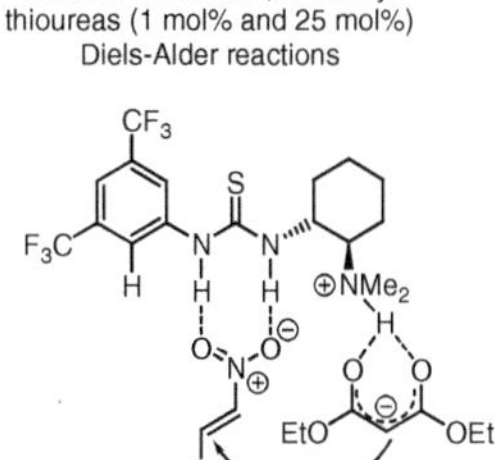

(**49**)
2003/2005, Takemoto
first bifunctional thiourea catalyst
asym. nitro-Michael addition (10 mol%)

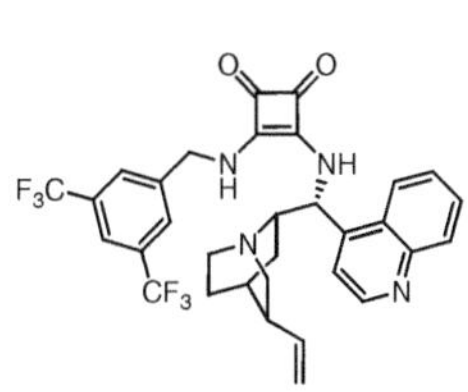

50
2008, Rawal
first squaramide catalyst
asym. nitro-Michael addition (0.5 mol%)

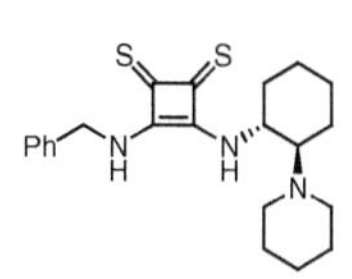

51
2017, Rawal
first thiosquaramide catalyst
asym. nitro-Michael addition (0.05 mol%)

Scheme 1.8 Milestones in hydrogen-bonding thiourea catalysis of neutral and anionic Lewis basic sites with (thio)urea derivatives.

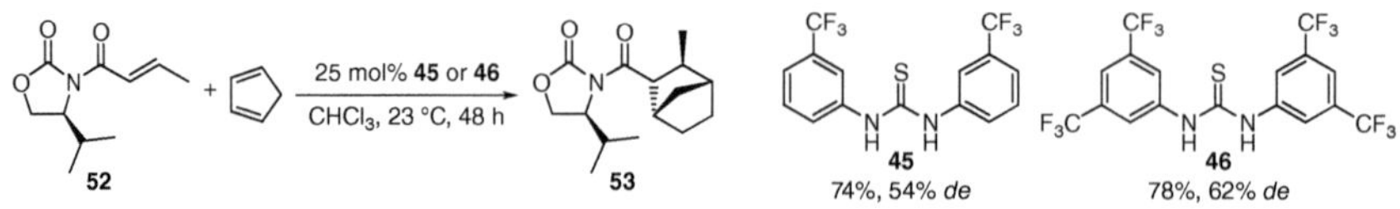

Scheme 1.9 Diastereoselective cycloaddition of *N*-acyloxazolidinone **52** and cyclopentadiene in the presence of **45** and **46**.

using 25 mol% of **45** supported the spectroscopic experiments and computational data [158]. More acidic **46** [143] eventually became a highly active privileged organocatalyst that activates neutral or anionic substrates through explicit double hydrogen bonding (Scheme 1.9) [142]. The 3,5-bis(trifluoromethyl)phenyl group became a very highly appreciated "privileged" anchor group in a large variety of organocatalysts not limited to hydrogen-bonding catalysis [158, 161–170] but also including, *inter alia*, proline [171, 172] and Brønsted acid catalysis [7, 173].

In pursuit of a tridentate chiral metal ligand in enantioselective Strecker reactions of *N*-allyl protected aldimines **54**, in 1998, Jacobsen's group identified in the course of high-throughput screening (HTS) of more than 130 tridentate Schiff bases thiourea catalysts [1] **47** and **48**, which were the first enantioselective thiourea organocatalysts (Scheme 1.10) [174]. Strecker products **55** were obtained as trifluoroacetyl-protected species (65–92%, 70–91% *ee*). Catalyst evolution and additional catalysts for asymmetric Strecker reactions are discussed in Section 1.2.2 including a detailed discussion of the substrate activation.

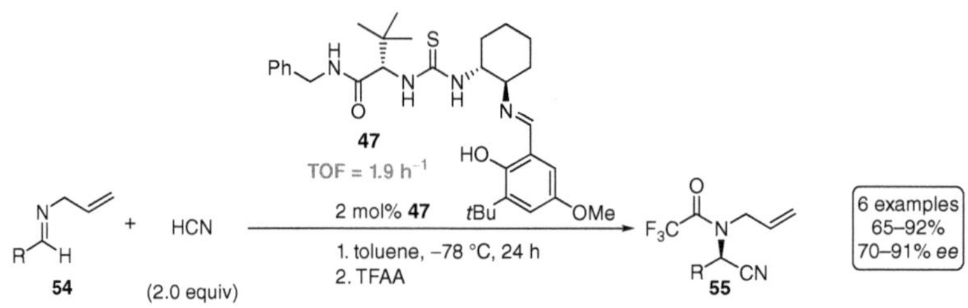

Scheme 1.10 Asymmetric Strecker reaction of aromatic and aliphatic *N*-allyl-protected aldimines catalyzed with **47**.

Structure–reactivity relationships of hydrogen-bonding catalysts in Diels–Alder reactions of cyclopentadiene and various dienophiles such as methyl vinyl ketones and (aza)chalcones, utilizing symmetrically *N,N'*-disubstituted thioureas with 1 mol% catalyst loading, were examined as early as 2003 [157]. The reaction rates determined by ^{1}H NMR qualitatively confirmed the conceptual ideas of Hine *et al.* for the aminolysis of epoxides (Table 1.1) [117], namely, that rigid

1 Jacobsen introduced the term "Schiff base catalyst" to demonstrate that the structure of this novel catalyst class originates from Schiff base ligands – originally developed for metal-based catalysis – and incorporates a Schiff base moiety. Notably, this term does not indicate that these catalysts operate as bases, but their high catalytic efficiencies result from the thiourea moiety as shown in Section 1.2.2. In the following, the scientifically established term "Schiff base (thio)urea" is used to describe this class of organocatalysts.

electron-withdrawing aromatic substituents lead to higher catalytic activity, as **45** (k_{rel} = 2–6; TOF = 52 h^{-1}) and **46** (k_{rel} = 5–8; TOF = 72 h^{-1}) were the most efficient catalysts, while dialkyl or electron-rich aromatic substituents are poorly activating (e.g., *N,N′*-diphenylthiourea **56**: k_{rel} = 1–2). Furthermore, product inhibition turned out to be very low because catalyst activity was still present even after 80% conversion. This emphasizes the entropy term in the formation of catalyst–substrate complexes and the weak enthalpic binding between the thiourea catalyst and the carbonyl group (~7 kcal mol^{-1} at rt in DCM) [158]. In addition, more conformationally flexible catalysts (e.g., **56** with a DFT (B3LYP/6-31G(d)) computed C_{aryl}–N rotational barrier of only ΔE_{rot} = 1.5 kcal mol^{-1}) perform poorly compared to conformationally more restricted derivative **46** (ΔE_{rot} = 3.4 kcal mol^{-1}) [157]. Moreover, electron-deficient substituents in meta- or para-position polarize the aromatic ortho-hydrogens, leading to intramolecular hydrogen-bonding interactions to the Lewis basic thiocarbonyl sulfur that hinders the rotation of the thiourea moiety and favors complexation entropically (Scheme 1.8). Substituents in ortho position, in contrast, lead to repulsive interactions with the thiocarbonyl group and the *syn–syn* conformation is entropically disfavored [130, 157]. Therefore, catalysts with substituents in meta- or para-position are more active than derivatives containing substituents in the ortho position. Schreiner's group highlighted the concept of thiourea organocatalysis at the very beginning of this research field and provided rough guidelines for (thio)urea catalyst design, which are still valid today and can readily be identified in every (thio)urea organocatalyst developed in the past 20 years [1]:

(1) The relative stabilization of the transition state must be larger than the stabilization of the starting material(s) and product(s);
(2) Bi- or multidentate catalyst–substrate complexes are favorable;
(3) There should be a balance between rigidity and flexibility in the catalyst structure to avoid entropy loss on the one hand and to retain adaptability to the substrate structure on the other hand;
(4) To prevent self-association, the catalyst should not incorporate strong hydrogen bond acceptors such as ester groups – the non-coordinating acidifying CF_3 group as in the 3,5-bis(trifluoromethyl)phenyl moiety of **46** appears to be ideal;
(5) To avoid product inhibition, the enthalpic binding interactions should be weak to allow low catalyst loadings;
(6) The catalyst should be water compatible or even catalytically active in water.

In 2003, Takemoto's group introduced the first bifunctional thiourea **49** for use in Michael addition reactions [161]. They postulated the simultaneous activation of the nitro olefins **57** by the N–H thiourea protons and malonates **58** by the amine (cf. Scheme 1.8), thereby obtaining α-chiral malonate derivatives **59** depicted in Scheme 1.11 (TOF = 0.7 h^{-1}) [175]. The authors performed kinetic studies on the Michael reaction and observed first-order kinetics for both reactants and the catalyst. Catalyst evolution and additional catalyst derivatives for asymmetric Michael reactions are discussed in Section 1.2.3, including a detailed discussion of the mode of substrate activation.

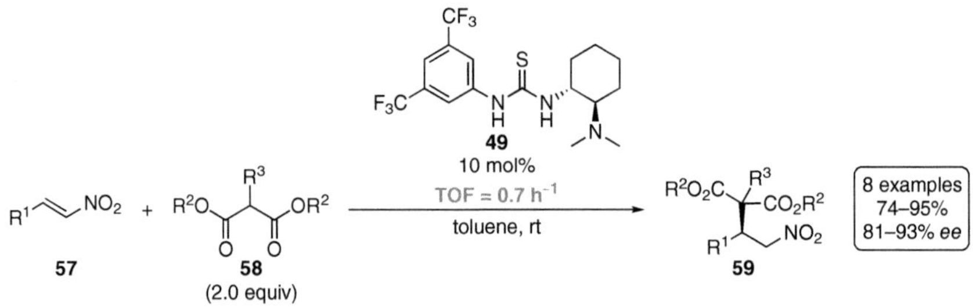

Scheme 1.11 Product range of the asymmetric Michael addition of dialkyl malonates to *trans*-β-nitrostyrenes catalyzed with **49**.

In 2006, Kotke and Schreiner applied the principle of anion-binding catalysis to high-yielding acid-free acetalizations [176]. They utilized various aromatic, aliphatic, unsaturated, and acid-labile aldehydes as well as ketones, which could be cleanly acetalized in the presence of 4.0–4.3 equivalents of methanol, ethanol, 1-propanol, 2-propanol, and 1,2-ethandiol as the alcohol components and as a solvent. The acetals were obtained in yields of 61–95% at low catalyst loading of only 0.01–1.0 mol% **46** – at that time the lowest catalyst loading in an organocatalytic reaction – at room temperature (Scheme 1.12). A limitation of the protocol, however, was the acetalization of electron-rich carbonyl compounds such as *p*-tolylaldehyde, cyclohexanone, and acetophenone **64** with long reaction times (92–250 h) and lower conversions (71–74%). In contrast, aldehydes were acetalized at much shorter reaction times (9–14 h) and in high yields (89–95%). Kotke and Schreiner proved the synthetic utility of this protocol when they isolated 67% of pure product from the acetalization of acid-labile TBS-protected *m*-hydroxybenzaldehyde **68** that was reported to react rather sluggishly under Brønsted [177] or Lewis acid catalysis [178]. Kotke and Schreiner found for the diethyl acetalization of *p*-chlorobenzaldehyde **65** and *n*-octanal TOFs of 632 h^{-1} or rather turn over number (TON) = 9800, and 577 h^{-1} (TON = 9700), respectively. These values are approximately ten times higher than those of previously reported thiourea-catalyzed reactions and were

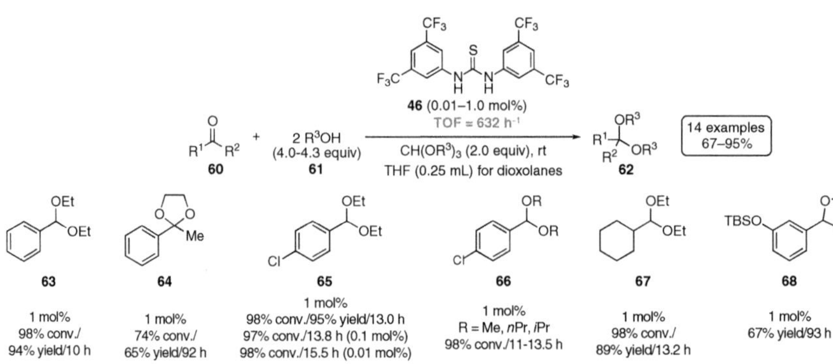

Scheme 1.12 Representative examples of the first anion-binding catalysis reaction described in the literature. Products was obtained under mild and acid-free conditions catalyzed by **46**.

significant, as the uncatalyzed reactions generally even after one week furnished <1% product. In addition, the authors found, on the one hand, no influence of the utilized alcohol but to the carbonyl component on the other hand. A competition experiment between benzaldehyde and acetophenone – using ethanol as well as triethyl orthoformate – resulted in a product mixture of 6.1 : 1 favoring the diethyl benzaldehyde acetal. This chemoselectivity is in line with the different reactivity of aldehydes and ketones under nucleophilic attack [176].

The initial working hypothesis was that the catalyst undergoes double hydrogen bonding to the carbonyl moiety – comparable to dienophile activation depicted in Scheme 1.8 – to increase the electrophilicity of the carbonyl carbon and, at the same time, to stabilize the increased negative charge on the carbonyl oxygen during the nucleophilic attack. However, utilizing thiols instead of alcohols in the presence of ethyl orthoester only provided the diethyl acetal, although thiols are much better nucleophiles [179]. Therefore, Kotke and Schreiner proposed a mechanism that starts with thiourea-assisted orthoester heterolysis, with **46** stabilizing multiple oxyanion intermediates. The stabilized alcoholate anion **69** rapidly undergoes nucleophilic attack onto the carbonyl compound and **46** binds the incipient hemiketal anion **70**. Subsequent nucleophilic addition provides the "double acetal **71**," which dissociates to alkyl formate **72**, the oxocarbenium ion **73**, and another **46**-stabilized alcoholate **69** that in turn attacks **73**, leading to the acetal product and releases the catalyst for subsequent turnover (Scheme 1.13). This mechanism is in in line with the well-established enzymatic mechanisms in which hydrogen-bonding interactions stabilize alkoxide intermediates [180].

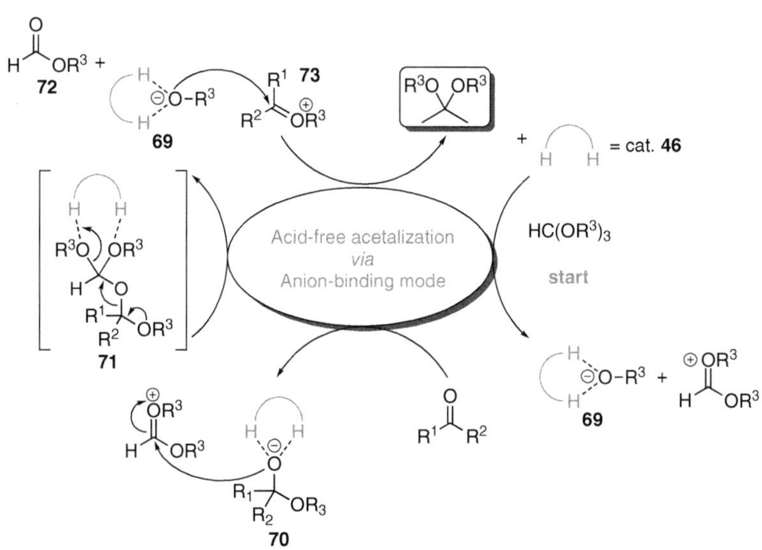

Scheme 1.13 Proposed acetalization mechanism catalyzed with **46** starting with thiourea-assisted heterolysis of an orthoester and subsequently multiple oxyanion stabilization through **46**-mediated anion binding.

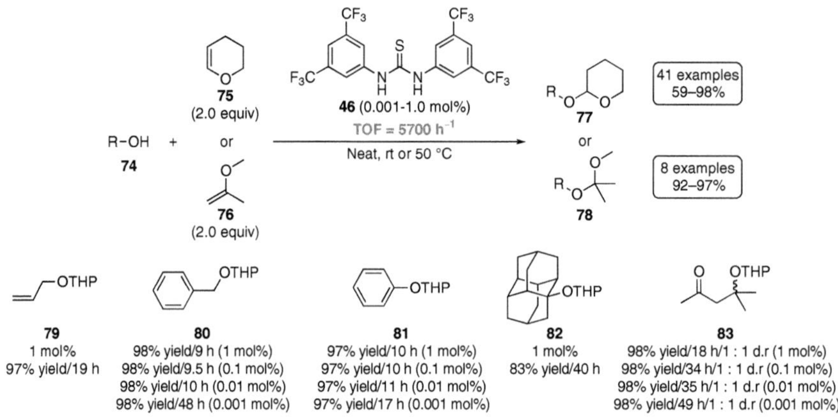

Scheme 1.14 Representative products of the THP and MOP protection of hydroxyl functionalities catalyzed with **46**.

In 2007, Kotke and Schreiner introduced the organocatalytic tetrahydropyran (THP) and 2-methoxypropene (MOP) protection of alcohols, phenols, and other hydroxyl containing substrates (Scheme 1.14) [182]. The tetrahydropyranylation catalyzed with **46** using 3,4-dihydro-2*H*-pyran **75** (DHP) as reactant and solvent was applicable to a broad spectrum of alcohols and furnished the corresponding tetrahydropyranyl ethers **77** under mild conditions and in yields of 59–98% at very low catalyst loading of only 0.001–1.0 mol% **46**. The THP protection of benzyl alcohol **80** at very low catalyst loadings down to 0.001 mol% emphasized the catalytic power of **46** and revealed a TON close to 100 000 and TOF of approximately 2000 h^{-1}. Phenol derivatives were also readily converted into the corresponding acetals at 50 °C. As shown for **81**, THP protection could be achieved with 0.001 mol% catalyst loading, resulting in 97% isolated product after 17 h. This indicates a TOF of approximately 5700 h^{-1}, which still marks the most efficient organocatalytic reaction utilizing hydrogen bonding to date. Scale-up experiments (50 mmol scale), e.g., in the case of phenol, also demonstrated that catalyst loadings of only 0.01–0.1 mol% are sufficient and feasible for preparative THP protection. Tertiary alcohols, which are generally difficult to protect as THP ethers due to steric hindrance and elimination as a common side reaction, could also be THP protected under these conditions. Remarkable is the tolerance of even most sterically hindered substrates diamantan-1-ol affording ether **82**. Because of the acid-free conditions, the protocol also tolerated acid-labile hydroxyl-functionalized substrates such as aldol reaction products (**83**; TOF = 2000 h^{-1}), β-hydroxy esters, epoxides, acetonides, cyanohydrins, oximes, and highly acid-sensitive TBS-protected alcohols without detectable side reactions in yields of 59–98%. Kotke and Schreiner applied their protocol to the alternative enol ether 2-methoxypropene **76** and prepared the corresponding MOP-ethers **78** in yields ranging from 92% to 97% at room temperature; it was noted that MOP is so reactive that the uncatalyzed reaction also proceeded albeit at lower rates (Scheme 1.14) [182].

A reasonable mechanistic entry into this reaction may start with the complexation of catalyst **46** with the alcohol substrate [182]. This double hydrogen-bonding-mediated coordination increases the alcohols' acidity as well as polarizability and

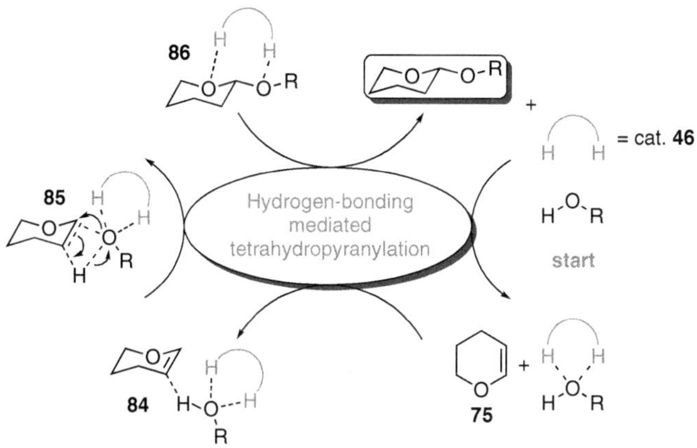

Scheme 1.15 Proposed mechanism for the **46**-mediated tetrahydropyranylation of alcohols.

hence its ability to form a subsequent ternary complex **84** with enol ether **75**. The catalyst remains attached during the polar addition through a highly polarized transition structure **85** and is finally released from the product complex **86** to initiate a new catalytic cycle (Scheme 1.15).

This mechanistic proposal indicates the departure from the often-implied concept of carbonyl or related functionalities such as nitroolefins activation through hydrogen bonding with (thio)urea derivatives and other hydrogen-bonding organocatalysts [3, 183]. Hence, this mechanistic alternative suggested either the hydrogen-bond-assisted generation of the free nucleophile (e.g., RO^- and CN^-) and subsequent anion binding or the stabilization of the active form of the nucleophile through hydrogen-bonding and polar interactions to the respective precursor (e.g., ROH, $HC(OR)_3$ [176], HCN, and TMSCN [184]). Kotke and Schreiner elucidated this mechanistic proposal utilizing DFT (B3LYP/6-31G(d,p)) and coupled cluster computations (CCSD(T)/cc-pVDZ,), which demonstrated that the catalyst preferentially stabilized the developing *oxyanion hole* in the transition state through double hydrogen-bonding without the formation of a charged alkoxide nucleophile. This conclusion was based on a comparative computational analysis of the uncatalyzed versus catalyzed model reaction of methanol with **75**. The stabilizing effect of **46** on the key transition structure was estimated to be 23 kcal mol^{-1}, resulting in a minimized barrier of only 17.7 kcal mol^{-1} (uncatalyzed: 45.2 kcal mol^{-1}): These values are in line with the experimentally found efficacy of **46** already at room temperature in contrast to the uncatalyzed THP protection, which showed no product formation under otherwise identical conditions. Additionally, the transition structure revealed that **46** helps pre-organizing the reactants and the overall geometric changes in going from the complexes to the transition structures; **46** is placed sideways and points away from the "R group" of the substrate making steric hindrance not a critical factor, as found experimentally [182]. Pápai, Varga, and coworkers reinvestigated the mechanism of the THP protection mechanisms utilizing DFT studies (ωB97X-D/6-311G(d,p)) and isotope labeling experiments [185]. In

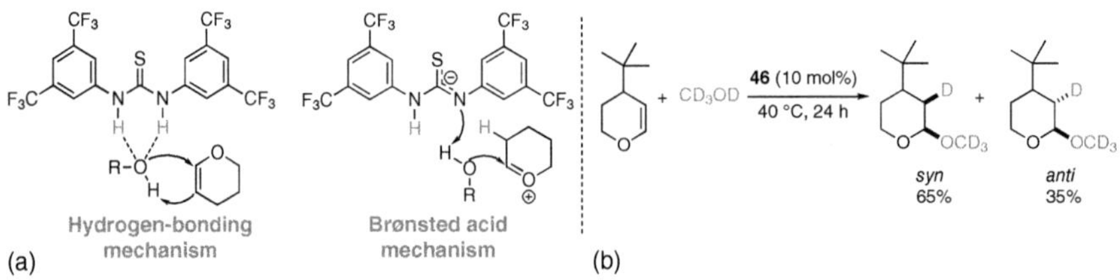

Scheme 1.16 (a) The two investigated possible mechanisms of the THP protection of alcohols and (b) isotope labeling experiment utilizing CD_3OD.

addition to the mechanism proposed by Schreiner and Kotke, the authors also examined an alternative reaction pathway with **46** acting as a Brønsted acid and protonating **75** (Scheme 1.16a). The computed transition structure for the Brønsted acid pathway was $\Delta\Delta G = 6.5\,\text{kcal mol}^{-1}$ lower in energy compared to the hydrogen-bonding mechanism. In isotope labeling experiments, utilizing CD_3OD and conformationally rigid DHP derivative 4-(*tert*-butyl)-3,4-dihydro-2*H*-pyran, a mixture of the *syn*- and *anti*-product was obtained in a 65 : 35 ratio, respectively, that supported a two-step mechanism *via* Brønsted acid mechanism (Scheme 1.16b). An additional hint for a Brønsted acid mechanism was obtained by utilizing 2-thiouracil that could not engage in double hydrogen bonding to the alcohol, however furnished THP-protected phenol **81** in similar yield (92% conversion, 12 h, 0.2 mol% catalyst loading) compared to **46** (100% conversion, 12 h, 1.0 mol% catalyst loading). In 2019, McGarrigle and coworkers utilized 2-thiouracil in the stereoselective glycosylation of galactals and proposed a mechanism where thiourea or 2-thiouracil engaged *via* general acid/base catalysis – similar to the mechanisms proposed by Pápai, Varga, and coworkers – but excluded a "simple" acid-catalyzed mechanism [186]. Utilizing acids with similar pK_a values such as Meldrum's acid, $Et_3N{\cdot}HCl$, or 4-nitrobenzoic acid does not lead to product formation [185–187].

In 2011, Schreiner and coworkers developed a cooperative catalyst system for the enantioselective cyanosilylation of aldehydes **87** (Scheme 1.17) [170]. They utilized thiourea **89** as the catalyst, benzoic acid as the cocatalyst, and TMSCN as the cyanide source and obtained acetyl-protected chiral cyanohydrins **88** in yields of 57–90% and *ee*-values of 8–88% utilizing 10 mol% **89** and cocatalyst each (TOF = $0.6\,h^{-1}$).

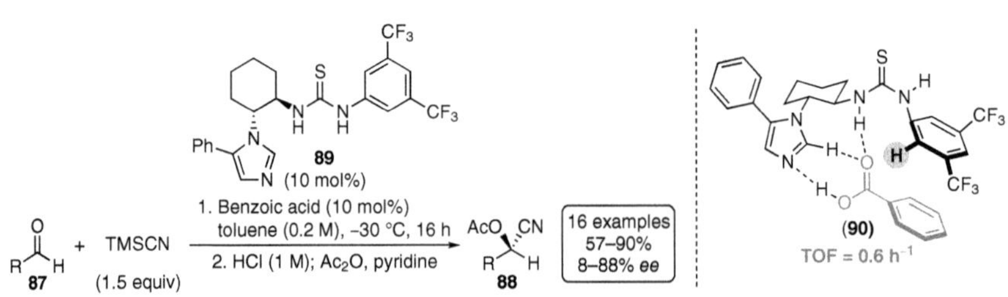

Scheme 1.17 Enantioselective cyanosilylation of aldehydes catalyzed utilizing the *in situ* formed cooperative catalyst (**90**).

The authors utilized a combination of NMR and ESI-MS techniques for mechanistic studies that reveal the presence of a hydrogen-bonded complex of **89** and benzoic acid that was also supported by DFT computations (M06/6-31G(d,p)). The complex (**90**) formed a well-defined chiral hydrogen-bonding environment, where one N–H proton coordinated the benzoic acid, whereas the second N–H proton formed hydrogen bonding to the aldehyde. Furthermore, the strong fixation of the aromatic aldehyde occurred *via* T-shaped π–π stacking interactions of the acidified ortho protons of the 3,5-bis(trifluoromethyl)phenyl moiety with the phenyl ring, which was supported by downfield shift of the ortho-protons in NMR experiments. This interplay of hydrogen bonding and π–π stacking interactions supported the formation of a chiral ternary complex preferring the observed stereoinduction of cyanide addition and emphasized the importance of the catalysts ortho-protons for substrate activation [170].

In 2012, Schreiner's group measured the pK_a values of various (thio)urea derivatives in dimethyl sulfoxide (DMSO) to examine the influence of CF_3 groups on the N–H protons' acidity (Scheme 1.18) [143]. The reason for choosing DMSO and not water as the solvent was that most organic reactions are carried out in organic solvents; the aqueous pK_a values are not useful because the differences to non-aqueous pK_a values are large and not even qualitatively meaningful [188]. The authors found that the pK_a values correlate well with the overall number of CF_3 groups attached to the aromatic rings; each CF_3 group decreases the pK_a by approximately 1.2 pK_a units (cf. **49** vs. **97**), and the combined effect of the four CF_3 groups in thiourea **46** resulted in a pK_a value of 8.5, which allows its deprotonation by organic amine bases such as *N*,*N*-diisopropylethylamine (DIPEA) [189]. Additionally, the pK_a values of thioureas were approximately 5 pK_a units lower compared to the corresponding urea derivatives (**94** vs. **46**).

The influence of the electron-withdrawing substituents on the ortho-protons and their engagement in substrate binding was investigated more closely by Schreiner's group in 2012, utilizing a combination of low-temperature IR spectroscopy, 2D NMR techniques, MS investigations, and DFT computations (M06/6-31+G(d,p)) [190]. The authors found strong evidence of **46** binding to δ-valerolactone both through the N–H protons and the ortho-proton. The IR spectra of a 1 : 1 mixture revealed a small

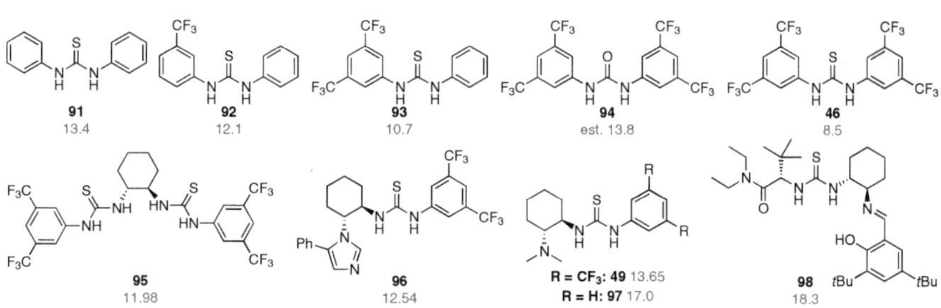

Scheme 1.18 Substituent effects on a selection of (thio)urea derivative pK_a values in DMSO. The pK_a value of **94** was estimated because of solubility issues. The pK_a value of **98** is an apparent value, a result of the similar acidity of the thiourea moiety and the phenolic OH.

$\nu_{C=O}$ band of free lactone and a large red-shifted band ($\Delta\nu = 32\,cm^{-1}$), indicating a hydrogen-bonding complex. In addition, ^{19}F-^{1}H HOESY experiments showed a cross-peak between the fluorine of the CF_3 group and the δ-methylene protons of the lactone, and the binary complex could also be identified by ESI-MS. [143].

As a further development of the thiourea catalysts, Rawal and coworkers introduced chiral cinchona-based squaramide derivatives as alternative (thio)carbonyl organocatalysts in 2008 (Scheme 1.19) [191]. Utilizing squaramide **50**, the authors performed the Michael addition of β-nitrostyrenes **99** and 2,4-pentanediones **100** – similar to the substrates used by Takemoto *et al.* in 2003 [161] – and obtained the products **101** in yields ranging from 65% to 98% and *ee*-values of 88–98%. Low catalyst loading of 0.5 mol% resulted in TOF = $24\,h^{-1}$, a 30 times higher value as compared to Takemoto's approach (TOF = $0.7\,h^{-1}$; cf. Scheme 1.11).

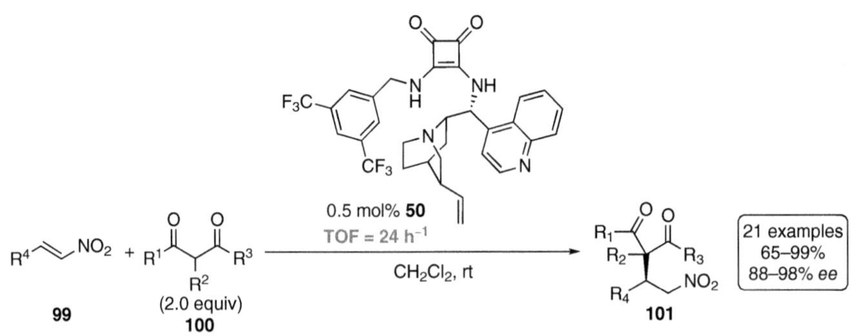

Scheme 1.19 Nitro Michael addition of 2,4-pentadienones to *trans*-β-nitroolefins catalyzed with squaramide **50**.

Four reasons for the higher catalytic activity of squaramides as compared to thioureas for this particular reaction were provided (Scheme 1.20) [192–194]:

(1) The nitrogen lone pairs are delocalized in both derivatives, restricting C–N bond rotation. The delocalization in squaramides takes place through the aromatic cyclobutenedione moiety; hence, the N–H acidity of squaramides is higher as compared to that of thioureas [143, 195].
(2) The squaramide scaffold is more rigid, which leads to reduced conformational flexibility [196].
(3) The squaramide moiety forms stronger hydrogen bonds due to the geometric structure of the cyclobutenedione ring that results in a larger distance and

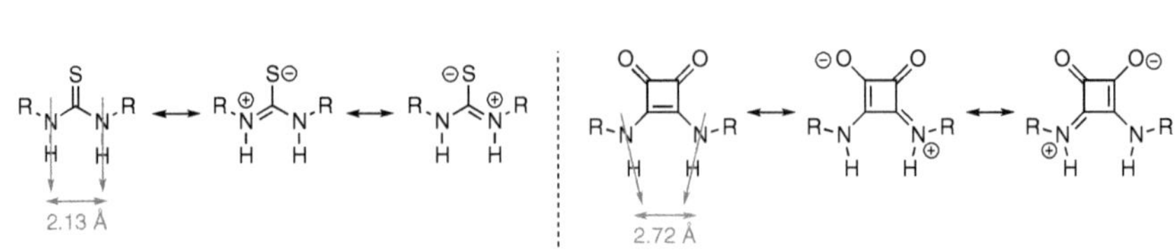

Scheme 1.20 Comparison of thiourea and squaramide skeletons regarding the zwitterionic species, the N–H-proton distance, and the orientation of the N–H-protons in thioureas (parallel) and squaramides (convergent).

convergent orientation of the N–H-protons with a larger H···X···H angle (X = Lewis base) [191, 192].

(4) Squaramides exhibit dual functionality because the carbonyl oxygens may act as Lewis bases [197–199].

Subsequently, squaramides were used in various organocatalyzed transformations, driving on concepts of double hydrogen bonding of neutral and anion-binding Lewis basic sites [192, 193, 200]. Almost ten years later, Rawal's group synthesized chiral thiosquaramide **51** and examined this derivate in Michael addition reactions of β-nitrostyrene derivatives **102** and barbiturates **103** (Scheme 1.21) [201]. Catalyst **51** provided Michael adducts **104** in ranges of 88–99% and *ee*-values of 72–98% at a catalyst loading of only 0.05 mol%. Similar to ureas and thioureas, thiosquaramides are more acidic than squaramides [202–204], which can be easily identified using TOF values: Compared to squaramide **50**, the TOF value of **51** ($3960\,h^{-1}$) is about 160 times higher and is in the order of the TOF value of **46** in the THP protection of alcohols (cf. Scheme 1.14) [182].

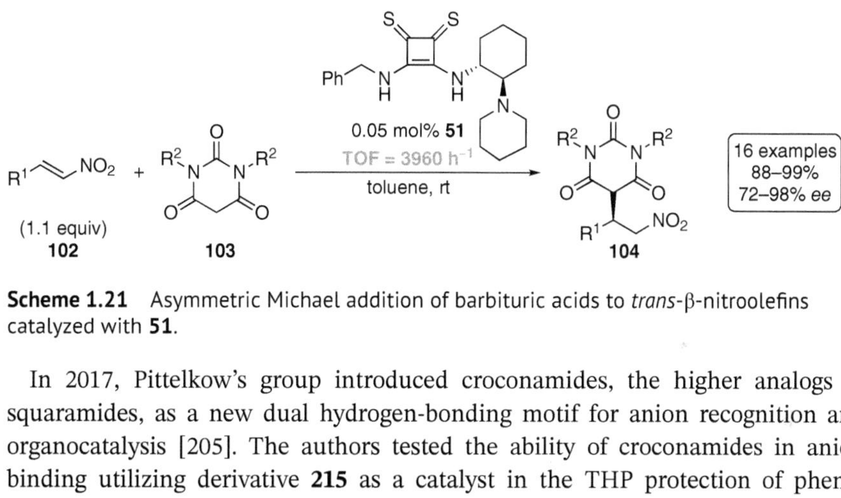

Scheme 1.21 Asymmetric Michael addition of barbituric acids to *trans*-β-nitroolefins catalyzed with **51**.

In 2017, Pittelkow's group introduced croconamides, the higher analogs of squaramides, as a new dual hydrogen-bonding motif for anion recognition and organocatalysis [205]. The authors tested the ability of croconamides in anion binding utilizing derivative **215** as a catalyst in the THP protection of phenol derivatives and measured rate constants for a series of substrates (Scheme 1.22).

At the same time Ho, Jolliffe, and coworkers employed quantum mechanical computations (G3(MP2)) to predict the pK_a values of several (thio)ureas, (thio)deltamides, (thio)squaramides, and croconamides [206]. Because many pK_a values of these hydrogen-bonding donors are known from the literature [195], the authors compared the computed values with the experimental data and observed differences of approximately 1.5 pK_a units. The croconamides were identified as the most acidic oxo-species; however, thiosquaramides were slightly more acidic (~2 pK_a units), but no investigations of thiocroconamides – which have not yet been reported – were made. Additionally, Jolliffe's group introduced deltamides as anion receptors [206]. The authors investigated the chloride-binding affinities and observed an unusual trend for the croconamide series: the binding affinities for *N*,*N*′-bisalkylcroconamides were higher than for *N*,*N*′-bisaryl derivatives. These results were explained by the high acidity of aryl-substituted derivatives that exist as

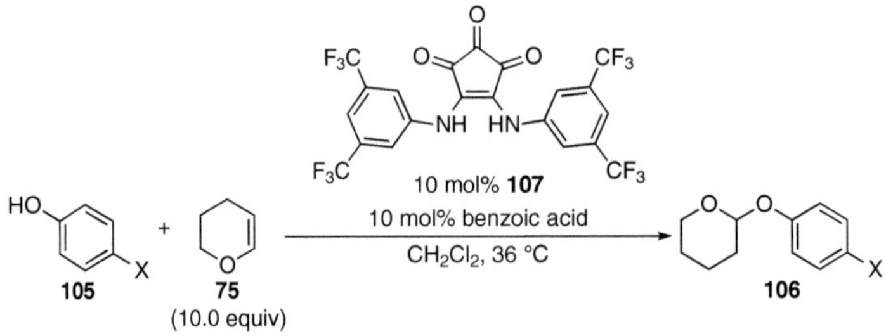

Scheme 1.22 THP protection of a series of phenols catalyzed with **215**.

partially deprotonated at neutral pH, thereby reducing their ability to bind anions. Because deltamides are less acidic than squaramides, but more acidic than ureas, *N,N′*-bisaryl derivatives resulted in higher affinities than *N,N′*-bisalkyl derivatives, and attachment of phenyl groups decreased the pK_a value and increased the anion affinity. Moreover, deltamides displayed lower chloride affinities than ureas and squaramides, whereas the affinities of all three species to the coplanar acetate anion are similar. In contrast to ureas and squaramides, where the anion-binding affinity correlates with anion basicity, deltamides also revealed higher affinities for tetrahedral $H_2PO_4^-$ than for coplanar carboxylates. The authors explained these results by the different structures of deltamides, where the DFT-computed distance of the N–H protons is approximately 3.5 Å and is therefore significantly larger than in ureas (2.13 Å) and squaramides (2.72 Å, Scheme 1.20). These results emphasize the strong correlation between the N–H geometry and the anion-binding affinity [206].

1.1.2 Evolution of Triazole-Based Catalysts

Non-covalent interactions involving C–H hydrogen bond donors have been known since the fundamental work by Kumler in 1935 [207], which took place at the same time when Pauling described "*traditional*" N–H and O–H hydrogen bond donors [181]. However, C–H hydrogen bonding was mentioned only occasionally in the following 30 years until Sutor utilized X-ray single-crystal analysis of 1,3,7,9-tetramethyluric acid **108** and described an intramolecular C–H· · ·O hydrogen bond (Scheme 1.23) [209–211]. Sutor argued that the distance of 3.00 Å between the methyl group (C_{11}) and the oxygen atom (O_{12}) was much shorter than the sum of the van der Waals radii of a methyl group and an oxygen atom (3.40 Å) [211]. In 1982, Taylor and Kennard investigated the crystal structures of 113 organic compounds (many of them are amino acid and nucleoside derivatives) with respect to C–H· · ·O hydrogen bonding and found strong evidence for this kind of NCI [133]. In 2008, a number of seminal studies on anion binding *via* C–H bonds emerged [208, 212–218]. In parallel, the 1,2,3-triazole unit was introduced in macrocyclic receptors, e.g., in the triazolophane **109** (Scheme 1.23) [208]. The C–H-mediated hydrogen-bonding motif was comprehensively reviewed [219–225], as where the triazoles utilized in anion recognition processes [226, 227].

Mancheño's group employed 1,2,3-triazol-based catalysts as alternative hydrogen-bonding donors in anion-binding catalysis [228] and carried out halide

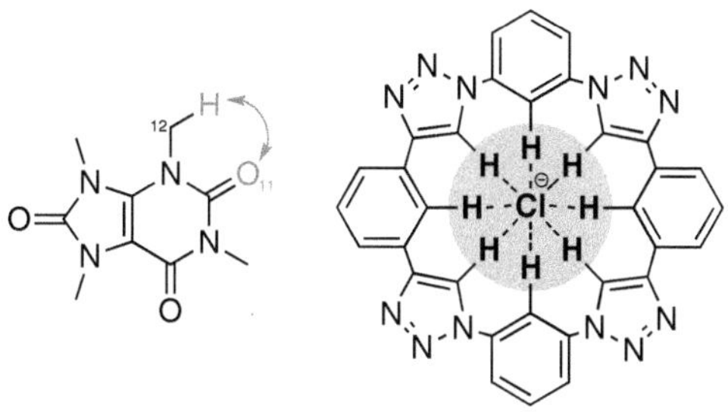

108 **109**

Scheme 1.23 C–H Hydrogen bonding in **108** and triazolophane-bound chloride anion in **109**.

anion-binding affinity and activity studies of a series of bis-triazole derivatives (Scheme 1.24) [229]. The authors synthesized a series of derivatives **110–120** with variation of the triazole substituent and the bridged phenyl moiety. Mancheño *et al.* performed chloride-binding studies and measured the highest chloride affinity for **114** that was substituted by the well-established 3,5-bis(trifluoromethyl)-phenyl moiety [190] as a triazole substituent and the 2-(methoxy-propan-2-yl)acetylene group as the substituent in *para* position for the phenyl unit [229, 230]. The importance of the ortho-protons of the 3,5-bis(trifluoromethyl)-phenyl moiety was also supported by the finding that the corresponding pentafluorophenyl derivative **113** bound chloride much less effectively than **114**. Furthermore, the authors observed a high preference for chloride binding ($k_{THF} = 1417\,M^{-1}$) as compared to bromide ($k_{THF} = 438\,M^{-1}$) and iodide ($k_{THF} = 319\,M^{-1}$). This was explained with the limited space inside the spherical binding pocket, which was further supported by the low binding affinity for the small, but linear cyanide ion ($k_{THF} = 394\,M^{-1}$).

Thereafter, Mancheño's group utilized bis-triazole **114** for alkylation of primary and secondary amines as well as thiophenol **127** and phenol **128** with 4-dimethylamino-*N*-triphenylmethyl-pyridinium chloride **121** (12–85%; 10 mol% catalyst loading; TOF = $0.12\,h^{-1}$) by binding of the counteranion and therefore activation of the electrophile (Scheme 1.25) [229, 230]. The high chloride affinity of the catalyst and the anion-binding mechanism were confirmed by anion selectivity studies utilizing each 10 mol% **114** and thiourea **46** in the presence of 20 mol% tetrabutylammonium bromide. The conversion dropped from 79% to 51% and from 82% to 37% for **114** and **46**, respectively (48 h).

In 2014, Mancheño and coworkers introduced the first helical triazole-based anion-binding catalyst with four triazole subunits and chiral 1,2-diaminocyclohexane backbone in a Reissert-type reaction of indole derivatives with silyl ketene acetals as nucleophiles [231–236] and in the dearomatization of pyrylium derivatives [237]. The evolution of the helical design and catalyst application are discussed in Section 1.2.3. The concept of triazole-based anion catalysis was applied to switchable catalysts as well [238]. In 2020, Dorel and Feringa reported a switchable stilbene-based catalyst with helical chirality that incorporates four triazole units [239]. This first-generation molecular motor-based [240] catalyst was utilized in

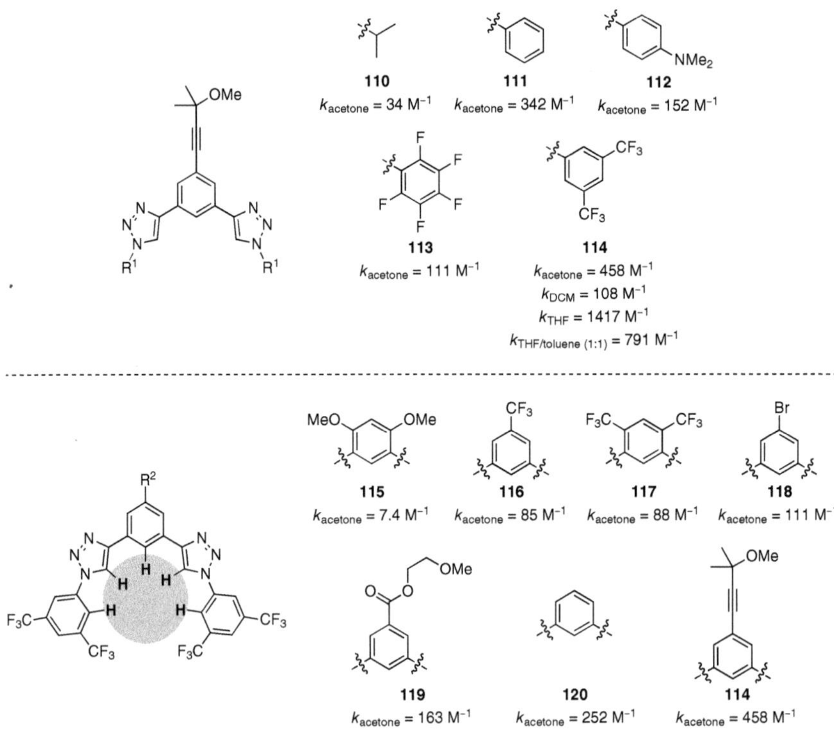

Scheme 1.24 Various 1,2,3-triazole-based catalysts and the corresponding chloride-binding affinity. The blue cycle indicates the spherical anion recognition pocket.

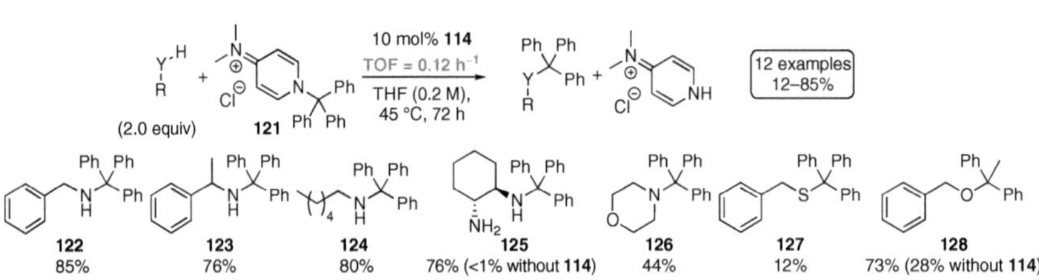

Scheme 1.25 Bistriazole **114**-catalyzed alkylation utilizing DMAP-derived alkylation reagent **121**.

the stereodivergent addition of silyl ketene acetals to oxocarbenium ions through chloride abstraction and binding with 10 mol% catalyst loading. The chirality of the catalyst was switched through light- and heat-driven rotation around the stilbene moiety, resulting in reversal of the enantiomeric ratio of the products. While Dorel and Feringa obtained a racemic mixture in the presence of the *trans* state, the (*M,M*)-*cis* and (*P,P*)-*cis* states gave up to 80 : 20 *e.r.* and 9 : 91 *e.r.*, respectively, of the opposite enantiomers, which means an Δ*ee* of 142%. Although the enantioselectivities were lower than those reported for non-switchable thiourea organocatalyst **223** (Section 1.2.3, Scheme 1.54, 92% *ee*; 0.1 mol% catalyst loading) [242], this work provided an exciting proof of concept [239].

1.1.3 Progress on Halogen-Binding-Based Catalysts

A novel kind of anion-binding catalysis utilizes halogen bonding instead of hydrogen bonding. The basic interaction dates back about 200 years when Colin reported an ammonia–iodine complex during his investigations of nitrogen iodine compounds such as NI_3 in 1814, only about two years after iodine was accidentally discovered and isolated [243, 244]. Because Colin did not reported a formula of the complex, Bineau investigated the formation of this complex further and reported a composition of the complex of three ammonia and one iodine molecule in 1844 [245]. About 20 years later, Guthrie reported this complex (**129**) as a 1 : 1 mixture during his study of the reaction with mercury, where he observed only nitrogen and mercury(I)-iodide in a 1 : 2 ratio [246], and it was not until 1950 that Mulliken described the compound as a charge transfer complex (Scheme 1.26) [247, 248]. Notwithstanding this realization, Benesi and Hildebrand examined the interaction between iodide and benzene and its alkyl-substituted derivatives in 1949. They reported a 1 : 1 complex of benzene and iodine (**130**) that formed in non-polar solvents such as carbon tetrachloride or *n*-heptane and determined the thermodynamic data of the complex [249]. Beginning in the 1950s, Hassel and coworkers published a series of X-ray structure analyses of halogen-bonding adducts, describing a linear orientation of the heteroatoms, e.g., in the adduct between dioxane and bromine (**131**) [250–254]. In the following years, the interest in this kind of non-covalent interactions considerably waned until Paolo and Sandorfy described complexes between fluorinated iodoalkanes and amines (such as **132**) [255]. In the 1970s and 1980s, Dumas introduced the term “halogen-bonding” during his research on the interactions between aliphatic halides and organic bases [256, 257].

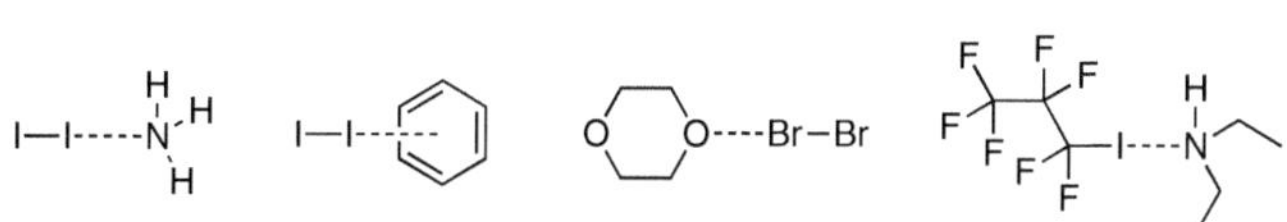

(129) (130) (131) (132)

Scheme 1.26 Some adducts of milestones in halogen-bonding adducts.

The concept of “halogen-bonding interactions” based on the σ-hole, which is well known for atoms of group 14 (tetrels), 15 (pnictogens), 16 (chalcogens), and 17 (halogens) elements when covalently bonded to electron-withdrawing substituents [258]. The electronic density of these atoms is not equally distributed but rather anisotropic. The σ-hole describes the region of positive molecular electrostatic potential (MEP) and is placed collinearly on the opposite side relative to the substituent. Because of σ-holes, in principle, “negatively charged” anions act as if they were “positively charged” [259, 260]. In the last few years, the concept of halogen bonding and σ-hole interactions has been comprehensively reviewed [243, 258, 259, 261–271]. In 2011, Huber’s group reported the first anion-binding additive utilizing double halogen bond donor species in a Ritter-type solvolysis of benzhydryl bromide **133** in acetonitrile (Scheme 1.27) [272]. The authors synthesized a series of bis-imidazolium derivatives and observed that only the double

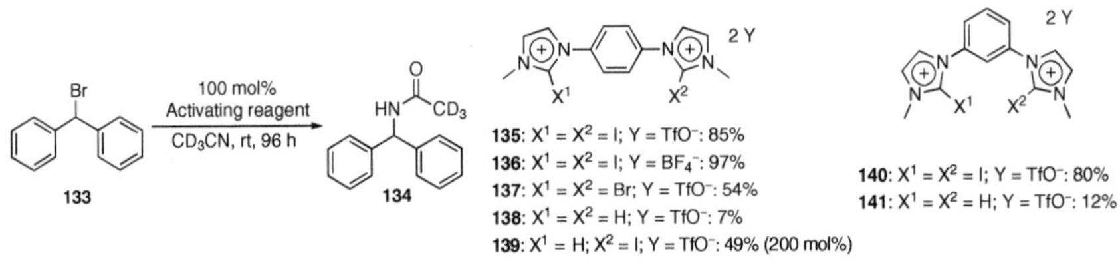

Scheme 1.27 First anion-binding activation induced by charged halogen-bonding catalysts.

halogen bond donors **135, 136,** and **140** – attached with two iodine substituents in 2-position of the imidazole moiety – accelerated the reaction well (>80% yield; 100 mol% loading), whereby the *para*-substituted derivative showed a slightly higher activity (**135** = 85%; **136** = 80%). The corresponding bromine derivative **137** gave only 54% yield, which indicated the importance of the halogen species. Hydrogen-bonding donors **138** and **141** did not provide significant conversion, and single halogen-bonding donor **139** only gave 49% conversion utilizing 200 mol% of the additive. Additionally, Huber and coworkers observed a slight influence of the counterion (**135** = TfO^- = 85%; **136** = BF_4^- = 97%). These results indicated the role that halogen-bonding plays in abstracting and binding the bromide, as also confirmed by isothermal calorimetric titrations [273].

In 2013, Huber's group synthesized a series of neutral halogen-bonding donors based on bridged 2,6-diiodo-3,4,5-trifluorophenyl moieties such as the tridentate variant **146** and utilized them as catalysts for substitution reactions between 1-chloroisochroman **142** with various silyl ketene acetals (10 mol% catalyst loading; TOF = 0.76^{-1}) [274]. The tridentate catalyst **146** (91%) was much more active than the fluorinated derivative **147** (<5%) or the bidentate derivatives **144** (37%; TOF = 0.31^{-1}) and **145** (14%; TOF = 0.12^{-1}), and also the well-established thiourea catalyst **46** resulted in significant lower product formation (12% at −78 °C; 28% at −40 °C). Because **144** should provide a better basis for possible future catalyst development, the authors tested the activity at −40 °C and obtained 52% of the product (Scheme 1.28). To prove the anion-binding mode in this reaction, the authors ran a reaction with 10 mol% **146** in the presence of 20 mol% TBACl and observed no conversion. This was the first anion-binding catalysis utilizing a neutral halogen-bonding donor catalyst and indicated that the number of iodine substituents is relevant for catalyst activity [274].

1.1.4 Miscellaneous Anion-Binding Catalysts

One of the most recent motifs for anion-binding catalysis is anion–π binding. Anion–π interactions are defined as favorable NCIs between anions and arenes' π-acidic faces [275–277] and were first proposed in 2002 based on gas-phase computations [278–280]. In 2013, again Matile's group described a series of π-acidic naphthalenediimides for anion-π-binding Kemp-type elimination [281]. A general base deprotonates 5-nitro-1,2-benzisoxazole **148** in the key step yielding *ortho*-hydroxybenzonitrile **149** as the product (Scheme 1.29). Utilizing tetrabutyl-ammonium hydroxide (TBAOH) as a base, up to 7606-fold rate

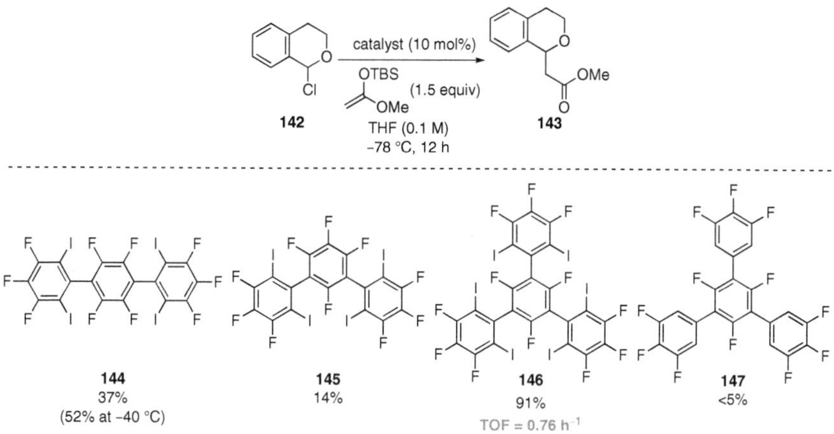

Scheme 1.28 Screening of neutral halogen-bonding derivatives in anion-binding catalysis through chlorine abstraction.

enhancements were observed, corresponding to a transition state stabilization by the naphthalenediimides derivatives of approximately $30\,\text{kJ mol}^{-1}$. Additionally, the authors investigated perylenediimides as a less π-acidic catalyst because of the expanded π-system to examine the π–π interactions in the reaction. The initial rates at high dilution utilizing perylenediimides were significantly lower and indicated the nearly irrelevant contributions of π–π interactions to anion–π catalysis.

Computational investigations using DFT methods (IEFPCM/M06-2X/def2-TZVP/-/IEFPCM/B97-D/6-311G(d,p)) as well as the independence of ground-state stabilization on increasing π-acidity supported this finding [282]. However, in 2014, a theoretical investigation by Lu and Wheeler (CM-M062X/def2-TZVP//PCM-B97-D/6-311g(d,p)) investigated the role of the anion–π interactions utilizing a smaller model system by omitting the linker to the carboxylic acid (**151**; Scheme 1.29c) [283]. The authors concluded that naphthalenediimides stabilizes the substrate complex **150·148** by anion–π binding at least as well as the transition structure, which even leads to a slightly increased activation barrier ($\Delta\Delta G = 0.3\,\text{kcal mol}^{-1}$) and that additional dispersion interactions such as π–π stacking stabilize **150·148** and furnish the high rate acceleration. In the following years, recent examples of anion–π-binding catalysis [284] and the application of non-covalent π–π-interactions for catalyst design were discussed [285].

In 2014, Berkessel *et al.* presented a pyridinium-based catalyst **153·BPh$_4$** for the alkylation of 1-chloroisochroman (48–92% yield) for which an X-ray structure analysis revealed an anion–π interaction, and an NMR titration indicated a 1 : 1 complex with a binding energy of $\Delta G = 13\,\text{kJ mol}^{-1}$ [286]. Using low catalyst loadings of 2–10 mol% resulted in TOF = $3.7\,\text{h}^{-1}$. The authors described that a non-coordinating anion is important for catalytic activity; otherwise, no anion abstraction was observed. Berkessel's group proposed a mechanism starting with the abstraction of chloride and formation of an ion pair **160**. Without the addition of silyl ketene acetal, this ion pair was observed after 24 h as crystalline species suitable for X-ray analysis, confirming the anion-binding mode of the reaction. By the addition of silyl ketene acetal, C–C bond formation occurred, and after

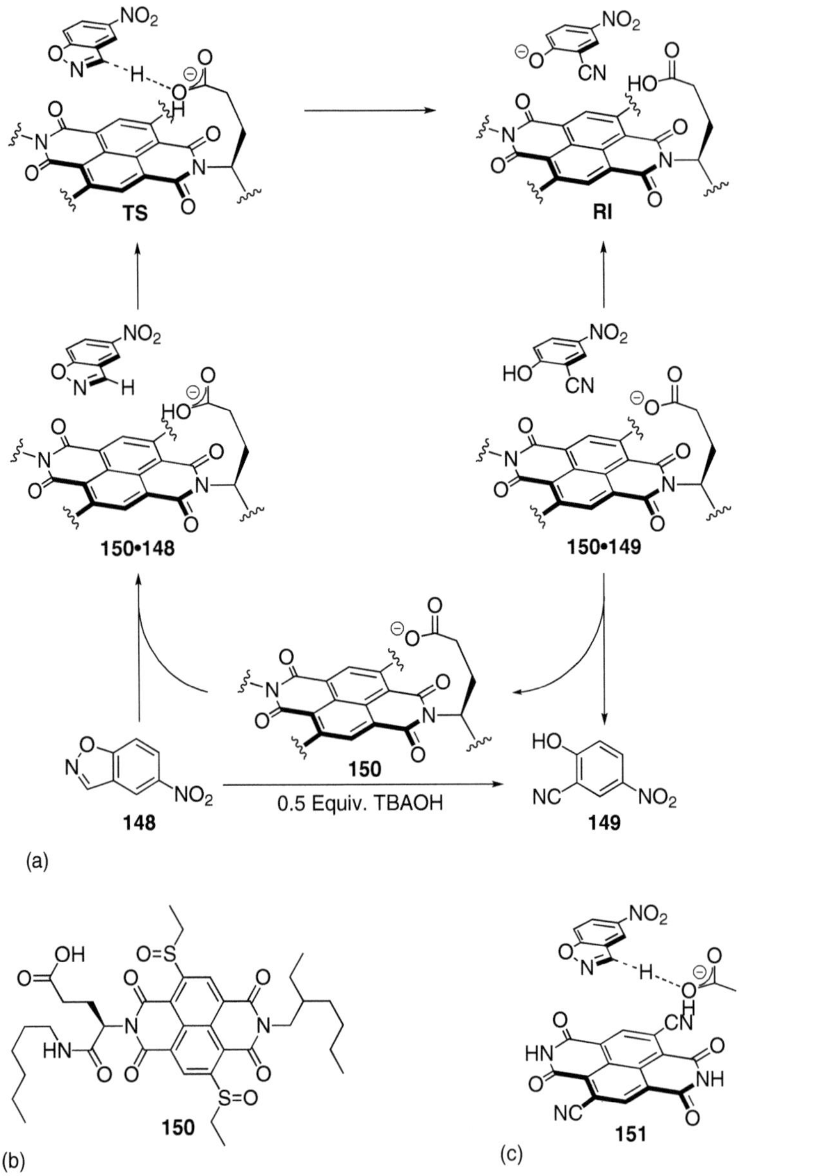

Scheme 1.29 (a) Anion–π-binding mediated Kemp-type elimination of 5-nitro-1,2-benzisoxazole **148**, (b) structure of the most active anion–π catalyst **150**, and (c) reduced model system used for computational investigations by Lu and Wheeler.

desilylation of intermediate **161**, the product and free catalyst were obtained (Scheme 1.30) [286].

For the sake of completeness, we must also mention that 1,1'-binaphthyl-based silanediol [287–289] and thiophosphoramide-based catalysts [290] have been utilized for anion-binding catalysis *via* hydrogen bonding. In 2013, the group of Mattson introduced silanediol **165** for anion-binding-mediated enantioselective acyl Mannich reactions of isoquinolines (Scheme 1.31). The authors proposed an anion abstraction mechanism and nucleophilic attack of silyl ketene acetals to the chiral ion pair and supported this suggestion by crystal structure analysis of the ion pair between **165** and isoquinoline hydrochloride, where **165** binds the chloride and forms a supramolecular complex [287]. Two years later, the same group synthesized derivative **166** that carries four phenyl substituents and showed higher activity (60% *ee* vs. 28% *ee*). Furthermore, Mattson's group suggested that NCIs such as π–π and π–cation interactions contribute to the stabilization of the transition state, which explained the higher stereoinduction of **166** compared to **165** [288].

In 2013, based on the work of Schreiner and coworkers from 2008, where the authors introduced the concept of cooperative catalysis of a Brønsted acid (mandelic acid) and thiourea **46** for Brønsted acid enhancement [291], Nagorny and coworkers utilized a similar catalytic system in [4+2] cycloadditions under mild reaction conditions [290] as compared to previous methods based on the use of highly ionic media, such as lithium perchlorate [292]. In catalyst screening utilizing cyclopentadiene and acrolein acetal **167** as a test reaction, thiophosphoramide **170** was found to be superior to the related thiourea **46** and squaramide **169** in terms of activity (Scheme 1.32). The authors utilized *para*-toluenesulfonic acid that supposedly forms a tetrahedral anion, which explains the high activity of **170** with its ability to form three strong hydrogen bonds to the anion, thus leading to a more stable complex than using **46** or **169**. This effect was support by NMR titration experiments and by the lower activation of HCl that was also activated by the thiourea derivative. With the optimized conditions in hand (6 mol% **170** + 3 mol% *p*-TSA), the authors obtained Diels–Alder products in yields ranging from 57% to 92% (Scheme 1.32). The anion-binding mode in this reaction differs from previously reported reactions in this chapter, where either the nucleophile or the electrophile is activated through anion binding [290].

The sheer number of reactions mirror that anion-binding catalysis, especially utilizing hydrogen-bonding catalysts such as (thio)urea derivatives, is an essential concept in organocatalysis. Moreover, the long known topic *organocatalysis* that had begun in 1832 with the report by Liebig and Wöhler about the well-known cyanide-catalyzed benzoin addition [293] has been very extensively reviewed over the past two decades. We emphasize this because the term "organocatalysis" is a translation of the old and neglected concept "organic catalysts" ("*die organischen Katalysatoren*"), coined by Langenbeck, who recognized the importance of "organic catalysts" and their relationship with the enzymes, with the first authoritative review on this topic as early as 1927 [294–301]. In section 1.2, we present the different modes of substrate activation in anion-binding catalysis. Selected examples from the literature will be used to demonstrate catalyst design concepts; experimental

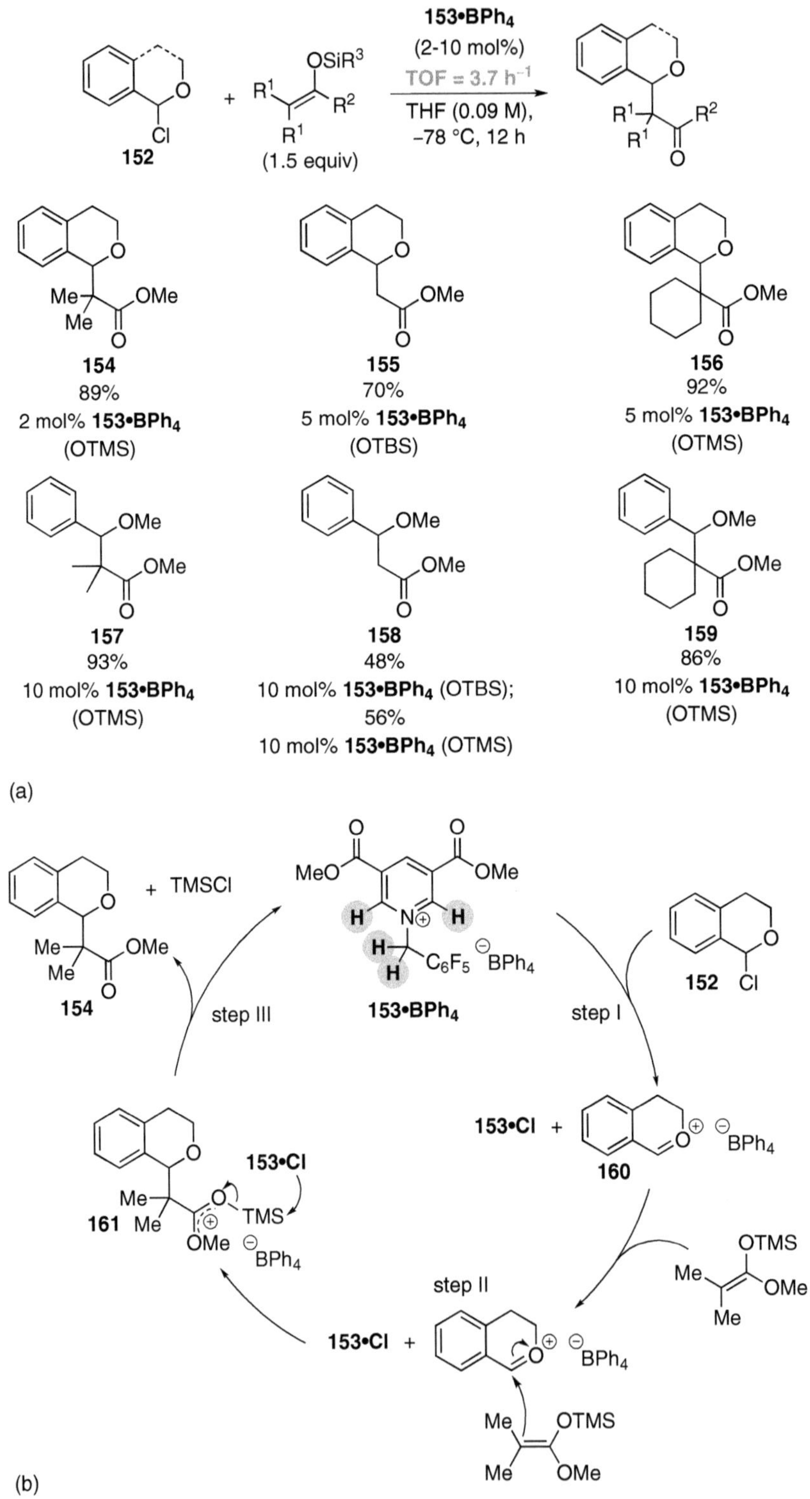

Scheme 1.30 (a) Products of the **153**·BPh_4-catalyzed substitution reaction; (b) proposed catalytic cycle with formation of the oxocarbenium ion **160,** followed by nucleophilic attack of the silyl ketene acetal. The important protons for anion recognition are highlighted.

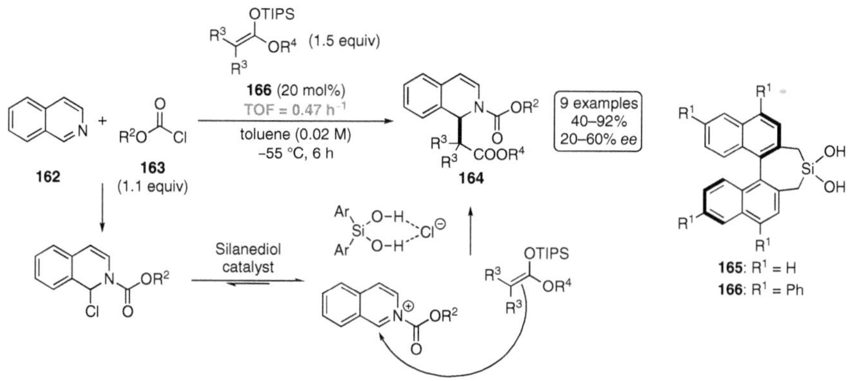

Scheme 1.31 The **166**-catalyed acyl-Mannich reaction utilizing silyl-ketene acetals as a nucleophile.

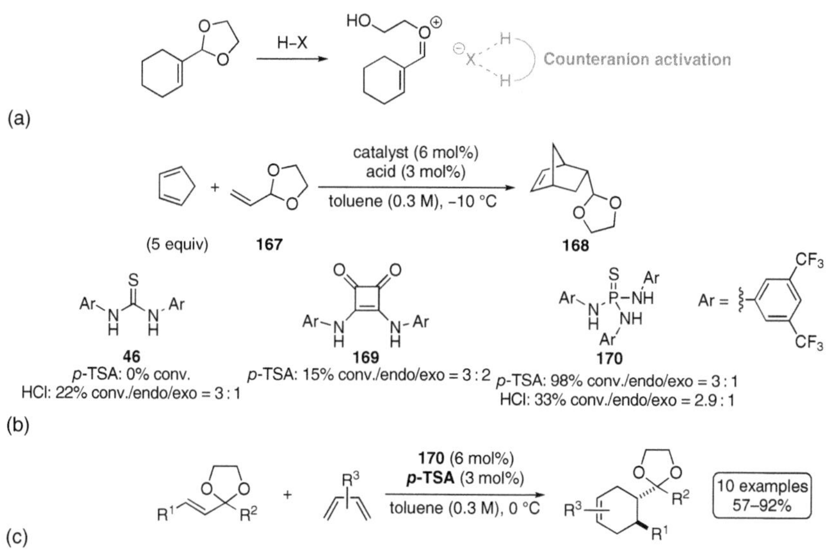

Scheme 1.32 (a) Proposed activation of the oxocarbenium ion in [4+2] cycloadditions *via* Brønsted acid enhancement; (b) comparison of three different hydrogen-bonding catalysts in the activation of the dienophile; and (c) [4+2] cycloadditions catalyzed with the **170**·*p*-TSA system.

details such as structure optimization studies, screening conditions, reaction conditions, and the typical substrate and product spectrum of the representative procedure; and some selected mechanistic studies.

1.2 Concepts in Anion-Binding Catalysis

1.2.1 Introduction

Controlling the stereochemistry of a chemical reaction through charged reagents or catalysts is the goal in asymmetric organic synthesis [302, 303] as every chemical reaction takes place with changes in polarization or charge [304–306]. Asymmetric induction using chiral catalysts is beneficial from economic and ecological viewpoints because the catalyst is utilized in a sub-stoichiometric amount (even though this is not a requirement for catalysis) [307, 308] and is reusable. Nature also catalyzes reactions this way when using enzymes [309]. Catalysts can be distinguished into metal-based or metal-free catalysts ("organocatalysts"). Further, the organocatalysis field can be divided into two main directions: (i) covalent organocatalysis, where the catalyst interacts with the substrate(s) because of covalent bonding, and (ii) non-covalent organocatalysis, where the catalyst interacts through NCIs such as anion-binding, hydrogen-bonding, halogen-bonding, chalcogenide-bonding, and/or dispersion interactions [310]. As anion-binding catalysis is part of non-covalent organocatalysis, this section will not review the field of covalent organocatalysis as it has been comprehensively reviewed elsewhere [311–319]. The two main parts in non-covalent organocatalysis are on the one hand hydrogen-bonding catalysis [6, 157], where the catalyst activates a neutral electrophile, and on the other hand ion pair catalysis, which is characterized by charged catalyst–substrate species. Furthermore, ion pair catalysis is classified into ion-binding catalysis and counterion-directed catalysis and is additionally sub-classified into anionic type and cationic type (Scheme 1.33) [9].

In 1926, Bjerrum introduced the physical principle of ion pairing, in which ion-pairing catalysis rests on and which derives directly from Coulomb's law [320, 321]. Based on his work, Anslyn and Dougherty defined that an ion pair is present when the electrostatic attraction is greater than the thermal energy $k_B T$ (k_B = Boltzmann constant; T = absolute temperature in K) that would be required for separation [322] because of Brownian motion [323]. The distance λ_B (Bjerrum length) defines the gap between two oppositely charged ions (q_1 and q_2), where the electrostatic interaction is inversely related to the dielectric constant of the medium (ε_r), the vacuum permittivity (ε_0), and the thermal energy $k_B T$. Utilizing Eq. (1.1), the distance where electrostatic interactions matter in catalysis can easily be calculated [321]. Therefore, solvation effects are important in ion pair catalysis.

$$\text{Bjerrum length}: \lambda_B = \frac{q_1 q_2}{4\pi\varepsilon_0\varepsilon_e} \frac{1}{k_B T} \tag{1.1}$$

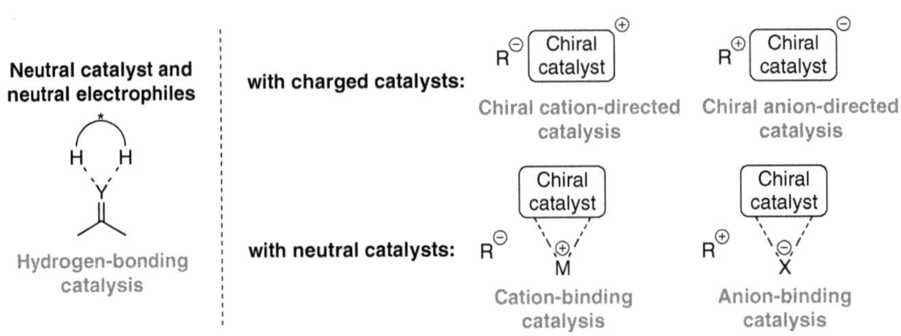

Scheme 1.33 General activation modes of hydrogen-bonding catalysis (left) and ion pair catalysis (right) with counterion-directed catalysis above and counterion-binding catalysis below. Source: Brak and Jacobsen [9].

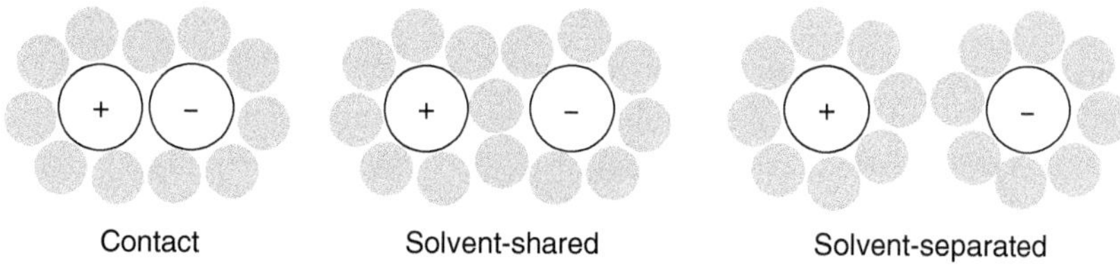

Scheme 1.34 Schematic presentation of different ion pairs . Source: Marcus [324].

Ion pairs are divided into three types that describe the association of the ions and the solvent molecules with one another [324, 325]. A *contact ion pair* exists if both anion and cation share one solvation shell without solvent molecules between them. If solvent molecules are present between the ions, there are two different terms: (i) *solvent shared*, if they share the same solvent shell but are separated through solvent molecules, and (ii) *solvent separated*, if each of the ions has its own solvent shell (Scheme 1.34). Consequently, in ion pair catalysis, the solvent has a significant influence on what type of ion pair exists. Generally, non-polar solvents favor *contact ion pairs* [326, 327], and therefore, in many (enantio)selective reactions, where the selectivity is directed largely by the catalyst, non-polar solvents are usually used.

The synthetic chemist's goal is utilizing small molecules as catalysts compared to nature's catalytic systems, e.g., enzymes [1]. Several intermolecular interactions between the catalyst and the substrate control the selectivity of a reaction, and highly directed catalyst–substrate interactions are required to provide transition-state stabilization. Generally, a distinction is made between four types of interactions as depicted in Scheme 1.35 [9]. Covalent catalysis such as iminium-based catalysis [328, 329] or metal-based Lewis acid catalysis [330, 331] has successfully been utilized for carbonyl activation in enantioselective addition reactions, and both rely on strong and directional interactions. Organic-based Lewis acid catalysis and Brønsted acid catalysis are based on weaker and less-directional interactions between the substrate and the catalyst [1, 11]. While these three types of catalytic systems obviously work well in enantioselective catalysis, ion-paring catalysis is the best strategy for enantioselective transformations when charged intermediates are present. Because ion pairing interactions are inherently less directional, the design of (an)ion-binding catalyst for (enantio)selective reactions combined with low catalyst loadings is a real

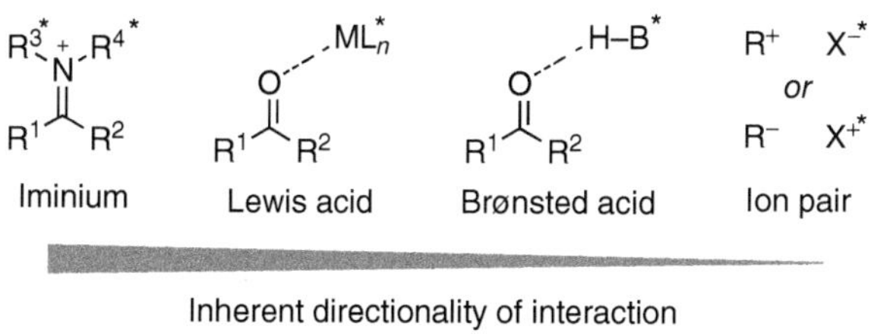

Scheme 1.35 Trend of directionality in common asymmetric catalyst–substrate interactions (L_n^* = chiral ligand; B*H = chiral Brønsted acid, X* = chiral counterion) [9]. Source: Based on Brak and Jacobsen [9].

challenge [9, 10]. However, in the past years, a large number of small molecules were described that operate through the general mechanism depicted in Scheme 1.33.

Note that the concepts of anion-binding catalysis and counterion-directed catalysis are quite different: in anion-binding catalysis, the catalyst binds the anion resulting in a supramolecular complex. In counterion-directed catalysis, such as the well-known asymmetric counterion-directed catalysis (ACDC) [12], the catalyst is an ion pair, e.g., **174**. In ACDC, the chiral counteranion of the catalyst induces stereochemistry, e.g., in the reduction of α,β-unsaturated aldehyde **175** depicted in Scheme 1.36 [332]. First, the formation of a chiral iminium–phosphoric acid ion pair occurs, followed by reduction with Hantzsch ester and formation of **176**. In contrast, the neutral (or positively charged) catalyst conveys the stereoinduction in asymmetric anion-binding catalysis, which is depicted for the Strecker reaction of aldimine **172**. Catalyst **171** binds the cyanide anion in a supramolecular complex. Moreover, catalyst and cyanide ion bind the aldimine *via* hydrogen-bonding interactions, and product **173** forms after nucleophilic attack (Scheme 1.36) [184].

One important characteristic of anion-binding catalysis is that the anion-binding motif of the catalyst only participates in the anion recognition process and does not engage in protonation or deprotonation, which distinguishes anion-binding catalysis from, e.g., guanidinium catalysis. Scheme 1.37 depicts an example of a guanidine-catalyzed asymmetric Strecker reaction utilizing C_2-symmetric **177** that emphasizes these differences [333]. Corey and Grogan postulated a mechanism that starts with the catalyst's coordination (**177·HCN**) and deprotonation of HCN, while coordination of imine **178** gives complex (**179**). After nucleophilic attack of the cyanide, the catalyst reprotonates the imine to furnish product **180**, however, not with the HCN proton but rather with the catalyst's proton. The mechanism was supported by computational studies using DFT methods (B3LYP/6-31G(d)) by Han and coworkers [334].

Most asymmetric anion-binding catalysts bear additional structural elements for substrate activation, e.g., dispersion interaction donors/acceptors [335, 336] or additional hydrogen bond donors/acceptors, and not only the "*anion recognition motif*." In the following sections, the catalyst applications are categorized by their mode of activation, and the evolution of the catalysts regarding conversion and stereoinduction will be discussed with some representative examples. The main activation modes in anion-binding catalysis can be separated into five mechanisms (Scheme 1.38):

(1) *Addition reaction catalyzed via anion binding* [176, 184]: The electrophile **E** is activated *via* a Lewis acid – formally the "leaving group" **M** of the nucleophile $\mathbf{Nu^-}$ – and the catalyst binds the nucleophile. Additionally, NCIs such as, e.g., a π-cation [337], and hydrogen-bonding [184] interactions take place between the catalyst and the electrophile. The nucleophilic attack occurs directly in the supramolecular complex (Scheme 1.38a).
(2) *Catalysis of S_N1-type reactions via anion abstraction and delivery of a cationic species* [242, 338]: The catalyst binds to the leaving group ($\mathbf{X^-}$) and intermolecular nucleophilic attack occurs (Scheme 1.38b). A similar activation mode takes

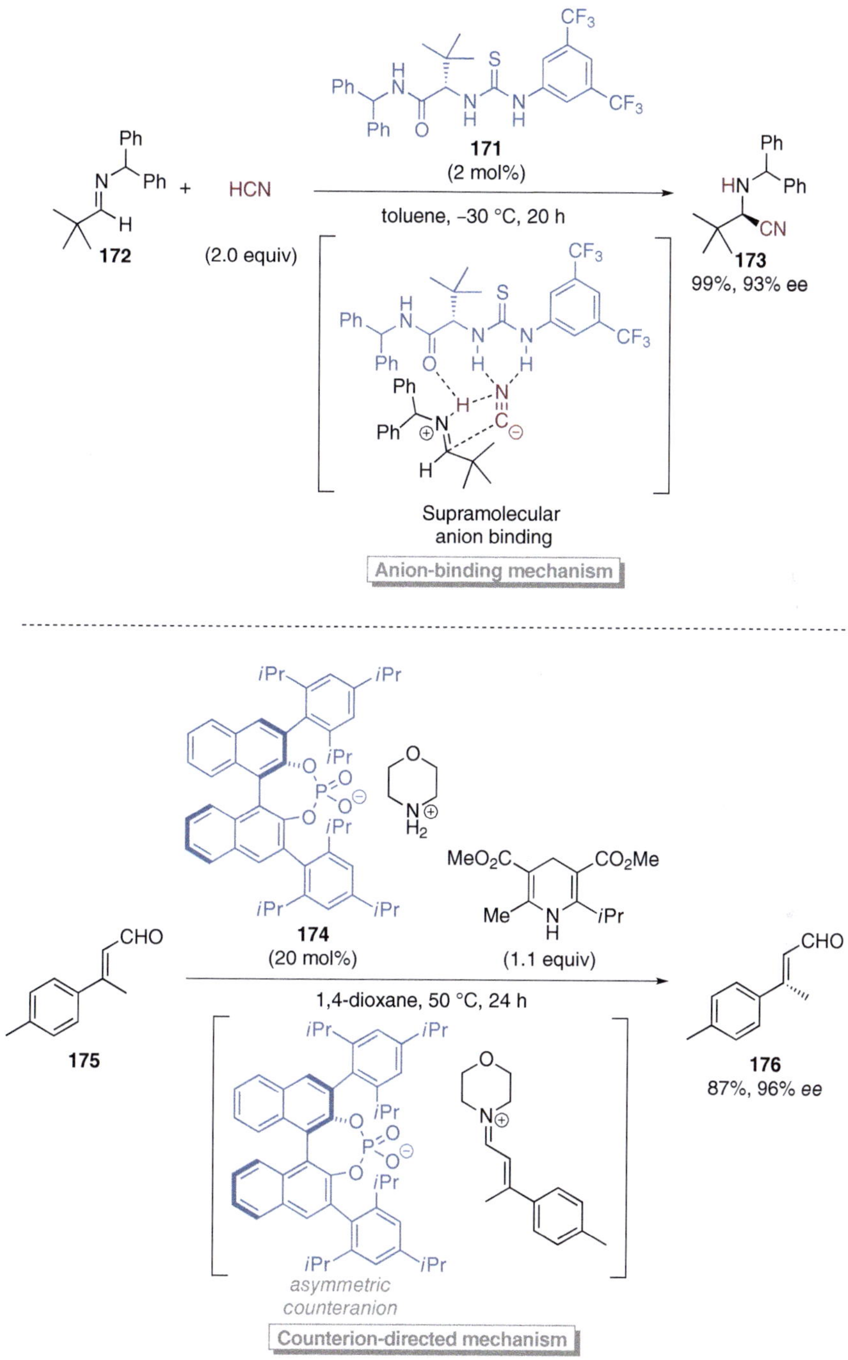

Scheme 1.36 Comparison of the anion-binding mode and the counterion-directed mechanism in asymmetric catalysis. The catalysts are shown in blue and the atoms of the HCN molecule in red for visualization of the anion-binding mode.

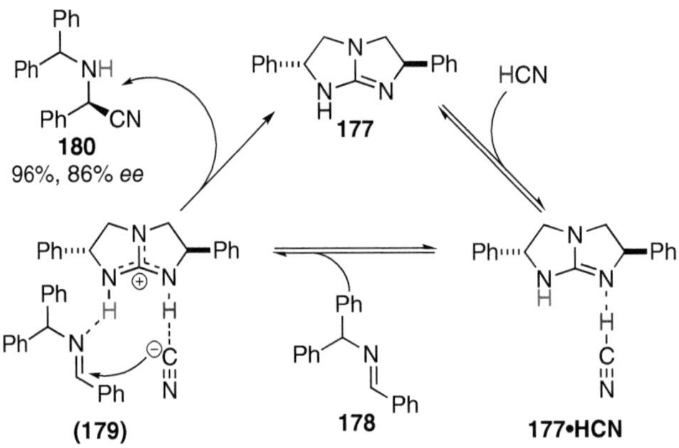

Scheme 1.37 Asymmetric Strecker reaction catalyzed by C_2-symmetric guanidine **177**.

place in the cationic cyclization of iminium [339] or allyl cations [340] *via* anion abstraction.

(3) *Cooperative catalysis with an achiral nucleophilic* ***cocatalyst*** [341]: The **cocatalyst** activates the electrophile **E** and the anion-binding catalyst recognizes the leaving group $\mathbf{X^-}$. Two mechanisms are described: either the direct nucleophilic attack that furnishes the product and the catalyst [342] or an anion exchange from $\mathbf{X^-}$ to $\mathbf{Nu^-}$ with subsequent nucleophilic attack on the electrophile (Scheme 1.38c) [337, 343].

(4) *Lewis acid enhancement catalysis by recognition of a Lewis base* [344, 345]: An initial Lewis acid $\mathbf{LB{-}LA_1}$ – typically trialkylsilyl triflates – dissociates and the anion-binding catalyst binds the Lewis basic part "LB" (triflate) as well as the Lewis acidic part "$\mathbf{LA_1}$" (trialkylsilyl), leading to an enhanced Lewis acid. The catalyst binds the electrophile, which also possesses a Lewis acidic part "$\mathbf{LA_2}$" – similar to "LA_1" – by NCIs. After the nucleophilic attack, the product and the enhanced Lewis acid catalyst – containing Lewis acidic part "$\mathbf{LA_2}$" – form (Scheme 1.38d).

(5) *Phase transfer catalysis via anion delivery*: The catalyst binds an anion $\mathbf{X^-}$ that derives from a soluble and an insoluble salt (in organic solvents) such as cesium fluoride [346, 347] or potassium fluoride [348] and forms a soluble ion pair. Additionally, the catalyst binds the electrophile through NCIs and the nucleophilic attack occurs directly in the supramolecular complex (Scheme 1.38e).

1.2.2 Anion-Binding Catalysis in Addition Reactions

In 1998, Sigman and Jacobsen utilized a library of polystyrene-bound Schiff bases and high-throughput screening (HTS) in the asymmetric Strecker reaction of *N*-allyl benzaldimine with TBSCN as the cyanide source and described the first asymmetric thiourea organocatalyst (Scheme 1.10) [174]. The basic goal of this research work was to find a new tridentate ligand for a chiral organometallic catalytic system; therefore, the authors used solid-phase synthesis and systematic structure

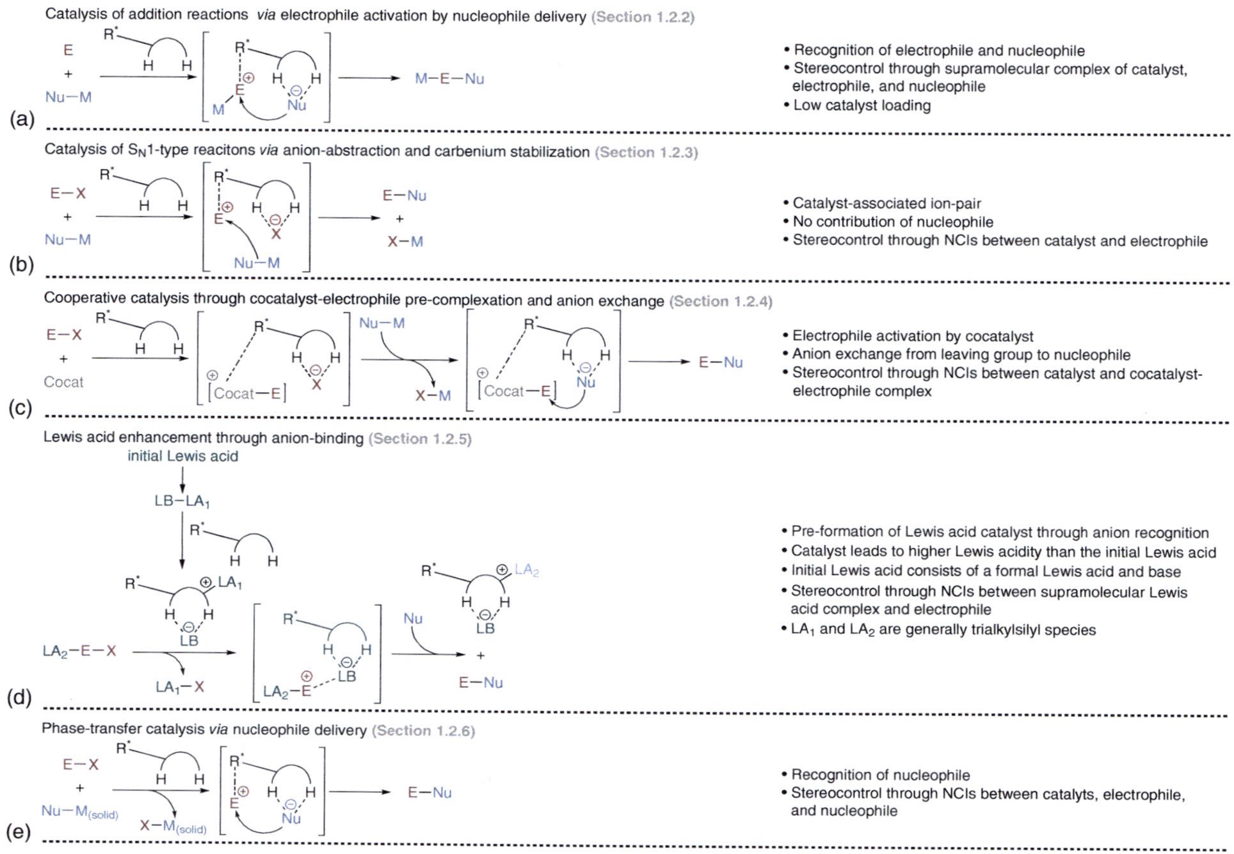

Scheme 1.38 The five different activation modes in asymmetric anion-binding catalysis: (a) Addition reactions *via* binding of the anionic nucleophile, (b) S_N1-type reactions by anion abstraction, (c) cooperative catalysis with anion exchange from the counterion to the nucleophile, (d) binding of a Lewis base and formation of a Lewis acid with enhanced acidity, and (e) phase transfer catalysis *via* nucleophile delivery.

optimization variations leading to high diversity of potential catalysts. The core structure of the tridentate ligand was based on a chiral amino alcohol and a salicylaldehyde derivative, which coordinates a metal ion. In order to implement the idea of solid-phase synthesis, the amino alcohol was replaced with chiral 1,2-diamines ((*R*,*R*)-1,2-diaminocyclohexane or (*R*,*R*)-1,2-diphenyl-ethylendiamine), which could be attached *via* a linker to a solid support. In addition, a second chiral element, an amino acid, was incorporated and linked *via* caproic acid (linker 1) to the polystyrene support and *via* a (thio)urea or a guanidine group (linker 2) to the chiral diamine (Scheme 1.39) [174].

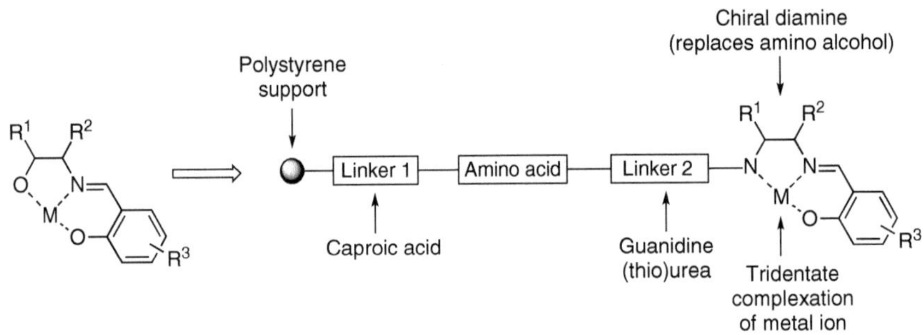

Scheme 1.39 Initial work stream for HTS optimization to identify chiral metal ligand for enantioselective metal-based Strecker reaction.

Sigman and Jacobsen performed iterative HTS optimization of the Strecker reaction enantioselectivity with three libraries consisting of 12 (library 1), 48 (library 2), and 132 (library 3) polymer-bound catalyst candidates. Library 1 (variation of the metal ion) led to the highest enantioselectivity when the pure, metal-free organocatalytic system (19% *ee*) was utilized rather than in the presence of a metal ion (11 complexes tested, highest enantioselectivity: Ru = 13% *ee*). In Library 2, the general influence of both linkers, the relative stereochemistry of the amino acid compared to the diamine, the amino acid unit, and the substitution pattern of the salicylaldehyde derivative were investigated. The authors found that linker 1 (caproic acid) leads to an unspecified background side reaction and was therefore removed, leading to the amino acid unit coupled directly to the polystyrene support. Furthermore, Sigman and Jacobsen observed the highest *ee*-values with the 3-*tert*-butyl-substituted salicylaldehyde and leucine as amino acid. The relative stereochemistry of the catalyst was important for enantioselectivity; with (*R*,*R*)-diamine derivatives and L-leucine (32% *ee*), the enantioselectivity was much higher than with D-leucine (5% *ee*). Linker 2 was also found to be crucial for enantioselectivity, and thiourea derivatives (55% *ee*) turned out to be superior to either urea-based systems (45% *ee*) or guanidine derivatives (21% *ee*). The focus of library 3 was on different non-polar amino acids, 1,2-diamino derivatives, and the substitution pattern of the salicylaldehyde unit. Sigman and Jacobsen found bulky L-*tert*-leucine and 3-*tert*-butyl-5-methoxysalicylaldehyde to provide the best results. Moreover, (*R*,*R*)-1,2-diaminocyclohexane was found to be slightly superior to (*R*,*R*)-1,2-diphenyl-ethylendiamine, and finally, Schiff base thiourea **48** was determined as the most efficient catalyst in terms of enantioselectivity [174]. The corresponding Schiff base thiourea **47** incorporating the important units of the polystyrene-bound derivative **48** was synthesized independently in solution and tested in the asymmetric Strecker reaction of aromatic as well as aliphatic *N*-allyl-imine derivatives. Utilizing 2 mol% of catalyst **47** and HCN as a cyanide source, trifluoroacetylated Strecker adducts were formed in the range from 65% to 92% yield and *ee*-values of 70–91% (cf. Scheme 1.10) [174]. Based on the core structure of catalyst **47**, Jacobsen and coworkers constructed a new optimization library of 70 polymer-bound Schiff bases incorporating seven amino acids with large α-substituents and ten new salicylaldehyde derivatives in 2000 [349]. After evaluating each library member in HCN [350] by adding to the *N*-allyl imine of pivaldehyde at 23 °C, polymer-bound 5-pivaloyl-substituded derivative **181** was found as the most efficient catalyst. Furthermore, Jacobsen's group synthesized the urea derivative **182** because of the higher solubility and easier preparation and observed comparable catalyst efficiency and utilized **182** in the asymmetric Strecker reaction of various aliphatic as well as aromatic aldimines and ketoimines (Scheme 1.40).

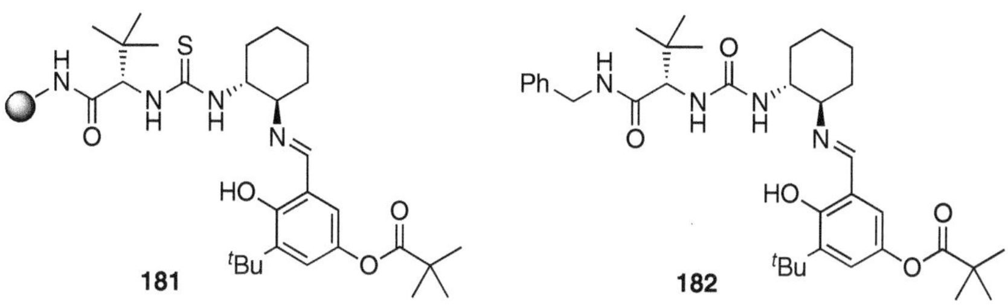

181

182

Scheme 1.40 More active Schiff base catalyst **181** found by HTS and the polymer-free urea derivative **182**.

In 2002, Vachal and Jacobsen utilized NMR spectroscopy, kinetic structure–activity relationships, and DFT studies (at the uncorrected B3LYP/6-31G(d,p) level of theory on a much simplified system to model the catalyst–imine and catalyst–product complexes in the gas phase) to elucidate the activation mode of Schiff base **182** [351]. Based on NOESY and ROESY NMR, the authors found that **182** exists in a well-defined secondary structure with the urea moiety in a *syn–syn* conformation. The kinetics of the hydrocyanation followed Michaelis–Menten kinetics, with first-order dependence on **182** and HCN, and saturation kinetics with respect to the imine substrate. Through alkylation of the urea nitrogens, or replacement of the urea moiety by a carbamate unit, the authors found that both urea N–H protons bind the imine substrate, similar to proposals by Wittkopp and Schreiner [158, 352]. Based on multiple NOE interactions between **182** and a series of *Z*-imines, such as 3,4-dihydroisoquinoline, Vachal and Jacobsen determined the orientation between the substrate and the catalyst. Cross-peaks of the catalyst in the free and bound state were essentially invariant and indicated that no significant change in conformation results from binding the substrate. The authors describe a clamp-like hydrogen-bonding interaction of both urea N–H to the imine-nitrogen, with the substrate being twisted. Furthermore, titration of the catalyst with *Z*-imines results in up-field shifts of the two urea protons to similar degrees, supporting this proposal. The gas-phase DFT studies underscored the results of the NMR measurements, even though the computed structure of the catalyst–product complex only provided one hydrogen bond as the most energetically favorable complex. The bridging interaction in the catalyst–substrate complex was found stronger (8.5 kcal mol^{-1} for urea; 10.0 kcal mol^{-1} for thiourea) than the single hydrogen bond event (5.0 and 6.3 kcal mol^{-1}, respectively). Based on these results, the authors describe the reversible formation of an imine–catalyst complex with **182** binding the imine through hydrogen bonding (cf. Scheme 1.8), with approximately 80% formation of the complex found in NMR experiments. Furthermore, the model of the catalyst–substrate complex provided information about the observed scope and stereoselectivity of the Schiff base (thio)urea-catalyzed Strecker reaction:

(1) The large group on the imine carbon is directed away from the catalyst and into the solvent. This explains why the Schiff base catalyst promoted hydrocyanations with high *ee*-values, regardless of the steric and electronic properties of the substrate.
(2) The small group (H for aldimines and Me for methylketoimines) is directed toward the catalyst, which indicates that ketoimines bearing larger substituents are poor substrates for the reaction, presumably because they could not adopt the optimal geometry.
(3) The N-substituent is also directed away from the catalyst. However, its size is restricted as a result of the requirement to access the *Z*-isomer of the imine.
(4) On the basis of the observed trend of stereoinduction, addition of HCN takes place over the diaminocyclohexane portion of the catalyst and away from the amino acid/amide portion.

Following the work of Kotke and Schreiner in anion-binding catalysis in 2006 (cf. Schemes 1.12 and 1.13 [176]), Zuend and Jacobsen re-examined the cyanide addition to imines utilizing computations, Hammett analyses, catalyst structure/activity relationships, and isotope labeling studies and concluded that non-covalent interactions are crucial for stereoinduction [184]. The authors utilized modified (thio)urea derivatives of catalyst **182** and instead of the diaminocyclohexane moiety various phenyl substituents were incorporated, especially the well-established 3,5-bis(trifluoromethyl)phenyl group [190]. Similar to the work of Vachal and Jacobsen in 2002, Zuend and Jacobsen observed first-order kinetics for imine and HCN; however, the value of 0.8 for the catalyst is different from the previous work utilizing **182** (*vide supra*). Utilizing "same excess" experiments, the authors excluded deactivation of the catalyst and that product inhibition was negligible. After the authors excluded the competitive uncatalyzed background reaction due to the high *ee*-values and the slow initial rate of approximately 5% compared to the catalyzed reaction, they proposed a small degree of catalyst dimerization that was also observed in previous work [353, 354]. A Hammett analysis for distinguishing the two possible reaction pathways that could be started by protonation of the imine (Scheme 1.41a) or by addition of cyanide to the imine (Scheme 1.41b) provided strong evidence for the former because negative ρ-values were obtained

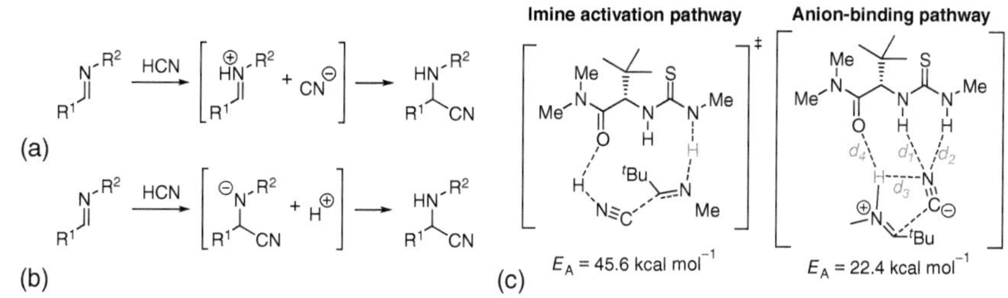

Scheme 1.41 Two possible mechanistic pathways of (thio)urea-catalyzed Strecker reaction *via* (a) an iminium ion or (b) an α-aminonitrile anion and (c) computed activation energies for both reaction pathways.

($\rho_{\text{thiourea}} = -2.7$; $\rho_{\text{urea}} = -2.5$). Additionally, utilizing uncorrected DFT gas-phase computations (B3LYP/6-31G(d)) on a reduced model system (cf. Scheme 1.41c) at 0 K, the authors found that the activation barrier for the iminium activation pathway was approximately 23 kcal mol^{-1} higher in energy than the anion-binding pathway and that the activation energy using urea catalyst is about 5.6 kcal mol^{-1} higher than that of the thiourea catalyst [184].

Based on these result, Zuend and Jacobsen investigated a mechanism based on the thiourea binding the cyanide and forming an ion pair complex with the iminium ion utilizing isotope labeling experiments [184]. The authors utilized DCN and obtained the N-deuterated Strecker product, which indicated that the N–H protons of the (thio)urea moiety do not acts as Brønsted acids as would be required in pathway **B** (Scheme 1.41c). Based on the computations noted above, the authors proposed a reaction pathway that includes proton transfer of HCN (or HNC) to the imine, and generation of a thiourea-bound cyanide–iminium ion pair (Scheme 1.41c). The experimental energy differences (Δ*G*) between the HCN and HNC protonation pathways were too low to be distinguished. Nevertheless, the outcome is the same, as the HNC protonation pathway convert into the HCN protonation pathway because of rearrangement of the cyanide ion during the reaction. This proposed mechanism was supported by the strong correlation between the experimental and the computed enantioselectivities for eight different thiourea catalysts. To elucidate the stereoinduction, correlation plots were constructed for different bond lengths vs. enantioselectivity. No trend was observed when plotting the sum of the computed thiourea–cyanide bond lengths ($d_1 + d_2$) vs. enantioselectivity, implying that the enantioselectivity cannot result from the stabilization of the cyanide. In contrast, a positive correlation was observed between the enantioselectivities and the computed imine N–H hydrogen bond distances to the cyanide anion and amide carbonyl ($d_3 + d_4$) (Scheme 1.41c). Therefore, the enantioselectivity can be ascribed to the stabilization of the iminium cation in the diastereomeric transition states of the ion pair rearrangement [184]. All in all, the authors showed that secondary design elements for the formation of NCIs are crucial for stereoinduction. Simultaneously, Jacobsen's group utilized thiourea **183** (2 mol%) in Strecker reactions of various imines and obtained the products **185** in yields ranging from 96% to 99% and *ee*-values of 73% to 99% [355]. Compared to the previously utilized Schiff base **182**, the catalyst loading was slightly higher, but the yields and *ee*-values significantly increased. The authors developed two different methods for the *in situ* generation of HCN utilizing KCN + AcOH or TMSCN + MeOH. The KCN/AcOH method gave the products with slightly decreased enantioselectivities at higher reaction temperatures, concentrations, and reaction rates (Scheme 1.42). Additionally, solid KCN is more practical than liquid TMSCN, and starting with the imine of pivaldehyde, Jacobsen's group obtained (*R*)-Boc-*tert*-leucine in 62–65% yield of crystalline product and 98–99% *ee* over four steps [355].

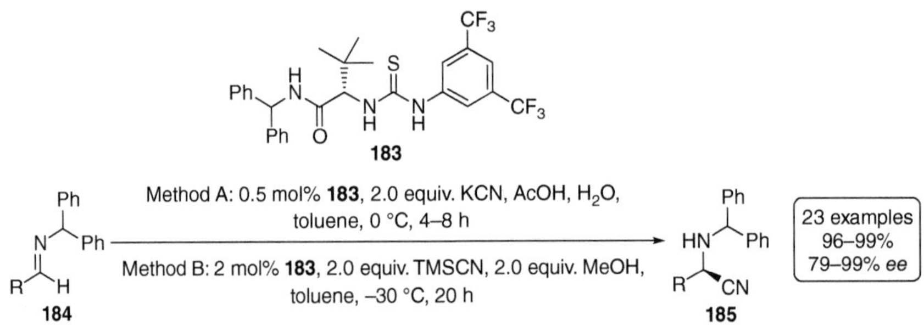

Scheme 1.42 The improved Strecker reaction of various aldimines with thiourea **183** with two different HCN sources.

In addition to the Strecker reaction, the Michael addition is a common and often used reaction in organic synthesis [356–358]. In 2003, Takemoto and coworkers introduced bifunctional catalyst **49** for the use in this kind of addition reaction (cf. Scheme 1.11) [161]. Takemoto's group synthesized a series of diaminocyclohexane-based thioureas (e.g., **49** and **189–192**) and screened them in the model Michael addition of diethyl malonate **186** to *trans*-β-nitrostyrene **187** at room temperature (Scheme 1.43). The authors identified tertiary amine-functionalized **49** as the most efficient catalyst in terms of catalytic activity (86% yield; 24 h) and enantioinduction (83% *ee*). Utilizing the chiral amine **189** that lacks the thiourea moiety, Takemoto's group obtained **188** in only 14% yield (24 h) and 35% *ee*. With achiral **190** lacking the tertiary amine group (requiring addition of 10 mol% triethylamine), the authors provided **188** in only 57% yield after 24 h. Using bulkier tertiary amine **191** and longer reaction time (48 h), the yield decreased (76%; 48 h) and the *ee*-value was slightly lower (87%). Comparing **49** and **192**, which lacks the two CF_3 groups, the advantage of the 3,5-bis(trifluoromethyl)phenyl motif was well recognized [157, 158, 190], as **192** provided the Michael adduct in only 58% yield and 80% *ee* within 48 h (Scheme 1.43) [161].

In 2005, Takemoto's group investigated the mechanism of the **49**-catalyzed Michael addition utilizing NMR kinetic studies and NMR titrations and postulated a reaction mechanism that starts with the deprotonation of the C–H-acidic 1,3-dicarbonyl compound by the tertiary amine and formation of complex (**193**), in which the six-membered enol of **187** is stabilized through interactions with the protonated tertiary amine group of **54** [175]. Subsequently, thiourea's N–H protons coordinate with the Michael acceptor through hydrogen-bonding interactions

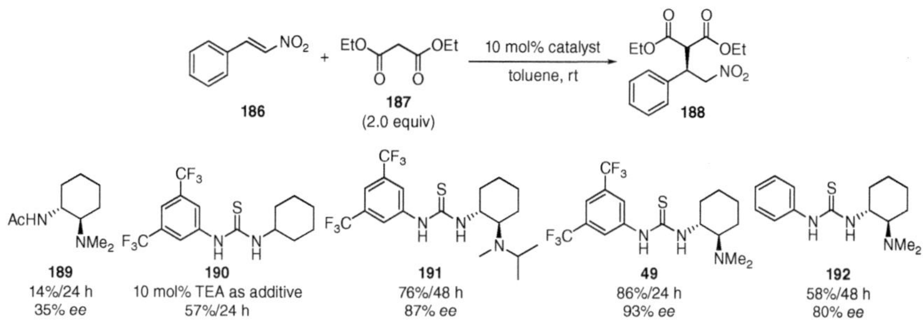

Scheme 1.43 Catalyst screening using chiral 1,2-diaminocyclohexane **413** and various thiourea derivatives in asymmetric Michael addition of diethyl malonate to *trans*-β-nitrostyrene.

resulting in the organized ternary complex (**194**) and a relative orientation that allows nucleophilic attack in an (*R*)-favored mode leading to complex (**195**). After final protonation, the catalyst–product complex dissociates and furnishes free catalyst and **188** (Scheme 1.44). Based on this proposed mechanism, Liu and coworkers performed DFT computations (B3LYP/6-31G(d)), utilizing diethyl malonate and simplified nitroethene as well as the corresponding urea of **49**

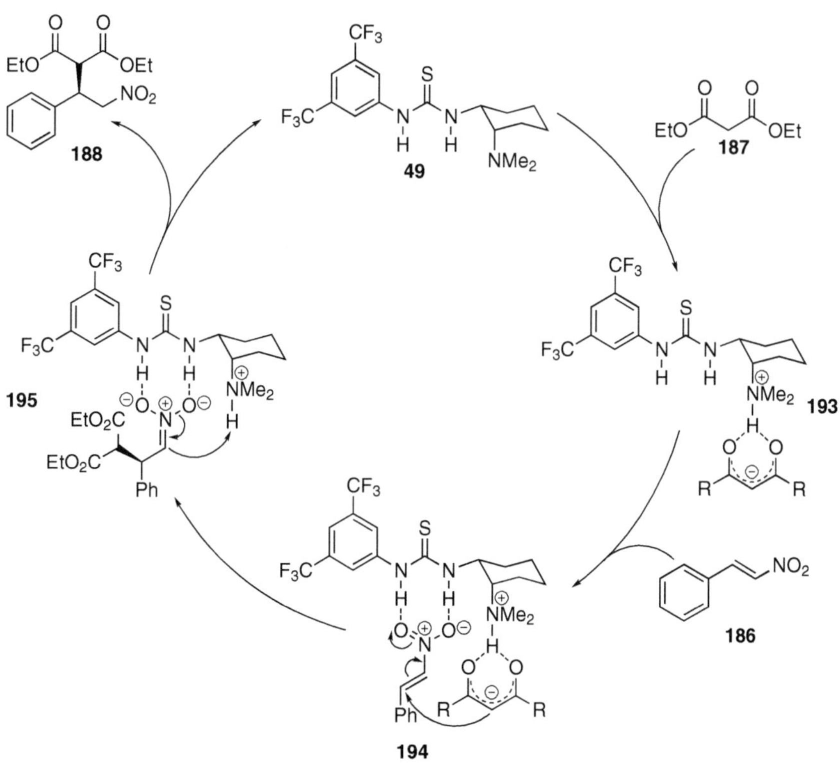

Scheme 1.44 Postulated mechanism of the enantioselective Michael addition between *trans*-β-nitrostyrene **186** and diethyl malonate **187** promoted by bifunctional **49**.

unfortunately lacking the 3,5-bis(trifluoromethyl)phenyl substituent, and proposed that the C–C bond formation step from **194** to **195** is enantiodetermining, while the reprotonation from catalyst's amine to the α-carbon in complex **195** was identified as the rate-determining step [359].

In the following years, thiourea **49** and its derivatives were well established in organic synthesis and were utilized in many asymmetric reactions [4, 360–363]. The synergetic activation of nucleophile and electrophile, leading to high conversion and enantioselectivity, was generally accepted. In 2006, Soós, Pápai, and coworkers reinvestigated the mechanism of the **49**-catalyzed Michael addition utilizing DFT methods (B3LYP/6-311++G(d,p)//B3LYP/6-31G(d)), with β-nitrostyrene as Michael acceptor and 2,4-pentadienone as Michael donor [364]. The authors investigated in addition to Takemoto's mechanism (hydrogen-bonding activation; pathway A) an alternative mechanism *via* anion binding of the enolate (pathway B). Both mechanisms afford the (*R*)-configured product; nevertheless, the transition structure for pathway B is preferred by $\Delta\Delta G = 2.7\,\text{kcal mol}^{-1}$ compared to pathway A because of the higher number of NCIs. This theoretical study reveals the alternative mechanism of the **49**-catalyzed Michael addition and provides strong evidence for the anion-binding mode. In 2019, Hirschi, Vetticatt, and coworkers utilized a combination of experimental ^{13}C kinetic isotope effects (KIEs) and DFT computations (B3LYP/6-31+G(d,p)/PCM (toluene)) for mechanistic investigations on the **49**-catalyzed Michael addition of diethyl malonate to β-nitrostyrene [365]. The authors found the lowest lying transition structure **196**, which is very similar to that described in previous work [366]. Moreover, the lowest lying transition structure **197** of the opposite enantiomer was $\Delta\Delta G = 2.4\,\text{kcal mol}^{-1}$ higher in energy because of the lack of the stabilizing hydrogen bond between the ortho C–H and the carbonyl oxygen. Furthermore, a similar transition structure **198** to that postulated by Takemoto's group (hydrogen bond activation of the nitroolefin) [175], but incorporating an additional hydrogen bond between ortho C–H and the carbonyl oxygen, was computed and found to be $\Delta\Delta G = 4.7\,\text{kcal mol}^{-1}$ higher in energy (Scheme 1.45). The authors also computed the transition structures for the deprotonation of the diethyl malonate and the reprotonation of the nitronate, and the corresponding KIEs for the key carbon atoms of the reactants for all computed transition structures. The authors found that the C–C bond forming step is the rate-determining as well as the enantiodetermining step of this reaction, which was inconsistent to the proposed mechanism by Liu and coworkers [359], who postulated the reprotonation of the nitronate as the rate-determining step. The work by Hirschi, Vetticatt, and coworkers confirmed the anion-binding mode of the **49**-catalyzed Michael addition.

1.2.3 Anion-Binding Catalysis in Substitution Reactions

(Thio)ureas are well known as potent halide-binding hosts in supramolecular chemistry, and the corresponding supramolecular complexes could be easily investigated *via* ^{1}H NMR and IR spectroscopy [82, 367]. Therefore, over the years, many halide-binding approaches utilizing (thio)ureas were developed, in particular,

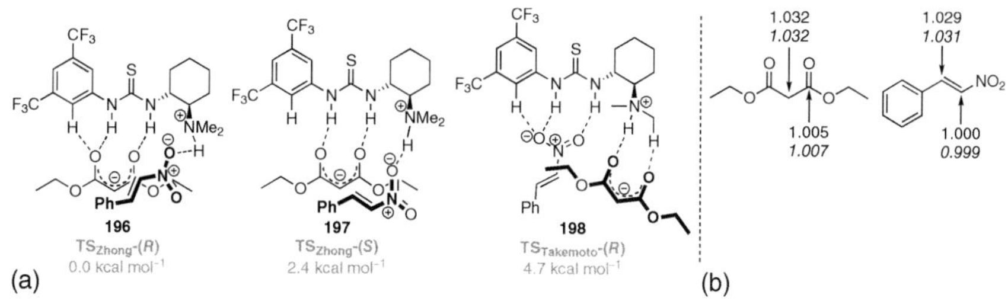

Scheme 1.45 (a) Three computed transition structures for the Michael addition of diethyl malonate and β-nitrostyrene with ΔG values related to **196** and (b) experimental KIE and computed (italics) KIE for C–C bond forming step as the rate-determining step.

for halide abstraction and substitution reactions [2, 9, 10]. In 2002, Wenzel and Jacobsen used a Schiff base catalyst in asymmetric Mannich reactions for the synthesis of β-aryl-β-amino acids and proposed an activation of the utilized *N*-Boc aldimines through hydrogen bonding [368], similar to the proposed activation mode in Strecker reactions by Vachal and Jacobsen (*vide supra*) [351]. In 2004, based on the work of Wenzel and Jacobsen, Taylor and Jacobsen presented the cyclization of indole derivatives **200** because of an asymmetric acetyl Pictet–Spengler reaction and suggested a similar activation mode [338]. The authors screened various thiourea derivatives, such as Schiff base **182** (cf. Scheme 1.40). Utilizing acetic anhydride, Taylor and Jacobsen did not observe product formation, even at high temperatures (Pictet–Spengler conditions). Switching to acetyl chloride as the acetylation reagent, the authors obtained the product in 65% and 59% *ee* catalyzed with 10 mol% **182**. After structure optimization of the catalyst, **199** was described as the most active one, bearing 2-methyl-5-phenylpyrrole moiety instead of the salicylaldehyde unit in **182** (70% yield; 93% *ee*). The authors described activation of the acyl-iminium ion by the thiourea's N–H protons and obtained Pictet–Spengler products **201** in yields ranging from 65% to 81% and *ee*-values of 86–95%, utilizing 5 or 10 mol% catalyst loading, whereby the imine substrate was generated *in situ* by condensation of the tryptamine derivatives with the corresponding aldehyde (Scheme 1.46). Furthermore, the products could be easily converted to tetrahydro-β-carbolines that are core structure elements in natural compounds [369, 370].

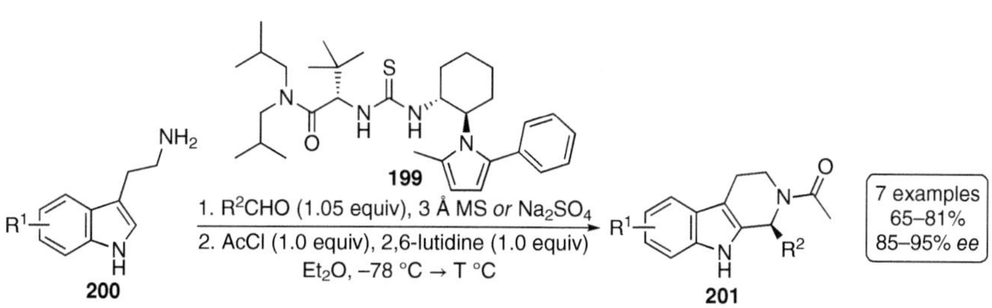

Scheme 1.46 Asymmetric acetyl Pictet–Spengler reaction catalyzed with 5 or 10 mol% **199**.

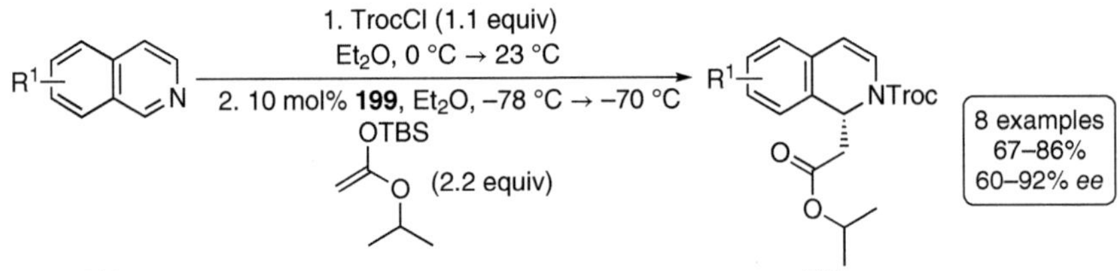

202 203

Scheme 1.47 Enantioselective acyl-Mannich reactions utilizing various substituted isoquinolines catalyzed with **199**.

In 2005, the same group utilized **199** in asymmetric acyl-Mannich reactions of isoquinolines **202** and obtained substituted dihydroisoquinolines **203** in yields ranging from 67% to 86% and *ee*-values of 60–92%, utilizing 10 mol% catalyst loading (Scheme 1.47) [371]. Additionally, Jacobsen's group observed in the acyl-Mannich as well as in the acyl Pictet–Spengler reactions a pronounced solvent effect, with diethyl ether providing the highest *ee*-values. The authors pointed out that the nature of the acyl-imine adduct is important in the reaction and deemed TrocCl the best acylation reagent, with *tert*-butyldimethylsilyl ketene acetal being the most reactive nucleophile. Bose, Spiegelman, and Manhas observed in the acylation of benzylimine with various acyl chlorides the formation of a covalent chloroamide in non-polar solvents, such as carbon tetrachloride [372]. Based on this work, and because of the strong leaving group effect of the acylation reagent and the high enantioinduction in non-polar solvents, such as diethyl ether, Jacobsen's group postulated the presence of the chloroamide structure, rather than the *N*-acylium chloride structure of the acyl-imine adduct [371].

In 2007, Jacobsen's group presented the enantioselective Pictet–Spengler-type cyclization of β-indolyl ethyl hydroxy lactams **205** utilizing TMSCl as a dehydrating agent to form *in situ* and irreversibly the corresponding chloride derivatives [373]. The hydroxy lactam substrates were synthesized either by imide reduction utilizing $NaBH_4$ or by imide alkylation with organolithium reagents. The authors obtained the cyclization products **206** in yields ranging from 51% to 94% and *ee*-values of 81–99%, utilizing pyrrole-based catalyst **204** (Scheme 1.48), which is the *N*-methylpentyl amide derivative of **199** (cf. Scheme 1.46) [373].

Jacobsen's group also investigated the mechanism using substituent-, counterion-, solvent-, and kinetic isotope effects and variable temperature ^{1}H NMR studies and

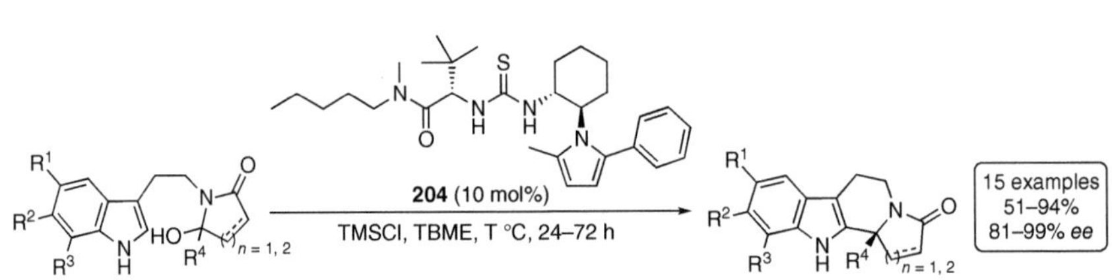

205 206

Scheme 1.48 Asymmetric Pictet–Spengler-type cyclization of hydroxy lactams *via in situ* generation of iminium ions catalyzed with **204**.

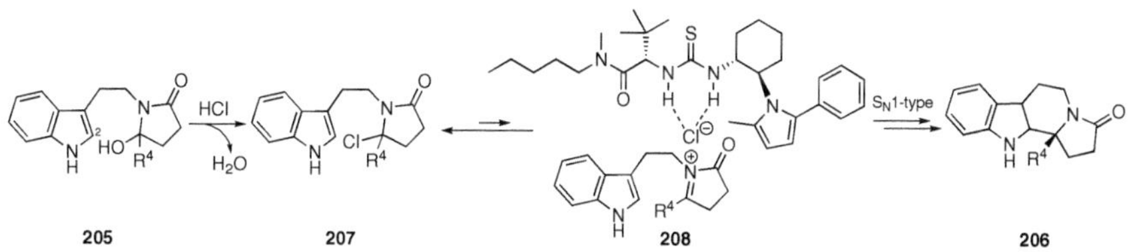

Scheme 1.49 Proposed mechanism for the enantioselective Pictet–Spengler-type cyclization of β-indolyl ethyl hydroxy lactams **205** with thiourea **204**.

suggested that the reaction starts with the TMSCl-induced formation of a chlorolactam, comparable to the chloroamides in the acyl Mannich reaction (cf. Scheme 1.47). Subsequently, the product forms through either an S_N2-type mechanism or a stepwise S_N1 route involving coordination of the chloride to the thiourea. Multiple observations indicate the stepwise mechanism (Scheme 1.49):

(1) The reaction rate was much higher with substituents stabilizing positive charge in the α-position.
(2) There is a strong enantioselectivity dependence on the counteranion (Cl, 97% *ee*; Br, 68% *ee*; I <5% *ee*), supporting the stepwise mechanisms *via* anion binding.
(3) No KIE was observed in reactions utilizing indole with deuterated C_2, ruling out the possibility of a rate-limiting deprotonation/rearomatization step.
(4) NMR experiments of mixtures of the catalyst with tetrabutylammonium chloride (TBAC) resulted in a downfield shift of 0.56 ppm of thiourea N–H protons, while bromide as well as iodide counterions appear at lower downfield shifts. These results indicate the anion-binding mode [373].

This reaction is the second reported example for anion-binding catalysis utilizing hydrogen-bonding organocatalysts, after Kotke's and Schreiner's fundamental work (cf. Schemes 1.12 and 1.13) [176]. Nevertheless, this is the first mechanistic proposal that suggests that a hydrogen-bonding organocatalyst binds an anion in an enantioselective reaction [373], whereby it is generally excepted that also the acetyl Pictet–Spengler (Scheme 1.46) as well as the acyl-Mannich reactions (Scheme 1.47) are catalyzed by anion-binding interactions [9, 10]. A similar chloride-binding concept was applied in many other anion-binding catalyzed reactions [2, 9, 10], e.g., Jacobsen and coworkers utilized a similar activation mode in the biomimetic cyclization of hydroxy lactams [339], which is based on the polycyclization of *N*-acyliminium ions of Dijkink and Speckamp of the 1970s [374, 375]. The idea was to convert the hydroxy lactams *in situ* into the corresponding chlorolactams and to use a bifunctional catalyst that activates the chlorolactams by anion abstraction while stabilizing the cationic intermediates **211** [339]. The authors utilized HCl for the *in situ* formation of the chlorolactams and 4 Å molecular sieve for water removal. Using **209** as a model substrate and starting from thiourea **212** (10%, 9% *ee*), the authors performed catalyst optimization and introduced a conformationally rigid pyrrolidine substituent with an additional stereogenic center (Scheme 1.50).

Catalyst **213** afforded product **210** in a slightly higher yield and increased enantioinduction (12%, 25% *ee*). Changing the phenyl-substituent to 1- and 2-naphthyl substituents, increased catalyst reactivity and stereoselectivity were observed

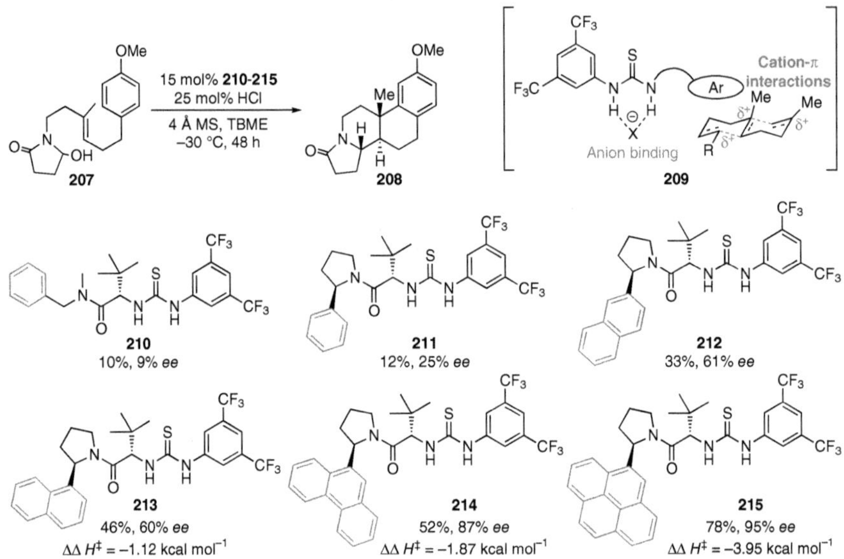

Scheme 1.50 Catalyst optimization for hydroxy lactam polycyclization and stabilization of the cationic intermediate **211**. For catalysts **215–217**, different activation enthalpies are given.

(46%, 60% *ee*; 33%, 61% *ee*, respectively). The 9-phenanthryl-substituted derivative **216** furnished **210** with a slightly decreased yield (52%) but increased enantioselectivity (87% *ee*). In the last optimization step, Jacobsen's group introduced the 4-pyrenyl substituent and obtained **210** with a good yield (78%) and a high *ee*-value (95% *ee*). Notably, the authors observed with all catalysts depicted in Scheme 1.50 the formation of one single diastereomer of 210, whereas performing the reaction without thiourea catalyst only monocyclic products were obtained [339].

With the optimized catalyst in hand, Jacobsen's group synthesized tetracyclic products **219** in yields ranging from 51% to 77% and *ee*-values of 89–94% and a typical TOF value of 0.07 h^{-1} (Scheme 1.51) [339]. The fact that the enantioinduction depends strongly on catalyst size, the authors noted that stabilizing cation–π interactions might play a key role in the intermediate as well as transition structure

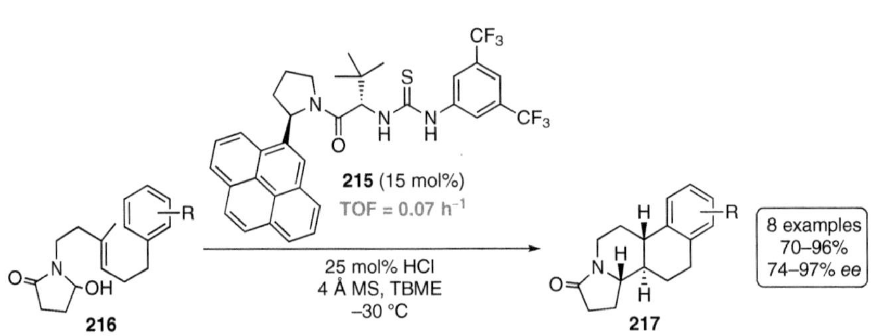

Scheme 1.51 Polycyclization of hydroxy lactams *via in situ* generation of the cationic intermediate catalyzed with **217**.

stabilization. Because catalysts **215–217** displayed a linear correlation between ln(*e.r.*) and reciprocal temperature over a 70 °C range, an Eyring analysis of the enantioselectivity revealed that the enantioselectivity was enthalpically controlled. Furthermore, the differential enthalpy increased obviously as the catalyst arene increased in size and was only attenuated slightly by increased differential entropy terms. These data support the importance of cation–π stabilization as the essential component in the mechanism [339].

Another example utilizing the *in situ* generation of an electrophile through chloride abstraction is the enantioselective addition to oxocarbenium ions generated from 1-chloroisochromans **220** [242]. Reisman, Doyle, and Jacobsen utilized 10 mol% of thiourea **223** and various silyl ketene acetals **221** as nucleophiles and obtained **222** in yields ranging from 70% to 96%, *ee*-values of 74–95%, and TOF = 1.45 h^{-1} (Scheme 1.52). The corresponding 1-chloroisochromans were prepared in a one-pot, two-stage procedure from the corresponding methyl acetals [242].

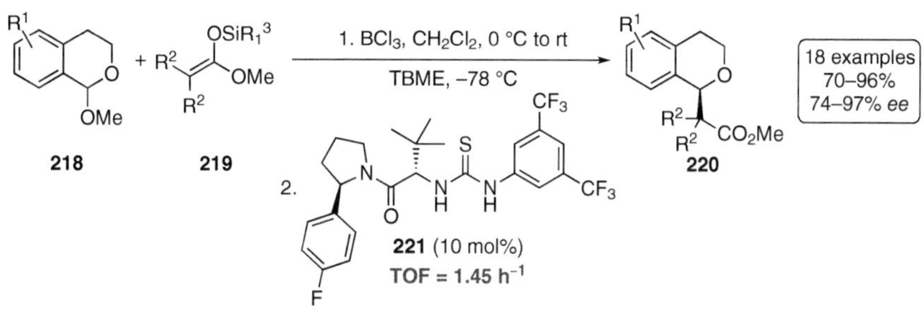

Scheme 1.52 Addition from various silyl ketene acetals to *in situ* generated oxocarbenium ions catalyzed with **223**.

In 2016, Jacobsen's group investigated the activation mode of the **223**-catalyzed chloride abstraction reaction [241, 376–378]. In "same-excess" experiments in the alkylation of 1-chloroisochroman, the authors did not observe catalyst deactivation through decomposition pathways or by product inhibition [376]. Additionally, the catalyst's reaction rate at high loading (>5 mol%) was found to be first order, while non-linear behavior was observed at low catalyst loading, and a positive non-linear relationship between product *ee* and catalyst *ee* was identified. The authors observed three head-to-tail catalyst dimers in 2D NOESY NMR studies, where each thiourea moiety forms hydrogen bonds to the amide oxygen, and that this agglomeration leads to an "off-cycle aggregation" event [376]. Furthermore, the authors obtained single crystals of dimeric catalyst complexes of $[(Z\text{-}\mathbf{223})_2\cdot\mathbf{Cl}]_2^-$ by addition of tetramethylammonium chloride. The X-ray single-crystal analysis shows the formation of a 4H–anion-binding mode with the four N–H protons of the thiourea moiety, and the authors suggested a cooperative and structurally similar chloride abstraction through two catalyst molecules *via* 4H–anion binding (Scheme 1.53) [376].

Jacobsen's group described pairwise catalyst combinations leading to four transition structures (*ZZ*–TS, *EE*–TS, *ZE*–TS, and *EZ*–TS) that apparently all make contributions to the overall reaction rate and enantioselectivity [241]. To suppress the (*Z*)-(*E*)-amide isomerization, the authors performed computational

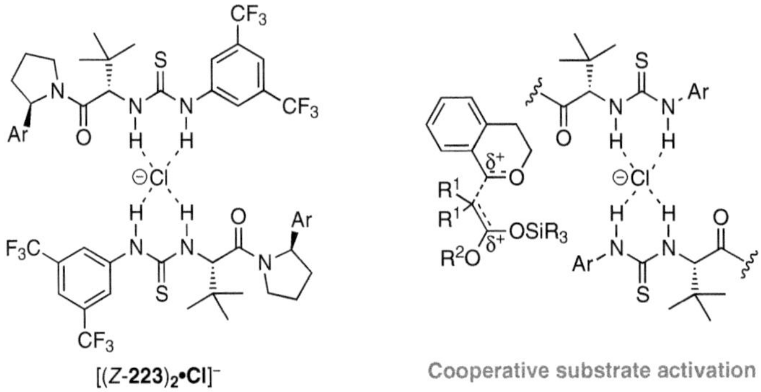

Scheme 1.53 Structures of the bis-thiourea coordinated chloride found in X-ray single-crystal analysis and the cooperative activation *via* the 4H–anion-binding mode.

analyses at the B3LYP/6-31G+(d,p)//B3LYP/6-31G(d) level (at 0 K, not taking zero-point vibrational energy corrections or dispersion interactions into account) of the relative energies of thiourea rotamers bearing a variety of substituents on the pyrrolidine moiety and identified the 2-aryl-2-methylpyrrolidine-derived thiourea **224** as a promising candidate. Jacobsen *et al.* identified in ^{1}H, ^{13}C, and 2D NOESY NMR experiments only the (*Z*)-amide rotamer and showed the improved activity compared to **224** with **223** in the oxocarbenium alkylation, where both yield and enantioselectivity increased (Scheme 1.54) [241]. Based on these results – the 4H–anion-binding mode and the benefit of the conformationally rigid 2-aryl-2-methyl-pyrrolidine moiety – Jacobsen's group synthesized bis-thiourea **225** that showed strict first-order behavior, high rate acceleration compared to **223**, and also a linear relationship between product *ee* and catalyst *ee* [378]. These results provided strong clues for a monocatalyst activation process of **225**, and much higher activity was shown in the formation of **226**, utilizing 0.1 mol% **225** (at 100 times lower catalyst loading, cf. Scheme 1.52), eight times shorter reaction time, and five times higher concentration (Scheme 1.54) [378].

In 2017, Jacobsen's group published the stereospecific β-glycosylation of various sugars catalyzed with thiourea derivatives [379]. The idea was to mimic the glycosyltransferase-catalyzed glycosylation [380]. As depicted in Scheme 1.55, the *cis*-1,2-*O*-glycosylation of α-mannosyl chloride **229** was utilized as a model reaction because the β-glycosidic linkage is strongly disfavored both sterically and electronically [381].

The glycosylation utilizing benzyl alcohol in the absence of any catalyst furnished the α-product predominantly (84 : 16 α:β) with very low yield (0.1%), whereas with 5 mol% **223** slightly higher reactivity, but no selectivity was observed (1%, 52 : 48 α:β). Nevertheless, this result showed that the thiourea catalyst could invert the selectivity, and bis-thiourea **231** leads to a very moderate yield and β-selectivity (15%, 20 : 80 α:β). To improve both the reactivity and selectivity, Jacobsen's group synthesized macrocyclic derivate **232** that forms **230** in good yield and β-selectivity (68%, 12 : 82 α:β). The introduction of a 2,3-dihydroindole instead of the pyrrolidine moiety

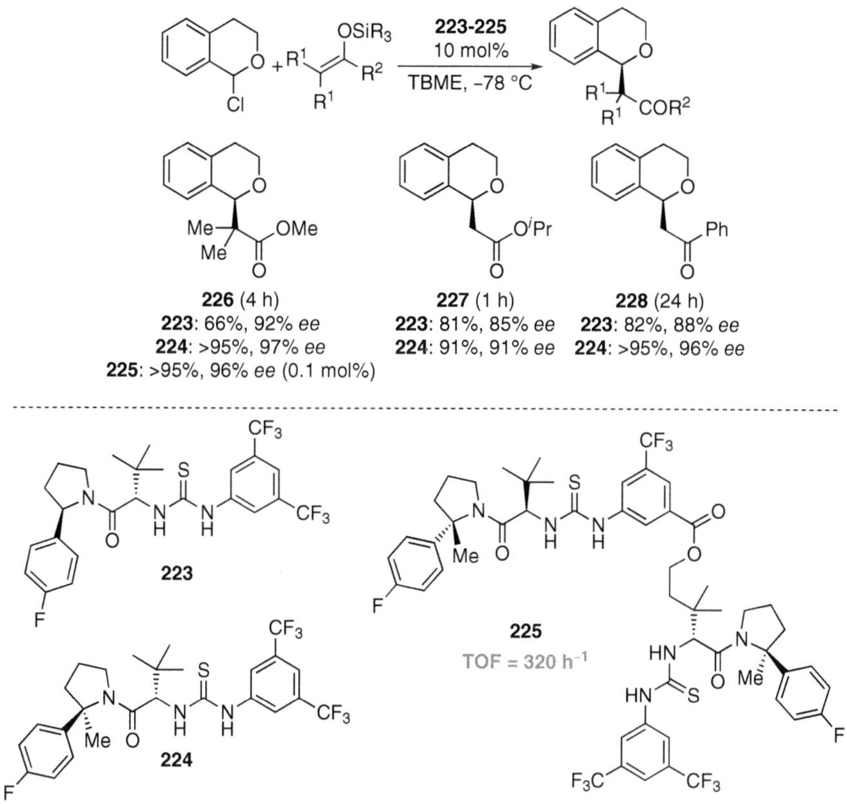

Scheme 1.54 Comparison of the enantioselective oxocarbenium alkylations utilizing thiourea **223**, conformational rigid **224**, and improved bis-thiourea **225**.

resulted in catalyst **233** that furnished the product in high yield as well as high β-selectivity (88%, 8 : 92 α:β) [379].

With the optimized catalysts in hand, the authors synthesized various disaccharides utilizing α-glycosyl chlorides **234** and obtained **235–240** in yields ranging from 64% to 88% and as predominantly β-linked products (Scheme 1.56) [379]. The thiourea activation of the leaving group (chloride) promotes both the S_N1 and S_N2 pathways, but nucleophile activation would exclusively support the S_N2 pathway. Overall, Jacobsen's group suggested an S_N2 mechanism based on the following observations:

(1) The products were obtained with a high degree of inversion.
(2) The reaction was insensitive to relative catalyst–substrate as well as nucleophile–electrophile stereochemical relationships.
(3) No limitations concerning the substrate scope could be observed.

The authors assumed an activation of the nucleophile by the catalyst's amide oxygen that acts as a Lewis base and found in DFT studies (M06-2X/6-31G(d) with PCM:benzene solvent inclusion, no temperature given) on the concerted

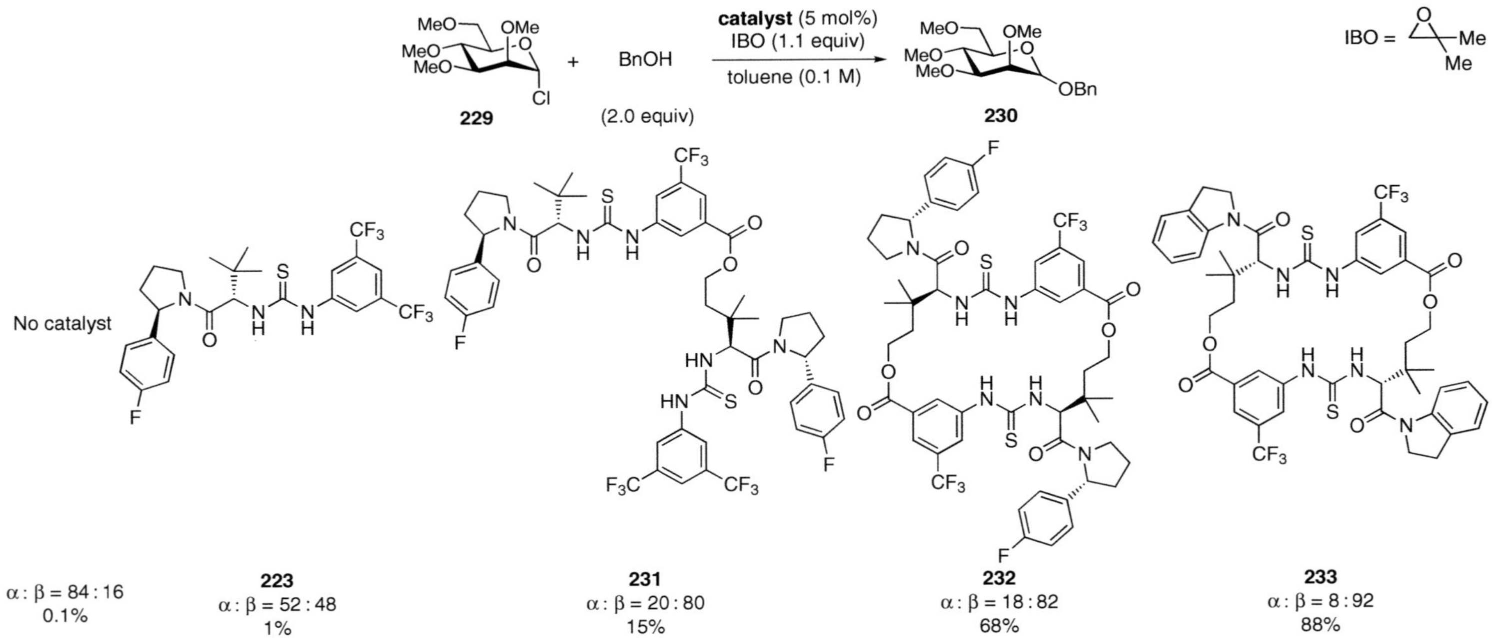

Scheme 1.55 Evolution of thiourea catalysts for the *cis*-1,2-*O*-glycosylation of mannosyl chloride utilizing benzyl alcohol as nucleophile and IBO as the HCl trapping agent.

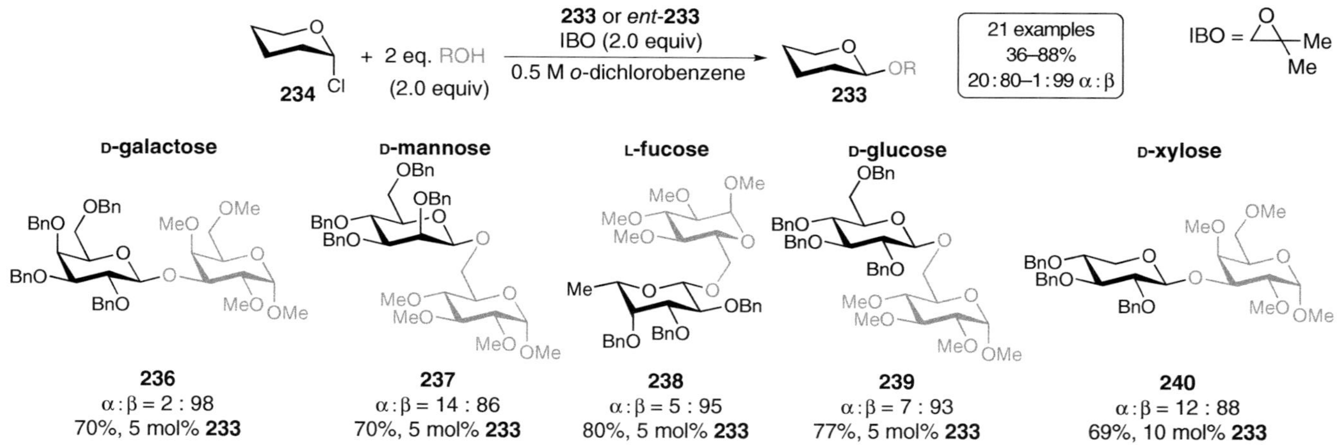

Scheme 1.56 Typical glycosylation products with high β-selectivity utilizing macrocyclic bis-thioureas **233** or *ent*-**233**.

glycosylation of glucosyl chloride by methanol a loose and asynchronous transition structure with hydrogen-bonding interactions between the methanol O–H and the amide oxygen that supported the S_N2 mechanism by both leaving group as well as nucleophile activation [379].

In 2014, on expanding from achiral triazole-based anion-binding catalyst **114** (Scheme 1.25) [229, 230], Mancheño's group introduced helical chiral derivatives with a chiral 1,2-diaminocyclohexane backbone [231]. The idea was to synthesize a flexible catalyst structure that in the "non-active mode" will be present in an equilibrium between linear and helical conformations, and that in the complexation mode to a chloride anion, the helical system is the dominant species. The authors synthesized a series of helical triazole-based anion-binding catalysts **243–247** and tested them in the Reissert-type reaction of quinolone **241** utilizing silyl ketene acetals as nucleophiles. Using bis-triazole-based catalyst **243** that cannot adopt helical chirality upon chloride binding, the observed enantioinduction was low (18% *ee*). Going on to tetra-triazole-based catalysts such as **243–247**, Mancheño's group obtained product **242** with increased *ee*-values (62–92% *ee*) and observed that the substitution pattern of the central phenyl ring was important with the 2-(methoxy-propan-2-yl)acetylene substituent as the most active one. The difference between catalysts **246** and **247**, where the connection of the phenyl groups by the triazole units was slightly changed, was marginal (90% *ee* and 92% *ee*, respectively) [231]. To validate the anion-binding mode of **247**, the authors performed NMR titration experiments with Troc-quinolinium chloride and observed that the eight hydrogens, which are highlighted in Scheme 1.57, form hydrogen bonds to the chloride [231]. Additionally, circular dichroism (CD) titration of **247** with TBAC showed conformational changes in the folding behavior of the flexible oligomer. Increased absorption bands as 250 and 265–280 nm in the positive as well as negative regions of the UV spectrum indicated catalysts' chloride binding and the formation of the helical chiral form. Furthermore, the authors synthesized a series of Reissert-type products utilizing 5 mol% of **247** and obtained products in yields ranging from 54% to 96% and *ee*-values of 44–96% [231].

1.2.4 Anion Binding in Cooperative Catalysis

Besides the high ability to coordinate with chloride (cf. Section 1.2.3), (thio)ureas are well-known receptors for Y-shaped anions, such as carboxylates [135, 138, 382, 383]. The topology of these coplanar anions allows a bidentate hydrogen-bonding mode with an N–H···O angle of approximately 170–175° that leads to anion stabilization [384, 385]. Dual thiourea/carboxylic acid catalyst systems have been utilized in many anion-binding mode-catalyzed reactions, e.g., the alcoholysis of styrene oxides [291], protio Pictet–Spengler reactions [169, 386, 387], [5+2] cycloadditions [388], and, in particular, kinetic resolution of primary amines [342, 343, 389–393], a topic that was reviewed by Seidel in 2014 [341]. The general concept of dual-catalyst kinetic resolution is based on the *in situ* formation of a chiral acylating reagent and consists of three components: a pyridinium species (most reactions use 4-dimethylaminopyridine (DMAP) [393] or its derivatives, e.g.,

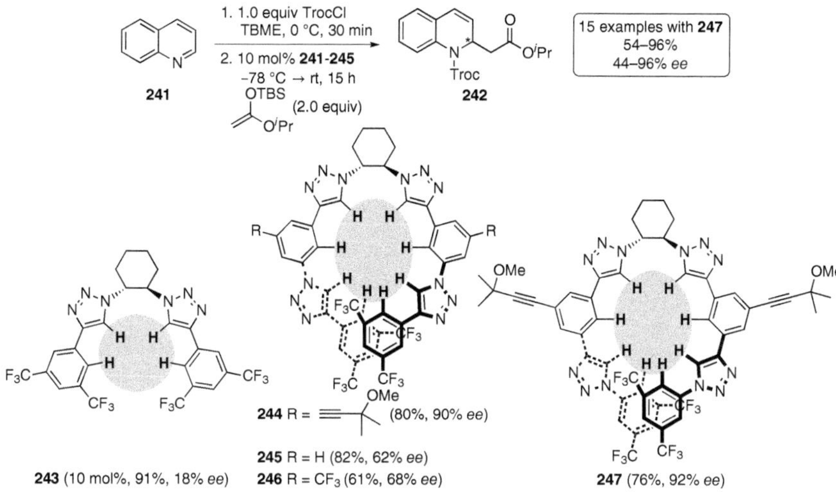

Scheme 1.57 Optimization of helical chiral catalyst structure in Reissert-type reactions. Hydrogens coordinating the chloride are highlighted.

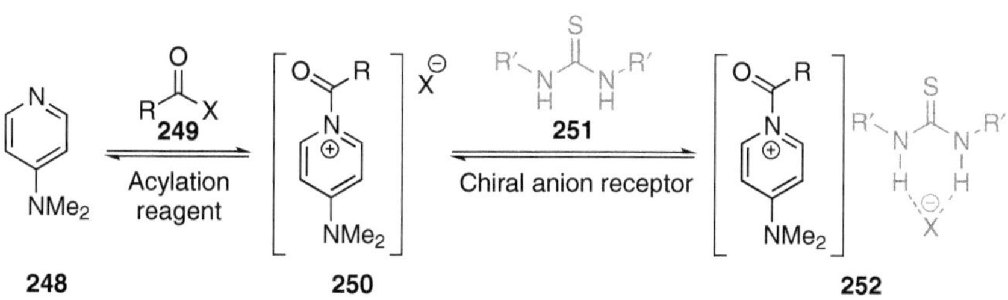

Scheme 1.58 Anion-binding concept for formation of chiral supramolecular ion pairs that serve as chiral acyl transfer species.

4-(pyrrolidino)pyridine (PPY) [390]), an achiral acylating reagent, and a chiral anion-binding/hydrogen-bonding catalyst. Mixtures of DMAP **248** and acylating reagents **249** exist in equilibrium with the corresponding achiral acylpyridinium salt **250** [394]. Chiral catalyst **251** binds the anion, leading to supramolecular chiral ion pair formation that affects the equilibrium between DMAP and its acylpyridinium salt because the supramolecular ion **252** is generally more soluble in organic solvents. Consequently, the substrates should rather react with the chiral supramolecular ion pair than with the acylpyridinium salt (Scheme 1.58).

Because the nature of the chiral supramolecular ion complex was at that time unknown, Seidel, Schreiner, and coworkers investigated the mechanism of the dual-catalysis anion-binding approach in the kinetic resolution of amines utilizing both experimental and computational approaches [393]. Based on the original study by Seidel's group in 2009 [342], Seidel, Schreiner, and coworkers utilized 1-phenylethylamine *rac*-**253** as a model substrate, DMAP as pyridinium species, and tested a series of chiral catalysts. Starting catalyst evolution with the "original" bis-thiourea **256** (s-factor = 8.5), amide–thiourea catalyst **257** (s-factor = 13.0),

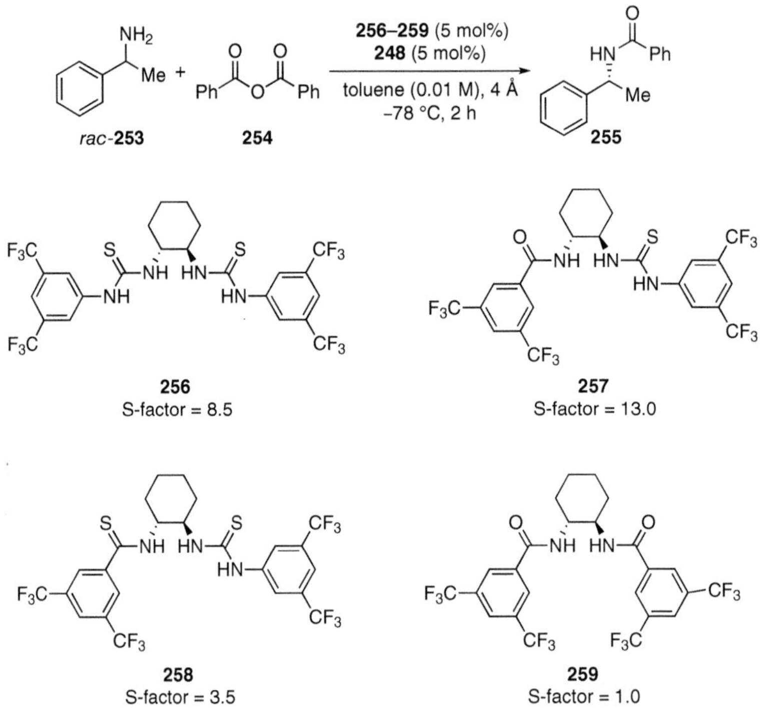

Scheme 1.59 Catalyst evolution for kinetic resolution of 1-phenylethanamine utilizing various catalysts with DMAP as cocatalyst.

which was also utilized in kinetic resolution of primary amines [341], was found to be the most selective catalyst. In the catalyst evolution study, Seidel, Schreiner, and coworkers found that the thiourea as well as the amide moieties are crucial for high selectivity because the corresponding thioamide–thiourea **258** (s-factor = 3.5) and diamide **259** (s-factor = 1.0) were much less selective (Scheme 1.59) [393]. Using various acylating reagents, the authors obtained the highest selectivities with benzoic anhydride but could not identify a trend regarding the electronic nature utilizing substituted benzoic anhydride derivatives ((4-CF_3–PhCO)$_2$O s-factor = 7.5; (4-Me–PhCO)$_2$O s-factor = 12.8; (4-MeO–PhCO)$_2$O s-factor = 4.8) [393]. Because the catalyst–substrate ion pairing is obviously crucial for the selectivity, toluene (s-factor = 13.0) was found as the solvent that furnished the highest *s*-value, as it favors (contact) ion pairs, whereas more polar solvents, such as ethyl acetate (s-factor = 1.5), resulted in solvent-shared or solvent-separated ion pairs (cf. Section 1.2.1) [326, 327].

Because deprotonation of the thiourea moiety would lead to an alternative ion pair, the authors performed deprotonation studies and identified hydrogen-bonding interactions between the catalyst's N–Hs and various amines, such as DMAP, Hünig's base, and 1-phenylethylamine in ^{1}H NMR experiments, but could not observe a deprotonated catalyst species, which should be easily identifiable by shifts in the ^{13}C NMR spectrum [143, 189]. These findings confirm the structure of supramolecular ion pair **252**. Only when the stronger base BEMP was utilized, the deprotonated catalyst could be identified in the ^{1}H and ^{13}C NMR spectra. Using DFT-based computations (ΔH_0, D_0, and ΔG values at M06/6-31G(d,p) including PCM solvent corrections for toluene at various temperatures), Seidel, Schreiner, and coworkers obtained for the ion pair of DMAP and benzoic anhydride substantial and negative dissociation energy ($D_0 = -6.8\,kcal\,mol^{-1}$; $D_{298} = -22.2\,kcal\,mol^{-1}$), which is consistent with the absence of NOE signals in NMR experiments; however, this could also be due to long proton–proton distances and fast exchange. Nevertheless, DOSY NMR spectroscopy equally did not reveal evidence of ion pair formation. Utilizing bis-thiourea **256**, the authors obtained a positive dissociation energy for ternary complex **252** (Scheme 1.58) in the gas-phase ($D_{195} = +10.4\,kcal\,mol^{-1}$) as well as in solution (toluene, $D_{195} = +3.6\,kcal\,mol^{-1}$), with **256** displaying (*Z,Z*)-oriented N–H protons. Additionally, Seidel, Schreiner, and coworkers identified that the benzoyl group is fixed in the ternary complex through π–π stacking with one of the thiourea aryl rings. Adding the (*R*)-configured amine, the lowest lying quaternary complex was found to coordinate the benzoate through double hydrogen bonding by one thiourea moiety, whereas the second thiourea unit coordinates the first thiourea moiety also through hydrogen bonding. Simultaneously, benzoate binds to the ortho and meta protons of the pyridinium cation. In this quaternary complex, the authors found threefold π–π stacking of one thiourea aryl, DMAP's pyridine, and the 1-phenylethylamine ring. Accordingly, quaternary complex formation was observed at −78 °C ($D_{195} = +19.0\,kcal\,mol^{-1}$; $D_{298} = +3.1\,kcal\,mol^{-1}$), whereas with the (*S*)-configured amine, the quaternary complex was less favorable ($D_{298} = +0.7\,kcal\,mol^{-1}$). Utilizing amide–thiourea catalyst **257**, the authors also found a (*Z,Z*)-oriented thiourea unit that binds the benzoate through double hydrogen bonding. Additionally, the acidified ortho-proton of the 3,5-bis(trifluoromethyl)phenyl moiety forms a hydrogen bond to one of the benzoate oxygens [190], and, similar to **256**, the amide binds *via* an N–H···S interaction to the thiourea. Benzoate forms hydrogen-bonds to DMAP's ortho proton, which itself is fixed by the two aryl groups of the catalyst, leading to a well-defined binding pocket. The dissociation energy is negative at 298 °C ($-0.5\,kcal\,mol^{-1}$) but positive at −78 °C ($+12.6\,kcal\,mol^{-1}$; $+5.2\,kcal\,mol^{-1}$ in toluene). These mechanistic studies using a combination of experimental as well as computational studies emphasize the high relevance of NCIs in anion-binding catalysis for the formation of a well-defined binding pocket that can furnish reactions with high stereoinduction [393].

1.2.5 Anion-Binding in Lewis Acid Enhancement Catalysis

In 2011, Schreiner's group utilized silicon–thiourea Lewis acid **260** in the House–Meinwald rearrangement of tri-substituted epoxides **261** and obtained the corresponding quaternary aldehydes **262** in yields of 43–88% (Scheme 1.60) [189]. Catalyst **260** forms by deprotonation of **46** and addition of $SiCl_4$ as evident from IR, NMR, and MS experiments. In blind experiments utilizing either **46** or $SiCl_4$, the authors did not observe product formation, which confirmed

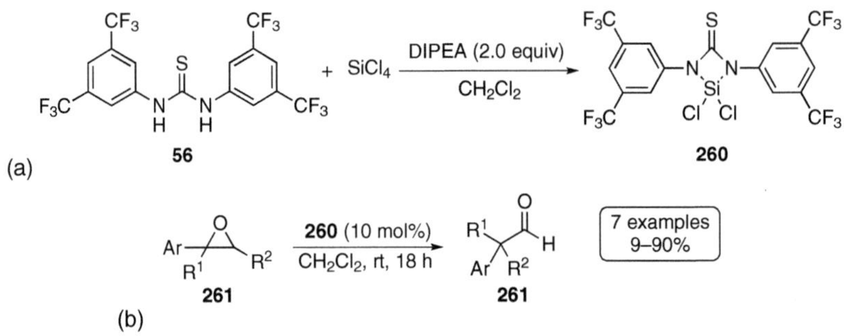

Scheme 1.60 (a) Synthesis of silicon-thiourea Lewis acid through deprotonation of **56** and (b) House–Meinwald rearrangement utilizing tri-substituted epoxides.

the increased Lewis acidity of **260**. When utilizing enantioenriched epoxides, synthesized by Shi epoxidation [395], the authors observed increased enantiopurity for the starting material as well as the product. They performed a negative control experiment spiking the reaction mixture with enantioenriched product. However, a corresponding autocatalytic process could be excluded as no chirality enhancement for the product was observed.

Schreiner and coworkers proposed that an epoxide coordinates the silicon first and forms the active catalyst species. Binding by another epoxide would consequently lead to diastereomeric transition structures and to diastereomeric matched and mismatched combinations. Therefore, the reaction was described as "*similar to a kinetic resolution of non-racemic starting materials*" [396]. This reaction utilized a thiourea-based complex that increased the silicon's Lewis acidity by covalent bond formation and not through anion binding. Nevertheless, this was a proof-of-concept study for Lewis acid enhancement utilizing (thio)ureas.

In 2017, Jacobsen's group took up this concept in the activation of triflates for the generation of oxocarbenium ions [344]. Using **263** as a model substrate in Mukaiyama aldol reactions, the authors screened various (thio)ureas and squaramides and observed that only squaramides catalyzed the aldol reaction, which was explained by squaramides' dual functionality [197–199]. Jacobsen's group employed squaramides with various dispersion energy donors and found 1-pyrenyl-substituted derivate **266** as the most active one (100% conversion, 88% *ee*, Scheme 1.61); it displays a structure similar to that of thiourea **217** (cf. Scheme 1.50). The importance of catalysts hydrogen-bonding donor motif was validated through the *N,N'*-dimethylated analog of **266** that promoted the aldol reaction only little and nearly without selectivity (43% conversion, 2% *ee*) [344].

Subsequently, Jacobsen's group utilized the **266**-silyl-triflate system in [4+3] cycloadditions of oxyallyl cations and furan derivatives **268** and obtained bicyclic **269** in yields of 55–98% and 66–96% *ee* as single diastereomers (Scheme 1.62) [344]. The authors performed kinetic analysis and found zero-order kinetics for furan derivatives **268**, first-order kinetics for the oxyallyl cation precursor **267** as well as squaramide catalyst **266**, and saturation kinetics for trialkyl silyl triflate (TESOTf). The kinetic data are consistent with a pre-equilibrium between **266** and TESOTf, and rate-limiting oxyallyl cation formation. To prove the formation of a

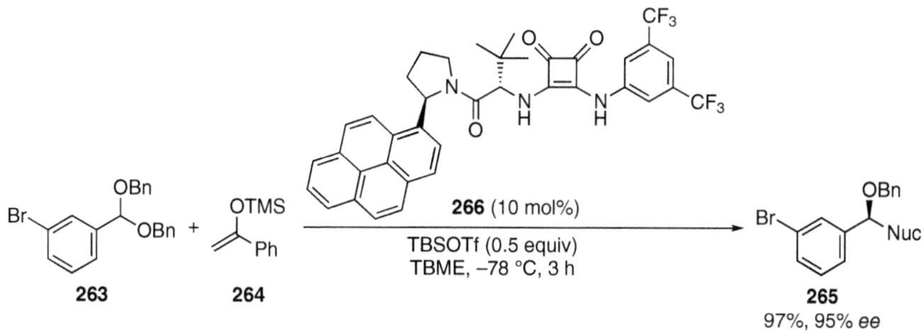

Scheme 1.61 Representative example of enantioselective aldol reaction utilizing TBSOTf as Lewis acid and catalyst **266**.

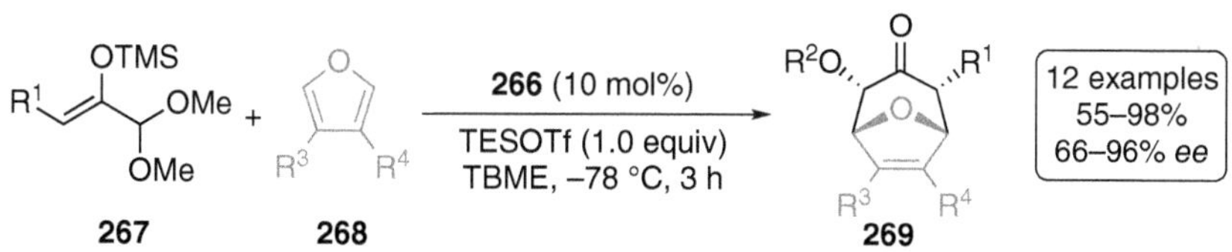

Scheme 1.62 Enantioselective [4+3] cycloaddition catalyzed with **266** and TESOTf.

pre-equilibrium, the authors preformed ^{1}H NMR experiments with NBu_4OTf and TESOTf and observed stable complexes with both triflate species. However, TESOTf was found to bind 4000 times as strongly as NBu_4OTf and forms simultaneous hydrogen bonds to the squaramides' N–Hs. While monitoring TESOTf addition to the squaramide catalyst utilizing IR spectroscopy, Jacobsen's group observed the disappearance of the absorbance attributed to the squaramide carbonyl groups. The authors proposed that complexation of **266** and trialkyl silyl triflates (R_3SiOTf) is more Lewis acid than (R_3SiOTf) alone because of the stabilization of the triflate anion through hydrogen-bonding interactions [344].

The authors suggested a catalytic cycle that starts with the resting state formation of the catalytically active enhanced Lewis acid **270** [344]. After the rate-determining ionization and generation of the oxyallyl cation, ion pair **271** forms. In a stepwise and enantiodetermining cycloaddition, the desired product forms and free enhanced Lewis acid **270** ensues (Scheme 1.63). Furthermore, the authors performed DFT studies (at the uncorrected M06-2X/6-31+G(d,p)//B3LYP/6-31G(d) level of theory) and identified transition structures leading to the formation of the major and minor products. Jacobsen's group described that the furan position near the catalyst's aromatic substituent leads to a stabilization of the major enantiomer trough NCIs, whereas the transition structure for the minor enantiomer lacks this stabilization [344].

1.2.6 Anion-Binding in Phase Transfer Catalysis

Fluorine incorporation in organic molecules leads to modified properties, such as lowering the pK_a of the neighboring groups and changing molecules' dipole moments. In particular, sp^3 C–F bonds have a large influence on metabolic

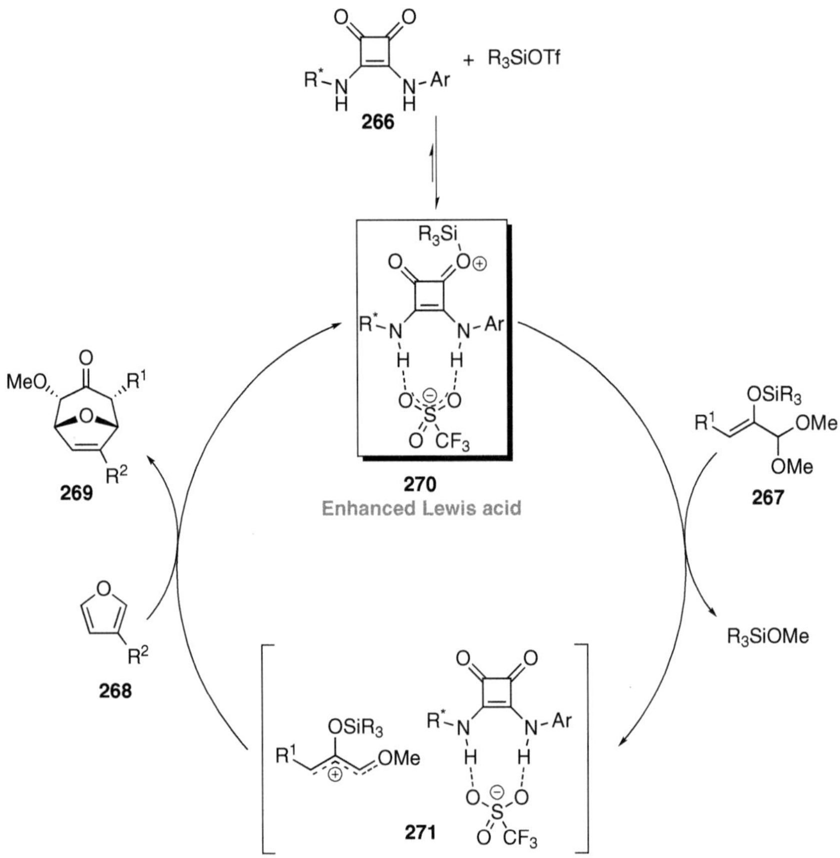

Scheme 1.63 Pre-equilibrium leading to the formation of the enhanced Lewis acid 268 and the proposed catalytic cycle for the enantioselective [4+3] cycloaddition.

stability, lipophilicity, and bioavailability – characteristics that are highly important in pharmaceutical science [397–402]. Gouverneur's group utilized urea-based phase transfer catalysts (PTC) [403–405] for the asymmetric nucleophilic fluorine incorporation [346–348] and utilized the well-known potent ability of ureas' fluoride recognition [384, 406–408]. Generally, fluorination is dominated by methods utilizing electrophilic fluorine species [409, 410] such as selectfluor [411]. Additionally, asymmetric PTC using organic electrophilic fluorine species are known [412, 413], whereas the anion-binding-mediated fluorination approach of Gouverneur's group based on cheap alkali-metal fluorides, such as cesium fluoride [346, 347] or potassium fluoride [348], mimic the nucleophilic enzymatic fluorination of *S*-adenosyl-L-methionine [414–416]. In 2016, Gouverneur and coworkers obtained single crystals by refluxing *N,N′*-bis(4-chloro)phenylurea and tetrabutylammonium fluoride (TBAF) in *n*-hexane. An X-ray crystal structure analysis of this complex revealed two different complexes with both urea moieties forming hydrogen bonds to the fluoride. The main difference of the two complexes is the twisted geometries as expressed through the interplanar angles of 40.6° and 76.3°. Gouverneur and coworkers synthesized tridentate and tetradentate catalysts

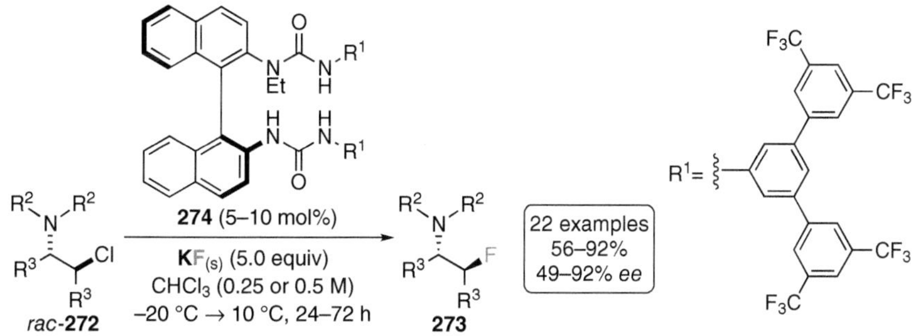

Scheme 1.64 Generation of β-fluoroamines obtained by asymmetric PTC fluorination with bis-urea **274**.

with two urea units, the privileged 3,5-bis(trifluoromethyl)phenyl motif [190], and 1,1′-binaphthalene-2,2′-diamine (BINAM) as the chiral backbone. The authors observed higher enantioinduction using tridentate derivatives as monoalkyl incorporation into one urea resulted in a preferred *anti-syn* conformation and formation of a well-defined binding pocket [346]. In 2019 [347], the same group utilized anion-binding-mediated PTC for the enantioselective synthesis of β-fluoroamines **273** that are a highly relevant substance class in medicinal chemistry [417–420]. The idea was the *in situ* formation of a prochiral aziridinium species **276** (Scheme 1.65) that subsequently underwent ring opening by the fluoride. The authors optimized the catalyst and identified **274** as the most active catalyst incorporating additional 3,5-bis(trifluoromethyl)phenyl substituents (cf. Scheme 1.64). Utilizing 5–10 mol% of bis-urea **274**, β-fluoroamines were obtained in yields of 63–92% and 49–92% *ee* [348].

The authors utilized molecular dynamics (MD) simulations and DFT computations (ΔG values at ωB97X-D3/(ma)-def2-TZVPP/COSMO($CHCl_3$)//M06-2X/def2-SVP(TZVPPD)/CPCM($CHCl_3$)) to gain insights into the reaction mechanism [348]. The MD simulations reveal a preferred *anti-syn* conformation of the catalysts. Utilizing DFT computations, 15 optimized transition structures for ring-opening of bisaryl-based aziridinium species were found, with the lowest lying transition structure leading to the major product being favored by 1.6 kcal mol^{-1}. The most favorable transition structure for both major and minor enantiomer reveals that the aziridinium ion N-substituents are pointing away from the catalytic pocket into the solvent, thereby helping to rationalize the indifference to this substituent (Scheme 1.64). Furthermore, Gouverneur and coworkers found cation–π interactions between the naphthyl ring and the aziridinium Cα–H protons. The authors described stronger cation–π interactions for transition structure leading to the major enantiomer since the distance compared with the transition structure for the minor enantiomer was shorter (2.26 Å vs. 2.41 Å, respectively). Additionally, unfavorable geometric distortions due to steric crowding contribute with about 1.0 kcal mol^{-1}. The authors proposed a mechanism that starts with the formation of the soluble and chiral fluoride from inorganic and insoluble potassium fluoride. Additionally, Gouverneur and coworkers suggested an equilibrium of the racemic β-chloroamines **272** and the reactive aziridinium chloride ion pair **276**. Ion pairs **275** and **276** underwent

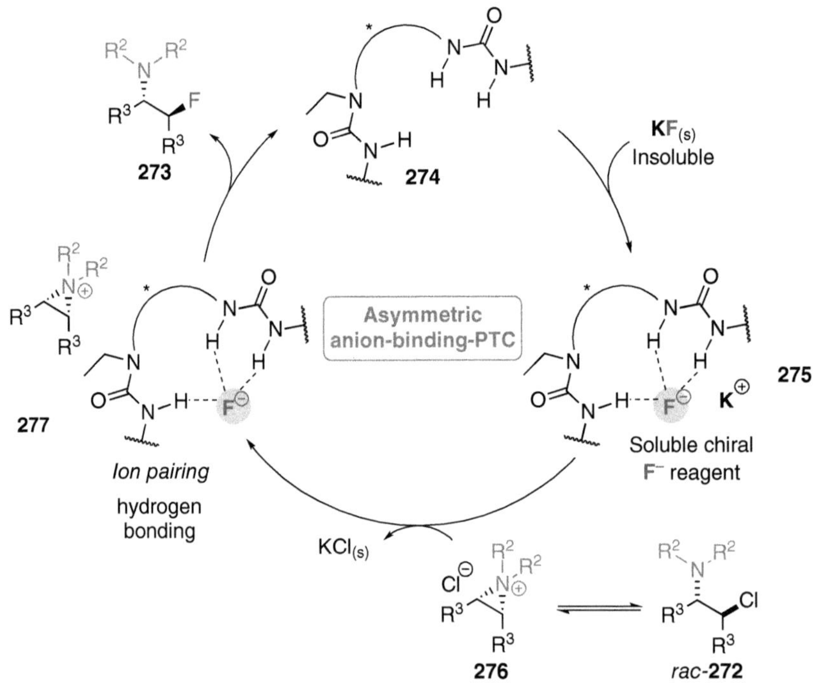

Scheme 1.65 Proposed mechanism of the asymmetric hydrogen-bonding PTC to furnish enantioenriched β-fluoroamines, utilizing KF, racemic β-chloroamines, and bis-urea **274**.

an ion-change process leading to the formation of insoluble potassium chloride and supramolecular anion pair **277**. Nucleophilic addition to the aziridinium ion furnishes the desired product **273** and free catalyst **274** (Scheme 1.65) [348].

1.3 Summary and Outlook

This chapter reviews the long evolution in anion-binding chemistry starting with the first observations in the 1940s and 1950s and ending today with highly efficient organocatalysts that activate and direct reactions through anion binding. Starting with unselective (thio)ureas that show low TOF values ($<1\,h^{-1}$), in the last years, organic chemist designed highly active anion-binding organocatalysts for asymmetric induction with TOF values up to $4000\,h^{-1}$, thereby underlining the success story of anion-binding catalysts – after initial ignorance and skepticism in the early years. After the foundational work in the late 1990s and early 2000s, the growing interest in this research field has been exponential, as demonstrated by the milestone achievements and guidelines for (thio)urea catalyst design summarized and presented here. The evident success of these design principles for anion-binding catalysts also triggered the development of novel catalyst classes over the past few years, such as those binding through C–H hydrogen bonds and σ-holes and incorporating anion-binding motifs into switchable catalysts. One crucial design element in asymmetric anion-binding catalysis is NCIs, including hydrogen-bonding, π–π as well as cation–π stacking, and dipole interactions. What is missing is the appreciation of London dispersion interactions (*via* dispersion energy donors, DEDs) as a design element. The inclusion of these interactions will finally complete all supposedly "weak" interactions that are at the heart of transition-state stabilization in a catalytic event.

The preferred use of non-polar solvents supports contact ion pair formation in intermediates and transition structures. In previous reviews, the authors divided organocatalysts in bound anion species, such as halide, enolate, and so on. In this chapter, we presented the five general activation modes in anion-binding catalysis utilizing selected representative examples, describing catalyst design principles, and presenting mechanistic proposals that go along with them:

(1) Recognition of the nucleophile in addition reactions and NCIs between the catalyst and the electrophile.
(2) Abstraction of the leaving group in S_N1-type reactions and formation of a chiral contact ion pair.
(3) Cooperative catalysis with an additional Brønsted acid forming a well-defined binding pocket to interact with the substrate.
(4) Lewis acid enhancement through activation by a Lewis base.
(5) Nucleophile delivery in phase transfer catalysis by nucleophile-binding and subsequent transfer into the organic medium.

Some important conceptual and practical points to be considered in future anion-binding-catalyzed reactions (and not only these) may include the following:

(1) To make catalyst reactivities and stereoselectivities more comparable, TOF values should be used and uncatalyzed reference reactions should always be reported.
(2) Typically, reactions are performed on a very small scale (0.1–0.2 mmol). To show the practicability of a new reaction, scale-up experiments (5–10 mmol) should be carried out to leave the "proof-of-concept" phase toward the challenging phase in which research should focus on broader, even large-scale applications.
(3) The product yields should be given and not just conversion of the starting materials because any side reaction also consumes the starting material. Furthermore – as we all know all too well – the work-up process is crucial for the isolation of a pure product. Only the yield determined for a sizeable amount of pure product matters when, for instance, it is to be used for (bio)medical applications.
(4) Novel applications in anion-binding catalysis should be investigated using experimental and theoretical studies to elucidate catalyst activation modes and principle mechanistic hypotheses.

Acknowledgment

We gratefully acknowledge the financial support by the German Research Foundation (Deutsche Forschungsgemeinschaft) and the State of Hesse for the many years of continuing support. We thank Raffael C. Wende and Bastian Bernhardt for valuable discussions.

References

1 Schreiner, P.R. (2003). *Chem. Soc. Rev.* 32: 289–296.
2 Zhang, Z. and Schreiner, P.R. (2009). *Chem. Soc. Rev.* 38: 1187–1198.
3 Pihko, P.M. (2004). *Angew. Chem. Int. Ed.* 43: 2062–2064.
4 Takemoto, Y. (2005). *Org. Biomol. Chem.* 3: 4299–4306.
5 Connon, S.J. (2006). *Chem. Eur. J.* 12: 5418–5427.
6 Taylor, M.S. and Jacobsen, E.N. (2006). *Angew. Chem. Int. Ed.* 45: 1520–1543.
7 Akiyama, T., Itoh, J., and Fuchibe, K. (2006). *Adv. Synth. Chem.* 348: 999–1010.
8 Doyle, A.G. and Jacobsen, E.N. (2007). *Chem. Rev.* 107: 5713–5743.
9 Brak, K. and Jacobsen, E.N. (2013). *Angew. Chem. Int. Ed.* 52: 534–561.
10 Beckendorf, S., Asmus, S., and García Mancheño, O. (2012). *ChemCatChem* 4: 926–936.
11 Pihko, P.M. (2009). *Hydrogen Bonding in Organic Synthesis.* Weinheim: Wiley VCH.
12 Mahlau, M. and List, B. (2013). *Angew. Chem. Int. Ed.* 52: 518–533.
13 Auvil, T.J., Schafer, A.G., and Mattson, A.E. (2014). *Eur. J. Org. Chem.* 2014: 2633–2646.
14 Phipps, R.J., Hamilton, G.L., and Toste, F.D. (2012). *Nat. Chem.* 4: 603–614.
15 Schramm, V.L. (2006). *Chem. Rev.* 106: 3029–3030.
16 Watson, J.D. and Crick, F.H.C. (1953). *Nature* 171: 737–738.
17 Jeffrey, G.A. and Saenger, W. (1994). *The Role of Hydrogen Bonding in the Structure and Function of the Nucleic Acids.* Berlin: Springer.
18 Fonseca Guerra, C., Bickelhaupt, F.M., Snijders, J.G., and Baerends, E.J. (2000). *J. Am. Chem. Soc.* 122: 4117–4128.
19 Fersht, A. (1999). *Structure and Mechanism in Protein Science.* New York: Macmillan Education.
20 Gao, J., Ma, S., Major, D.T. *et al.* (2006). *Chem. Rev.* 106: 3188–3209.
21 Carey, P.R. (2006). *Chem. Rev.* 106: 3043–3054.
22 Boehr, D.D., Dyson, H.J., and Wright, P.E. (2006). *Chem. Rev.* 106: 3055–3079.
23 Bruice, T.C. (2006). *Chem. Rev.* 106: 3119–3139.
24 Antoniou, D., Basner, J., Núñez, S., and Schwartz, S.D. (2006). *Chem. Rev.* 106: 3170–3187.
25 Kirby, A.J. (1996). *Angew. Chem. Int. Ed.* 35: 706–724.
26 Kohen, A. and Klinman, J.P. (1998). *Acc. Chem. Res.* 31: 397–404.
27 Lindskog, S. (1997). *Pharmacol. Ther.* 74: 1–20.
28 Bruice, T.C. and Benkovic, S.J. (2000). *Biochemistry* 39: 6267–6274.
29 Williams, D.H., Stephens, E., O'Brien, D.P., and Zhou, M. (2004). *Angew. Chem. Int. Ed.* 43: 6596–6616.
30 Evans, M.J. and Cravatt, B.F. (2006). *Chem. Rev.* 106: 3279–3301.
31 Christianson, D.W. (2006). *Chem. Rev.* 106: 3412–3442.

32 Pauling, L. (1960). *The Nature of the Chemical Bond*. Ithaca, NY: Cornell University Press.

33 Pimentel, G.C. and McClellan, A.L. (1960). *The Hydrogen Bond*. San Francisco: W. H. Freemann.

34 Jeffrey, G.A. and Jeffrey, G.A. (1997). *An Introduction to Hydrogen Bonding*. New York: Oxford University Press.

35 Prins, L.J., Reinhoudt, D.N., and Timmerman, P. (2001). *Angew. Chem. Int. Ed.* 40: 2382–2426.

36 Steiner, T. (2002). *Angew. Chem. Int. Ed.* 41: 48–76.

37 Hunter, C.A. (2004). *Angew. Chem. Int. Ed.* 43: 5310–5324.

38 Riordan, J., McElvany, K., and Borders, C. (1977). *Science* 195: 884–886.

39 Riordan, J.F. (1979). *Mol. Cell. Biochem.* 26: 71–92.

40 Riley, K.E. and Hobza, P. (2013). *Acc. Chem. Res.* 46: 927–936.

41 Meyer, E.A., Castellano, R.K., and Diederich, F. (2003). *Angew. Chem. Int. Ed.* 42: 1210–1250.

42 Dessent, C.E.H. and Müller-Dethlefs, K. (2000). *Chem. Rev.* 100: 3999–4022.

43 Wear, M.A., Kan, D., Rabu, A., and Walkinshaw, M.D. (2007). *Angew. Chem. Int. Ed.* 46: 6453–6456.

44 Warshel, A., Sharma, P.K., Kato, M. *et al.* (2006). *Chem. Rev.* 106: 3210–3235.

45 Kraut, D.A., Sigala, P.A., Pybus, B. *et al.* (2006). *PLoS Biol.* 4: 201–519.

46 Lau, E.Y., Newby, Z.E., and Bruice, T.C. (2001). *J. Am. Chem. Soc.* 123: 3350–3357.

47 Hopmann, K.H. and Himo, F. (2006). *Chem. Eur. J.* 12: 6898–6909.

48 Hopmann, K.H. and Himo, F. (2006). *J. Phys. Chem. B* 110: 21299–21310.

49 de Vries, E.J. and Janssen, D.B. (2003). *Curr. Opin. Biochem.* 14: 414–420.

50 Jormakka, M., Törnroth, S., Abramson, J. *et al.* (2002). *Acta Crystallogr. D* 58: 160–162.

51 Hedstrom, L. (2002). *Chem. Rev.* 102: 4501–4524.

52 Product inhibition and substrate inhibition are effects also known in enzyme catalysis that can reduce catalytic efficiency. Generally, catalytic systems (natural or artificial) based on covalent interactions are more sensitive towards inhibitions than non-covalent systems utilizing weak interactions:Garcia-Junceda, E. (2008). *Multi-Step Enzyme Catalysis*. Weinheim: Wiley-VCH.

53 Vigny, M., Bon, S., Massoulié, J., and Leterrier, F. (1978). *Eur. J. Biochem.* 85: 317–323.

54 Blackmond, D.G., Armstrong, A., Coombe, V., and Wells, A. (2007). *Angew. Chem. Int. Ed.* 46: 3798–3800.

55 Sheldon, R.A. (2008). *Chem. Commun.* 44: 3352–3365.

56 Murakami, Y., Kikuchi, J.-i., Hisaeda, Y., and Hayashida, O. (1996). *Chem. Rev.* 96: 721–758.

57 Kirby, A.J. and Hollfelder, F. (2009). *From Enzyme Models to Model Enzymes*. Cambridge: Royal Society of Chemistry.

58 Breslow, R. (2005). Artificial Enzymes. In: *Artificial Enzymes* (ed. R. Breslow), 1–35. Weinheim: Wiley-VCH.

59 Breslow, R. (1982). *Science* 218: 532–537.

60 Breslow, R. (1995). *Acc. Chem. Res.* 28: 146–153.
61 Pascal, R. (2003). *Eur. J. Org. Chem.* 2003: 1813–1824.
62 Egorova, K.S. and Ananikov, V.P. (2017). *Organometallics* 36: 4071–4090.
63 Pauling, L. (1948). *Nature* 161: 707–709.
64 Pauling, L. (1946). *Chem. Eng. News* 24: 1375–1377.
65 Zhang, X. and Houk, K.N. (2005). *Acc. Chem. Res.* 38: 379–385.
66 Jencks, W.P. (1980). *Acc. Chem. Res.* 13: 161–169.
67 Jencks, W.P. (1976). *Acc. Chem. Res.* 9: 425–432.
68 Warshel, A. (1981). *Biochemistry* 20: 3167–3177.
69 Johnson, K.A. and Goody, R.S. (2011). *Biochemistry* 50: 8264–8269.
70 Michaelis, L. and Menten, M. (1913). *Biochem. Z.* 49: 333–369.
71 Bruice, T.C. (2002). *Acc. Chem. Res.* 35: 139–148.
72 Fox, K.M. and Karplus, P.A. (1994). *Structure* 2: 1089–1105.
73 Kawai, Y., Inaba, Y., and Tokitoh, N. (2001). *Tetrahedron: Asymmetry* 12: 309–318.
74 Kawai, Y., Inaba, Y., Hayashi, M., and Tokitoh, N. (2001). *Tetrahedron Lett.* 42: 3367–3368.
75 Meah, Y. and Massey, V. (2000). *Proc. Natl. Acad. Sci. U.S.A.* 97: 10733–10738.
76 Bartlett, P.D. and Dauben, H.J. Jr., (1940). *J. Am. Chem. Soc.* 62: 1339–1344.
77 Swain, C.G. (1948). *J. Am. Chem. Soc.* 70: 1119–1128.
78 Lund, H. (1958). *Acta Chem. Scand.* 12: 298–302.
79 Bufalini, J. and Stern, K.H. (1961). *J. Am. Chem. Soc.* 83: 4362–4366.
80 Tsuboi, M. (1952). *Bull. Chem. Soc. Jpn.* 25: 60–66.
81 Tsubomura, H. (1955). *J. Chem. Phys.* 23: 2130–2133.
82 Allerhand, A. and v. R. Schleyer, P. (1963). *J. Am. Chem. Soc.* 85: 1233–1237.
83 Simmons, H.E. and Park, C.H. (1968). *J. Am. Chem. Soc.* 90: 2428–2429.
84 Park, C.H. and Simmons, H.E. (1968). *J. Am. Chem. Soc.* 90: 2429–2431.
85 Park, C.H. and Simmons, H.E. (1968). *J. Am. Chem. Soc.* 90: 2431–2432.
86 Bell, R.A., Christoph, G.G., Fronczek, F.R., and Marsh, R.E. (1975). *Science* 190: 151–152.
87 Busschaert, N., Caltagirone, C., Van Rossom, W., and Gale, P.A. (2015). *Chem. Rev.* 115: 8038–8155.
88 Molina, P., Zapata, F., and Caballero, A. (2017). *Chem. Rev.* 117: 9907–9972.
89 He, Q., Vargas-Zúñiga, G.I., Kim, S.H. *et al.* (2019). *Chem. Rev.* 119: 9753–9835.
90 Schmidtchen, F.P. and Berger, M. (1997). *Chem. Rev.* 97: 1609–1646.
91 Kim, S.K. and Sessler, J.L. (2010). *Chem. Soc. Rev.* 39: 3784–3809.
92 Park, I.-W., Yoo, J., Adhikari, S. *et al.* (2012). *Chem. Eur. J.* 18: 15073–15078.
93 Gunnlaugsson, T., Glynn, M., Tocci, G.M. *et al.* (2006). *Coord. Chem. Rev.* 250: 3094–3117.
94 Mutihac, L., Lee, J.H., Kim, J.S., and Vicens, J. (2011). *Chem. Soc. Rev.* 40: 2777–2796.
95 Teresa Albelda, M., Frías, J.C., García-España, E., and Schneider, H.-J. (2012). *Chem. Soc. Rev.* 41: 3859–3877.
96 Wichmann, K., Antonioli, B., Söhnel, T. *et al.* (2006). *Coord. Chem. Rev.* 250: 2987–3003.

97 Moyer, B.A. (2013). *Ion Exchange and Solvent Extraction: Volume 21, Supramolecular Aspects of Solvent Extraction*. Boca Raton: CRC Press.
98 Davis, A.P. (2006). *Coord. Chem. Rev.* 250: 2939–2951.
99 Davis, A.P., Sheppard, D.N., and Smith, B.D. (2007). *Chem. Soc. Rev.* 36: 348–357.
100 Matile, S., Vargas Jentzsch, A., Montenegro, J., and Fin, A. (2011). *Chem. Soc. Rev.* 40: 2453–2474.
101 Davis, J.T., Okunola, O., and Quesada, R. (2010). *Chem. Soc. Rev.* 39: 3843–3862.
102 Gale, P.A. (2011). *Acc. Chem. Res.* 44: 216–226.
103 Behera, H. and Madhavan, N. (2017). *J. Am. Chem. Soc.* 139: 12919–12922.
104 Busschaert, N. and Gale, P.A. (2013). *Angew. Chem. Int. Ed.* 52: 1374–1382.
105 Gokel, G.W. and Barkey, N. (2009). *New J. Chem.* 33: 947–963.
106 Lankshear, M.D. and Beer, P.D. (2006). *Coord. Chem. Rev.* 250: 3142–3160.
107 Gimeno, N. and Vilar, R. (2006). *Coord. Chem. Rev.* 250: 3161–3189.
108 Custelcean, R. (2014). *Chem. Soc. Rev.* 43: 1813–1824.
109 Vilar, R. (2003). *Angew. Chem. Int. Ed.* 42: 1460–1477.
110 Galbraith, E. and James, T.D. (2010). *Chem. Soc. Rev.* 39: 3831–3842.
111 Nishiyabu, R., Kubo, Y., James, T.D., and Fossey, J.S. (2011). *Chem. Commun.* 47: 1106–1123.
112 Hudnall, T.W., Chiu, C.-W., and Gabbaï, F.P. (2009). *Acc. Chem. Res.* 42: 388–397.
113 Izzo, I., Licen, S., Maulucci, N. *et al.* (2008). *Chem. Commun.* 44: 2986–2988.
114 Caballero, A., Zapata, F., White, N.G. *et al.* (2012). *Angew. Chem. Int. Ed.* 51: 1876–1880.
115 Hine, J., Ahn, K., Gallucci, J.C., and Linden, S.M. (1984). *J. Am. Chem. Soc.* 106: 7980–7981.
116 Hine, J., Linden, S.M., and Kanagasabapathy, V.M. (1985). *J. Am. Chem. Soc.* 107: 1082–1083.
117 Hine, J., Linden, S.M., and Kanagasabapathy, V.M. (1985). *J. Org. Chem.* 50: 5096–5099.
118 Errera, J. and Mollet, P. (1936). *Nature* 138: 882–882.
119 Wolf, K.L., Prahm, H., and Harms, H. (1937). *Z. Phys. Chem.* 36B: 237.
120 Sidgwick, N.V. (1927). *Electronic Theory of Valency*. New York: Oxford University Press.
121 Hine, J., Hahn, S., and Miles, D.E. (1986). *J. Org. Chem.* 51: 577–584.
122 Hine, J. and Ahn, K. (1987). *J. Org. Chem.* 52: 2083–2086.
123 Hine, J., Hahn, S., Miles, D.E., and Ahn, K. (1985). *J. Org. Chem.* 50: 5092–5096.
124 Hine, J. and Ahn, K. (1987). *J. Org. Chem.* 52: 2089–2091.
125 Kelly, T.R., Meghani, P., and Ekkundi, V.S. (1990). *Tetrahedron Lett.* 31: 3381–3384.
126 Blake, J.F. and Jorgensen, W.L. (1991). *J. Am. Chem. Soc.* 113: 7430–7432.
127 Blake, J.F., Lim, D., and Jorgensen, W.L. (1994). *J. Org. Chem.* 59: 803–805.
128 Severance, D.L. and Jorgensen, W.L. (1992). *J. Am. Chem. Soc.* 114: 10966–10968.
129 Etter, M.C. and Baures, P.W. (1988). *J. Am. Chem. Soc.* 110: 639–640.

130 Etter, M.C. and Panunto, T.W. (1988). *J. Am. Chem. Soc.* 110: 5896–5897.

131 Etter, M.C. and Reutzel, S.M. (1991). *J. Am. Chem. Soc.* 113: 2586–2598.

132 Dannecker, W. and Kopf, J. (1979). *Acta Cryst.* 8: 429–432.

133 Taylor, R. and Kennard, O. (1982). *J. Am. Chem. Soc.* 104: 5063–5070.

134 Etter, M.C. (1991). *J. Phys. Chem. A* 95: 4601–4610.

135 Smith, P.J., Reddington, M.V., and Wilcox, C.S. (1992). *Tetrahedron Lett.* 33: 6085–6088.

136 Wilcox, C.S., Adrian, J.C. Jr.,, Webb, T.H., and Zawacki, F.J. (1992). *J. Am. Chem. Soc.* 114: 10189–10197.

137 Wilcox, C.S., Kim, E.-i., Romano, D. *et al.* (1995). *Tetrahedron* 51: 621–634.

138 Fan, E., Van Arman, S.A., Kincaid, S., and Hamilton, A.D. (1993). *J. Am. Chem. Soc.* 115: 369–370.

139 Curran, D.P. and Kuo, L.H. (1995). *Tetrahedron Lett.* 36: 6647–6650.

140 Curran, D.P. and Kuo, L.H. (1994). *J. Org. Chem.* 59: 3259–3261.

141 Wittkopp, A. (1997). Wasserstoffbrückenbindungen als dirigierendeWechselwirkung in der organischen Synthese: Thioharnstoffe als Katalysatoren. Diploma Thesis. University of Göttingen, Germany.

142 Kotke, M. and Schreiner, P.R. (2009). (Thio)urea Organocatalysts. In: *Hydrogen Bonding in Organic Synthesis* (ed. P.M. Pihko), 141–351. Weinheim: Wiley VCH.

143 Jakab, G., Tancon, C., Zhang, Z. *et al.* (2012). *Org. Lett.* 14: 1724–1727.

144 Busschaert, N., Karagiannidis, L.E., Wenzel, M. *et al.* (2014). *Chem. Sci.* 5: 1118–1127.

145 Minami, T., Liu, Y., Akdeniz, A. *et al.* (2014). *J. Am. Chem. Soc.* 136: 11396–11401.

146 Gómez, D.E., Fabbrizzi, L., Licchelli, M., and Monzani, E. (2005). *Org. Biomol. Chem.* 3: 1495–1500.

147 Bonizzoni, M., Fabbrizzi, L., Taglietti, A., and Tiengo, F. (2006). *Eur. J. Org. Chem.* 2006: 3567–3574.

148 Jiménez Blanco, J.L., Bootello, P., Benito, J.M. *et al.* (2006). *J. Org. Chem.* 71: 5136–5143.

149 Custelcean, R. (2008). *Chem. Commun.* 44: 295–307.

150 Succaw, G.L., Weakley, T.J.R., Han, F., and Doxsee, K.M. (2005). *Cryst. Growth Des.* 5: 2288–2298.

151 Yonova, P.A. and Stoilkova, G.M. (2004). *J. Plant Growth Regul.* 23: 280–291.

152 Richter, C.P. (1945). *JAMA* 129: 927–931.

153 Bhogala, B.R., Captain, B., Parthasarathy, A., and Ramamurthy, V. (2010). *J. Am. Chem. Soc.* 132: 13434–13442.

154 Harris, K.D.M. (1997). *Chem. Soc. Rev.* 26: 279–289.

155 Harris, K.D.M. (2007). *Supramol. Chem.* 19: 47–53.

156 Wittkopp, A. and Schreiner, P.R. (2000). *The Chemistry of Dienes and Polyenes*, vol. 2. New York: John Wiley & Sons.

157 Wittkopp, A. and Schreiner, P.R. (2003). *Chem. Eur. J.* 9: 407–414.

158 Schreiner, P.R. and Wittkopp, A. (2002). *Org. Lett.* 4: 217–220.

159 Jai-Nhuknan, J., Karipides, A., Hughes, J., and Cantrell, J. (1997). *Acta Crystallogr. C* 53: 455–457.

160 Kotke, M. (2010). Hydrogen-Bonding (Thio)urea Organocatalysts in Organic Synthesis: State of the Art and Practical Methods for Acetalization, Tetrahydropyranylation, and Cooperative Epoxide Alcoholysis. Dissertation, Universitätsbibliothek Gießen (University of Gießen, Germany, published online).
161 Okino, T., Hoashi, Y., and Takemoto, Y. (2003). *J. Am. Chem. Soc.* 125: 12672–12673.
162 Ooi, T., Miki, T., Taniguchi, M. *et al.* (2003). *Angew. Chem. Int. Ed.* 42: 3796–3798.
163 Sohtome, Y., Tanatani, A., Hashimoto, Y., and Nagasawa, K. (2004). *Tetrahedron Lett.* 45: 5589–5592.
164 Herrera, R.P., Sgarzani, V., Bernardi, L., and Ricci, A. (2005). *Angew. Chem. Int. Ed.* 44: 6576–6579.
165 Honjo, T., Sano, S., Shiro, M., and Nagao, Y. (2005). *Angew. Chem. Int. Ed.* 44: 5838–5841.
166 Wang, J., Li, H., Yu, X. *et al.* (2005). *Org. Lett.* 7: 4293–4296.
167 Vakulya, B., Varga, S., Csámpai, A., and Soós, T. (2005). *Org. Lett.* 7: 1967–1969.
168 Lan, Q., Wang, X., and Maruoka, K. (2007). *Tetrahedron Lett.* 48: 4675–4678.
169 Klausen, R.S. and Jacobsen, E.N. (2009). *Org. Lett.* 11: 887–890.
170 Zhang, Z., Lippert, K.M., Hausmann, H. *et al.* (2011). *J. Org. Chem.* 76: 9764–9776.
171 Hayashi, Y., Gotoh, H., Hayashi, T., and Shoji, M. (2005). *Angew. Chem. Int. Ed.* 44: 4212–4215.
172 Marigo, M., Wabnitz, T.C., Fielenbach, D., and Jørgensen, K.A. (2005). *Angew. Chem. Int. Ed.* 44: 794–797.
173 Akiyama, T., Morita, H., Itoh, J., and Fuchibe, K. (2005). *Org. Lett.* 7: 2583–2585.
174 Sigman, M.S. and Jacobsen, E.N. (1998). *J. Am. Chem. Soc.* 120: 4901–4902.
175 Okino, T., Hoashi, Y., Furukawa, T. *et al.* (2005). *J. Am. Chem. Soc.* 127: 119–125.
176 Kotke, M. and Schreiner, P.R. (2006). *Tetrahedron* 62: 434–439.
177 Karimi, B., Hazarkhani, H., and Maleki, J. (2005). *Synthesis*: 279–285.
178 Ranu, B.C., Jana, R., and Samanta, S. (2004). *Adv. Synth. Chem.* 346: 446–450.
179 Mayr, H., Bug, T., Gotta, M.F. *et al.* (2001). *J. Am. Chem. Soc.* 123: 9500–9512.
180 Tantillo, D.J. and Houk, K.N. (2002). *J. Comput. Chem.* 23: 84–95.
181 Pauling, L. (1935). *J. Am. Chem. Soc.* 57: 2680–2684.
182 Kotke, M. and Schreiner, P.R. (2007). *Synthesis* 2007: 779–790.
183 Li, H., Wang, Y., Tang, L., and Deng, L. (2004). *J. Am. Chem. Soc.* 126: 9906–9907.
184 Zuend, S.J. and Jacobsen, E.N. (2009). *J. Am. Chem. Soc.* 131: 15358–15374.
185 Madarász, Á., Dósa, Z., Varga, S. *et al.* (2016). *ACS Catal.* 6: 4379–4387.
186 Bradshaw, G.A., Colgan, A.C., Allen, N.P. *et al.* (2019). *Chem. Sci.* 10: 508–514.
187 Balmond, E.I., Coe, D.M., Galan, M.C., and McGarrigle, E.M. (2012). *Angew. Chem. Int. Ed.* 51: 9152–9155.
188 Rossini, E., Bochevarov, A.D., and Knapp, E.W. (2018). *ACS Omega* 3: 1653–1662.

189 Hrdina, R., Müller, C.E., Wende, R.C. *et al.* (2011). *J. Am. Chem. Soc.* 133: 7624–7627.

190 Lippert, K.M., Hof, K., Gerbig, D. *et al.* (2012). *Eur. J. Org. Chem.* 2012: 5919–5927.

191 Malerich, J.P., Hagihara, K., and Rawal, V.H. (2008). *J. Am. Chem. Soc.* 130: 14416–14417.

192 Alemán, J., Parra, A., Jiang, H., and Jørgensen, K.A. (2011). *Chem. Eur. J.* 17: 6890–6899.

193 Chauhan, P., Mahajan, S., Kaya, U. *et al.* (2015). *Adv. Synth. Chem.* 357: 253–281.

194 Storer, R.I., Aciro, C., and Jones, L.H. (2011). *Chem. Soc. Rev.* 40: 2330–2346.

195 Ni, X., Li, X., Wang, Z., and Cheng, J.-P. (2014). *Org. Lett.* 16: 1786–1789.

196 Prohens, R., Tomàs, S., Morey, J. *et al.* (1998). *Tetrahedron Lett.* 39: 1063–1066.

197 Tomàs, S., Prohens, R., Vega, M. *et al.* (1996). *J. Org. Chem.* 61: 9394–9401.

198 Davis, P.A., Draper, S., M., Dunne, G., and Ashton, P. (1999). *Chem. Commun.* 35: 2265–2266.

199 Quiñonero, D., Prohens, R., Garau, C. *et al.* (2002). *Chem. Phys. Lett.* 351: 115–120.

200 Zhao, B.-L., Li, J.-H., and Du, D.-M. (2017). *Chem. Rec.* 17: 994–1018.

201 Rombola, M., Sumaria, C.S., Montgomery, T.D., and Rawal, V.H. (2017). *J. Am. Chem. Soc.* 139: 5297–5300.

202 Busschaert, N., Elmes, R.B.P., Czech, D.D. *et al.* (2014). *Chem. Sci.* 5: 3617–3626.

203 Elmes, R.B.P., Busschaert, N., Czech, D.D. *et al.* (2015). *Chem. Commun.* 51: 10107–10110.

204 Ho, J., Zwicker, V.E., Yuen, K.K.Y., and Jolliffe, K.A. (2017). *J. Org. Chem.* 82: 10732–10736.

205 Jeppesen, A., Nielsen, B.E., Larsen, D. *et al.* (2017). *Org. Biomol. Chem.* 15: 2784–2790.

206 Zwicker, V.E., Yuen, K.K.Y., Smith, D.G. *et al.* (2018). *Chem. Eur. J.* 24: 1140–1150.

207 Kumler, W.D. (1935). *J. Am. Chem. Soc.* 57: 600–605.

208 Li, Y. and Flood, A.H. (2008). *Angew. Chem. Int. Ed.* 47: 2649–2652.

209 Sutor, D.J. (1960). *Nature* 188: 47–48.

210 Sutor, D.J. (1963). *Acta Crystallogr. A* 16: 97–104.

211 Sutor, D.J. (1963). *J. Chem. Soc.*: 1105–1110.

212 Zhu, S.S., Staats, H., Brandhorst, K. *et al.* (2008). *Angew. Chem. Int. Ed.* 47: 788–792.

213 Li, Y. and Flood, A.H. (2008). *J. Am. Chem. Soc.* 130: 12111–12122.

214 Juwarker, H., Lenhardt, J.M., Pham, D.M., and Craig, S.L. (2008). *Angew. Chem. Int. Ed.* 47: 3740–3743.

215 Meudtner, R.M. and Hecht, S. (2008). *Angew. Chem. Int. Ed.* 47: 4926–4930.

216 Yoon, D.-W., Gross, D.E., Lynch, V.M. *et al.* (2008). *Angew. Chem. Int. Ed.* 47: 5038–5042.

217 Maeda, H., Mihashi, Y., and Haketa, Y. (2008). *Org. Lett.* 10: 3179–3182.

218 Berryman, O.B., Sather, A.C., Hay, B.P. *et al.* (2008). *J. Am. Chem. Soc.* 130: 10895–10897.

219 Nishio, M., Umezawa, Y., Fantini, J. *et al.* (2014). *Phys. Chem. Chem. Phys.* 16: 12648–12683.

220 Desiraju, G.R. (1996). *Acc. Chem. Res.* 29: 441–449.

221 Nishio, M., Umezawa, Y., Hirota, M., and Takeuchi, Y. (1995). *Tetrahedron* 51: 8665–8701.

222 Nishio, M. (2004). *CrystEngComm* 6: 130–158.

223 Nishio, M., Umezawa, Y., Honda, K. *et al.* (2009). *CrystEngComm* 11: 1757–1788.

224 Takahashi, O., Kohno, Y., and Nishio, M. (2010). *Chem. Rev.* 110: 6049–6076.

225 Nishio, M. (2011). *Phys. Chem. Chem. Phys.* 13: 13873–13900.

226 Hua, Y. and Flood, A.H. (2010). *Chem. Soc. Rev.* 39: 1262–1271.

227 McDonald, K.P., Hua, Y., Lee, S., and Flood, A.H. (2012). *Chem. Commun.* 48: 5065–5075.

228 Zurro, M. and García Mancheño, O. (2017). *Chem. Rec.* 17: 485–498.

229 Beckendorf, S., Asmus, S., Mück-Lichtenfeld, C., and García Mancheño, O. (2013). *Chem. Eur. J.* 19: 1581–1585.

230 Asmus, S., Beckendorf, S., Zurro, M. *et al.* (2014). *Chem. Asian J.* 9: 2178–2186.

231 Zurro, M., Asmus, S., Beckendorf, S. *et al.* (2014). *J. Am. Chem. Soc.* 136: 13999–14002.

232 García Mancheño, O., Asmus, S., Zurro, M., and Fischer, T. (2015). *Angew. Chem. Int. Ed.* 54: 8823–8827.

233 Fischer, T., Bamberger, J., and García Mancheño, O. (2016). *Org. Biomol. Chem.* 14: 5794–5802.

234 Fischer, T., Duong, Q.-N., and García Mancheño, O. (2017). *Chem. Eur. J.* 23: 5983–5987.

235 Duong, Q.-N., Schifferer, L., and García Mancheño, O. (2019). *Eur. J. Org. Chem.* 2019: 5452–5461.

236 Zurro, M., Asmus, S., Bamberger, J. *et al.* (2016). *Chem. Eur. J.* 22: 3785–3793.

237 Fischer, T., Bamberger, J., Gómez-Martínez, M. *et al.* (2019). *Angew. Chem. Int. Ed.* 58: 3217–3221.

238 Eichstaedt, K., Jaramillo-Garcia, J., Leigh, D.A. *et al.* (2017). *J. Am. Chem. Soc.* 139: 9376–9381.

239 Dorel, R. and Feringa, B.L. (2020). *Angew. Chem. Int. Ed.* 59: 785–789.

240 Koumura, N., Zijlstra, R.W.J., van Delden, R.A. *et al.* (1999). *Nature* 401: 152–155.

241 Lehnherr, D., Ford, D.D., Bendelsmith, A.J. *et al.* (2016). *Org. Lett.* 18: 3214–3217.

242 Reisman, S.E., Doyle, A.G., and Jacobsen, E.N. (2008). *J. Am. Chem. Soc.* 130: 7198–7199.

243 Wang, H., Wang, W., and Jin, W.J. (2016). *Chem. Rev.* 116: 5072–5104.

244 Colin, M. (1814). *Ann. Chim.* 91: 252–272.

245 Bineau, M.A. (1844). *C. R. Acad. Sci.* 55: 762–767.

246 Guthrie, F. (1863). *J. Am. Chem. Soc.* 16: 239–244.

247 Mulliken, R.S. (1950). *J. Am. Chem. Soc.* 72: 600–608.

248 Mulliken, R.S. (1952). *J. Phys. Chem. A* 56: 801–822.

249 Benesi, H.A. and Hildebrand, J.H. (1949). *J. Am. Chem. Soc.* 71: 2703–2707.

250 Hassel, O. and Hvoslef, J. (1954). *Acta Chem. Scand.* 8: 873–873.

251 Hassel, O. and Stromme, K. (1958). *Acta Chem. Scand.* 12: 1146–1147.

252 Hassel, O. and Stromme, K. (1959). *Acta Chem. Scand.* 13: 1781–1786.

253 Hassel, O. and Strømme, K. (1959). *Acta Chem. Scand.* 13: 275–280.

254 Hassel, O. and Rømming, C. (1962). *Q. Rev. Chem. Soc.* 16: 1–18.

255 Paolo, T.D. and Sandorfy, C. (1974). *Can. J. Chem.* 52: 3612–3622.

256 Dumas, J.M., Peurichard, H., and Gomel, M. (1978). *J. Chem. Res.*: 54–55.

257 Dumas, J.M., Gomel, M., and Guerin, M. (1983). Molecular Interactions Involving Organic Halides. In: *Halides, Pseudo-Halides and Azides* (eds. S. Patai and Z. Rappoport), 985–1020. Weinheim: Wiley VCH.

258 Lim, J.Y.C. and Beer, P.D. (2018). *Chem* 4: 731–783.

259 Clark, T. (2013). *WIREs Comput. Mol. Sci.* 3: 13–20.

260 Bauzá, A. and Frontera, A. (2015). *Angew. Chem. Int. Ed.* 54: 7340–7343.

261 Scholfield, M.R., Zanden, C.M.V., Carter, M., and Ho, P.S. (2013). *Protein Sci.* 22: 139–152.

262 Politzer, P., Lane, P., Concha, M.C. *et al.* (2007). *J. Mol. Model.* 13: 305–311.

263 Politzer, P. and Murray, J.S. (2013). *ChemPhysChem* 14: 278–294.

264 Erdélyi, M. (2012). *Chem. Soc. Rev.* 41: 3547–3557.

265 Wolters, L.P., Schyman, P., Pavan, M.J. *et al.* (2014). *WIREs Comput. Mol. Sci.* 4: 523–540.

266 Beale, T.M., Chudzinski, M.G., Sarwar, M.G., and Taylor, M.S. (2013). *Chem. Soc. Rev.* 42: 1667–1680.

267 Metrangolo, P., Neukirch, H., Pilati, T., and Resnati, G. (2005). *Acc. Chem. Res.* 38: 386–395.

268 Clark, T., Hennemann, M., Murray, J.S., and Politzer, P. (2007). *J. Mol. Model.* 13: 291–296.

269 Clark, T. (2017). *Faraday Discuss.* 203: 9–27.

270 Politzer, P., Murray, J.S., Clark, T., and Resnati, G. (2017). *Phys. Chem. Chem. Phys.* 19: 32166–32178.

271 Breugst, M. and Koenig, J.J. (2020). *Eur. J. Org. Chem.* 2020: 5473–5487.

272 Walter, S.M., Kniep, F., Herdtweck, E., and Huber, S.M. (2011). *Angew. Chem. Int. Ed.* 50: 7187–7191.

273 Walter, S.M., Kniep, F., Rout, L. *et al.* (2012). *J. Am. Chem. Soc.* 134: 8507–8512.

274 Kniep, F., Jungbauer, S.H., Zhang, Q. *et al.* (2013). *Angew. Chem. Int. Ed.* 52: 7028–7032.

275 Schottel, B.L., Chifotides, H.T., and Dunbar, K.R. (2008). *Chem. Soc. Rev.* 37: 68–83.

276 Frontera, A., Gamez, P., Mascal, M. *et al.* (2011). *Angew. Chem. Int. Ed.* 50: 9564–9583.

277 Frontera, A., Quiñonero, D., and Deyà, P.M. (2011). *WIREs Comput. Mol. Sci.* 1: 440–459.

278 Quiñonero, D., Garau, C., Rotger, C. *et al.* (2002). *Angew. Chem. Int. Ed.* 41: 3389–3392.
279 Alkorta, I., Rozas, I., and Elguero, J. (2002). *J. Am. Chem. Soc.* 124: 8593–8598.
280 Mascal, M., Armstrong, A., and Bartberger, M.D. (2002). *J. Am. Chem. Soc.* 124: 6274–6276.
281 Zhao, Y., Domoto, Y., Orentas, E. *et al.* (2013). *Angew. Chem. Int. Ed.* 52: 9940–9943.
282 Zhao, Y., Beuchat, C., Domoto, Y. *et al.* (2014). *J. Am. Chem. Soc.* 136: 2101–2111.
283 Lu, T. and Wheeler, S.E. (2014). *Org. Lett.* 16: 3268–3271.
284 Zhao, Y., Cotelle, Y., Liu, L. *et al.* (2018). *Acc. Chem. Res.* 51: 2255–2263.
285 Neel, A.J., Hilton, M.J., Sigman, M.S., and Toste, F.D. (2017). *Nature* 543: 637–646.
286 Berkessel, A., Das, S., Pekel, D., and Neudörfl, J.-M. (2014). *Angew. Chem. Int. Ed.* 53: 11660–11664.
287 Schafer, A.G., Wieting, J.M., Fisher, T.J., and Mattson, A.E. (2013). *Angew. Chem. Int. Ed.* 52: 11321–11324.
288 Wieting, J.M., Fisher, T.J., Schafer, A.G. *et al.* (2015). *Eur. J. Org. Chem.* 2015: 525–533.
289 Attard, J.W., Osawa, K., Guan, Y. *et al.* (2019). *Synthesis* 51: 2107–2115.
290 Borovika, A., Tang, P.-I., Klapman, S., and Nagorny, P. (2013). *Angew. Chem. Int. Ed.* 52: 13424–13428.
291 Weil, T., Kotke, M., Kleiner, C.M., and Schreiner, P.R. (2008). *Org. Lett.* 10: 1513–1516.
292 Lee, J.H., Kim, W.H., and Danishefsky, S.J. (2010). *Tetrahedron Lett.* 51: 1252–1253.
293 Wöhler, F. and v. Liebig, J. (1832). *Liebigs Ann.* 3: 249–282.
294 Langenbeck, W. (1927). *Ber. Dtsch. Chem. Ges.* 60: 930–934.
295 Langenbeck, W. (1928). *Angew. Chem.* 41: 740–745.
296 Langenbeck, W. (1929). *Liebigs Ann.* 469: 16–25.
297 Langenbeck, W. (1932). *Angew. Chem.* 45: 97–99.
298 Langenbeck, W. (1949). *Die organischen Katalysatoren und ihre Beziehungen zu den Fermenten (Organic Catalysts and their Relation to the Enzymes).* Berlin: Springer.
299 Langenbeck, W. (1958). *Tetrahedron* 3: 185–196.
300 v. Liebig, J. (1860). *Liebigs Ann.* 113: 246–247.
301 v. Liebig, J. (1870). *Liebigs Ann.* 153: 1–47.
302 Seebach, D. (1990). *Angew. Chem. Int. Ed.* 29: 1320–1367.
303 Morrill, L.A., Susick, R.B., Chari, J.V., and Garg, N.K. (2019). *J. Am. Chem. Soc.* 141: 12423–12443.
304 Simons, J.P. (1987). *J. Phys. Chem. A* 91: 5378–5387.
305 Aldegunde, J., de Miranda, M.P., Haigh, J.M. *et al.* (2005). *J. Phys. Chem. A* 109: 6200–6217.
306 Zhan, C.-G., Bentley, J., and Chipman, D.M. (1998). *J. Chem. Phys.* 108: 177–192.

307 Ostwald, W. (1894). *Z. Phys. Chem.* 15: 705–707.

308 Roberts, M.W. (2000). *Catal. Lett.* 67: 1–4.

309 Palmer, T. and Bonner, P.L. (2011). The Chemical Nature of Enzyme Catalysis. In: *Enzymes*, 2e (eds. T. Palmer and P. L. Bonner), 189–221. Sawston: Horwood Publishing Ltd.

310 List, B. (2007). *Chem. Rev.* 107: 5413–5883. special issue on organocatalysis.

311 Abbasov, M.E. and Romo, D. (2014). *Nat. Prod. Rep.* 31: 1318–1327.

312 Wende, R.C. and Schreiner, P.R. (2012). *Green Chem.* 14: 1821–1849.

313 Seayad, J. and List, B. (2005). *Org. Biomol. Chem.* 3: 719–724.

314 Berkessel, A. and Gröger, H. (2006). *Asymmetric Organocatalysis: From Biomimetic Concepts to Applications in Asymmetric Synthesis.* Weinheim: Wiley VCH.

315 Marcia de Figueiredo, R. and Christmann, M. (2007). *Eur. J. Org. Chem.* 2007: 2575–2600.

316 Marqués-López, E., Herrera, R.P., and Christmann, M. (2010). *Nat. Prod. Rep.* 27: 1138–1167.

317 MacMillan, D.W.C. (2008). *Nature* 455: 304–308.

318 Dalko, P.I. and Moisan, L. (2001). *Angew. Chem. Int. Ed.* 40: 3726–3748.

319 Dalko, P.I. and Moisan, L. (2004). *Angew. Chem. Int. Ed.* 43: 5138–5175.

320 Szwarc, M. (1969). *Acc. Chem. Res.* 2: 87–96.

321 Bjerrum, N. (1926). *K. Dan. Vidensk. Selsk. Mat.-Fys. Medd.* 7: 3–48.

322 Anslyn, E.V. and Dougherty, D.A. (2006). *Modern Physical Organic Chemistry.* Sausalito: University Science Books.

323 Einstein, A. (1905). *Ann. Phys.* 322: 549–560.

324 Marcus, Y. (1985). Solvation of Ion Pairs. In: *Ion Solvation* (ed. Y. Marcus), 218–244. Chichester: Wiley.

325 Marcus, Y. and Hefter, G. (2006). *Chem. Rev.* 106: 4585–4621.

326 Sadek, H. and Fuoss, R.M. (1954). *J. Am. Chem. Soc.* 76: 5897–5901.

327 Winstein, S., Clippinger, E., Fainberg, A.H., and Robinson, G.C. (1954). *J. Am. Chem. Soc.* 76: 2597–2598.

328 Erkkilä, A., Majander, I., and Pihko, P.M. (2007). *Chem. Rev.* 107: 5416–5470.

329 Huang, Y., Walji, A.M., Larsen, C.H., and MacMillan, D.W.C. (2005). *J. Am. Chem. Soc.* 127: 15051–15053.

330 Yamashita, Y., Yasukawa, T., Yoo, W.-J. *et al.* (2018). *Chem. Soc. Rev.* 47: 4388–4480.

331 Romea, P. and Urpi, F. (2013). Stereoselective Acetate Aldol Reactions. In: *Modern Methods in Stereoselective Aldol Reactions* (ed. R. Mahrwald), 1–81. Weinheim: Wiley-VCH.

332 Mayer, S. and List, B. (2006). *Angew. Chem. Int. Ed.* 45: 4193–4195.

333 Corey, E.J. and Grogan, M.J. (1999). *Org. Lett.* 1: 157–160.

334 Li, J., Jiang, W.-Y., Han, K.-L. *et al.* (2003). *J. Org. Chem.* 68: 8786–8789.

335 Goerigk, L., Kruse, H., and Grimme, S. (2011). *ChemPhysChem* 12: 3421–3433.

336 Wagner, J.P. and Schreiner, P.R. (2015). *Angew. Chem. Int. Ed.* 54: 12274–12296.

337 Birrell, J.A., Desrosiers, J.-N., and Jacobsen, E.N. (2011). *J. Am. Chem. Soc.* 133: 13872–13875.

338 Taylor, M.S. and Jacobsen, E.N. (2004). *J. Am. Chem. Soc.* 126: 10558–10559.
339 Knowles, R.R., Lin, S., and Jacobsen, E.N. (2010). *J. Am. Chem. Soc.* 132: 5030–5032.
340 Kutateladze, D.A., Strassfeld, D.A., and Jacobsen, E.N. (2020). *J. Am. Chem. Soc.* 142: 6951–6956.
341 Seidel, D. (2014). *Synlett* 25: 783–794.
342 De, C.K., Klauber, E.G., and Seidel, D. (2009). *J. Am. Chem. Soc.* 131: 17060–17061.
343 De, C.K., Mittal, N., and Seidel, D. (2011). *J. Am. Chem. Soc.* 133: 16802–16805.
344 Banik, S.M., Levina, A., Hyde, A.M., and Jacobsen, E.N. (2017). *Science* 358: 761–764.
345 Wendlandt, A.E., Vangal, P., and Jacobsen, E.N. (2018). *Nature* 556: 447–451.
346 Pupo, G., Ibba, F., Ascough, D.M.H. *et al.* (2018). *Science* 360: 638–642.
347 Roagna, G., Ascough, D.M.H., Ibba, F. *et al.* (2020). *J. Am. Chem. Soc.* 142: 14045–14051.
348 Pupo, G., Vicini, A.C., Ascough, D.M.H. *et al.* (2019). *J. Am. Chem. Soc.* 141: 2878–2883.
349 Sigman, M.S., Vachal, P., and Jacobsen, E.N. (2000). *Angew. Chem. Int. Ed.* 39: 1279–1281.
350 HCN was generated by the method of Ziegler. Alternatively, generation of HCN in situ by reaction of TMSCN and MeOH afforded identical results: Ziegler, K. (1932). In: *Organic Syntheses Collective*, Volume 1 (eds. H. Gilman and H. Blatt), p. 314. New York: Wiley
351 Vachal, P. and Jacobsen, E.N. (2002). *J. Am. Chem. Soc.* 124: 10012–10014.
352 Wittkopp, A. (2001). Organokatalyse von Diels–Alder-Reaktionen durch neutrale Wasserstoffbrückendonoren in organischen und wässrigen Lösungsmitteln. Dissertation, University of Göttingen, Germany, published online (Germany).
353 Tárkányi, G., Király, P., Varga, S. *et al.* (2008). *Chem. Eur. J.* 14: 6078–6086.
354 Oh, S.H., Rho, H.S., Lee, J.W. *et al.* (2008). *Angew. Chem. Int. Ed.* 47: 7872–7875.
355 Zuend, S.J., Coughlin, M.P., Lalonde, M.P., and Jacobsen, E.N. (2009). *Nature* 461: 968–970.
356 Wadhwa, P., Kharbanda, A., and Sharma, A. (2018). *Asian J. Org. Chem.* 7: 634–661.
357 Enders, D., Lüttgen, K., and Narine, A.A. (2007). *Synthesis* 2007: 959–980.
358 Enders, D., Wang, C., and Liebich, J.X. (2009). *Chem. Eur. J.* 15: 11058–11076.
359 Zhu, R., Zhang, D., Wu, J., and Liu, C. (2006). *Tetrahedron: Asymmetry* 17: 1611–1616.
360 Hoashi, Y., Yabuta, T., and Takemoto, Y. (2004). *Tetrahedron Lett.* 45: 9185–9188.
361 Hoashi, Y., Yabuta, T., Yuan, P. *et al.* (2006). *Tetrahedron* 62: 365–374.
362 Takemoto, Y. (2010). *Chem. Pharm. Bull.* 58: 593–601.
363 An overview of numerous application is discussed in Kotke, M., Schreiner, P. R. (2009). (Thio)urea Organocatalysts. In: *Hydrogen Bonding in Organic Synthesis* (ed. P. M. Pihko), 141–351. Weinheim: Wiley VCH.

364 Hamza, A., Schubert, G., Soós, T., and Pápai, I. (2006). *J. Am. Chem. Soc.* 128: 13151–13160.
365 Izzo, J.A., Myshchuk, Y., Hirschi, J.S., and Vetticatt, M.J. (2019). *Org. Biomol. Chem.* 17: 3934–3939.
366 Tan, B., Lu, Y., Zeng, X. *et al.* (2010). *Org. Lett.* 12: 2682–2685.
367 Nishizawa, S., Bühlmann, P., Iwao, M., and Umezawa, Y. (1995). *Tetrahedron Lett.* 36: 6483–6486.
368 Wenzel, A.G. and Jacobsen, E.N. (2002). *J. Am. Chem. Soc.* 124: 12964–12965.
369 Bentley, K.W. (2004). *Nat. Prod. Rep.* 21: 395–424.
370 Bentley, K.W. (2006). *Nat. Prod. Rep.* 23: 444–463.
371 Taylor, M.S., Tokunaga, N., and Jacobsen, E.N. (2005). *Angew. Chem. Int. Ed.* 44: 6700–6704.
372 Bose, A.K., Spiegelman, G., and Manhas, M.S. (1971). *Tetrahedron Lett.* 12: 3167–3170.
373 Raheem, I.T., Thiara, P.S., Peterson, E.A., and Jacobsen, E.N. (2007). *J. Am. Chem. Soc.* 129: 13404–13405.
374 Dijkink, J. and Speckamp, W.N. (1977). *Tetrahedron Lett.* 18: 935–938.
375 Dijkink, J. and Speckamp, W.N. (1978). *Tetrahedron* 34: 173–178.
376 Ford, D.D., Lehnherr, D., Kennedy, C.R., and Jacobsen, E.N. (2016). *J. Am. Chem. Soc.* 138: 7860–7863.
377 Ford, D.D., Lehnherr, D., Kennedy, C.R., and Jacobsen, E.N. (2016). *ACS Catal.* 6: 4616–4620.
378 Kennedy, C.R., Lehnherr, D., Rajapaksa, N.S. *et al.* (2016). *J. Am. Chem. Soc.* 138: 13525–13528.
379 Park, Y., Harper, K.C., Kuhl, N. *et al.* (2017). *Science* 355: 162–166.
380 Lairson, L.L., Henrissat, B., Davies, G.J., and Withers, S.G. (2008). *Annu. Rev. Biochem.* 77: 521–555.
381 Nigudkar, S.S. and Demchenko, A.V. (2015). *Chem. Sci.* 6: 2687–2704.
382 Albert, J.S. and Hamilton, A.D. (1993). *Tetrahedron Lett.* 34: 7363–7366.
383 Kelly, T.R. and Kim, M.H. (1994). *J. Am. Chem. Soc.* 116: 7072–7080.
384 Boiocchi, M., Del Boca, L., Gómez, D.E. *et al.* (2004). *J. Am. Chem. Soc.* 126: 16507–16514.
385 Amendola, V., Esteban-Gómez, D., Fabbrizzi, L., and Licchelli, M. (2006). *Acc. Chem. Res.* 39: 343–353.
386 Lee, Y., Klausen, R.S., and Jacobsen, E.N. (2011). *Org. Lett.* 13: 5564–5567.
387 Klausen, R.S., Kennedy, C.R., Hyde, A.M., and Jacobsen, E.N. (2017). *J. Am. Chem. Soc.* 139: 12299–12309.
388 Burns, N.Z., Witten, M.R., and Jacobsen, E.N. (2011). *J. Am. Chem. Soc.* 133: 14578–14581.
389 Klauber, E.G., De, C.K., Shah, T.K., and Seidel, D. (2010). *J. Am. Chem. Soc.* 132: 13624–13626.
390 Klauber, E.G., Mittal, N., Shah, T.K., and Seidel, D. (2011). *Org. Lett.* 13: 2464–2467.
391 Min, C., Mittal, N., De, C.K., and Seidel, D. (2012). *Chem. Commun.* 48: 10853–10855.
392 Mittal, N., Sun, D.X., and Seidel, D. (2012). *Org. Lett.* 14: 3084–3087.

393 Mittal, N., Lippert, K.M., De, C.K. *et al.* (2015). *J. Am. Chem. Soc.* 137: 5748–5758.

394 Lutz, V., Glatthaar, J., Würtele, C. *et al.* (2009). *Chem. Eur. J.* 15: 8548–8557.

395 Wang, Z.-X., Tu, Y., Frohn, M. *et al.* (1997). *J. Am. Chem. Soc.* 119: 11224–11235.

396 Kagan, H. and Fiaud, J. (1988). *Top. Stereochem.* 18: 21.

397 Müller, K., Faeh, C., and Diederich, F. (2007). *Science* 317: 1881–1886.

398 Purser, S., Moore, P.R., Swallow, S., and Gouverneur, V. (2008). *Chem. Soc. Rev.* 37: 320–330.

399 Ojima, I. (2009). *Fluorine in Medicinal Chemistry and Chemical Biology*. New York: John Wiley & Sons.

400 Gouverneur, V. and Müller, K. (2012). *Fluorine in Pharmaceutical and Medicinal Chemistry: From Biophysical Aspects to Clinical Applications*. Singapur: World Scientific.

401 Jeschke, P. (2004). *ChemBioChem* 5: 570–589.

402 Gillis, E.P., Eastman, K.J., Hill, M.D. *et al.* (2015). *J. Med. Chem.* 58: 8315–8359.

403 Nelson, A. (1999). *Angew. Chem. Int. Ed.* 38: 1583–1585.

404 Ooi, T. and Maruoka, K. (2007). *Angew. Chem. Int. Ed.* 46: 4222–4266.

405 Maruoka, K. (2008). *Asymmetric Phase Transfer Catalysis*. Weinheim: Wiley VCH.

406 Esteban-Gómez, D., Fabbrizzi, L., and Licchelli, M. (2005). *J. Org. Chem.* 70: 5717–5720.

407 Jose, D.A., Kumar, D.K., Ganguly, B., and Das, A. (2004). *Org. Lett.* 6: 3445–3448.

408 Pfeifer, L., Engle, K.M., Pidgeon, G.W. *et al.* (2016). *J. Am. Chem. Soc.* 138: 13314–13325.

409 Yang, X., Wu, T., Phipps, R.J., and Toste, F.D. (2015). *Chem. Rev.* 115: 826–870.

410 Auria-Luna, F., Mohammadi, S., Divar, M. *et al.* (2020). *Adv. Synth. Chem.* 362: 5275–5300.

411 Nyffeler, P.T., Durón, S.G., Burkart, M.D. *et al.* (2005). *Angew. Chem. Int. Ed.* 44: 192–212.

412 Wang, X., Lan, Q., Shirakawa, S., and Maruoka, K. (2010). *Chem. Commun.* 46: 321–323.

413 Rauniyar, V., Lackner, A.D., Hamilton, G.L., and Toste, F.D. (2011). *Science* 334: 1681–1684.

414 O'Hagan, D., Schaffrath, C., Cobb, S.L. *et al.* (2002). *Nature* 416: 279–279.

415 Zhu, X., Robinson, D.A., McEwan, A.R. *et al.* (2007). *J. Am. Chem. Soc.* 129: 14597–14604.

416 O'Hagan, D. and Deng, H. (2015). *Chem. Rev.* 115: 634–649.

417 Morgenthaler, M., Schweizer, E., Hoffmann-Röder, A. *et al.* (2007). *ChemMedChem* 2: 1100–1115.

418 Briggs, C.R.S., O'Hagan, D., Howard, J.A.K., and Yufit, D.S. (2003). *J. Fluor. Chem.* 119: 9–13.

419 Sofia, M.J., Bao, D., Chang, W. *et al.* (2010). *J. Med. Chem.* 53: 7202–7218.

420 Sato, K., Hoshino, K., Tanaka, M. *et al.* (1992). *Antimicrob. Agents Chemother.* 36: 1491–1498.

2

Anion Recognition and Binding Constant Determination

Edward G. Sheetz[1], David Van Craen[2], and Amar H. Flood[1]

[1] *Indiana University, Department of Chemistry, 800 East Kirkwood Avenue, Bloomington, IN 47405, USA*
[2] *TU Dortmund University, Department of Chemistry and Chemical Biology, Otto-Hahn Str. 6, 44227 Dortmund, Germany*

2.1 Introduction to Supramolecular Chemistry and Binding Constant Determination

2.1.1 Chapter Overview

Supramolecular chemistry is the study of intermolecular association, and it describes the reversible non-covalent reaction between two or more molecules [1]. It includes the sub-fields of host–guest chemistry and molecular recognition and extends to applications in sensing and catalysis, which is the topic of this book. Molecular recognition is specifically relevant in anion-assisted catalysis, where the mechanism usually involves the activation of a cationic substrate toward nucleophilic attack following recognition of the counter-anion by a catalyst (the receptor) and the formation of a contact ion pair. The cation and anion can be formed *in situ*, as intermediates, during the course of the reaction. Other catalytic systems, however, can also feature anion binding in various steps of the reaction [2–4]. We will contend first with the factors controlling the ion-binding equilibria of importance to anion-assisted catalysis and second with the methods to determine the equilibrium constants. Exemplary experiments have been identified in the text to help connect these methods to the studies of catalysis conducted by the following authors: Connon, Feringa, García Mancheño, Goldfuss, Gouverneur, Huber, Jacobsen, Lacour, Leigh, Maeda, Matile, Mattson, Kozlowski, Mukherjee, Nagorny, Ooi, Seidel, Schreiner, Vicario, Merino, Xu, Yoon, Yuan, and Zhang.

The anion-assisted catalysis developed by Jacobsen is exemplary (Figure 2.1a) [5]. The accepted mechanism [6] involves formation of an ion-paired intermediate that is activated toward stereospecific nucleophilic attack passing through a transition state that also displays some degree of ion pairing. Of the underlying equilibria (Figure 2.1b,c) leading to the activated intermediate, we will focus on anion binding and ion pair formation. Dissociation of the thiourea dimer can also be treated like any supramolecular process with its identification and inclusion in a binding model.

Anion-Binding Catalysis, First Edition. Edited by Olga García Mancheño.

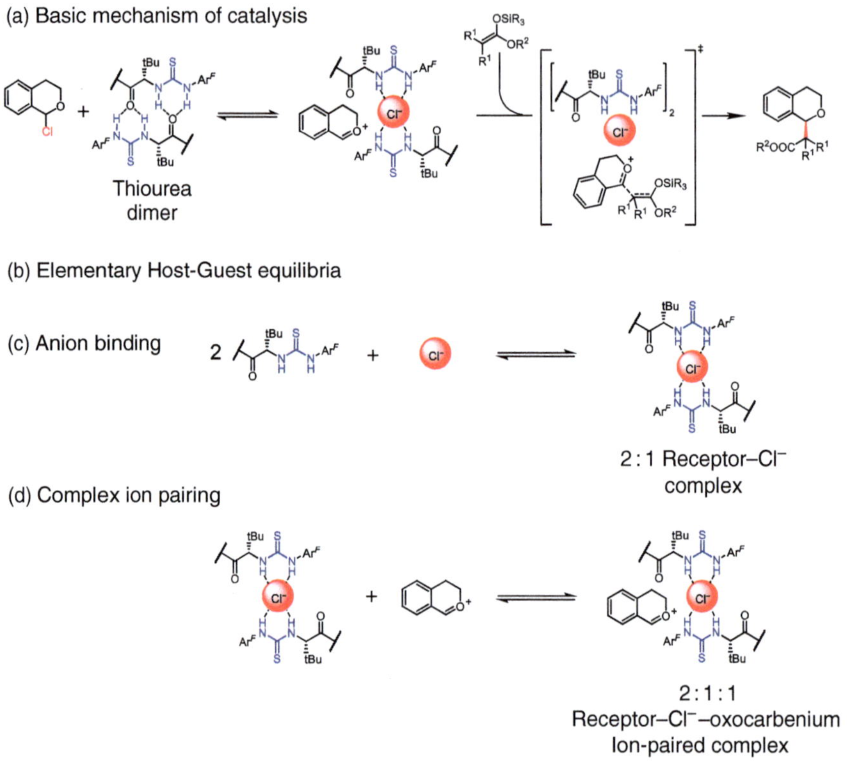

Figure 2.1 (a) Basic mechanism of anion-assisted catalysis based on Jacobsen's thiourea-catalyzed enantioselective alkylation of α-chloroethers. (b) Elementary host–guest equilibria for (c) the anion binding equilibrium and (d) ion pairing between the anion complex and the oxocarbenium cation. Source: Based on Ford *et al.* [5].

To this end, we will outline common strategies to measure the various association reactions and common concerns for the accurate determination of equilibrium constants, K_a.

2.1.2 Supramolecular Chemistry and Its Connection to Anion-Assisted Catalysis

In supramolecular catalysis, the ***association complexes*** that form during the course of the reaction (Figure 2.1b,c) can impact the overall yield and rate of product formation as well as the regio- and stereochemistry present in the product. Knowledge of the elementary processes of receptor–anion binding, e.g. Figure 2.2a, the equilibrium stabilities (K_a), and the potential structures (Figure 2.2b) of the complexes helps guide the intuition and understanding of the catalytic event. The simplest elementary case is the 1 : 1 association between the host (H) and guest (G) as defined in Eqs. (2.1) and (2.2):

$$H + G \leftrightarrow HG \tag{2.1}$$

$$K_a = \frac{[HG]_{eq.}}{[H]_{eq.} \cdot [G]_{eq.}} \tag{2.2}$$

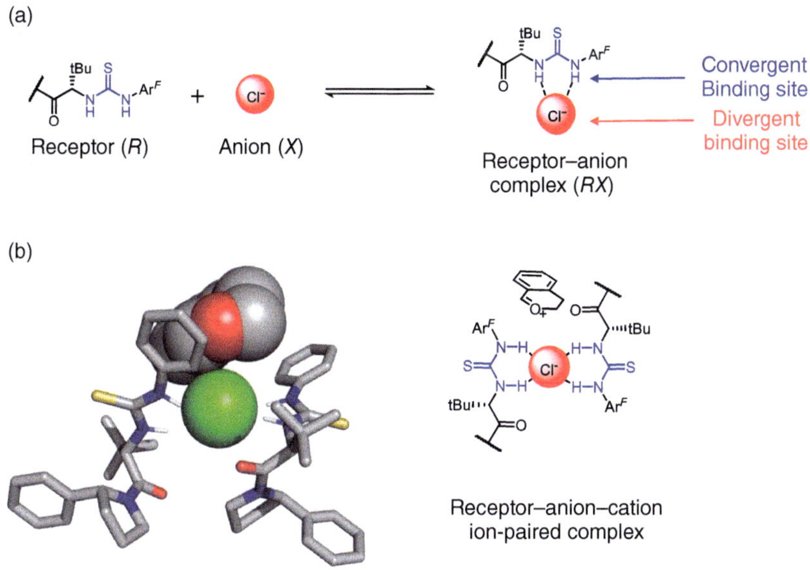

Figure 2.2 (a) Elementary 1 : 1 reaction between the receptor and the chloride anion. (b) Molecular model of the 2 : 1 : 1 complex formed between thioureas, chloride, and oxocarbenium cation.

In many of the receptor–anion cases described herein, the receptor is the host, and the anion is the guest. However, Many of the resulting complexes are multi-molecule assemblies, involving some combination of anion, receptor, and cation. In Jacobsen's example (see molecular model in Figure 2.2b), the species called the ion-pair complex involves multiple molecules, two receptors, one anion, and the oxocarbenium cation. Nevertheless, this multi-component assembly can be broken down to the elementary reactions. As such, they will always reliably conform to the principles (and intuition) of supramolecular chemistry. Thus, we will consider the anion recognition event (Figure 2.1c) and the ion-pairing event (Figure 2.1d) as separate elementary reactions. Whether they are separable in the catalytic reaction flask when cooperativity is expressed is another question. Nevertheless, they are usually always amenable to model studies and optimization.

Receptors are defined by convergent binding sites, e.g. –OH groups [7], thiourea NH donors [5], and triazole CH groups [8], that are directed inward toward the anion, while anionic guests and often their counter-cations have divergent binding sites that accommodate receptor interactions from many directions (Figure 2.2a). This concept helps identify the functional groups that dictate the strength and geometrical preferences of non-covalent interactions. For example, an $NH\cdots Cl^-$ hydrogen bonding interaction (Figure 2.3a) is ideal at 180°, although sub-optimal angles as low as 110° can also be stabilizing albeit with attenuated binding strength (Figure 2.3b) [9].

The architecture of the receptor, i.e. geometry and size, is the key attribute that governs its size and shape complementarity for guests. A more favorable binding occurs when the geometrical complementarity for a guest is highest, i.e.

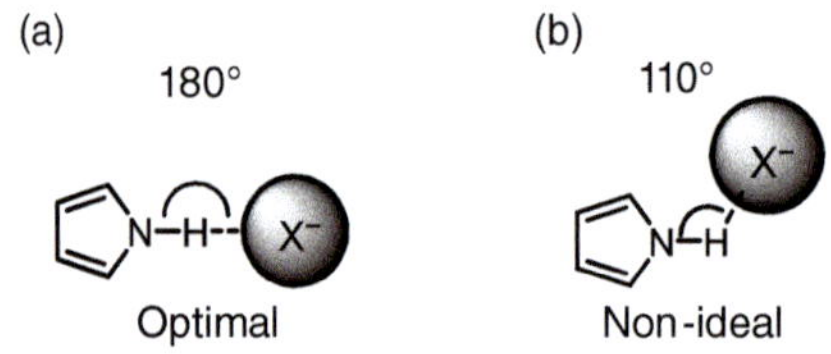

Figure 2.3 (a) Ideal (180°) and (b) less optimal (110°) H-bond angles for pyrrole–anion contacts. Source: Based on Dannenberg [9].

the lock-and-key principle [10], and it is best seen across a homologous series. High binding affinity and selectivity is conferred when the conformation of the receptor is preorganized for binding before associating with the guest. Thus, the conformational preferences of the ligand receptor are also important. This role of organization is exemplified by a recent aryl-triazole cage [11]; however, less organized receptors are often used in anion-assisted catalysis and changes in their organization always need to be considered. As an example, García Mancheño's receptor framework is built on aryl-triazoles where the large 5-Debye dipoles on the triazoles dictate that there are three conformations accessible in solution (Figure 2.4). The *syn-anti* conformation is most favored in solution, but the *syn–syn* is the one that is ideal for binding chloride anions. Thus, the receptor is not preorganized for binding, and the penalty of forming the *syn–syn* conformation is paid by the benefit of anion binding. It has been shown that preorganizing these into the anti–anti conformations boosts affinity [12].

Molecular recognition is most ideal when it is able to achieve strong and selective binding of one guest among others. In a catalyzed reaction, one typically knows the constituents of the reaction vessel, which provides a basis for the guest-targeted design of a ligand. This situation also allows for rationalization of observations. In Jacobsen's example, the anion is generated *in situ* and is not expected to face much competition from non-participating anions, and in García Mancheño's example, the substrate itself is a chloride salt, so competing anions are actively excluded in this system. The existence of ion-pairing equilibria between anions and cations as well

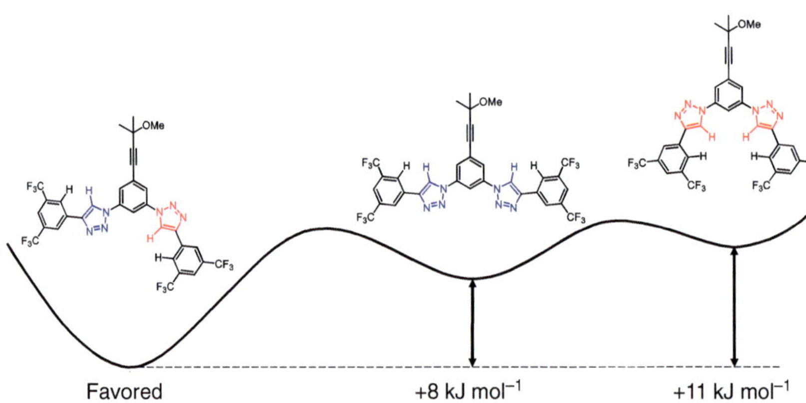

Figure 2.4 Conformation preferences of García Mancheño's aryl-triazole catalyst for the *syn–anti*, *syn–syn*, and *anti–anti* conformers (molecular mechanics, MMFF, and gas phase).

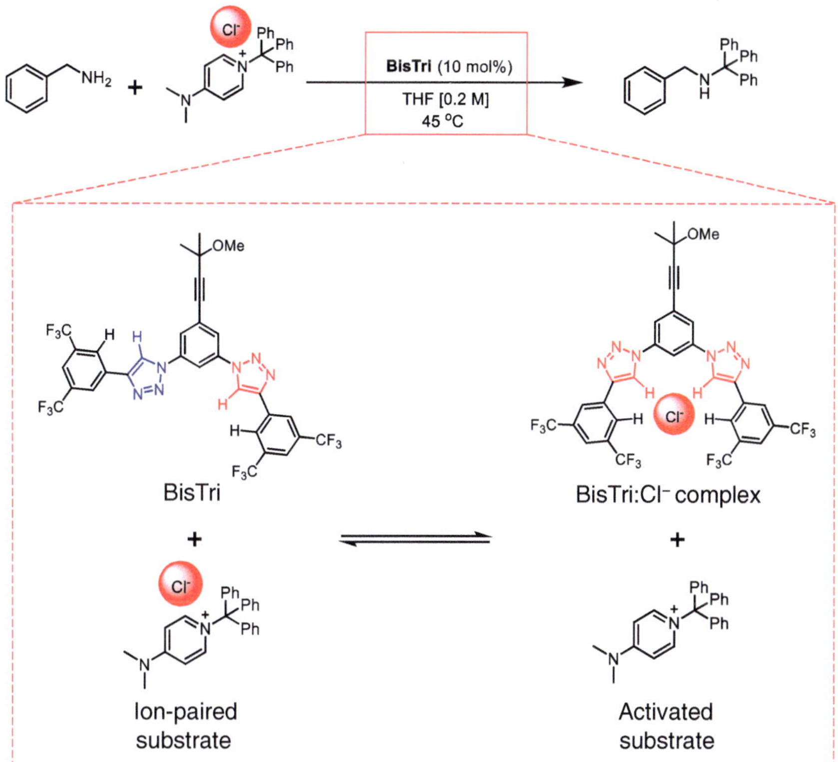

Figure 2.5 García Mancheño's bistriazole-catalyzed alkylation. Box shows anion-binding equilibrium that activates cationic substrate (ion-pairing equilibrium omitted).

as between the anionic complex and cations can add complexity to the selectivity picture. In Jacobsen's case, ion pairing between oxocarbenium and the complex is integral to the mechanism [6], while in García Mancheño's example, anion binding mitigates ion pairing to activate a cationic substrate (Figure 2.5).

Characterization of the structure and stability of the receptor–anion complex is key to understanding its role in the catalyzed reaction. The crystal structures of complexes can verify the expected non-covalent contacts and reveal the unexpected ones. Yet most catalysis occurs in the solution and crystals, while they look spectacular, do not always translate to solution, and must serve only as a starting point. The computational models of target complexes can also provide structures, allow one to test intuition, and help guide hypotheses about expected structures. For example, the crystal structures [6] of various thiourea receptors with chloride anions (Figure 2.6) can confirm ideas about NH· · ·Cl^- contacts and suggest possible stoichiometries in the case of 1 : 1 and 2 : 1 complexes (Figure 2.6). Other crystal structures in the same study showed stoichiometries and geometries that are less likely to be present in the solution. One such structure shows a unit cell with a 4 : 2 stoichiometry where two receptors are interacting with two chloride anions, but only one is engaging all of its hydrogen bond donors. Additionally, the other two receptors are not engaged at all

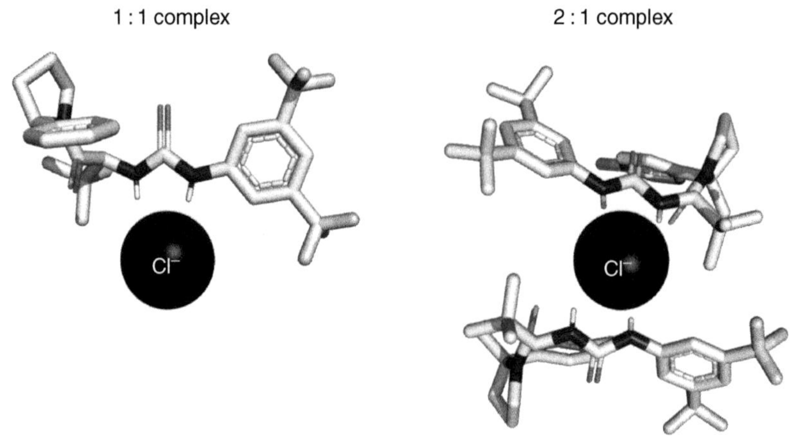

Figure 2.6 X-ray crystallographic structures of several of Jacobsen's thiourea catalysts engaged with chloride in various geometries and stoichiometries. These are the structures identified that neatly model the structures identified in solution.

with the anion but rather are forming hydrogen bonds with other receptors in the crystal.

Crystal structures can also help show the conformational preferences of empty receptors, some imparting lower reorganization penalties upon binding. For example, Goldfuss and coworkers showed cyclodiphosphazane receptors adopt either in–in conformations favorable to binding or in–out conformations depending on their substituents [13]. Beyond these initial models, estimation of the solution phase geometries by various NMR spectroscopies, determination of the dominant reaction stoichiometries, and measurement of binding constants by various titration methods are crucial to understanding the actual active species that are important to anion-assisted catalysis.

2.1.3 Brief History of Advances in Supramolecular Anion Binding

Anion recognition [14] took substantially longer to develop than the exploration of cation recognition that began with crown ether chemistry [15]. This situation emerges because anions have weaker binding affinities in solution than is seen with cations. This situation arises from their larger size, often twice the size, which makes them more charge diffuse. There is also negligible covalency to their binding. Thus, the idea of coordinative saturation, while an effective design parameter for transition metals, is often poorly defined in anion-assisted catalysis. Anions also have various geometries and protonation states, which need to be accommodated (or leveraged).

Anion binding is usually produced in receptors that bear cationic and/or hydrogen-bonding sites, although other contacts such as halogen and chalcogen

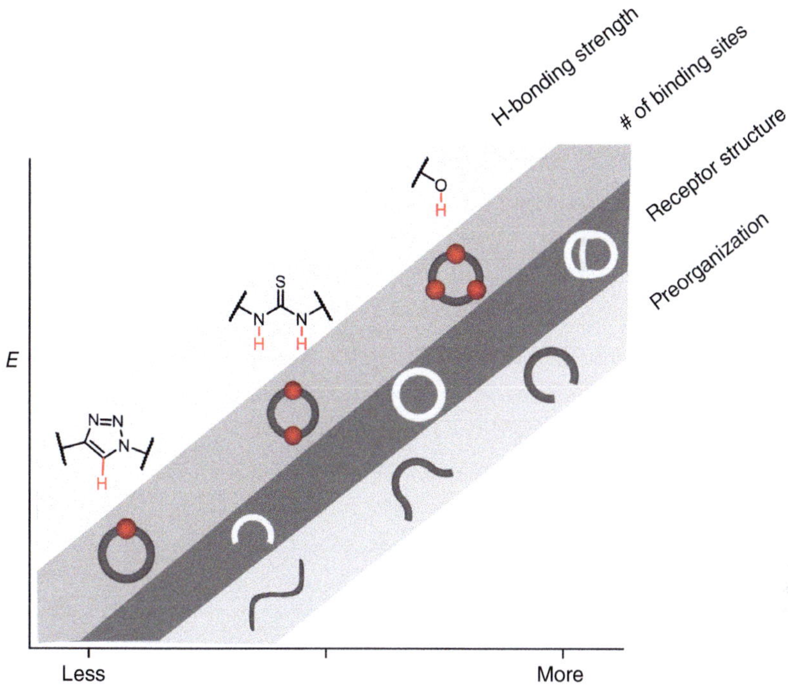

Figure 2.7 Correlation plot between binding energy (Δ*E*) and synthetic design features: H-bond strength of the binding sites (triazole, thiourea, and their total number), the structure of the receptor spanning from chelate to macrocycle to cage, and the extent of a receptor's preorganization.

bonding have recently emerged [16]. Some binding sites are known to be complementary to specific anions, thereby offering an intrinsic source of selectivity, e.g. ureas are complementary to phosphates, sulfates, and carboxylates [17]. Connon and coworkers, for instance, use a thiourea-based receptor with good complementarity to the two oxygen atoms in an anhydride-derived enolate to catalyze an enantioselective Tamura cycloaddition [18]. To boost binding energy (ΔE) of guests, there are various strategies available and amenable to synthetic receptor design (Figure 2.7). A preorganized receptor with multiple strong bonding sites affording a high degree of structural complementarity is ideal for binding. Common frameworks are chelates and macrocycles (Figure 2.8), but cages and foldamers have also been developed.

In recent work, binding was shown to be modulated by the incorporation of switchable functionalities. These efforts have translated into a catalyst that is based on an acid-responsive rotaxane from Leigh [29] and a photoswitchable CH–hydrogen bonding receptor catalyst from Feringa and Dorel [19]. In addition, physical parameters that are external to the molecular designs can be optimized to confer tight binding conditions on the receptor–anion system, such as use of high concentrations (>10 mM), non-competitive solvents, e.g. polarity equal or less than that of CH_2Cl_2, and low temperatures. A receptor's charge also plays a

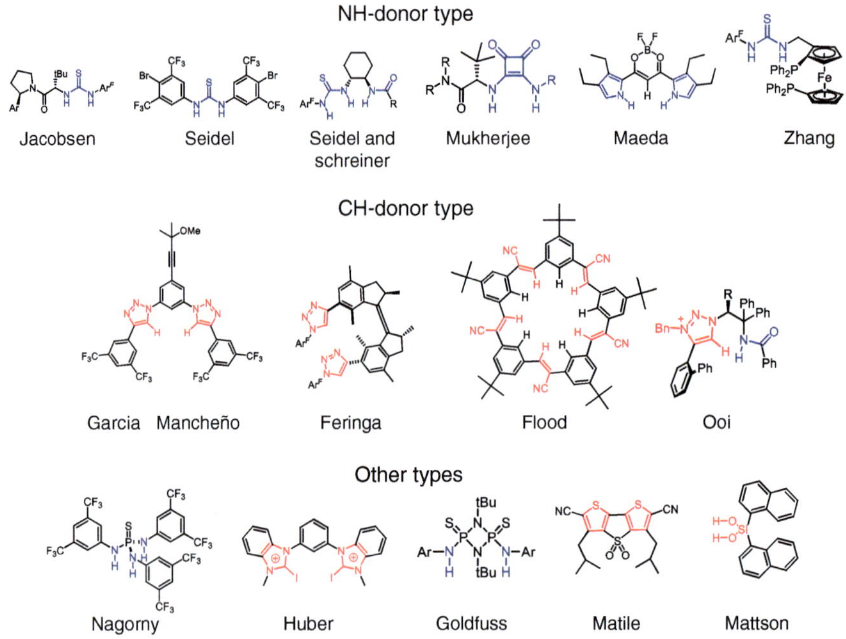

Figure 2.8 Variety of receptors used for anion-assisted catalysis and anion binding using various stabilizing units. (Refs. [3, 6–8, 13, 19–28]).

major role in introducing additional competing equilibria. For instance, cationic receptors carry a counter-anion that can have competing anion-binding equilibria. For neutral receptors, they pick up a charge when bound to anions, which can introduce additional ion-pairing equilibria.

2.1.4 Predicting the Model of Association and Simulating the Expected Species Distribution Profiles and Binding Curves

Every study of host–guest chemistry starts with an educated guess of the complementary binding partners, such as receptor and anion. The structure of the anion is usually known while that of the organic receptor needs to be modeled. The receptors are normally composed of a primary backbone architecture, and their conformational characteristics are important to assess the scope of structural complementarity and preorganization. The selectivity profiles of flexible receptors are often governed by different properties than rigid receptors [30]. For example, calixpyrrole macrocycles change shape to conform to its guests [31], thus favoring the charge-dense and small fluoride anions [32] more tightly than other anions. In contrast, rigid triazolophane macrocycles [33] cannot change shape and therefore offer size-selective binding that is matched to chloride despite being less charge dense than fluoride.

These preliminary assessments of the binding partners help inform the expected model of the binding equilibria, stoichiometries of the complexes, the potential binding strengths, and the extent of binding expected when both receptor and anion are present in a solution. The simplest model is 1 : 1 binding between receptor and anion. Nagorny's phosphoramides are good examples of 1 : 1 binding governed by their size, geometry, and binding site complementarity to the mesylate anion [20]. Higher order species with two receptors around one anion [5] (2 : 1, Figure 2.1) or two anions on the same receptor [34] (1 : 2) or a ternary ion-paired complex with a counter-cation (e.g. 2 : 1 : 1 in Figure 2.1) [6] can be considered in a similar way. The thioureas (Figures 2.1 and 2.2) likely achieve 2 : 1 binding because there is room for two such receptors and together they confer greater stabilization. Once the possible binding partners and stoichiometries constituting the model are identified, previous literature is usually available to provide an estimate of the equilibrium constants. It is critical, in such comparisons, that the solvent be selected to match or at least be taken into account when they do not; see more in the following.

The set of association reactions and equilibrium constants can be used to estimate the various species in solution. The model is the description of the elementary equilibrium reactions present in solution. We advocate using software to generate an expected titration curve. The software Hyss [35] is freely accessible and enables almost any combination of equilibria to be simulated. Thus, it is straightforward to estimate the extent of complexation between receptor and anion, that is, the position of equilibrium, simply by knowing their concentrations. This knowledge provides insights into the species present in solution. For example, if only 50% of the anion is bound during the catalyzed reaction, then it might only be 50% effective in achieving the desired enantio-, regio-, or disastereoselectivity of the reaction. Knowledge, therefore, of how to increase binding from 50% to >99% will have a direct impact on the reaction metrics. A criterion for knowing the extent of complexation between receptor and anion is whether the binding affinity, solvent, concentrations, equivalents, and the presence of competing anions favor strong, medium, or weak binding conditions. In practice, the set of possible association reactions, the stoichiometry of their complexation reactions, the structural models of any receptor–anion, ion pair, and receptor–anion–cation complexes, and the value of the equilibrium constants of each reaction are essential characteristics that need to be assessed in order to help understand and then optimize any anion-assisted catalysis.

2.2 Equilibrium Constants, Binding Curves, Titration Conditions, and Errors

2.2.1 Physical Origins of Equilibrium Binding Constants

Equilibrium binding constants, sometimes called association constants, are the key information of receptor–anion systems, and their routine determination date back to the crown ethers in the late 1960s [36]. The mass balance equation describes an equilibrium reaction and the position of the equilibrium expressed in the normal way (Figure 2.9).

To get an idea of the expected course of a titration, it is a good practice to simulate titration curves. There is a range of free software available for this purpose,

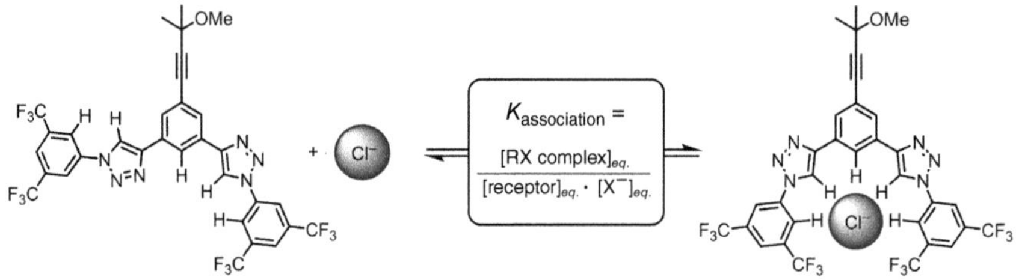

Figure 2.9 Formation of a 1 : 1 receptor–anion complex between an aryl-triazole chelate and the chloride anion [8]. Source: Based on Beckendorf *et al.* [8].

e.g. Hyss. After installation, a quick simulation of the 1 : 1 association (Figure 2.10) can be initiated by running a new simulation with no protons present. After entering symbols for host (H) and guest (G), selecting an initial value for the association constant (given the symbol b to refer to multiple equilibria), log $\beta = 3$, and concentrations for host (1 mM) and guest (0–2 mM), a binding curve is simulated out to 2 equiv of added guest. These simulations can allow a wide variety of situations to be explored, including ion pairing by having a model with three species for receptor, anion, and cation (see also Figure 2.17).

2.2.2 Explanation of the Basis for Titration Techniques and Binding Curves

Equilibrium constants are determined using titrations in which aliquots of a solution of the salt is added to a known concentration of the receptor. For a 1 : 1 binding event, typical initial aliquots involve addition of 0.1 equiv steps up to addition of 1.0 equiv after which 0.2, 0.5, or even 1, 2, and 5 equiv steps may be used to span out to addition of 10 (or more) equiv. Use of a dilution-corrected titration to keep the receptor's concentration constant helps rule out competing factors but is not needed if the species present are relatively few [37]. One option for dilution correction is to use separate tubes (Figure 2.11a). A second option (Figure 2.11b) is to use a single NMR tube and spike the anion solution with a known amount of receptor to replenish the change in concentration arising from dilution. In practice, changes in concentrations can be routinely addressed during the fitting process using standard analysis software. Thus, the third and simplest method (Figure 2.11c) is to add the anion from a highly concentrated stock solution to minimize dilution effects. Initial titrations are often conducted using this last method to first survey the possible species present in solution and their approximate binding strengths. When quantitative analyses are being conducted, it is also crucial to measure enough data points during the titration and to perform titrations multiple times to determine reliable uncertainties as outlined by Thordarson [37].

In a standard titration curve, the spectroscopic response of the system is plotted as a function of the added guest. One often selects a spectral feature that correlates with the amount of receptor–anion complex produced in solution. The response can be

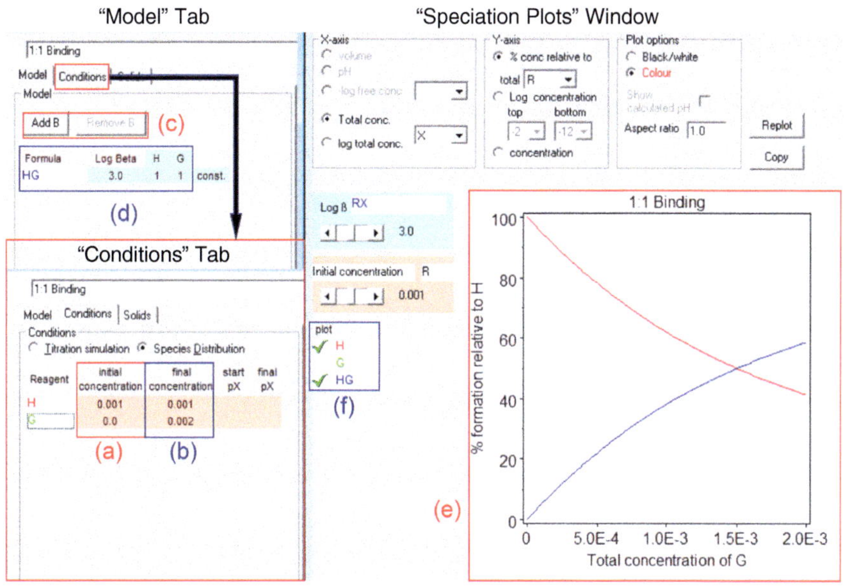

Figure 2.10 Screenshots from the Hyss modeling program featuring a typical 1 : 1 binding simulation. (a) Initial titration concentrations for host and guest (M). Generally, the initial titrant concentration is 0 M. (b) The final titration concentrations for host and guest (M) generally affects how many equivalents of guest are added relative to the host. (c) Add or remove equilibria from model, e.g. a 1 : 1 binding system would only have one equilibrium, while a 2 : 1 binding system would have both a 1 : 1 and 2 : 1 binding equilibrium. (d) Stoichiometries and binding constants (log β) to be used in the simulation. Host and guest labels can be customized. (e) Simulated speciation plot constructed from the conditions specified in the "Model" and "Conditions" tab. Y-axis corresponds to the % formation of a given species relative to the host, and the X-axis is the total concentration of the added titrant. (f) Legend for speciation plot. Species can be added to or removed from the plot here. Source: Hyperquad.

normalized from 0 to 1 for ease of comparison between signals and is often assumed to correlate with 0% to 100% conversion to the product, although this need not always be the case.

2.2.3 Hirose's Rule and Picking the Right Concentration, Solvent, and Technique

Performing titrations in the most sensitive concentration regime is critical for generating accurate binding constants. Hirose teaches us that failure to comply with this guidance will yield inaccurate binding constants that can deviate by many orders of magnitude [38]. As a good rule of thumb, the titration curve needs to have some curvature. This condition can be assessed visually by inspecting the raw titration plots and ensuring that it has curvature in it. The condition can also be met by ensuring that the concentration of host matches (or is within a factor of 10) of the dissociation

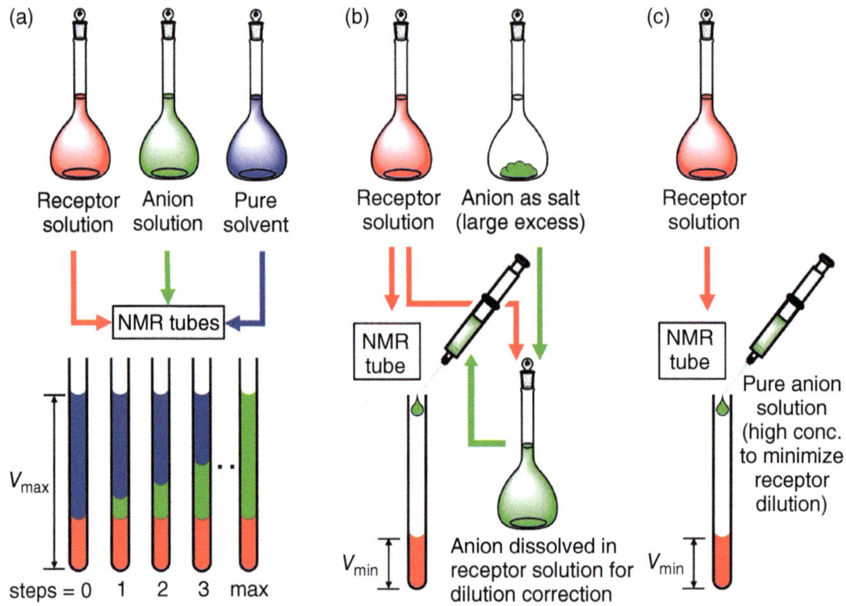

Figure 2.11 Schematic procedures for titrations. (a) Titration technique using separate NMR tubes for each data point. (b) Titration technique using a single NMR tube. (c) Direct titration method often used for initial surveys and to produce data that can be modeled using modern approaches.

constant, $K_d = 1/K_a$, defined as the inverse of the association constant and having unit of concentration. This last one is what we call Hirose's Rule. In addition, the degree of complexation represented by the titration must also span 20–80% [38–40].

Curvature is easily seen visually (Figure 2.12). The tight-binding regime is accessed at higher concentrations and leads to titration curves with no curvature (Figure 2.12a). While not useful for measuring binding constants, tight binding conditions are helpful for locking in specific species at their equivalence points and therefore helps with stoichiometry and structure determination. Qualitatively, a 1 : 1 receptor–anion complex should achieve full complexation at 1 equiv of both receptor and anion, which corresponds to the 1 : 1 receptor–anion stoichiometry. The medium binding regime is more accurate for determining binding constants and is accessed at lower concentrations where both the products and reactants are present in similar amounts. In the medium binding regime, 1 equiv of both receptor and anion should contain similar amounts of products and reactants, whose ratio defines the equilibrium constant [33, 41–43].

Representative 1 : 1 titration curves simulated using Hyss software [35] show the difference in binding regimes (Figure 2.12). In these plots, the percent formation of the receptor–anion complex is plotted against the equivalents of the added guest. The first plot (Figure 2.12a) shows the case of a titration simulated at NMR concentrations with the receptor at 1 mM. Each simulated curve corresponds to binding constants ranging from $K_a = 10^2$ to $10^6\ M^{-1}$. For simulations generated under medium

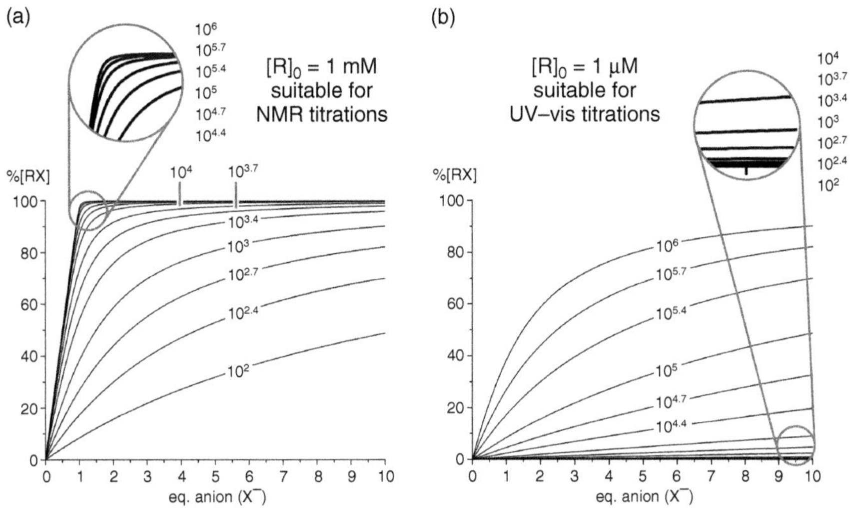

Figure 2.12 Effect of different host concentrations (a) 1 mM or (b) 1 μM on the plot of [RX] vs. equivalents of the added anionic guest for different binding constants. [RX] is directly related to the observed spectroscopic property, e.g. chemical shift. Curves generated using Hyss [35] and the parameters listed in the figure. Source: Based on Alderighi *et al.* [35].

binding conditions ($K_a = 10^2$ to $10^4\ M^{-1}$), we see curvature in the plots and we see that the degree of complexation at 1 equiv is 8.4% - 73%, i.e. much less than 100%. For plots generated with strong binding and thus having dissociation constants much lower than 1 mM, e.g. 100, 10, and 1 μM, the plots do not show any curvature and are nearly superposable. Only by reducing the concentration of the receptor by 3 orders of magnitude to 1 μM (Figure 2.12b), do we see curvature in the simulations corresponding to the higher binding constants. Therefore, finding conditions where there is curvature in the expected data allows accurate binding constant determination. The binding affinities for the weaker complexes, however, cannot be readily determined under these conditions because the degree of complexation does not span the requisite 20–80%.

The potential pitfalls of determining a binding constant are outlined here. In the tight binding regime, the solution composition favors products. Consequently, there is very little reactant, and its corresponding spectroscopic signal is too low in intensity that it is not possible to generate an accurate measure of its equilibrium concentration. (One of us made this mistake in their first paper in the field of anion recognition [33], where the binding constant was determined at a high host concentration (100 mM) to be $1.1 \times 10^5\ M^{-1}$. Only after learning of Hirose's rule [41] and multi-equilibrium fitting [42] was the equilibrium constant determined accurately [43] to be $5.5 \times 10^6\ M^{-1}$.)

A good guess of the initial binding constant for the system of interest can save time to find the right concentration region for the titration [37, 38]. However, starting at 1 mM provides a good general indication of how to alter the conditions in subsequent titrations.

The chosen concentration also has some influence on the method of analysis. Routine NMR spectroscopy is favored for higher concentrations but with many scans can reach down to 10 μM if required, and UV–visible (UV–vis) spectroscopy spans the lower regime (1–100 μM). NMR, however, is always the powerhouse technique of choice as it provides detailed structural information. Building on ^{1}H NMR spectra at each equivalence point, nuclear Overhauser effect (NOE) experiments can be performed to determine proximity of binding partners, such as the contacts between a receptor–anion complex and its cationic substrate. NMR spectroscopy can also provide diffusion coefficients to estimate increasing size of the species in solution upon complexation [44] Other techniques, including isothermal titration calorimetry (ITC) [21, 45] and fluorescence spectroscopy [22, 46], are appropriate for binding constant determination but fall outside the scope of this chapter.

2.2.4 Error Determination

Precise estimation of uncertainties and errors is essential to the accurate determination of binding constants. While it is straightforward to generate errors from the global fitting of the model to the data using related software, these errors may not necessarily reflect the major source of uncertainty in the measurement. These errors reflect how well the raw data conform to the model of binding and are akin to the errors generated when conducting least-squares regression analysis of a straight-line fit; if the primary data have errors, these will not necessarily emerge in the fitting. Experimental errors associated with sample preparation are best accounted for from conducting duplicates as suggested by Thordarson [37]. For more complicated systems that involve multiple equilibria, it is wise to perform these duplicates at different concentrations to ensure binding constants obtained are in fact "constant," i.e. they do not change if the concentration of the experiment was altered. Comparing binding constants from different techniques is also very useful in this respect. Finally, following standard procedures is also important when reporting the significant figures in both the value and the uncertainty.

2.3 Experimental Techniques: NMR Spectroscopy

2.3.1 When to Use

NMR spectroscopy is the "gold standard" for binding investigations and can be applied routinely for concentrations above 0.1 mM. NMR studies provide information on both binding strength and structural information about the nature of the complex. For instance, one can identify which protons are affected most by binding (and any attendant conformation changes), NOE experiments can be used to identify through-space coupling, and diffusion experiments [47] help confirm which signals arise from the same complex in solution. All these analyses help when proposing the structure of the complex present in solution. Multiple nuclei can be used other than ^{1}H, for instance, ^{19}F-NMR is gaining more use these days [23].

Variable temperature measurements are very helpful to match the actual reaction conditions applied in catalytic processes. Maeda and coworker were able to show a binding event at −50 °C between the dipyrrolyldiketone-based receptor and *N*-acyl quinolinium chloride, which is also the substrate of the catalytic reaction [24]. These findings can be combined with any crystal structures and computational results to provide a consistent picture of binding.

2.3.2 Slow Exchange vs. Fast Exchange

Binding processes that are under fast or slow exchange on the NMR timescale are easy to distinguish. During the titration, slow exchange is indicated by the growth of a unique set of signals for the host–guest complex and intensity loss in the signals of the free receptor (Figure 2.13). The chemical shift positions are not affected. The intensities of the signals assigned to the host and the complex are directly related to the concentrations of both at equilibrium, i.e. $[HG]_{eq}$ and $[H]_{eq}$, allowing direct determination of the association constant K_a by employing Eq. (2.1). However, instances when complexation is under slow exchange often emerge under tight binding conditions, which preclude accurate determination of binding constants as noted above in Section 2.2.3.

The majority of receptor–anion complexation events display fast exchange signatures [37] as indicated by shifting proton signals upon addition of anion. Their chemical shift represents the linear combination of the signal position of pure complex and free host according to their respective mole fractions (Figure 2.13).

Determination of the association constant K_a needs to be obtained by fitting the data of the complexation-induced shifts in the averaged peak positions (δ_{obs}) vs. the equivalents of guest added (equiv). There are two main approaches to fitting the titration curves. The first, and simplest, relies on exact analytical expressions derived by algebraic manipulation for specific binding models. These equations are

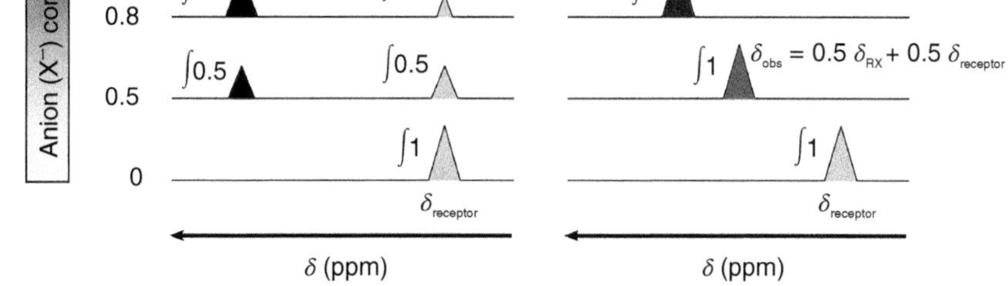

Figure 2.13 Difference between slow and fast exchanging binding behavior observed by NMR spectroscopy.

known [37] and readily available for fitting using BindFit [48] for 1 : 1, 2 : 1, and 1 : 2 receptor–anion stoichiometries. The complexity of the equations increases with higher order complexes such as ion pairing into a 2 : 1 : 1 receptor–anion–cation species or with addition of thiourea dimerization equilibria [6]. In these cases, methods are needed that allow the elementary equilibria to be used as input to constrain the data, e.g. the stepwise formation of 1 : 1 and 2 : 1 and then the formation of the ion-paired species, 2 : 1 : 1, would involve three elementary reactions.

The fitting of data to a complex binding model can be undertaken using non-linear regression methods. We recommend the use of software to automate the fitting, such as, HypNMR [49], which is related to the simulation software, Hyss. When using this software, the input data are the chemical shifts of the peaks paired with the concentration of added anion. It is important to include as many unique proton signals as possible (or as many wavelengths if using UV–vis data, see below). This approach helps ensure that the data include proton signals that are sensitive to different species and thus they can represent the totality of the various equilibria that are present in solution. For example, some protons that are engaged in hydrogen bonding to anions will be more sensitive to complexation events while protons on any counter-cations will better reflect the ion pairing. The ideal signals are those few that show serpentine shifts with clear inflection points corresponding to formation and then consumption of different species (see Figure 2.14). If there is not enough representation of the reactants or products in the data used for fitting, then the equilibrium constants obtained may have a higher degree of uncertainty associated with them.

2.3.3 Determination of the Underlying Equilibria

Before fitting the data, all species formed during the course of the titration need to be identified in order to use the correct set of reactions in the data fitting. If all species are in slow exchange, it is trivial to identify each of them, but then, the solutions need to be diluted to ensure medium binding conditions. Under fast exchange, it may be possible to increase concentrations to help drive tight binding conditions so that the formation of specific complexes can be maximized at their stoichiometric equivalence points. These stoichiometric points often occur upon saturation of the host by the guest. Another useful indication of stoichiometry is the observation of inflection points during the course of an NMR titration. These inflection points give the necessary evidence to identify the emergence of a species, even if it is not immediately possible to verify its exact structure. Inflection points represent turning points in the shifting behavior, e.g. from downfield to upfield, and are often called serpentine.

The most common saturation and inflection points occur upon addition of guest at 0.5 equiv for the 2 : 1 receptor–anion complex, 1 equiv for 1 : 1 (and 2 : 2) complexes, and 2 equiv for a 1 : 2 complex. For example, during the titration of an aryl-triazole macrocycle (Figure 2.14) with tetrabutylammonium chloride, there is an inflection point at 0.5–0.6 equiv (orange boxes), followed by a saturation of the chemical shift from 1.2 equiv onward (blue boxes) [43]. These two observations provide strong evidence for the formation of a 2 : 1 receptor–anion sandwich-type complex, followed

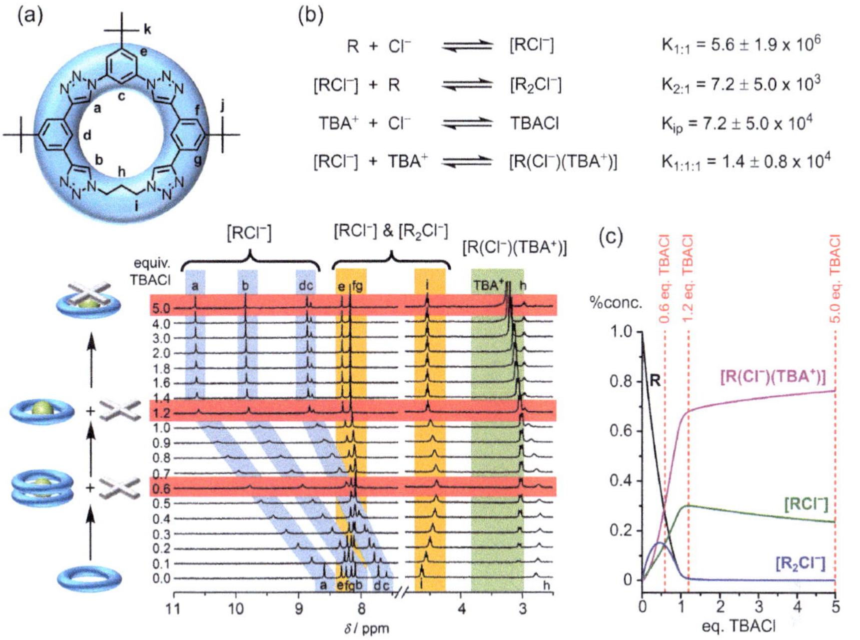

Figure 2.14 (a) Exemplary titration of a propylene-bridged aryl-triazole macrocycle with tetrabutylammonium chloride at 0.5 mM in CD_2Cl_2 at 298 K [43] with the corresponding stacked spectra, (b) the binding model shown as a set of equilibria, and (c) simulation of the concentration profile with Hyss. Source: Modified from Hua *et al.* [43].

by the formation of a 1 : 1 species. Another observation is the small shift of the tetrabutylammonium signal (Figure 2.14, green box) resulting in an inflection point at 1 equiv consistent with an ion-paring event between the 1 : 1 species and the counter-cation to form a 1 : 1 : 1 receptor–anion–cation. Confirmation of ion pairing was derived from measurements of the diffusion coefficient using the signals assigned to the cation. Under the conditions of the titration (0.5 mM), the binding constants revealed that ~50% of the 1 : 1 : 1 complex was present with 1 equiv of salt added (Figure 2.14c).

2.3.4 Software for Non-linear Regression Fitting

Free software for fitting binding curves in a non-linear fashion is the software BindFit [48]. The software allows fitting of different binding models such as 1 : 1, 1 : 2, and 2 : 1 host–guest complexes but complicated systems, such as receptor–anion–cation complexes (i.e. 1 : 1 : 1 or 2 : 1 : 1), cannot be analyzed yet [50]. Aggregation processes, such as those investigated for Seidel's and Schreiner's thiourea receptors [25], can be analyzed separately but not in combination with the host–guest binding. BindFit can be used with NMR and UV–vis data alike. The input parameters are the observed property like the chemical shift (absorbance or molar absorptivity for UV–vis) and host and guest concentrations for each data point. These data are compiled into an Excel sheet and uploaded to the program. The software shows multiple outputs such

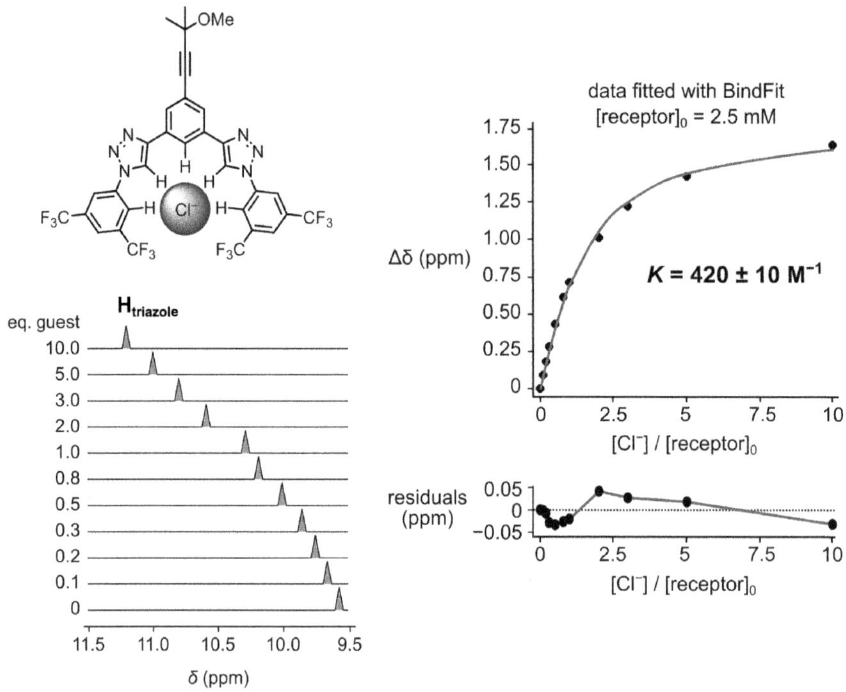

Figure 2.15 Exemplary titration of an aryl-triazole receptor with chloride performed by García Mancheño. The model NMR spectrum was generated from the original published data showing only the shifting triazole proton signal. The data were analyzed with BindFit with a 1 : 1 model. The result is in agreement with the published binding constant ($K = 450 \pm 40\ M^{-1}$) [8]. Source: Beckendorf *et al.* [8].

as the binding constant, the actual curve fitted to the data points, and the R^2 value as well as a residual plot (ideally random scatter) to help assess the quality of the fit. The fitting processes are documented, and the underlying equilibria are published by Thordarson [37, 50] who developed and maintains the program. The published raw data for chloride binding with García Mancheño's aryl-triazole chelate [8] were fitted with BindFit as an example (Figure 2.15), resulting in an association constant that matches the published value.

A commercial alternative is HypNMR [51, 52]. HypNMR2008 was applied, for example, by Lacour and coworkers to determine association constants for the binding of chloride with an aryl-triazole-based receptor [53]. It requires an input file that defines the binding model, which can be a collection of elementary equilibrium reactions of any number. Thus, complicated cases involving ion-pairing events, such as the one emerging with the aryl-triazole macrocycle (Figure 2.14), can also be analyzed by the software. The file can easily be produced using HySS [35]. Defining which species are NMR active and how many of the signals are to be tracked needs to be done before the data points can be included. Good initial guesses of the association constants are necessary for a successful fitting process. HypNMR delivers binding constants, R^2 values, a residual map, and estimates for the chemical shifts of intermediates, which all help to determine the quality of the fitting process.

2.3.5 Common Issues

NMR spectroscopy is not suitable for the determination of very high binding constants according to Hirose's rule (Section 2.2.3) [38]. Furthermore, ion pairing is more likely to emerge and thereby complicate analysis at the higher concentrations (*vide infra*). Lowering the concentration is a typical solution, either repeating the NMR titration using more scans or moving to UV–vis spectroscopy. Another issue that can arise is intermediate exchange. This situation is recognized by peak broadening and shifting but becomes most problematic when accompanied by a change in signal intensity. As a general rule, changing the concentration might help. In those cases when there are multiple equilibria with different exchange rates, extra care has to be taken during any fitting analysis. In all these problematic cases, initial NMR titrations can be used as a good starting point for qualitative comparison of the binding strengths of hosts before solving the problem. Changing the method of analysis to UV–vis spectroscopy [37] can often help circumvent these problems. More often, the two methods are paired as they can offer complementary information.

2.4 Experimental Techniques: UV–Vis Spectroscopy

2.4.1 When to Use

Traditionally, UV–vis titrations required chromophores that absorb in unique regions of the spectrum where peaks can be assigned to either the receptor or the complex. However, modern analysis techniques largely obviate this requirement. Nevertheless, these characteristics are often fulfilled by the presence of π-based chromophores common to anion receptors. Anion receptors with fluorophores are rarer, although the advice given here is broadly applicable to emission spectroscopy as well.

UV–vis spectroscopy offers sensitivity down to ~1 μM for measuring higher binding affinities. The one difference to NMR spectroscopy is that UV–vis spectroscopy has much less structural information. Nevertheless, UV–vis titrations can still provide information on the number of species and equilibria in solution. When the titration data and the complete model of binding are coupled with the proper analysis software, UV–vis spectroscopy can be used to determine accurate binding affinities for even highly complex systems.

2.4.2 Physical Origins of Optical Phenomena

The key quantitative element of UV–vis titrations is the relationship of absorption to concentration reflected in the Beer–Lambert law (Eq. 2.2):

$$A = \varepsilon\, b\, c \tag{2.3}$$

where A is the absorbance, ε is the molar absorptivity of the chemical compound in question, b is the optical path length, and c is the concentration of the chemical compound in question. With multiple species present during a titration, the total

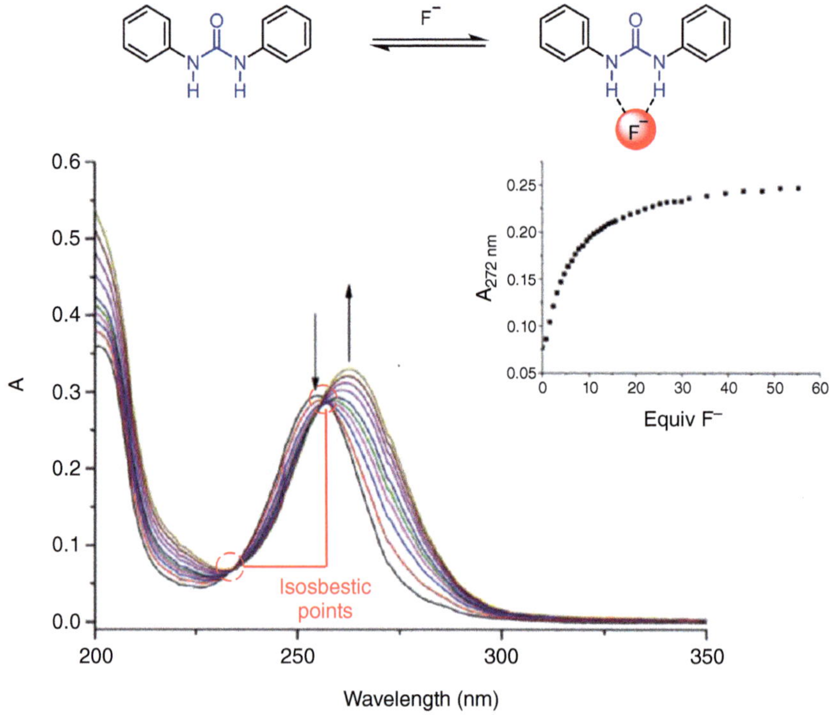

Figure 2.16 UV–vis absorbance data and titration binding curve (absorbance from 272 nm) for the binding of fluoride by a thiourea receptor (log $K_{1:1} = 4.29 \pm 0.01$, 8.0 μM thiourea, MeCN, 298 K). Source: Pfeifer *et al.* [47].

absorbance at any single wavelength is the summation of the absorbances of the host, guest, and complex present at equilibrium. Thus, the measured absorption is a linear combination of the contributing species, and their change through the titration follows the receptor–anion equilibrium. Again, modern software helps substantially with deconvoluting these individual contributions.

Several common phenomena characterize UV–vis titrations. Isosbestic points, corresponding to one wavelength where the absorbance remains constant across the titration, can be seen in some cases. These cases are associated with the conversion from an initial and usually a pure solution of the receptor to a final solution that is dominated by a single species. They are common to 1 : 1 complexation but can also be seen for other assemblies. The presence or absence of an isosbestic point can be helpful in diagnosing the binding behavior of simple host–guest systems, e.g. see the near-isosbestic behavior in 1 : 1 binding between Gouverneur's thiourea receptor and fluoride (Figure 2.16) [47].

2.4.3 Software for Non-linear Regression Analysis of UV–Vis Titrations

Several types of software are available for determining binding constants from UV–vis titrations using non-linear regression analysis. Much like for NMR, Bindfit

[48] can be used for non-linear regression analysis of simple 1 : 1, 2 : 1, and 1 : 2 host–guest binding systems. The raw absorbance data are provided as an input into a template, and the analysis proceeds much the same as for NMR titrations. The fitting provides binding constants, calculated molar absorptivity plots for each complex, and residual values for each titration point. While Bindfit is a powerful and free tool for analyzing simple binding equilibria, it cannot be used to analyze higher order equilibria as mentioned earlier.

For more complex binding systems, software such as SIVVU [54] and HypSpec 2014 [55] can be used, and they operate in a manner similar to HypNMR described above (Section 2.3.4). By using the entire wavelength region, more data points are provided for the fitting process. Thus, there is no limit on the number of equilibria that can be used to create a binding model to analyze more complex systems and to include ion pairing. Additionally, some of the output contains much more detailed error analysis tools and can provide calculated molar absorptivity plots for each complex present in the fitting. These molar absorptivity plots also need to conform to the expectation of chemical logic, e.g. a 2 : 1 species should absorb twice as much as a 1 : 1 receptor–anion species.

2.4.4 Common Issues

The issues that can arise during UV–vis titrations are similar to those in an NMR titration, but they can be harder to diagnose. Several common issues are highlighted here. One common issue seen in systems with simple binding stoichiometries is the absence of an isosbestic point. This situation most commonly arises if the titration is not dilution corrected. Second, and a more pressing concern, is the presence of an equilibrium that was not accounted for by the model. These equilibria can come in the form of ion pairing or a higher order species. When unanticipated equilibria arise, it is best to revisit the binding model and determine whether these species could be relevant. The best approach is to repeat the titration at a different concentration an order of magnitude higher or lower by both NMR and UV–vis. This change often raises or lowers the relevance of one of the species (Figure 2.17). Global fitting of both data sets should resolve a single set of equilibrium constants to recapitulate the titrations conducted at higher and lower concentrations.

2.5 Underappreciated Concerns in Binding Constant Determination: Multiple Binding Equilibria

2.5.1 When to Expect Additional Equilibria

When a receptor–anion system displays multiple species in solution, and not just a simple 1 : 1, the binding constants become more challenging to determine. Even though it can be tempting to approximate the binding curve a 1 : 1 event on account of its greater simplicity, this route should never be taken on account of the fact that the model is grossly inaccurate and will lead to wrong interpretations. This situation is particularly relevant in the field of anion-assisted catalysis, where various

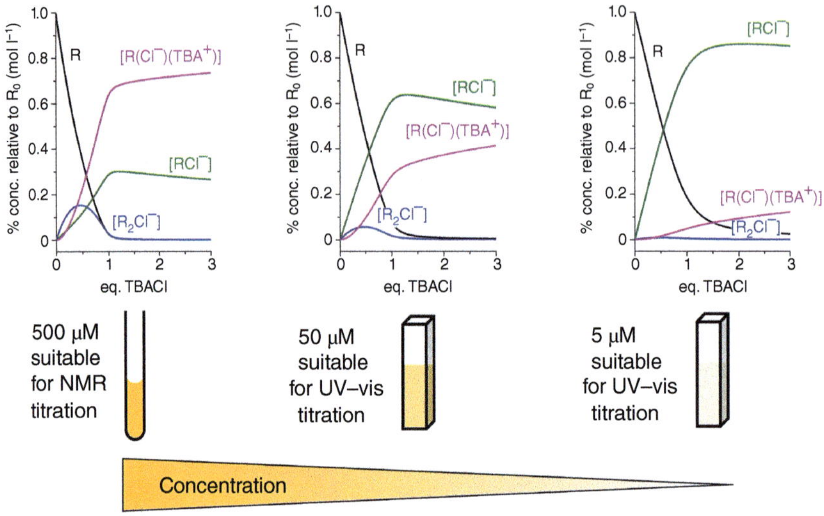

Figure 2.17 Using the concentration as a tool to dilute out one species of a complicated multi-equilibrium system. (a) and (b) All species need to be addressed in the fitting process. (c) Reducing the concentration to 5 μM reduces the abundance of the sandwich complex $[R_2Cl^-]$ to less than 5%, which can be reasonably neglected for the fitting process conducted at the lower concentration.

ion-pairing equilibria, e.g. cationic substrate and anion, can play a key role in reaction outcomes.

2.5.2 How to Diagnose Additional Equilibria

It is important to be open minded to species other than 1 : 1 and, once identified, to include them in the modeling. Good practice starts with Einstein's quote, "A model should be as simple as it can be but no simpler," and ends with a model that also follows chemical logic. One approach is to consider each possible species as a hypothesis and seek to confirm or refute their existence.

Common complex stoichiometries are 1 : 1, 2 : 2, 2 : 1, and 1 : 2 ratios for receptor and anion, but others can occur. In addition, ion pairing is a common and important process for anion-assisted catalysis. Thus, receptor–anion–cation species with 1 : 1 : 1 or 2 : 1 : 1 stoichiometries might be present. Moreover, some receptors can exist as dimers. The stoichiometry observed is defined by the demands of the structure and electronics of receptor, anion, and cation as described in the introductory remarks (Sections 2.1.2–2.1.4). Building an accurate model can be advanced, therefore, by considering the totality of the evidence to hand. This includes computational modeling, crystal structures, analogous systems, and the raw titration data. Clear indications for higher ordered complexes are the appearance of saturation and inflection points upon addition of specific equivalents of guest during NMR or UV–vis titrations.

Mass spectrometry can also help investigate the various species present in solution [2, 17, 56]. Electrospray ionization mass spectrometry (ESI-MS) is the method of choice because receptor–anion complexes survive the mild ionization method, and they already bear a charge. As a result, care needs to be taken to ensure that no additional acids or bases (salts) are added to the sample, and direct injection procedures should be followed to avoid cross-contamination from other samples. However, most ESI-MS spectra typically only allow for a qualitative measure of the possible complexes present by simply varying the amount of guest (e.g. a sub-stoichiometric, a stoichiometric, and an excess of guest). That is, the intensity of a peak may not always correspond to its relative concentration in solution. It should also be noted that ion-paired complexes, such as a 1 : 1 : 1 receptor:anion:cation complex, would not appear using ESI-MS on account of the fact that its overall charge is neutral. The observed complexes provide further evidence for the underlying speciation especially if the ESI-MS data match the solution evidence from NMR or UV–vis spectroscopy.

Evaluation of different possible models during the data fitting can also help provide further evidence for or against multiple binding equilibria. When a specific model provides a poor fit to the titration data, the poor fit is often a clear warning sign that something is not right in the model. Poor fits reveal themselves in residual maps with clear trends (ideally, residual maps should show random scattering) and high uncertainties being produced in the fitting process (R^2) as well as impossible outputs such as exceedingly high binding constants that are not physically meaningful. For NMR fitting, the estimated chemical shift of an intermediate is produced during the analysis, and these chemical shifts should always be inspected to ensure that they do not deviate too far outside of acceptable values. For instance, an aromatic proton should not deviate too far from 6.5–10 ppm unless rationalized from the structure of the complex formed. Such deviations can emerge with the formation of hydrogen bonds to anions. Thus, the output from the fitting should also be chemically meaningful. Sometimes, the initial guesses for the association constants are not good enough. This problem can be checked by varying the initial guess to either a higher or a lower value to observe whether a better fit is achieved. The final fitted values for the equilibrium constants should ideally be stable (invariant) in this respect and should not be sensitive to minor changes in the initial guess. Another consideration is that the titration was not performed carefully, so the data obtained are inaccurate. Duplicate titrations are advised in all cases. After these issues are ruled out, then the only remaining reason for a poor fit is that the model is not correct and cannot be applied to the obtained data.

2.5.3 How to Account for Additional Equilibria

Multiple binding equilibria need to be fitted to obtain accurate fits for the data observed. However, expanded models can lead to over fitting. Additional parameters will always lower the mathematical error, but they may introduce inaccuracies that do not necessarily reflect the underlying equilibria. Thus, simpler models are often selected on account of being chemically and physically meaningful even though they might produce slightly poorer fits. When an expanded model is used in the fitting process, the necessary evidence should be given in support of each additional species and equilibrium.

As noted, variation of the concentration can help solve issues with multiple binding equilibria. If the magnitude of the binding constants of the elementary reactions underlying the receptor–anion or receptor–anion–cation complexes is different enough, then the concentration of the titration can be lowered to dilute out the weaker binding species. Trace amounts of any species (<5%) can often be neglected to simplify the model for data fitting. Practice shows that the inclusion of the minor species often does not alter the quality of the fit.

The next example shows how multiple binding equilibria of the propylene-bridged aryl-triazole receptor seen earlier (Figure 2.14) were characterized by performing tetrabutylammonium chloride titrations at three different concentrations [43]. First, tetrabutylammonium chloride was known to be capable of ion pairing in dichloromethane. The value of its equilibrium constant was known [57] or could easily be determined by a dilution study [43, 58–63]. Any anion binding needs to compete with the cation; thus, this equilibrium is always included as a background reaction with a fixed value for its equilibrium constant. Second, the sandwich 2 : 1 receptor–anion complex can be diluted out by reducing the concentration to 5 μM (Figure 2.17). For this process, changing the technique from NMR to UV–vis spectroscopy is necessary to compensate for the lower concentration. This study results in reasonably accurate estimates of binding constants for the dominant 1 : 1 process (K_1) in solution. Third, with the 1 : 1 equilibrium accounted for, K_1 can be used as a fixed value when analyzing the data recorded at a higher concentration to determine the equilibrium constant of the weaker species, e.g. 2 : 1 complex (K_2 or β_2). Fourth, the same process was followed to establish the equilibrium constant for the formation of the 1 : 1 : 1 ion-paired species. Finally, this is an iterative process. The objective is a single set of binding constants that fit the data acquired over a range of concentrations. See the work of Yuan for an example that explores the link between the strength of ion pairing, binding affinity, and reactivity [64].

2.6 Underappreciated Concerns in Binding Constant Determination: Ion Pairing

2.6.1 When to Expect Ion Pairing

Any system that contains an anion must have a corresponding cation to balance its charge. The role of the counter-cation should always be addressed. The two ways that the counter-cation can influence anion binding systems is either through competition or by emergence of a new equilibrium producing some sort of receptor–anion–cation complex. Both processes were described in the previous section (Section 2.5.3), and competition from ion pairing of salts added exogenously can be assessed independently [43, 58–63]. When ion-pairing equilibria are not properly addressed, binding constants can be underestimated by orders of magnitude, and abnormal binding behavior can become impossible to diagnose.

The principles of structural complementarity between host and guest also govern the extent of ion pairing, and as a consequence, they can be quite powerful in some cases. A notable example comes from Gale, Schmidtchen, and Sessler who demonstrated that differently sized tetraalkylammonium cations can have substantial effects on the chloride binding of calix[4]pyrroles, owing to the ability of smaller cations to interact closely with the bowl of a chloride-bound calix[4]pyrrole complex [65].

2.6.2 Role of Solvent and Concentration in Ion Pairing

The strength and impact of ion-pairing equilibria are also governed by two external factors, solvent and concentration. According to Coulomb's law, electrostatic interactions are the dominant force in ion pairing. Electrostatic interactions are strongest in low dielectric media, so ion-pairing equilibria are most relevant in solvents with low dielectric constants, e.g. toluene ($\varepsilon = 2.4$), chloroform ($\varepsilon = 4.8$), and up to dichloromethane ($\varepsilon = 8.9$). In higher dielectric media, e.g. acetonitrile ($\varepsilon = 37.5$), dimethylsulfoxide ($\varepsilon = 46.7$), and water ($\varepsilon = 80.1$), electrostatic interactions are screened and ion pairing becomes less relevant [60]. Changing solvent to shut down ion pairing is one strategy to simplify the binding model, but this strategy is not always possible on account of solubility issues or other factors. However, even ion pairs that have low affinities for each other can demonstrate tight ion pairing at high concentrations. This situation can be solved by lowering the titration concentration. If these strategies are ineffective and ion pairing is present, the ion-pairing equilibria must be taken into account. Non-linear regression analysis software (Sections 2.3.4 and 2.4.2) can allow for the deconvolution of ion-pairing equilibria from the receptor–anion equilibria.

2.6.3 How to Diagnose Ion Pairing

Shifts in the NMR peak positions of the balancing cation often indicate ion pairing. Care must be taken, however, to include the role of any background ion-pairing equilibria before considering additional complexation events.

One observation that could indicate ion pairing is the emergence of different binding constants for the same process when titrations are conducted at different concentrations. In this case, there is likely an equilibrium that is not accounted for. Ion-pairing equilibria are often the missing piece in low dielectric solvents.

In some cases, when tight pairing of the organic cation with a receptor–anion complex is indicated, two NMR techniques are useful. NOE cross-peaks can be observed between the receptor and the cation that would not be possible unless they are associated together. This approach also helps to determine the possible relative spatial locations of the two species. Diffusion NMR may also be used. When the receptor–anion complex and cation form a tight ion pair, they diffuse together as one single entity and their diffusion coefficients are identical. One key example (Figure 2.18) involves the 2 : 2 : 2 stoichiometry between cyanostar macrocycles, bisulfate anions (HSO_4^-), and tetrabutylammonium cations [44].

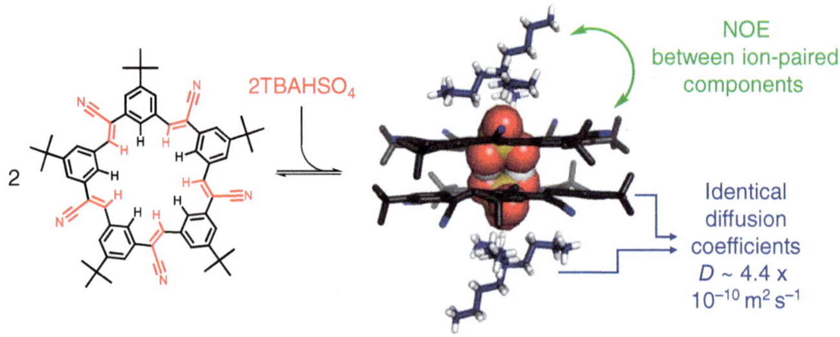

Figure 2.18 Cyanostar:bisulfate:tetrabutylammonium 2 : 2 : 2 complex. Ion-paired structure exists at high fidelity in low dielectric solvent (chloroform), as identified by diffusion NMR analysis and NOESY experiments. Source: Modified from Fatila *et al.* [44].

2.7 Underappreciated Concerns in Binding Constant Determination: Kinetic Processes

In rare cases, kinetic effects can be relevant in supramolecular anion-binding equilibria. Kinetic signatures are important to recognize as an indication that equilibrium conditions are not met. In the majority of systems in the literature, the timescale of most experimental methods used to determine binding constants is faster than the rate of complexation. However, for receptors and guests with especially large steric bulk near the binding site, reaching equilibrium can be slow enough as to cause issues during titrations. These situations are easy to diagnose from time-dependent peak shifting, and they can be remedied by letting the system reach equilibrium at each titration point.

2.8 Connecting Equilibrium Constants to Structures and Catalysis

Once accurate binding models and equilibrium constants have been obtained, a picture can be created of which species are present in solution and their relative amounts can be determined from the concentration of all species. This knowledge can be connected to the role of these species on the course of the catalytic reaction, whether they are beneficial or not. To consider ways to suppress or elevate different species requires consideration of how the structures of the complexes emerge from the specific architectures of receptor, anion, and cation. Often crystal structures and computational models return to assist with this evaluation and thus provide guidance as to which non-covalent contacts to tune up or down during iterative synthetic design of the receptor.

2.9 Conclusion

The principles of supramolecular chemistry enable the design and understanding of receptors capable of selective, high-affinity association of anions and ion pairing for the purpose of anion-assisted catalysis. The connection between the catalysis and the design principles is an accurate binding model describing the total set of species and equilibria present in solution. The correct binding model is an absolute necessity for accurate binding constant determination and, by extension, of improved catalyst design. Simulation software, such as Hyss, can help explore and identify binding models that match all of the observed binding behaviors. Once an accurate model is obtained, titrations using NMR and/or UV–vis spectroscopies can be conducted and subjected to quantitative fitting analyses. Care must be taken in the titrations to adhere to good practices: follow Hirose's rule of medium binding conditions to conduct titrations at relevant concentrations, use duplicates to estimate uncertainties and errors, and address any ion-pairing effects. These titrations can be processed using software that allows multiple spectroscopic signals to be followed using non-linear regression methods, such as HypNMR, HypSpec2014, Bind Fit, and SIVVU, to obtain accurate binding constants and help with estimating error values. With accurate binding constants and stoichiometries in hand, mechanistic interpretations of anion-binding events within studies of anion-assisted catalysis become more straightforward to either support or refute, thereby improving mechanistic analysis of these catalytic systems.

Acknowledgment

This work was supported by the Chemical Sciences, Geosciences, and Biosciences Division, Office of Basic Energy Sciences, Office of Science, U.S. Department of Energy (DEFG02-09ER16068). We thank Rachel Fadler for help with editing the manuscript.

References

1 Steed, J.W. and Atwood, J.L. (2009). *Supramolecular Chemistry*, 2e. John Wiley & Sons, Ltd.
2 Gu, Y., Wang, Y., Yu, T.-Y. *et al.* (2014). Rationally designed multifunctional supramolecular iminium catalysis: direct vinylogous Michael addition of unmodified linear dienol substrates. *Angew. Chem. Int. Ed.* 53: 14128–14131.
3 Kumar, V. and Mukherjee, S. (2013). Synergistic Lewis base and anion-binding catalysis for the enantioselective vinylogous addition of deconjugated butenolides to allenoates. *Chem. Commun.* 49: 11203–11205.

4 Ohmatsu, K., Kiyokawa, M., and Ooi, T. (2011). Chiral 1,2,3-triazoliums as new cationic organic catalysts with anion-recognition ability: application to asymmetric alkylation of oxindoles. *J. Am. Chem. Soc.* 133: 1307–1309.
5 Ford, D.D., Lehnherr, D., Kennedy, C.R., and Jacobsen, E.N. (2016). Anion-abstraction catalysis: the cooperative mechanism of α-chloroether activation by dual hydrogen-bond donors. *ACS Catal.* 6: 4616–4620.
6 Ford, D.D., Lehnherr, D., Kennedy, C.R., and Jacobsen, E.N. (2016). On- and off-cycle catalyst cooperativity in anion-binding catalysis. *J. Am. Chem. Soc.* 138: 7860–7863.
7 Schafer, A.G., Wieting, J.M., Fisher, T.J., and Mattson, A.E. (2013). Chiral silanediols in anion-binding catalysis. *Angew. Chem. Int. Ed.* 52: 11321–11324.
8 Beckendorf, S., Asmus, S., Mück-Lichtenfeld, C., and García Mancheño, O. (2013). "Click" bis-triazoles as neutral C–H···anion-acceptor organocatalysts. *Chem. Eur. J.* 19: 1581–1585.
9 Dannenberg, J.J. (1998). An introduction to hydrogen bonding; by George A. Jeffrey (University of Pittsburgh). Oxford University Press: New York and Oxford, 1997. *J. Am. Chem. Soc.* 120: 5604–5604.
10 Fischer, E. (1894). Einfluss der Configuration auf die Wirkung der Enzyme. *Ber. Dtsch. Chem. Ges.* 27: 2985–2993.
11 Liu, Y., Zhao, W., Chen, C.-H., and Flood, A.H. (2019). Chloride capture using a C–H hydrogen-bonding cage. *Science* 365: 159–161.
12 Lee, S., Hua, Y., Park, H., and Flood, A.H. (2010). Intramolecular hydrogen bonds preorganize an aryl-triazole receptor into a crescent for chloride binding. *Org. Lett.* 12: 2100–2102.
13 Wolf, F.F., Neudörfl, J.-M., and Goldfuss, B. (2018). Hydrogen-bonding cyclodiphosphazanes: superior effects of 3,5-$(CF_3)_2$-substitution in anion-recognition and counter-ion catalysis. *New J. Chem.* 42: 4854–4870.
14 Gale, P.A. and Dehaen, W. (2010). *Anion Recognition in Supramolecular Chemistry*. Berlin: Springer.
15 Pedersen, C.J. (1967). Cyclic polyethers and their complexes with metal salts. *J. Am. Chem. Soc.* 89: 7017–7036.
16 Vargas Jentzsch, A. and Matile, S. (2013). Transmembrane halogen-bonding cascades. *J. Am. Chem. Soc.* 135: 5302–5303.
17 Wang, Y., Yu, T.-Y., Zhang, H.-B. *et al.* (2012). Hydrogen-bond-mediated supramolecular iminium ion catalysis. *Angew. Chem. Int. Ed.* 51: 12339–12342.
18 Collar, A.G., Trujillo, C., and Connon, S.J. (2019). Highly enantio- and diastereoselective catalytic asymmetric tamura cycloaddition reactions. *Chem. Eur. J.* 25: 7270–7274.
19 Dorel, R. and Feringa, B.L. (2020). Stereodivergent anion binding catalysis with molecular motors. *Angew. Chem. Int. Ed.* 59: 785–789.
20 Borovika, A., Tang, P.-I., Klapman, S., and Nagorny, P. (2013). Thiophosphoramide-based cooperative catalysts for Brønsted acid promoted ionic Diels–Alder reactions. *Angew. Chem. Int. Ed.* 52: 13424–13428.

21 Walter, S.M., Kniep, F., Rout, L. *et al.* (2012). Isothermal calorimetric titrations on charge-assisted halogen bonds: role of entropy, counterions, solvent, and temperature. *J. Am. Chem. Soc.* 134: 8507–8512.

22 Benz, S., López-Andarias, J., Mareda, J. *et al.* (2017). Catalysis with chalcogen bonds. *Angew. Chem. Int. Ed.* 56: 812–815.

23 Ohmatsu, K., Hamajima, Y., and Ooi, T. (2012). Catalytic asymmetric ring openings of meso and terminal aziridines with halides mediated by chiral 1,2,3-triazolium silicates. *J. Am. Chem. Soc.* 134: 8794–8797.

24 Hirata, G. and Maeda, H. (2018). Pyrrole-based anion-responsive π-electronic molecules as hydrogen-bonding catalysts. *Org. Lett.* 20: 2853–2856.

25 Mittal, N., Lippert, K.M., De, C.K. *et al.* (2015). A dual-catalysis anion-binding approach to the kinetic resolution of amines: insights into the mechanism via a combined experimental and computational study. *J. Am. Chem. Soc.* 137: 5748–5758.

26 Zhao, C., Sojdak, C.A., Myint, W., and Seidel, D. (2017). *J. Am. Chem. Soc.* 139: 10224.

27 Zhao, Q., Wen, J., Tan, R. *et al.* (2014). *Angew. Chem. Int. Ed.* 53: 8467.

28 Lee, S., Chen, C.-H., and Flood, A.H. (2013). *Nat. Chem.* 5: 704.

29 Eichstaedt, K., Jaramillo-Garcia, J., Leigh, D.A. *et al.* (2017). Switching between anion-binding catalysis and aminocatalysis with a rotaxane dual-function catalyst. *J. Am. Chem. Soc.* 139: 9376–9381.

30 McDonald, K.P., Hua, Y., Lee, S., and Flood, A.H. (2012). Shape persistence delivers lock-and-key chloride binding in triazolophanes. *Chem. Commun.* 48: 5065–5075.

31 Kim, D.S. and Sessler, J.L. (2015). Calix[4]pyrroles: versatile molecular containers with ion transport, recognition, and molecular switching functions. *Chem. Soc. Rev.* 44: 532–546.

32 Wu, Y.-D., Wang, D.-F., and Sessler, J.L. (2001). Conformational features and anion-binding properties of calix[4]pyrrole: a theoretical study. *J. Organomet. Chem.* 66: 3739–3746.

33 Li, Y. and Flood, A.H. (2008). Pure C–H hydrogen bonding to chloride ions: a preorganized and rigid macrocyclic receptor. *Angew. Chem. Int. Ed.* 47: 2649–2652.

34 He, Q., Kelliher, M., Bähring, S. *et al.* (2017). A bis-calix[4]pyrrole enzyme mimic that constrains two oxoanions in close proximity. *J. Am. Chem. Soc.* 139: 7140–7143.

35 Alderighi, L., Gans, P., Ienco, A. *et al.* (1999). Hyperquad simulation and speciation (HySS): a utility program for the investigation of equilibria involving soluble and partially soluble species. *Coord. Chem. Rev.* 184: 311–318.

36 Jensen, W.B. (2002). Holleman-Wiberg's inorganic chemistry (edited by Wiberg, Nils). *J. Chem. Ed.* 79: 944.

37 Thordarson, P. (2011). Determining association constants from titration experiments in supramolecular chemistry. *Chem. Soc. Rev.* 40: 1305–1323.

38 Hirose, K. (2012). Quantitative analysis of binding properties. In: *Analytical Methods in Supramolecular Chemistry*, 2e, 27–66.

39 Wilcox, C.S., Schneider, H.-J., and Dürr, H. (1991). *Frontiers in Supramolecular Organic Chemistry and Photochemistry*. Wiley-VCH Verlag GmbH: Weinheim.

40 Weber, G. (1965). *Molecular Biophysics*. New York: Academic Press.

41 Li, Y., Griend, D.A.V., and Flood, A.H. (2009). Modelling triazolophane–halide binding equilibria using Sivvu analysis of UV–vis titration data recorded under medium binding conditions. *Supramol. Chem.* 21: 111–117.

42 Li, Y. and Flood, A.H. (2008). Strong, size-selective, and electronically tunable C–H···halide binding with steric control over aggregation from synthetically modular, shape-persistent [34]triazolophanes. *J. Am. Chem. Soc.* 130: 12111–12122.

43 Hua, Y., Ramabhadran, R.O., Uduehi, E.O. *et al.* (2011). Aromatic and aliphatic CH hydrogen bonds fight for chloride while competing alongside ion pairing within triazolophanes. *Chem. Eur. J.* 17: 312–321.

44 Fatila, E.M., Twum, E.B., Karty, J.A., and Flood, A.H. (2017). Ion pairing and co-facial stacking drive high-fidelity bisulfate assembly with cyanostar macrocyclic hosts. *Chem. Eur. J.* 23: 10652–10662.

45 Jungbauer, S.H. and Huber, S.M. (2015). Cationic multidentate halogen-bond donors in halide abstraction organocatalysis: catalyst optimization by preorganization. *J. Am. Chem. Soc.* 137: 12110–12120.

46 Farney, E.P., Chapman, S.J., Swords, W.B. *et al.* (2019). Discovery and elucidation of counteranion dependence in photoredox catalysis. *J. Am. Chem. Soc.* 141: 6385–6391.

47 Pfeifer, L., Engle, K.M., Pidgeon, G.W. *et al.* (2016). Hydrogen-bonded homoleptic fluoride–diarylurea complexes: structure, reactivity, and coordinating power. *J. Am. Chem. Soc.* 138: 13314–13325.

48 Pall Thordarson. Online tools for supramolecular chemistry research and analysis. http://supramolecular.org. (accessed 1 March 2021).

49 Hyperquad Limited. HypNMR2018: Equilibrium constants from averaged chemical shifts. http://www.hyperquad.co.uk/hypnmr.htm. (accessed 1 March 2021).

50 Brynn Hibbert, D. and Thordarson, P. (2016). The death of the Job plot, transparency, open science and online tools, uncertainty estimation methods and other developments in supramolecular chemistry data analysis. *Chem. Commun.* 52: 12792–12805.

51 Frassineti, C., Ghelli, S., Gans, P. *et al.* (1995). Nuclear magnetic resonance as a tool for determining protonation constants of natural polyprotic bases in solution. *Anal. Biochem.* 231: 374–382.

52 Frassineti, C., Alderighi, L., Gans, P. *et al.* (2003). Determination of protonation constants of some fluorinated polyamines by means of ^{13}C NMR data processed by the new computer program HypNMR2000. Protonation sequence in polyamines. *Anal. Bioanal.Chem.* 376: 1041–1052.

53 Bosmani, A., Pujari, S.A., Besnard, C. *et al.* (2017). Stereoselective and enantiospecific mono- and bis-C–H azidation of Tröger bases: insight on bridgehead iminium intermediates and application to anion-binding catalysis. *Chem. Eur. J.* 23: 8678–8684.

54 Douglas A. Vander. Sivvu. http://sivvu.org/. (accessed 1 March 2021).

55 Hyperquad Limited. Stability constants, binding constants, pK determination. http://www.hyperquad.co.uk/index.htm. (accessed 1 March 2021).

56 Juste-Navarro, V., Prieto, L., Delso, I. *et al.* (2017). A case study of thiourea-assisted iminium formation by hydroxyl anion binding: kinetic, spectroscopic and computational evidences. *Adv. Synth. Catal.* 359: 4122–4128.

57 Svorstol, I., Hoiland, H., and Songstad, J. (1984). Solvent properties of dichloromethane. III. Conductivity studies of some tetraalkylammonium-, tetraphenylarsonium- and bis(triphenylphosphine)iminium salts in dichloromethane. *Acta Chem. Scand.* 38b: 885–893.

58 Ramabhadran, R.O., Liu, Y., Hua, Y. *et al.* (2014). An overlooked yet ubiquitous fluoride congenitor: binding bifluoride in triazolophanes using computer-aided design. *J. Am. Chem. Soc.* 136: 5078–5089.

59 Qiao, B., Sengupta, A., Liu, Y. *et al.* (2015). Electrostatic and allosteric cooperativity in ion-pair binding: a quantitative and coupled experiment–theory study with aryl–triazole–ether macrocycles. *J. Am. Chem. Soc.* 137: 9746–9757.

60 Liu, Y., Sengupta, A., Raghavachari, K., and Flood, A.H. (2017). Anion binding in solution: beyond the electrostatic regime. *Chem* 3: 411–427.

61 Fatila, E.M., Pink, M., Twum, E.B. *et al.* (2018). Phosphate–phosphate oligomerization drives higher order co-assemblies with stacks of cyanostar macrocycles. *Chem. Sci.* 9: 2863–2872.

62 Parks, F.C., Liu, Y., Debnath, S. *et al.* (2018). Allosteric control of photofoldamers for selecting between anion regulation and double-to-single helix switching. *J. Am. Chem. Soc.* 140: 17711–17723.

63 Qiao, B., Leverick, G.M., Zhao, W. *et al.* (2018). Supramolecular regulation of anions enhances conductivity and transference number of lithium in liquid electrolytes. *J. Am. Chem. Soc.* 140: 10932–10936.

64 Kang, K., Lohrman, J.A., Nagarajan, S. *et al.* (2019). Convergent ditopic receptors enhance anion binding upon alkali metal complexation for catalyzing the ritter reaction. *Org. Lett.* 21: 652–655.

65 Sessler, J.L., Gross, D.E., Cho, W.-S. *et al.* (2006). Calix[4]pyrrole as a chloride anion receptor: solvent and Countercation Effects. *J. Am. Chem. Soc.* 128: 12281–12288.

3

(Thio)urea and Squaramide-Catalyzed Anion-Binding Catalysis with Halogen Anions

Matthew A. Horwitz and Véronique Gouverneur

University of Oxford, Department of Chemistry, Chemistry Research Laboratory, 12 Mansfield Road, Oxford OX1 3TA, UK

3.1 Introduction

The synthetic possibilities offered by anion-binding catalysis with halogen anions have grown exponentially since the advent of this research area [1–4]. The conceptual evolution of this line of research began with the development of catalytic asymmetric methods that relied on noncovalent interactions between organocatalysts and electrophiles. In this regard, this chapter details the establishment of the accepted S_N1-type mechanism that has been proposed in many systems, whereby the catalyst activates a halogen-containing electrophile through binding with the halide; experimental support for this activation manifold is detailed below in the mechanistic discussions. Early examples in *N*-acyliminium chemistry followed by numerous other electrophiles demonstrate the usefulness of anion-binding catalysis and its application to many reaction subtypes including enantioselective variants. The progressive interrogation of motifs on (thio)urea and squaramide hydrogen bond donor (HBD) catalysts allowed not only a broader range of catalytic asymmetric transformations but also a more mature understanding from a physical organic chemistry perspective. The recent development of asymmetric catalytic methods where the catalyst is associated with the nucleophile through hydrogen-bonding motifs is also described.

3.2 History and Background

Although attempts to probe and characterize the interaction between hydrogen bond donors and halide anions were initially met with practical difficulties because of the insolubility of ionic compounds in common organic solvents, the utilization of quaternary ammonium salts enabled foundational spectroscopic investigations to proceed. In an early study by Lund, infrared spectral shifts were observed by combining tetrabutylammonium halide salts with various O–H and N–H proton donors [5]. Subsequent developments established that specific noncovalent interactions between halides and hydrogen bond donors could be quantified using

Anion-Binding Catalysis, First Edition. Edited by Olga García Mancheño.

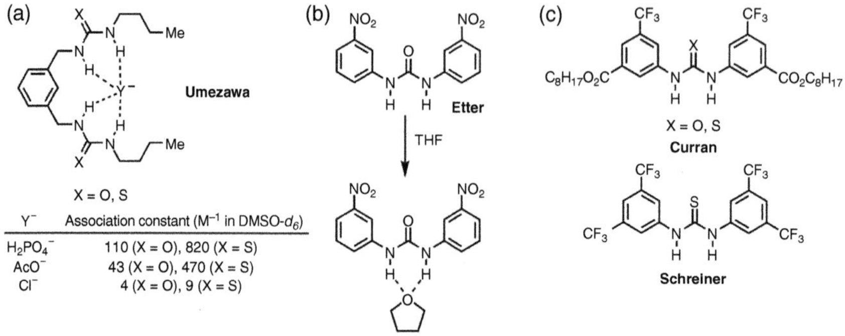

Figure 3.1 (Thio)ureas in (a) anion recognition, (b) complexation, and (c) catalysis. Sources: (a) Based on Nishizawa *et al.* [9], (b) Adapted from Etter and Panuto [10], (c) Based on Kelly *et al.* [11].

macroscopic observables, such as absorbance in infrared spectroscopy [6, 7]. This line of investigation led to the elucidation of noncovalent binding trends. Allerhand and von Ragué Schleyer were able to use spectral shifts to study the relative binding strengths of halide anions in hydrogen-bonding interactions, which led to the observed order of $\Delta\nu_{XH\cdots Cl^-} > \Delta\nu_{XH\cdots F^-} > \Delta\nu_{XH\cdots Br^-} > \Delta\nu_{XH\cdots I^-}$ (the deviation from the correct relative order of binding strengths, $F^- > Cl^- > Br^- > I^-$, was attributed to aggregation effects) [8]. A major advance came with the discovery that ureas and thioureas can act as anion receptors whose association constants can be calculated from changes in ^{1}H NMR chemical shifts (Figure 3.1a) [9]. These studies underscored the dually important aspects of basicity and geometry in determining the selectivity of a host for a given guest. The investigations of Etter in utilizing diarylureas for molecular recognition (Figure 3.1b) furthered the field's interest in this type of HBD scaffold [10, 12]. Hydrogen-bonding interactions between (thio)ureas and halide anions were later investigated from a computational standpoint, which led to the conclusion that fluoride binds most strongly to the donors surveyed, followed sequentially by chloride and bromide [13].

In parallel with these efforts to understand, quantify, and generalize noncovalent interactions between halide anions and hydrogen bond donors, noncovalent interactions were beginning to play an increasingly conspicuous role in the field of catalysis. Early promises of (thio)urea HBDs for applications in catalysis were established by Curran in the context of diastereoselective allylation reactions [14] and Claisen rearrangements [15], as well as by Schreiner [16–18], who used the Diels–Alder reaction to investigate noncovalent interactions (Figure 3.1c); prior art by Kelly [11] established that biphenylenediols are also competent HBD catalysts for this transformation. These studies would form the basis for asymmetric catalysis with (thio)ureas and squaramides, which soon exploited HBD–halide interactions. Over time, the field would conceptually bifurcate into two general approaches: catalyst association with the electrophile (Figure 3.2a) and catalyst association with the nucleophile (Figure 3.2b).

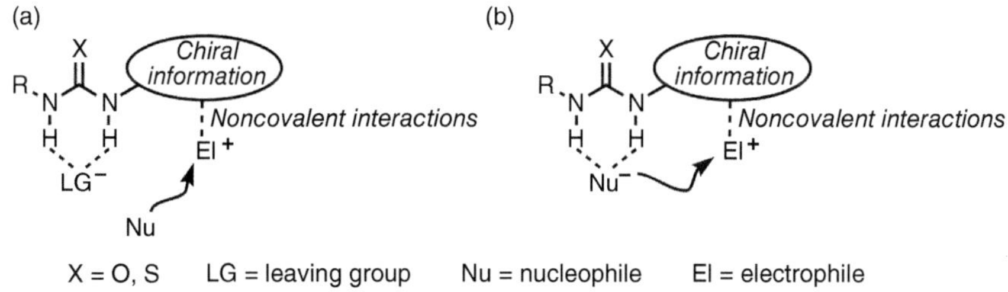

Figure 3.2 Modes of association in asymmetric catalysis with HBDs. (a) Catalyst association with electrophile. (b) Catalyst association with nucleophile.

3.3 Asymmetric Catalysis by Catalyst Association with the Electrophile

3.3.1 Examples Utilizing the *N*-Acyliminium Chloride Ion Pair

3.3.1.1 Pictet–Spengler Reaction and Variants

In 2004, Taylor and Jacobsen disclosed an asymmetric acyl-Pictet–Spengler reaction that provided *N*-acyl tetrahydro-β-carboline products using a chiral diaminocyclohexane-derived thiourea organocatalyst (Scheme 3.1a) [19]. Although the mechanism had not been elucidated at the time of publication, subsequent developments have proven the pivotal role that (thio)urea–chloride interactions played in enabling this transformation. Subsequent work by the same group resulted in the development of an asymmetric Pictet–Spengler reaction that used hydroxy lactams as precursors for *N*-acyliminium ions (Scheme 3.1b) [20]. In this report, alkyl-substituted hydroxylactams were found to be more reactive than their unsubstituted counterparts (in the formation of **5a** vs. **5b**, for example), which suggested that an S_N1-type mechanism was more plausible than an S_N2 mechanism (Scheme 3.1c). In this framework, the action of TMSCl with a hydroxylactam (**4**) led to the formation of HCl *in situ*, allowing irreversible formation of a chlorolactam (**6**). The asymmetric induction was proposed to originate from a *N*-acyliminium chloride–thiourea complex (**7**), which could cyclize via either of the two pathways (*A*, leading to **8**, or *B*, leading to **9**; Scheme 3.1c) to give the enantioenriched β-carboline product.

As evidence of the direct synthetic utility of these transformations, the total syntheses of (+)-harmicine (**10**) [20] and (+)-yohimbine (**12**) were completed [21]. While in the former case, the target was accessed through a single step consisting of a $LiAlH_4$ reduction of **5a** (Scheme 3.2a), the latter target required a nine-step sequence to elaborate the tetrahydro-β-carboline alkaloid core into **12** (Scheme 3.2b).

Moving beyond intramolecular reactions, an intermolecular addition of indoles to *N*-acyliminium ions was developed soon afterward (Scheme 3.3) [22]. During the optimization of this reaction, new catalyst motifs were explored in order to obtain highly enantioenriched products (**15**).

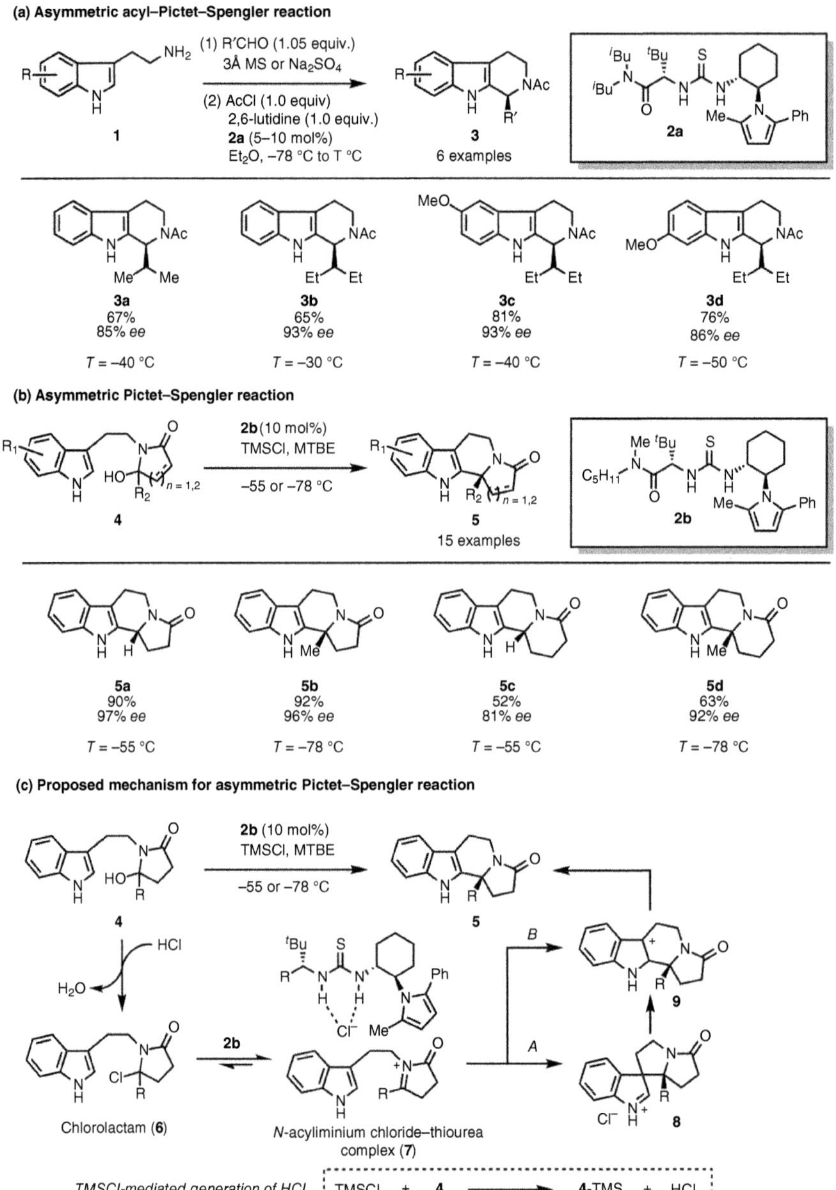

Scheme 3.1 Chiral thiourea-catalyzed asymmetric acyl-Pictet–Spengler and Pictet–Spengler reactions. Source: Based on Taylor and Jacobsen [19].

By incorporating a Schiff base into the diaminocyclohexane scaffold, the authors were able to develop catalyst **2c**, which provided high levels of enantioselectivity in the intermolecular addition. While in the case of electron-rich indoles the combination of TMSCl and catalytic H_2O was sufficient to generate the *N*-acyliminium species (via a chlorolactam intermediate, as illustrated in Scheme 3.1c) and allows

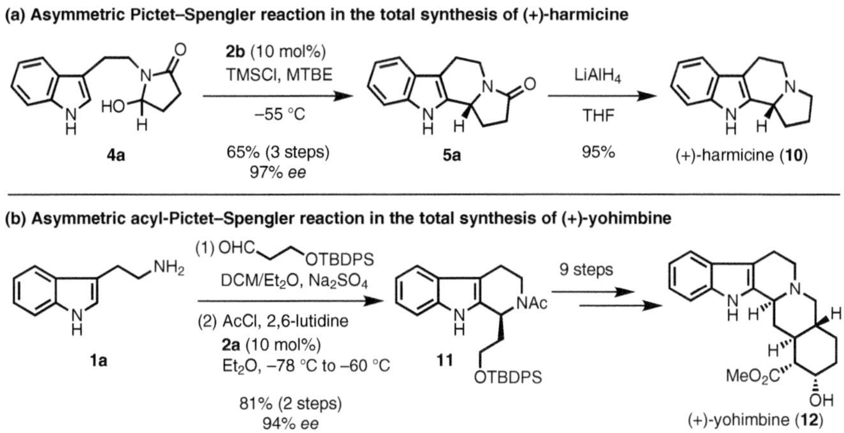

Scheme 3.2 Synthetic applications of the asymmetric Pictet–Spengler and acyl-Pictet–Spengler reactions. Source: Based on Raheem *et al.* [20].

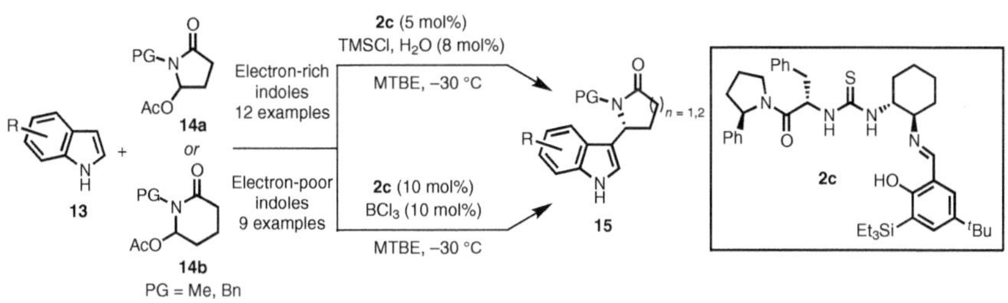

Scheme 3.3 Enantioselective intermolecular addition of indoles to *N*-acyliminium ions. Source: Based on Peterson and Jacobsen [22].

high yields for the reaction, extending the scope to electron-deficient indole nucleophiles required the use of BCl_3. In the former case (Scheme 3.3, top line), TMSCl was proposed to aid in the formation of the chlorolactam intermediate by facilitating the regeneration of HCl from HOAc, producing TMSOAc. Taken together, these developments highlighted the potential of asymmetric catalysis using HBD binding of halogen anions, which continued to evolve through other reaction subtypes.

3.3.1.2 Intramolecular Cyclizations with Other (Hetero)aromatic Nucleophiles

Expanding beyond indole nucleophiles, Jacobsen's group explored cyclization reactions of pyrroles onto pendant *N*-acyliminium electrophiles using chiral thiourea HBD catalysts (from **16** in Scheme 3.4) [23]. Cyclization at C2 occurred with high enantioselectivities using catalyst **2b** with succinimide-derived *N*-acyliminium ions (providing **17a** for example); enantioselectivities using glutarimide-derived electrophiles (as in **17b**) were more modest. Upon protection of the pyrrole with a sterically encumbering triisopropylsilyl (TIPS) protecting group, it was possible to redirect the regioselectivity of the reaction. However, while alkylated hydroxylactams could be employed to yield regio- and enantioselective cyclization products such as **18a**, *N*-TIPS-protected pyrrolohydroxylactams lacking alkylation on the

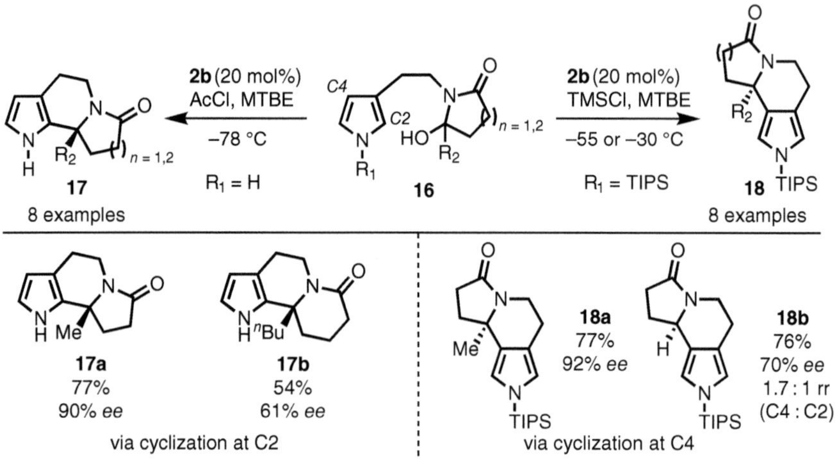

Scheme 3.4 Enantio- and regioselective asymmetric cyclization of pyrroles with pendant *N*-acyliminium ions. Source: Based on Raheem *et al.* [23].

lactam resulted in a mixture of C2 and C4 cyclization products (as with **18b**). Of further note is the fact that **18a** can be formed through an uncatalyzed background reaction in a non-regioselective manner (3 : 1 rr of reaction at C4:C2), but the utilization of catalyst **2b** improves the regioselectivity to >50 : 1, favoring cyclization at C4 while simultaneously allowing enantioselectivity in the process. Cyclization reactions of furans onto pendant *N*-acyliminium ions have also been explored using chiral xanthene-derived thiourea organocatalysts with good conversion but low levels of enantioselectivity [24].

In order to access more complex product classes through asymmetric anion-binding catalysis, enantioselective polycyclization reactions were pursued as a logical extension [25, 26]. In this line of investigation, *N*-acyliminium ions (derived from **19**) were employed in conjunction with chiral thiourea HBD catalyst **2d** to carry out cationic polycyclizations to form **20** (Scheme 3.5a). In this system, the enantioselectivity was found to have a strong dependence on the size of the extended arene present in the catalyst, which the authors proposed to be a consequence of cation–π interactions in the dominant transition-state structure. This hypothesis was supported by Eyring analysis (Scheme 3.5b), which found that enantioselectivity was enthalpically controlled and highlighted the critical role of strong noncovalent interactions (specifically, cation–π interactions arising from the pyrene of **2d**). Prior investigations into the thermodynamics of noncovalent association have similarly implicated the enthalpy of association as the primary manifestation of positive cooperativity [27–29].

3.3.1.3 Intramolecular and Intermolecular aza-Sakurai Reaction

Anion-binding catalysis in the context of *N*-acyliminium chemistry was further advanced by the development of the enantioselective aza-Sakurai reaction. Intramolecular cyclizations using this reaction have been reported to give bicyclic amides, carbamates, and ureas in an enantioenriched form (Scheme 3.6a) [30]. In

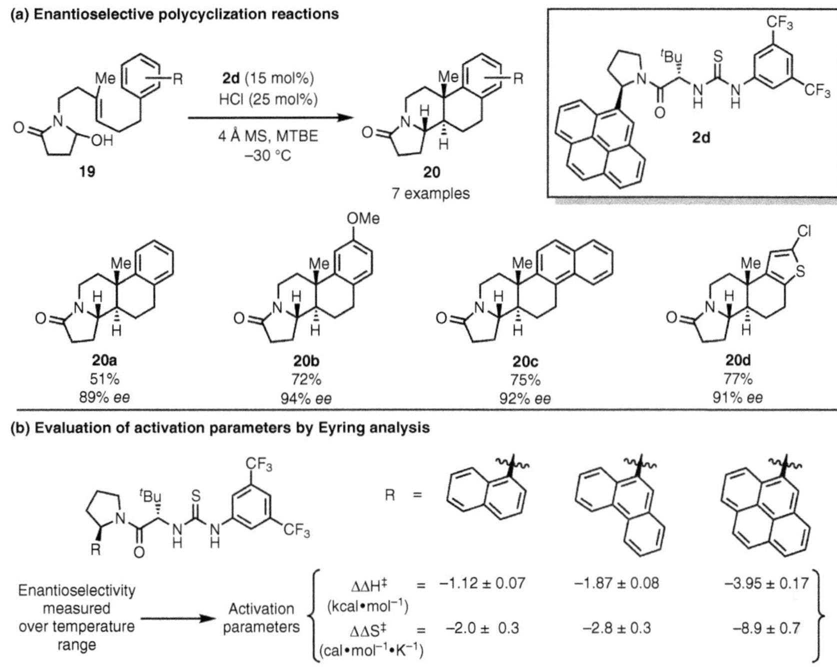

Scheme 3.5 Enantioselective cationic polycyclizations enabled by halide-binding thiourea catalyst **2d**.

this work, the authors found that the catalyst (**2e**) played a critical role in activating allylsilane using the thiourea moiety as a Lewis base.

Experiments comparing the relative rates of starting material consumption with a series of silylated substrates (**21-I**, **21-II**, and **21-III**) demonstrated that with a thiourea catalyst, the optimal rate was observed when the smallest and least electron-rich silane (**21-III**) was used (Scheme 3.6b). When the thiourea motif in the catalyst was replaced with a urea, it was found that the bulkier, electron-rich silane **21-II** was consumed the fastest, offering strong support for the proposal that the thiourea in **2e** plays a key role in the nucleophilic activation of the (*Z*)-allylsilane moiety (Scheme 3.6c). Two of the products of this reaction, **ent-22a** and **22d**, were transformed into the alkaloids (−)-tashiromine (**24**) and (+)-epi-lupinine (**26**), respectively, through an identical Lemieux–Johnson oxidation/$LiAlH_4$ reduction sequence (Scheme 3.7).

In a subsequent intermolecular variant, chiral squaramide **2f** was employed in the construction of α-allyl amino esters using α-chloro glycinate starting materials (**27**) in combination with allylsilane or allylstannane nucleophiles (**28**) (Scheme 3.8) [31]. From these starting materials, highly enantioenriched products (**29**) were obtained with excellent diastereocontrol in cases where multiple stereocenters were formed. It was proposed that the mechanism could either proceed via (i) a concerted S_N2 reaction where enantioconvergence is achieved through rapid racemization of the α-chloro glycinate, resulting in a dynamic kinetic resolution, or (ii) a stepwise S_N1

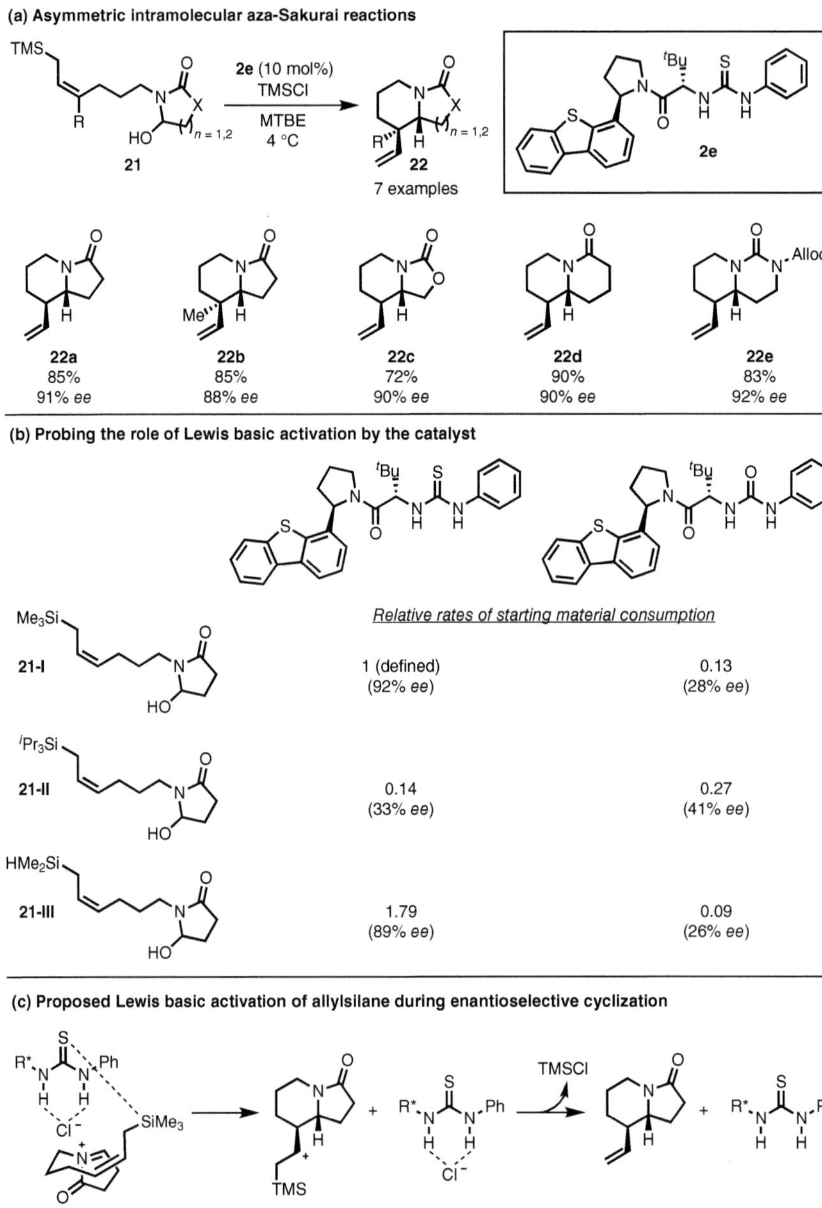

Scheme 3.6 Enantioselective aza-Sakurai cyclizations catalyzed by chiral thiourea HBD catalyst. Source: Based on Park *et al.* [30].

reaction where the catalyst assists in the ionization of the α-chloro glycinate. Computations were performed on the uncatalyzed reaction, which indicated that the S_N2 pathway is over 18.9 kcal mol^{-1} lower in energy than the S_N1 pathway in the absence of a HBD catalyst. Both the (*E*)- and (*Z*)-crotyltributylstannanes leading to **29e** provided the same major diastereomer (in 8 : 1 dr for the (*E*)-isomer and 4 : 1

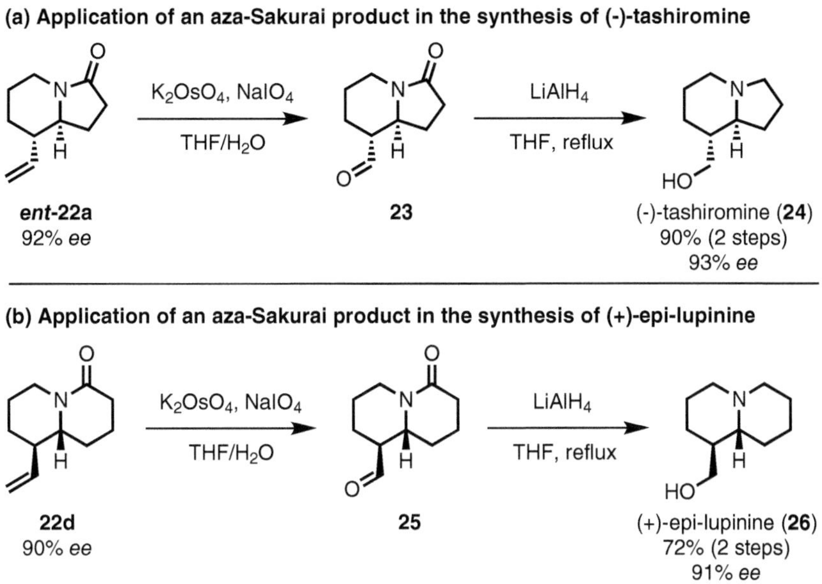

Scheme 3.7 Applications of aza-Sakurai products in alkaloid synthesis.

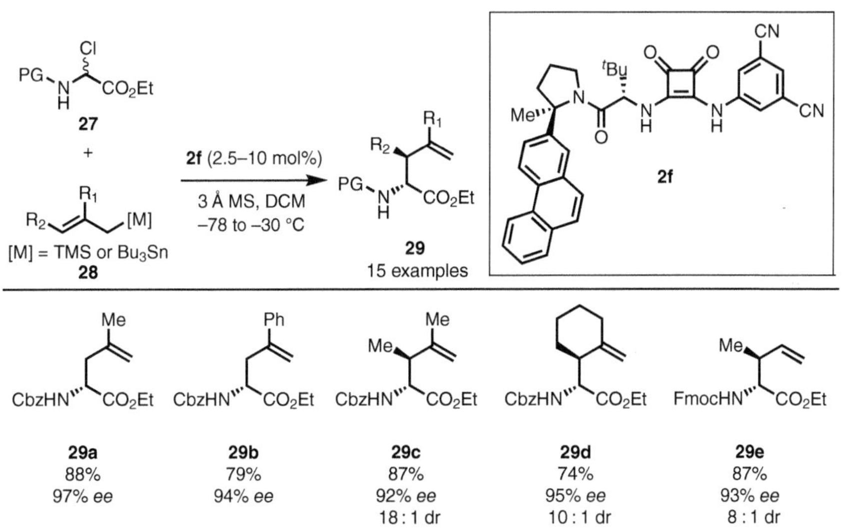

Scheme 3.8 Enantioselective synthesis of α-allyl amino esters under chiral squaramide catalysis. [M] = Me_3Si (TMS) or Bu_3Sn. Source: Based on Bendelsmith *et al.* [31].

dr starting for the (*Z*)-isomer). Analogous stereochemical outcomes were observed when using allylsilane pro-nucleophiles, which led the authors to propose that these reactions occurred through open transition states. Of further note, chiral squaramide catalysts have found a central niche in anion-binding catalysis – outside the scope

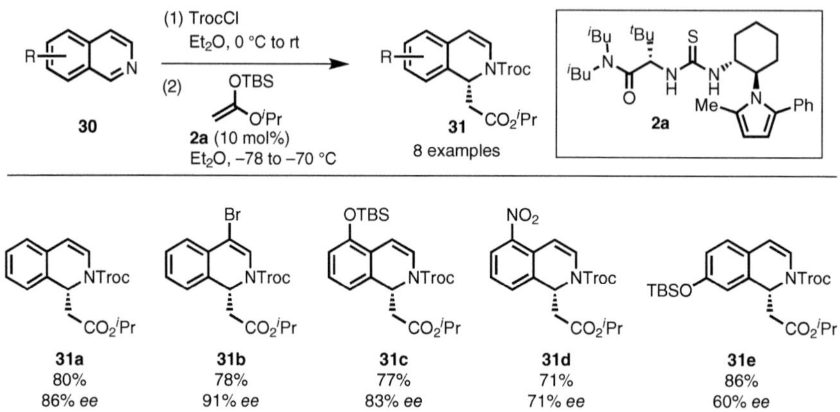

Scheme 3.9 Enantioselective acyl-Mannich reaction of isoquinolines with chiral thiourea HBD catalyst. Source: Based on Taylor *et al.* [34].

of halide binding – in the context of Lewis acid enhancement [32] and enantioconvergent S_N1 reactions [33].

3.3.1.4 Mannich Reaction and Variants

In a parallel line of investigation, *N*-acyliminium ions have been deployed as electrophiles in Mannich reactions and related variants. In an early example, acylation of isoquinolines (**30**) with 2,2,2-trichloroethoxycarbonyl chloride (TrocCl) and exposure to **2a** provided access to a *N*-acyliminium chloride–thiourea ion pair, which was found to react with enol silane nucleophiles in a Mannich reaction to form **31** in moderate to high enantioselectivity (Scheme 3.9) [34]. In this work, catalyst variants bearing differentially substituted pyrrole units (with 2,5-dimethyl, 2,5-diphenyl, or 2-methyl-5-phenyl substitution) were synthesized, and **2a** was found to provide the products in significantly higher enantioselectivity than the other variants studied. The authors partially rationalized this outcome using the crystal structure of the catalyst, which demonstrated the spatial proximity of the 5-phenyl substituent of the pyrrole to the thiourea-binding site of the catalyst, offering an explanation for the observed enantioinduction.

The concept of using isoquinolinium chloride ions as electrophiles in asymmetric Mannich reactions was subsequently applied to develop a C–H functionalization sequence for tetrahydroisoquinoline derivatives (Scheme 3.10) [35]. In this framework, tetrahydroisoquinolines were converted to α-chloro amines using a ruthenium photocatalyst and carbon tetrachloride as an oxidant; subsequent introduction of a halogen-binding thiourea catalyst enabled the formation of iminium ions that were used in asymmetric Mannich reactions. Using this strategy, tetrahydroisoquinolines (**32**) were converted to Mannich products (**33**) in moderate to high enantioselectivity (as in **33a–33c**). Substitution on the aromatic ring of the tetrahydroisoquinoline has a large impact on the enantioselectivity of the reaction, with electron-rich substrates performing poorly (leading to **33d**, for example).

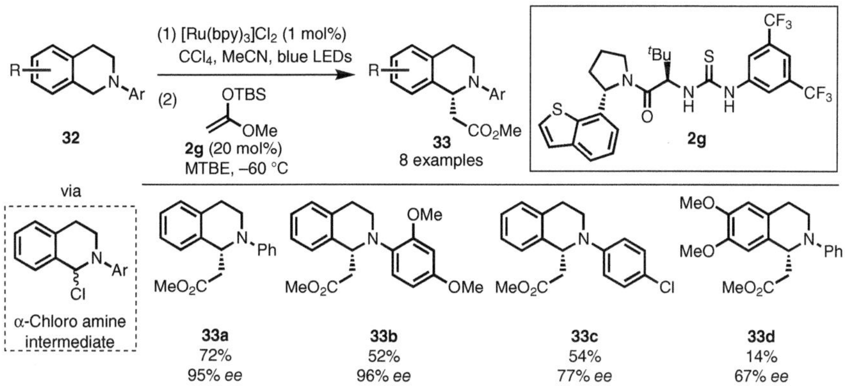

Scheme 3.10 Merging photoredox catalysis with anion-binding catalysis to achieve C–H functionalization. Source: Based on Bergonzini *et al.* [35].

The broader applicability of the asymmetric Mannich reaction can be seen in developments that moved beyond the isoquinolinium core. Using *N*-acyliminium ions derived from α-chloroglycinate substrates, the Jacobsen and Roche groups demonstrated that catalyst **2h** was capable of enabling enantioselective Mannich reactions with 1,3-dicarbonyl pro-nucleophiles (**34**) to form substituted α-amino ester products (**35**) (Scheme 3.11a) [36].

While examples furnishing adjacent stereocenters resulted in diastereomeric mixtures at or near parity, the enantioselectivities observed were high for 1,3-diarylketone pro-nucleophiles. Utilization of 1,3-diketones with aliphatic groups instead of arenes was found to decrease the yield and enantioselectivity of the reaction (comparing **35a** and **35b**, for example). In the case of β-ketoester pro-nucleophiles, benzhydryl esters led to significantly better results than their aliphatic ester analogs (comparing **35h** and **35i**, for example). Mechanistically, the authors proposed that the tertiary amine present in the catalyst may be involved in either (i) deprotonating the α-proton of the 1,3-dicarbonyl or (ii) deprotonating an iminium ion to form a thiourea-bound iminoester. Additional investigations have implicated hyperconjugation ($n_N \rightarrow \sigma^*_{C\text{–}Cl}$) as the primary driving force for the intrinsic reactivity of the α-chloroglycinates used in this study (Scheme 3.11b) [37].

3.3.1.5 Petasis-Type Reactions

With the extension of asymmetric halide-binding catalysis to the Mannich reaction, it was natural that this catalysis concept was soon adapted in the context of the closely related Petasis reaction. Takemoto's group reported that catalyst **2i** (Scheme 3.12) can be used for the direct addition of styrenylboronic acids to quinoline-derived *N*-acyliminium ions (formed from **36**) [38]. Recognizing the importance of forming a reactive "ate" intermediate from the boronic acid in this reaction, the authors introduced a primary alcohol in the catalyst scaffold, which resulted in a dramatic improvement in the enantioselectivity of the reaction (90% *ee* vs. 9% *ee*, which was obtained with a catalyst variant lacking the primary alcohol moiety). With this catalyst, the Petasis products (**37**) could be obtained in good

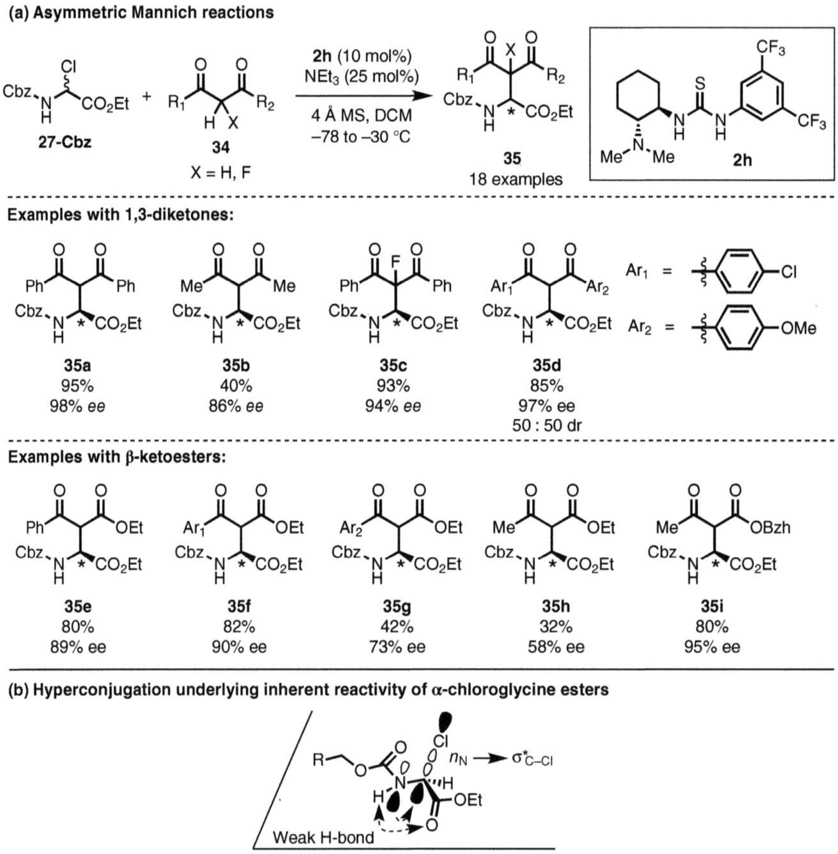

Scheme 3.11 Asymmetric Mannich reactions with 1,3-diketone or β-ketoester nucleophiles. Source: Based on Wasa *et al.* [36], Samanta and Roche [37].

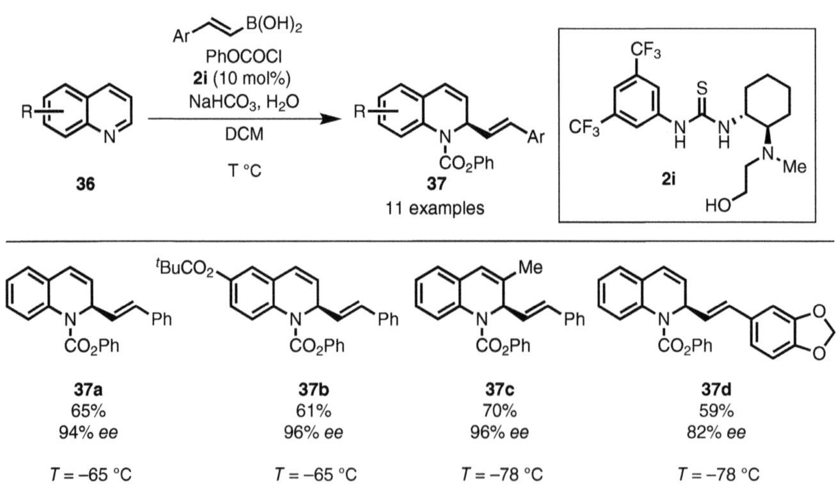

Scheme 3.12 Asymmetric Petasis reactions developed by Takemoto's group.

yield and excellent enantioselectivity. Utilization of H_2O and $NaHCO_3$ as additives improved the yield and enantioselectivity, which was attributed to more facile catalyst regeneration.

3.3.2 Examples Utilizing Electrophiles Other than *N*-Acyliminium Ion Precursors

3.3.2.1 Utilization of Oxocarbenium and Pyrone Intermediates

Research in the area of halide-binding by chiral (thio)ureas and squaramides profited enormously from developments beyond the *N*-acyliminium ion as an electrophile. One of the first adaptations with a novel electrophile was the enantioselective addition of silyl ketene acetals to oxocarbenium ions (Scheme 3.13a) [39]. Here, the methyl acetal starting material (**38**) was converted to a 1-chlorochroman intermediate, which acted as a substrate for anion-binding catalysis and resulted in oxocarbenium formation upon treatment with **2j**. Following an asymmetric addition of the silyl ketene acetal, the ester products (**39**) could be obtained in a highly enantioenriched form in most cases.

The authors found that addition of 10 mol% NBu_4Cl completely inhibited the reaction, which was consistent with a chloride-binding mode for the catalyst. While S_N2 kinetics were observed in this reaction, which indicated the participation of the silyl ketene acetal nucleophile in the rate-determining step, the mechanism was postulated to involve an oxocarbenium chloride–thiourea complex as the reactive electrophile.

In a subsequent study, catalyst **2j** was found to exist as an aggregate of inactive dimers, which had to dissociate in order to reach the rate-determining transition state in a 2 : 1 catalyst–substrate complex (Scheme 3.13b) [40]. While the authors proposed that this 2 : 1 catalyst–substrate complex in the transition state might possess a 4H-binding mode (Scheme 3.14b), they noted that a 2H-binding mode may be a viable alternative. This catalyst was also demonstrated to promote substrate ionization, leading to racemization through an S_N1 pathway and forward reaction through an oxocarbenium chloride–thiourea ion pair [42]. While the pyrrolidine amide of **2j** was found to exist as an interconverting mixture of (*E*)- and (*Z*)-rotamers, the catalyst could be "locked" into the (*Z*)-amide conformer by inclusion of a 2-methyl substituent [43]. Further mechanistic investigations into the asymmetric addition of silyl ketene acetals to oxocarbenium ions resulted in the development of catalyst **2k** (Scheme 3.14) [41].

With the aim of preventing the formation of nonproductive dimeric catalyst aggregates (Scheme 3.14a) and favoring a 4H-binding mode of cooperative catalysis (Scheme 3.14b), Jacobsen's group identified an appropriate tether to link together two chiral thiourea units. The novel bis-thiourea catalyst **2k** demonstrated significant improvement over monomeric analogs in terms of the reaction rate. Consequently, the catalyst loading could be greatly reduced (Scheme 3.14c) for improved throughput upon scale-up.

Pyrones were also shown to be competent electrophiles in halide-binding catalysis with arylpyrrolidine thiourea organocatalyst **2l** (Scheme 3.15) [44]. In this work,

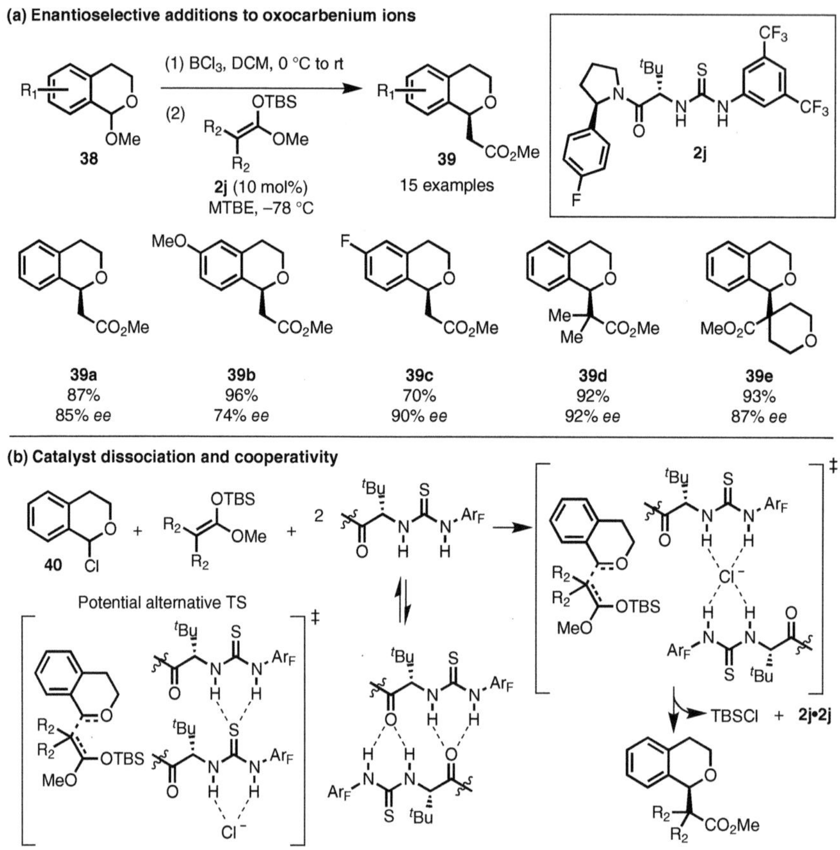

Scheme 3.13 Enantioselective additions of silyl ketene acetals to oxocarbenium ions. Source: Based on Reisman *et al.* [39], Based on Ford *et al.* [40].

the authors utilized a Brønsted acid for the desilylation of a γ-pyrone precursor (**42**), which was proposed to enter into an anion-binding catalysis cycle via bromide abstraction by the thiourea of **21**. Addition of the indole (**41**) to the pyrone led to **43** in a highly enantioselective manner when the authors employed catalyst **21** featuring an extended arene. Interestingly, the leaving group of **42** influenced the outcome of the reaction. Replacement of the bromine in **42** with chlorine was found to decrease the enantioselectivity (from 92% to 84% *ee* in one example), and utilization of a benzoate leaving group resulted in only traces of product. Furthermore, N-methylation on the indole afforded a racemic cycloadduct, which led the authors to conclude that the enantiodetermining step involved privileged interactions of the catalyst with the leaving group of pyrone as well as the N–H of the indole. They also proposed that attractive interactions (either π–π or C–H–π) enabled by the use of the extended arene in **21** were responsible for the high level of enantioselectivity in the reaction.

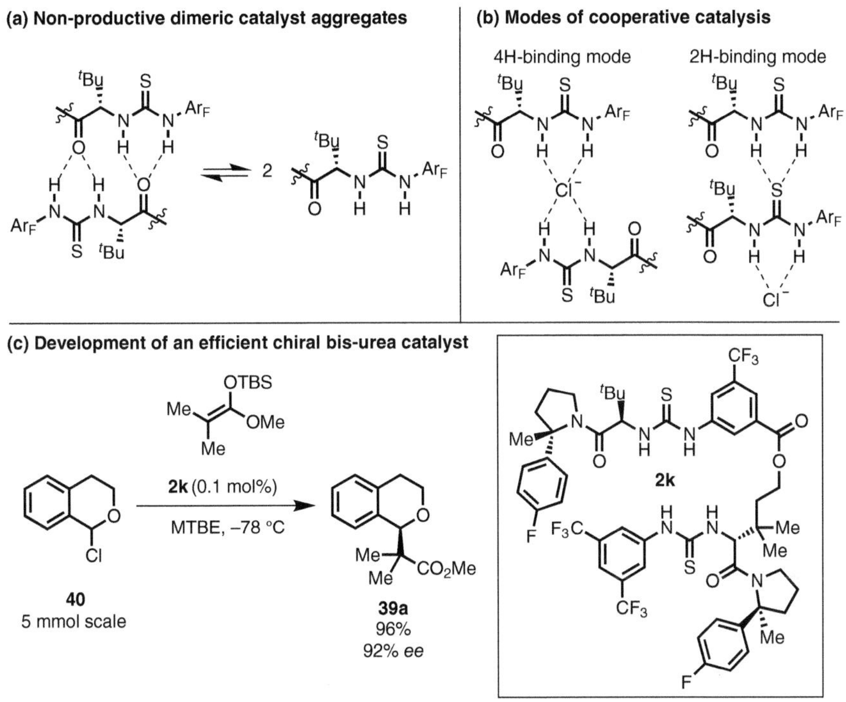

Scheme 3.14 Development of bis-urea organocatalysts with improved efficiency. Source: Based on Kennedy *et al.* [41].

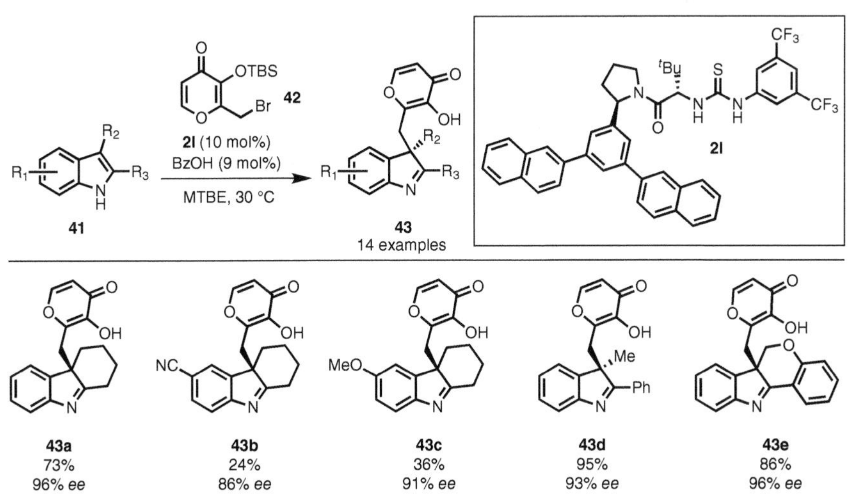

Scheme 3.15 Enantioselective addition of indoles to pyrones under thiourea catalysis. Source: Based on Yeung *et al.* [44].

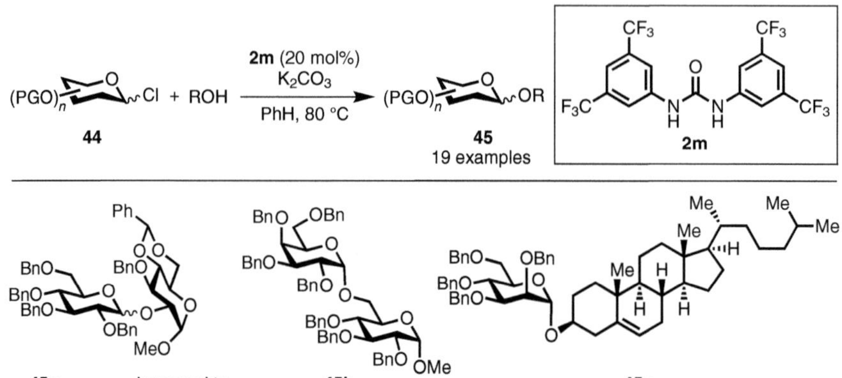

45a 90% α/β = 2 : 1 — Improved to 90%, α/β = 10.1 : 1 (+1.5 equiv TTMPP)

45b 99% α only

45c 98% α only

Scheme 3.16 Urea-catalyzed Koenigs–Knorr glycosylation reactions. Source: Based on Sun *et al.* [54].

3.3.2.2 Glycosylation Reactions Utilizing HBD–Halide Binding

The exploration of broader applications of HBD–halide interactions has fueled the development of novel glycosylation reactions. A foundational example was the application of Schreiner's urea (**2m**) to the Koenigs–Knorr glycosylation reaction of glycosyl chlorides (**44**) with alcohols (or a glycosyl donor) to give **45** (Scheme 3.16) [54]. While diastereoselectivity remained a problem in some cases (**45a**, for example), many products (such as **45b** and **45c**) could be formed with a high degree of stereoselectivity for the α-product. For cases where the α/β selectivity was low, the authors reported that tri-(2,4,6-trimethoxyphenyl)-phosphine (TTMPP) could be included to improve the selectivity in favor of the α-product, although the mechanism for this phenomenon was not fully elucidated in this report.

The application of halide-binding catalysis to the problem of stereospecific glycosylation reactions resulted in the development of the macrocyclic bis-thiourea catalyst **2n** (Scheme 3.17a) [46]. Using glycosyl chloride electrophiles (**47**) bearing an α-chloride substituent at the anomeric position, this system could be used to obtain products **48** in a highly stereospecific manner, resulting in *trans*-1,2- (**48a**), *cis*-1,2- (**48b**), and 2-deoxy-β-glycosides. In order to remove HCl formed during the reaction, isobutylene oxide (IBO) was identified as a suitable additive. The 2-azidogalactose derivative **48c** was formed in high diastereoselectivity and subsequently elaborated to trisaccharide **48d** through the same conditions. The Lewis basic amide carbonyls of the catalyst (**2n**) were found to be critical to both the yield and diastereoselectivity of the reaction; replacing them with isopropyl esters resulted in a decrease in both metrics. Furthermore, density functional theory (DFT) calculations (M06-2X/6-31G*/PCM: benzene) predicted that in a concerted glycosylation of glycosyl chloride with methanol, the carbonyl of the indoline amide in **2n** would form a hydrogen bond with the methanol hydroxyl moiety. This hypothesis was supported by the $^{12}C/^{13}C$ kinetic isotope effect (KIE) studies using polarization transfer, wherein the observed KIEs for C1, C2, C4, and C5 were

(a) Stereospecific glycosylation reactions

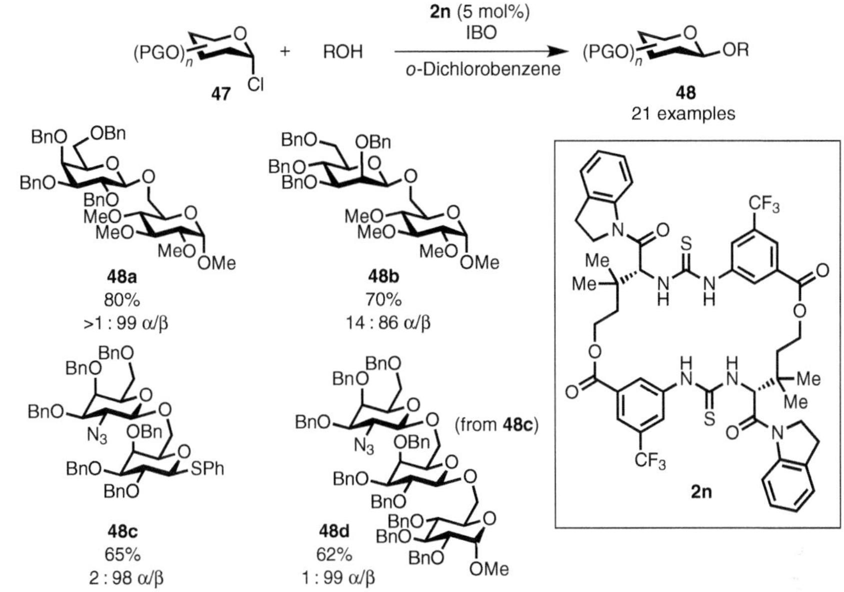

(b) Mechanistic elucidation using carbon KIEs

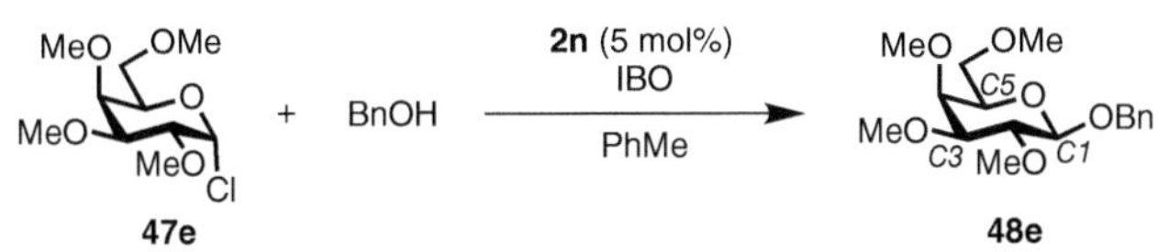

		C1	C2	C4	C5
Experimental KIE (DEPT-55)		1.000	1.006	1.000	1.008
S_N2 (predicted)	PBE0-D3(BJ)	0.999	1.006	0.999	1.007
	B3LYP-D3(BJ)	0.993	1.006	0.999	1.007
	M06-2X	1.008	1.011	0.998	1.005
S_N1 (predicted)	PBE0-D3(BJ)	0.981	1.007	1.000	1.011

Scheme 3.17 Stereospecific glycosylation reactions catalyzed by macrocyclic bis-thioureas. Source: Based on Park *et al.* [46], Kwan *et al.* [47].

found to closely match those predicted by PBE0-D3(BJ)/6-31G*/PCM for a S_N2 mechanism (Scheme 3.17b) [47]. The scope of the electrophile was later extended to glycosyl phosphates using non-macrocyclic bis-thioureas as anion-binding catalysts [48–50].

3.3.2.3 Utilization of Non-heteroatom-Stabilized Carbocations as Electrophiles

In a significant departure from prior art using nitrogen and oxygen-stabilized carbocations as intermediates, Jacobsen's group developed an enantioselective α-alkylation of aldehydes using a bifunctional primary amine/thiourea halide-binding catalyst (**2o**) with diarylbromomethane electrophiles (Scheme 3.18a) [51]. Replacement of the primary amine in the catalyst with a tertiary amine (in the

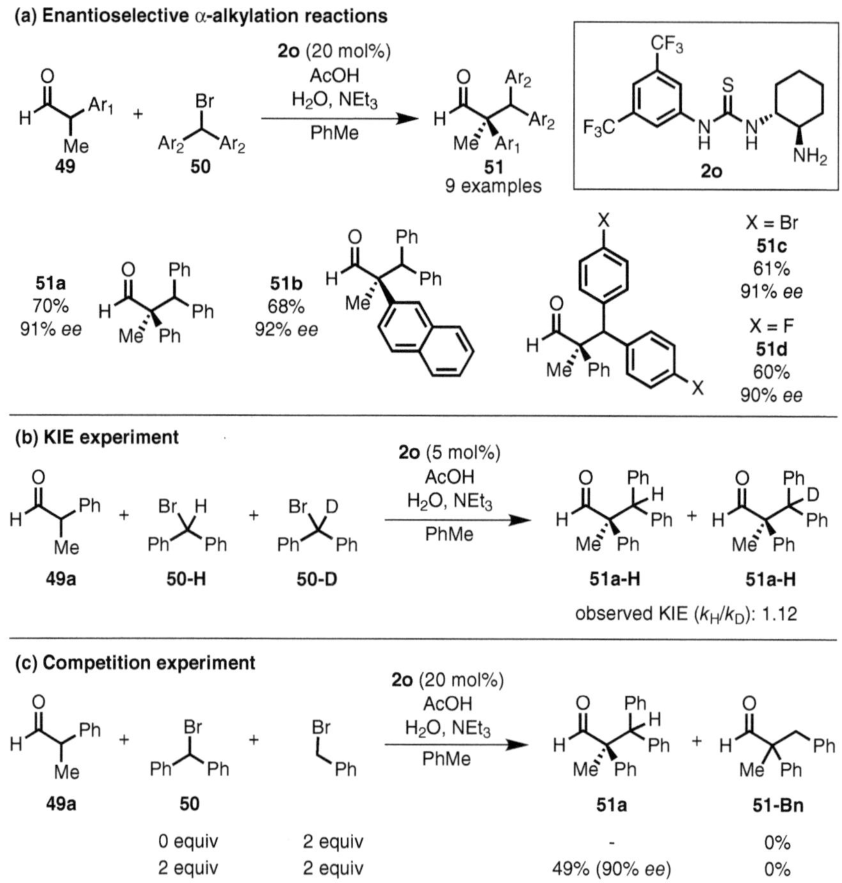

Scheme 3.18 Enantioselective α-alkylation of aldehydes using bifunctional primary amine/thiourea catalyst. Source: Based on Brown *et al.* [51].

form of a dimethylamino group) was found to completely shut down the reaction; the thiourea group was also essential for high reactivity and enantioselectivity. To distinguish between the two potential mechanisms, which were postulated to be a S_N1- or S_N2-type substitution, a kinetic isotope study was performed by employing a diarylbromomethane deuterated at the benzhydryl position. The secondary kinetic isotope effect (k_H/k_D) of 1.12 observed (Scheme 3.18b) indicated that a sp^3 to sp^2 rehybridization was occurring in the transition state, which is in agreement with a S_N1 pathway. Further evidence for this mechanism came in the form of a Hammett study, which found that electron-donating substituents on the electrophile increased the reaction rate ($\rho = -1.95$).

Competition experiments with benzyl bromide offered additional support for this mechanistic proposal (Scheme 3.18c). Notably, alkylations using enantioenriched *p*-chlorobenzhydryl chloride were found to proceed with high (95%) stereospecificity, an observation that was explained by the exceptional reactivity of benzhydryl

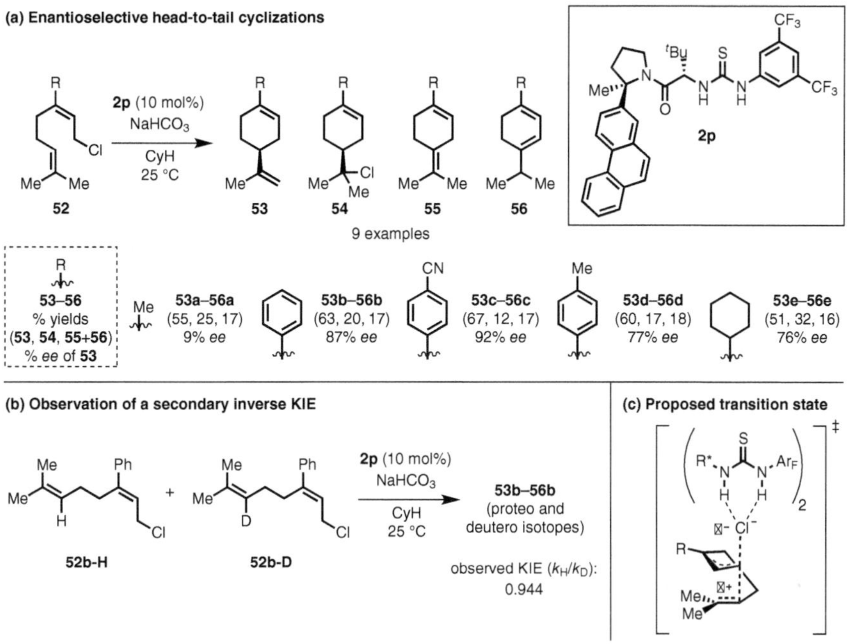

Scheme 3.19 Enantioselective thiourea-catalyzed tail-to-head cyclizations. Source: Based on Kutateladze *et al.* [52].

cations with enamines, which has been established to be high enough to outcompete racemization.

Tail-to-head cyclizations of neryl chloride derivatives were later developed using chiral thiourea HBD catalyst **2p** (Scheme 3.19a) [52]. While utilization of neryl chloride (**52a**, R = Me) led to a non-stereoselective cyclization (giving **53a–56a**), the inclusion of an aromatic ring on the nucleophilic olefin enabled cyclization to occur with high levels of enantioselectivity. Utilization of the (*E*)-alkene isomer of **52a** (geranyl chloride) resulted in low conversion and inefficient formation of the cyclized products relative to uncyclized by-products. A small secondary inverse KIE (k_H/k_D = 0.944) was observed for the vinylic C—H (or C—D) bond of the nucleophilic olefin (Scheme 3.19b). This evidence, in combination with a Hammett analysis and DFT, was used to support the hypothesis that π-participation by the nucleophilic olefin promoted the ionization of the substrate and prevented the formation of direct elimination products (Scheme 3.19c).

3.4 Asymmetric Catalysis by Catalyst Association with the Nucleophile

While the majority of (thio)urea and squaramide-catalyzed anion-binding catalysis involving halides has been developed by utilizing the association of an electrophile with a chiral catalyst through ion pairing, there has been momentum toward developing a complementary approach based on utilizing HBD-bound halides as nucleophiles. In the framework of asymmetric catalysis by association of an HBD with a nucleophile, there are two general trends that have emerged. In the first strategy, HBDs have been employed as phase-transfer catalysts in order to ensure that any nucleophile brought into solution during the reaction has a chiral environment (provided by the catalyst), an approach pioneered by Gouverneur's group (Section 3.4.1). The second strategy, formulated by Jacobsen's group, features homogeneous catalysis and a catalyst-bound nucleophile generated *in situ* from a suitable pro-nucleophile (Section 3.4.2).

3.4.1 Catalyst-Nucleophile Association in Phase-Transfer Catalysis

3.4.1.1 Investigation of Hydrogen-Bonded Fluoride: Structure and Reactivity

In recent investigations into the possibility of using HBD-bound halides as nucleophiles, a particular emphasis has been placed on fluoride. In a foundational structural investigation, fluoride was found to form hydrogen bonds with 1,3-bis(4-nitrophenyl)urea in a 1 : 1 complex [53]. Upon addition of a second equivalent of fluoride, the urea underwent deprotonation with the formation of HF_2^-. Subsequent work employing alcohols as HBDs in a series of fluoride–alcohol complexes demonstrated that reactivity and selectivity (for an S_N2 reaction over an E2 reaction) were modulable by changing the properties of the alcohol HBD [54]. The development of the HBD–fluoride interaction in anion-binding catalysis was further advanced through an investigation of urea–fluoride, amide–fluoride, and squaramide–fluoride complexes, which led to the conclusions that (i) dimeric urea–fluoride complexes $[U_2F]^-$ partially dissociate to the reactive species $[UF]^-$ and (ii) diarylurea–tetrabutylammonium fluoride (TBAF) complexes could be used for the formation of C—F bonds, where the substituents on the urea directly impact the selectivity of the reaction [55]. In a model system, it was found that while the reaction of **57** to form **58** and **59** was much slower using a diarylurea–TBAF complex than using the commercially available TBAF trihydrate (Scheme 3.20), utilization of an electron-deficient diarylurea–TBAF complex was able to kinetically favor the S_N2 product (**58**) over the E2 product (**59**).

3.4.1.2 Development of Hydrogen-Bonding Phase-Transfer Catalysis (HBPTC)

The development of urea–fluoride complexes presented new opportunities in asymmetric catalysis by offering a way to utilize the fluoride anion as a nucleophile rather than a base, and by employing chiral urea HBDs, it became possible to conceive applications in enantioselective fluorination. Furthermore, by utilizing urea HBDs as solid–liquid phase-transfer agents, it was realized that background reactions could be minimized if non-soluble alkali metal fluoride salts were used directly in stereoselective fluorination reactions. Gouverneur's group formulated this approach, that they coined hydrogen-bonding phase-transfer catalysis (HBPTC), originally using the enantioselective fluorination of *in situ* formed episulfonium ions as a model transformation (Scheme 3.21a) [56]. It was initially postulated that (i) "bis – tetradentate" or (ii) "cooperative – bidentate" binding

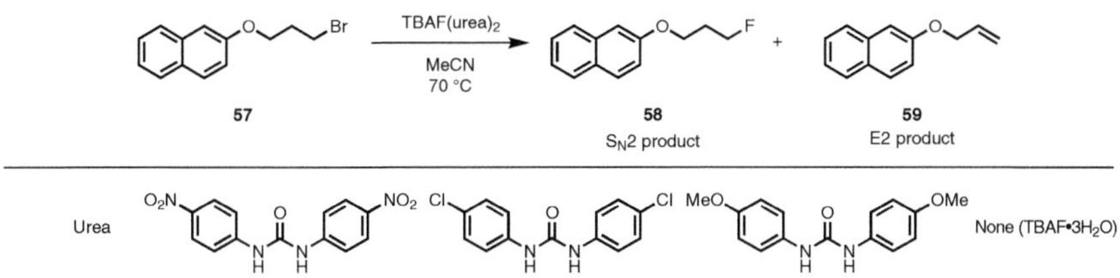

k_2 (S_N2) (x 10^{-5} $M^{-1}s^{-1}$)	3.2	5.83	12.7	375
k_2 (E2) (x 10^{-5} $M^{-1}s^{-1}$)	0.38	0.84	2.45	235
k_2 (S_N2)/k'_2 (E2)	8.4	6.9	5.2	1.6

Scheme 3.20 Modulating propensity of fluoride for S_N2 vs. E2 using diarylurea TBAF complexes.

modes (Scheme 3.21b) may be operative, drawing on 2H/4H-binding modes of cooperative thiourea catalysis. However, molecular dynamics on a tetradentate catalyst revealed the importance of the tridentate interaction (iii) between this class of catalysts and fluorides; such a coordination mode became possible when one of the urea motifs adopted a *syn–anti*-conformation, which ultimately allowed the authors to reach the final optimized catalyst structure. Furthermore, it was found that this catalyst was able to interact with the substrate through cation–π and C–H–π interactions, which allowed high enantioselectivities for stilbene-derived substrates. In the proposed mechanism, the bis-urea catalyst **61a** was used to deliver fluoride from solid cesium fluoride (CsF) to episulfonium precursors **60** by initial generation of an episulfonium bis-urea–bromide ion pair, which was proposed to undergo an anion exchange with fluoride to form a new ion pair enabling enantioselective fluoride delivery (Scheme 3.21c).

A subsequent NMR investigation found that the hydrogen-bonding network involved differential contributions from each N—H bond of the catalyst, which could be determined using quantitative ^1H-^{19}F heteronuclear Overhauser spectroscopy (HOESY) analysis of **61b**•TBAF in DCM-d_2 (Scheme 3.22a) [57]. Additionally, proton diffusion experiments (^1H DOSY) indicated the existence of a minor 2 : 1 catalyst–fluoride complex [(**61a**)$_2$•CsF; Scheme 3.22b], which was identified using exchange spectroscopy (EXSY/NOESY) and rotating frame Overhauser spectroscopy (ROESY). At room temperature, the monomeric species **61a**•CsF exists in equilibrium with (**61a**)$_2$•CsF, favoring the former in a 85 : 15 ratio.

This type of catalysis was subsequently advanced in a report that utilized potassium fluoride (KF) for the enantioselective fluorination of aziridinium ions (Scheme 3.23) [58]. In this work, racemic amino chlorides and bromides (**63**) served as precursors of *meso*-aziridinium ions, which could be efficiently desymmetrized using catalyst **61c** to obtain enantioenriched β-amino fluorides **64**. Conditions were developed to use either KF or CsF, and there was remarkable consistency in the enantioselectivities, although certain substrates gave marginally higher yields with one salt over the other. Using this methodology, fluorinated analogs of diphenidine (**64b**), lefetamine, and MT-45 were prepared in a highly enantioenriched form. Importantly, aziridinium precursors bearing distinct nitrogen

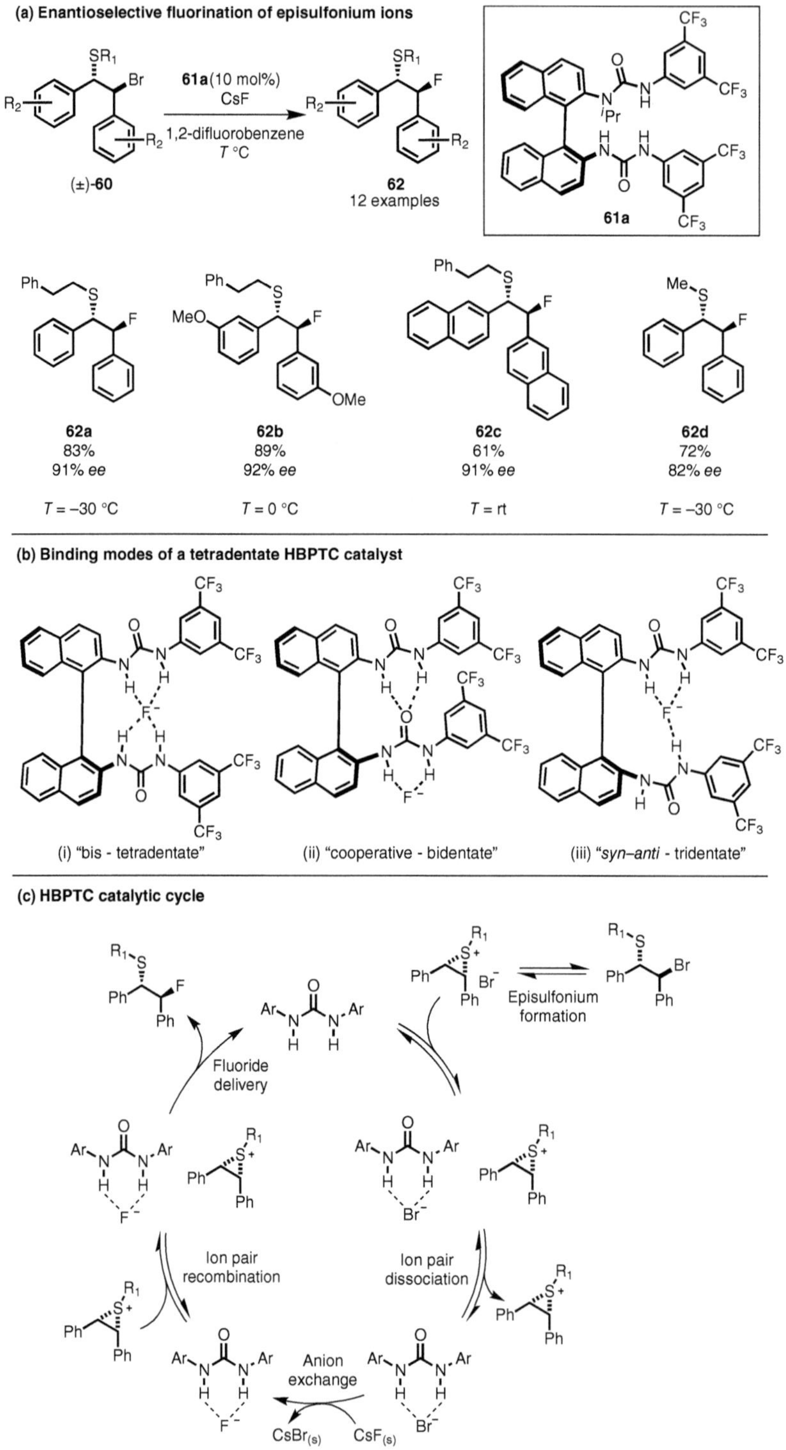

Scheme 3.21 Enantioselective fluorination of episulfonium ions under HBPTC. Source: Pupo *et al.* [56].

(a) Determination of H–F distances via quantitative HOESY analysis

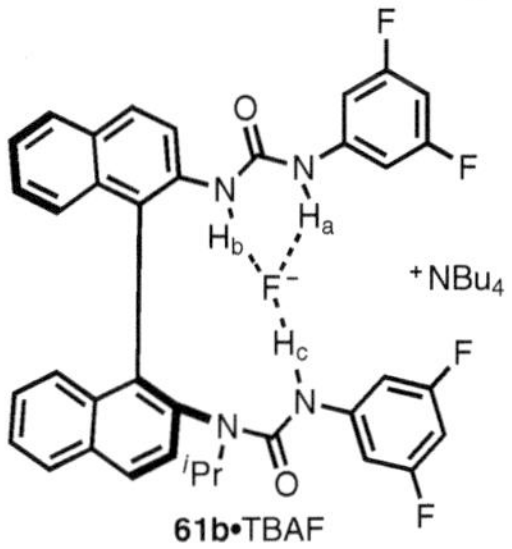

61b•TBAF

	Distance (Å)	$^{1h}J_{NH\cdots F^-}$ (Hz)
$NH_a\cdots F^-$	1.83	60
$NH_c\cdots F^-$	1.91	48
$NH_b\cdots F^-$	2.05	35

(b) Identification of a 2 : 1 catalyst–fluoride complex

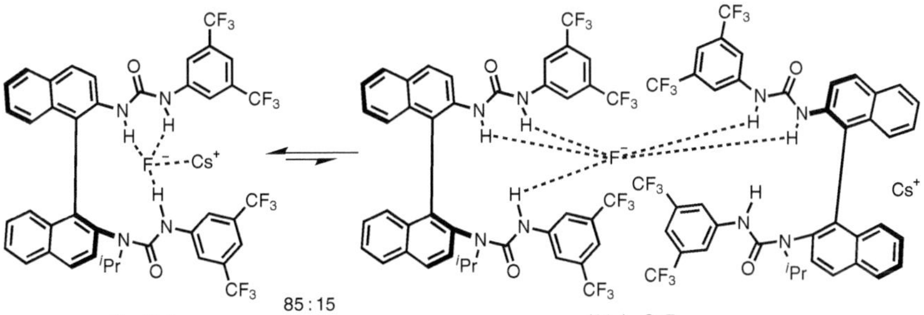

61a•CsF | **61a•CsF : (61a)$_2$•CsF** at 298 K | **(61a)$_2$•CsF**

Scheme 3.22 NMR investigations of HBPTC catalysts. Spectra acquired in DCM-d_2. Source: Ibba *et al.* [57].

substituents (which would generate a diastereomeric mixture of aziridinium ions) also provided the β-amino fluoride product (**64c**) in high enantioselectivity. Examples using non-stilbene-derived amino chlorides (leading to **64d** and **64e**, for example) gave lower enantioselectivities, underscoring the importance of the non-covalent catalyst–substrate interactions initially proposed for the fluorination of episulfonium ions. As a measure of scalability, the reaction to make **64b** was carried out on 50 g scale, resulting in 31.1 g product (66% yield) and 94% *ee*.

While these initial reports utilized transiently generated ionic intermediates, it was subsequently demonstrated that ionic reactants can also be used in this approach. Using catalyst **61a** and CsF, azetidinium ions **65** could be desymmetrized through a ring-opening fluorination reaction to form γ-fluoroamines **66** in a highly enantioselective manner (Scheme 3.24) [59]. Importantly, it was found that **65** could be used as a mixture of diastereomers; computations suggested that this phenomenon originates from the fact that the exocyclic N-substituents of the azetidinium are oriented away from the catalyst in the enantiodetermining transition state. Using this synthetic approach to γ-fluoroamines, the authors demonstrated that it was possible to access highly enantioenriched fluorinated lorcaserin (**67**). Outside of HBPTC with fluoride, urea HBDs have also been employed as phase-transfer catalysts for iodocyclization reactions, where they were proposed to bring *N*-iodosuccinimide (NIS) into solution at cryogenic temperatures by coordinating to the imide oxygen [60].

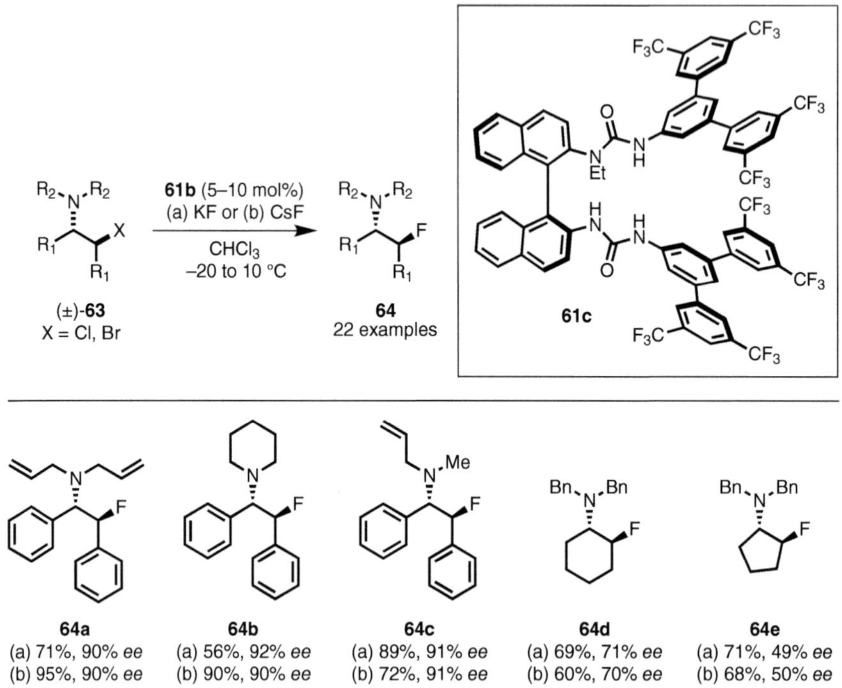

Scheme 3.23 Enantioselective fluorination of aziridinium ions under HBPTC. Source: Pupo *et al.* [58]. Licensed under CC BY 4.0.

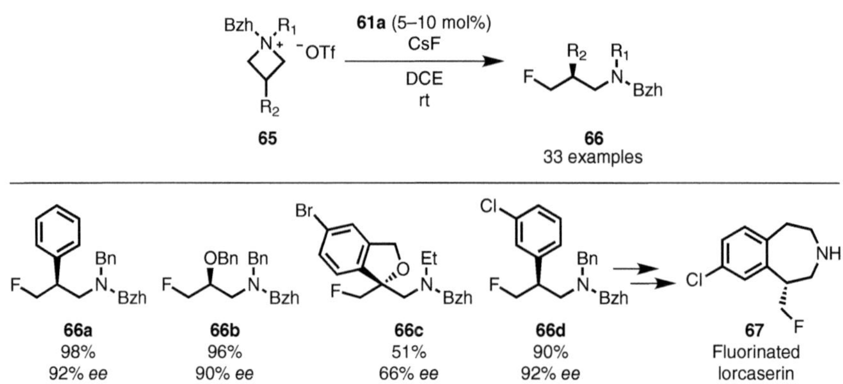

Scheme 3.24 Enantioselective fluorination of azetidinium ions under HBPTC. Source: Based on Roagna *et al.* [59].

3.4.1.3 Development of Acyl-Transfer Catalysis with Hydrogen-Bonded Fluoride

The exploration of other possibilities for applying hydrogen-bonded fluoride as a nucleophile in desymmetrization reactions led to a promising development in acyl-transfer catalysis. Using phase-transfer catalyst **69**, Connon's group demonstrated that *meso* anhydrides (**68**) can be desymmetrized with moderate

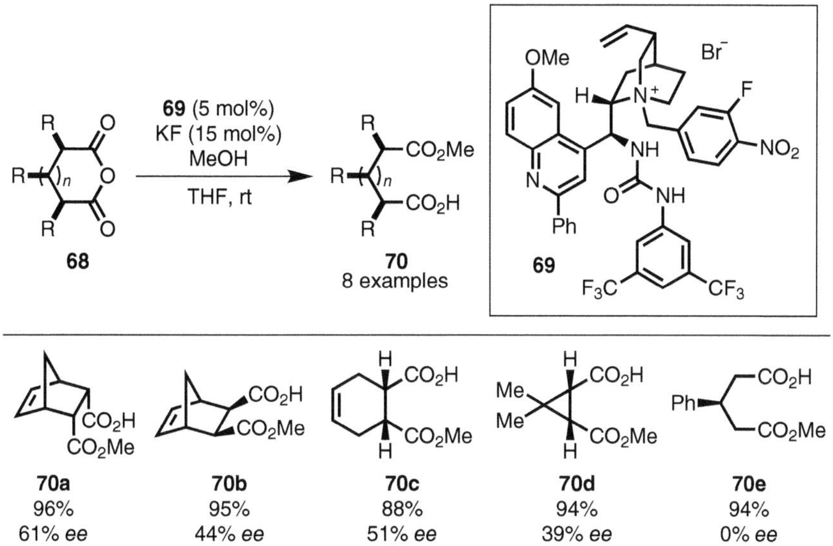

Scheme 3.25 Enantioselective acyl-transfer catalysis. Source: Based on Craig *et al.* [61].

enantioselectivity using bicyclic anhydride starting materials and methanol as a terminal nucleophile (Scheme 3.25) [61]. Monocyclic anhydride starting materials provided racemic or nearly racemic products (**70e**, for example), which the authors attributed to a significant background methanolysis resulting from the relative lack of steric hindrance. By performing ^{19}F NMR experiments, the authors were able to support the intermediacy of an acyl fluoride in the mechanism. The concept of utilizing fluoride for catalysis has also been investigated outside the scope of hydrogen bonding for the Steglich rearrangement [62].

3.4.2 Catalyst–Nucleophile Association in Homogeneous Catalysis

A logical advance in halide-binding catalysis with novel electrophiles came with the development of an enantioselective oxetane-opening reaction (Scheme 3.26a) [63]. Employing the chiral squaramide HBD catalyst **2q**, Jacobsen's group has demonstrated that oxetanes (**71**) could be opened with a bromide nucleophile to obtain enantioenriched ring-opened products (**72**). This reaction was postulated to proceed by silylation of the oxetane, leading to a transient oxonium species, which could be opened by a squaramide-bound bromide nucleophile. Verification of the reversible nature of the silylation, as well as the assignment of the bromide delivery as the enantiodetermining step, was made by carrying out a one-pot competitive KIE, which revealed a primary ^{12}C/^{13}C KIE of 1.126 (Scheme 3.26b). The association of the catalyst with TMSBr was proposed to not only result in the squaramide-bound bromide but also to accentuate the Lewis acidity of TMSBr as a silylation reagent. The synthetic value of this reaction was demonstrated by the application to the synthesis of pretomanid (**73**), an approved drug for the treatment of tuberculosis (Scheme 3.26c).

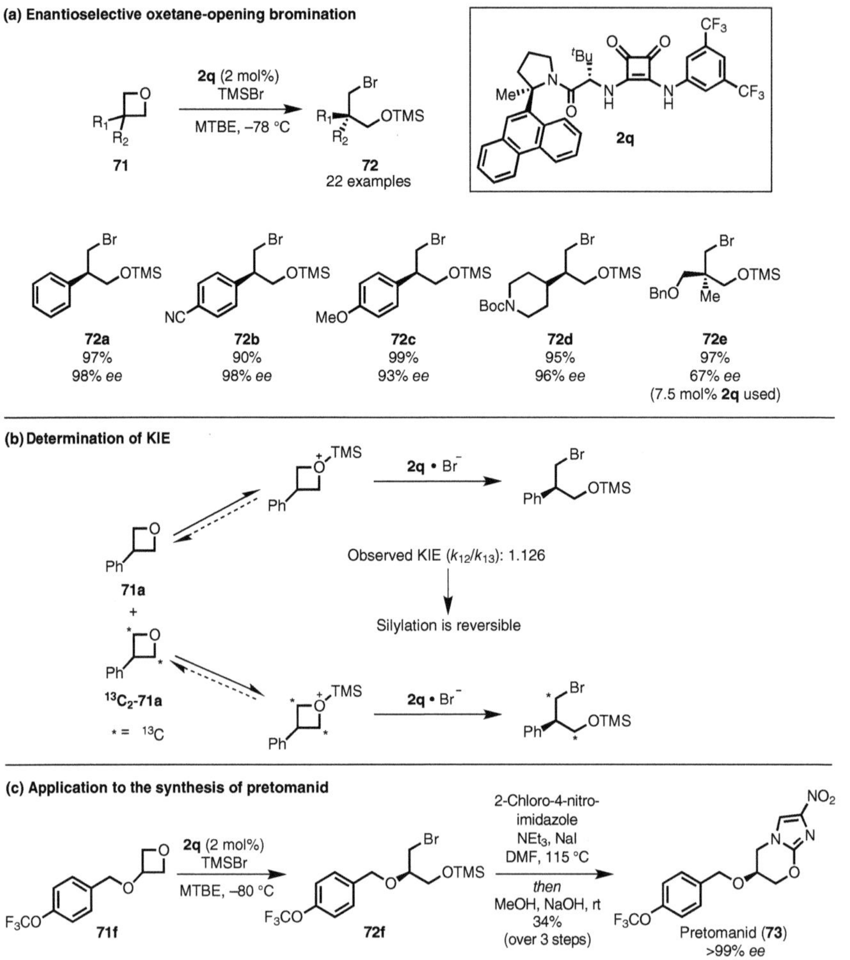

Scheme 3.26 Enantioselective oxetane-opening reactions using a chiral squaramide HBD catalyst. Source: Based on Strassfeld *et al.* [63].

3.5 Conclusions and Outlook

Anion-binding catalysis with halogen anions has been developed in broad directions. Future efforts in the field may take advantage of the foundations laid by the work discussed in this chapter. In the context of reactions that rely on catalyst–electrophile interactions, there is certainly room for growth in terms of the types of electrophiles that can be used. To expand the limits of the methodology in this way, it is almost always necessary for new catalyst scaffolds to be explored and interrogated with the tools of physical organic chemistry. Conversely, reactions that take advantage of catalyst–nucleophile interactions are still relatively underdeveloped, being primarily limited to examples with fluoride as a nucleophile. Future work to apply novel HBD-bound nucleophiles with less reactive electrophilic partners will aid in the development of this research area.

Acknowledgments

The authors thank Dr. Francesco Ibba, Dr. Jeroen Sap, Sebastiano Ortalli, and Frank Nightingale for assistance in the revision of this chapter.

References

1 Doyle, A.G. and Jacobsen, E.N. (2007). Small-molecule H-bond donors in asymmetric catalysis. *Chem. Rev.* 107: 5713–5743.

2 Zhang, Z. and Schreiner, P.R. (2009). (Thio)urea organocatalysis – what can be learnt from anion recognition? *Chem. Soc. Rev.* 38: 1187–1198.

3 Brak, K. and Jacobsen, E.N. (2013). Asymmetric ion-pairing catalysis. *Angew. Chem. Int. Ed.* 52: 534–561.

4 Visco, M.D., Attard, J., Guan, Y., and Mattson, A.E. (2017). Anion-binding catalyst designs for enantioselective synthesis. *Tetrahedron Lett.* 58: 2623–2628.

5 Lund, H. (1958). Infrared spectral properties of hydrogen bonding systems containing anions. *Acta Chem. Scand.* 12: 298–302.

6 Bufalini, J. and Stern, K.H. (1961). Interactions in ionic solutions. The effect of electrolytes on the infrared spectra of some hydrogen-bonded compounds. *J. Am. Chem. Soc.* 83: 4362–4366.

7 Hyne, J.B. and Levy, R.M. (1962). The effect of added electrolyte on hydrogen-bonding equilibrium in dilute solutions of *t*-butyl alcohol in carbon tetrachloride. *Can. J. Chem.* 40: 692–700.

8 Allerhand, A. and von Rague Schleyer, P. (1963). Halide anions as proton acceptors in hydrogen bonding. *J. Am. Chem. Soc.* 85: 1233–1237.

9 Nishizawa, S., Bühlmann, P., Iwao, M., and Umezawa, Y. (1995). Anion recognition by urea and thiourea groups: remarkably simple neutral receptors for dihydrogenphosphate. *Tetrahedron Lett.* 36: 6483–6486.

10 Etter, M.C. and Panunto, T.W. (1988). 1,3-bis(*m*-nitrophenyl)urea: an exceptionally good complexing agent for proton acceptors. *J. Am. Chem. Soc.* 110: 5896–5897.

11 Kelly, T.R., Meghani, P., and Ekkundi, V.S. (1990). Diels–Alder reactions: rate acceleration promoted by a biphenylenediol. *Tetrahedron Lett.* 31: 3381–3384.

12 Etter, M.C., Urbanczyk-Lipkowska, Z., Zia-Ebrahimi, M., and Panunto, T.W. (1990). Hydrogen bond-directed cocrystallization and molecular recognition properties of diarylureas. *J. Am. Chem. Soc.* 112: 8415–8426.

13 Jose, D.A., Singh, A., Das, A., and Ganguly, B. (2007). A density functional study towards the preferential binding of anions to urea and thiourea. *Tetrahedron Lett.* 48: 3695–3698.

14 Curran, D.P. and Kuo, L.H. (1994). Altering the stereochemistry of allylation reactions of cyclic α-sulfinyl radicals with diarylureas. *J. Org. Chem.* 59: 3259–3261.

15 Curran, D.P. and Kuo, L.H. (1995). Acceleration of a dipolar Claisen rearrangement by hydrogen bonding to a soluble diaryl urea. *Tetrahedron Lett.* 36: 6647–6650.

16 Schreiner, P.R. and Wittkopp, A. (2002). H-bonding additives act like Lewis acid catalysts. *Org. Lett.* 4: 217–220.

17 Wittkopp, A. and Schreiner, P.R. (2003). Metal-free, noncovalent catalysis of Diels–Alder reactions by neutral hydrogen bond donors in organic solvents and in water. *Chem. Eur. J.* 9: 407–414.

18 Schreiner, P.R. (2003). Metal-free organocatalysis through explicit hydrogen bonding interactions. *Chem. Soc. Rev.* 32: 289–296.

19 Taylor, M.S. and Jacobsen, E.N. (2004). Highly enantioselective catalytic acyl-Pictet–Spengler reactions. *J. Am. Chem. Soc.* 126: 10558–10559.

20 Raheem, I.T., Thiara, P.S., Peterson, E.A., and Jacobsen, E.N. (2007). Enantioselective Pictet–Spengler-type cyclizations of hydroxylactams: H-bond donor catalysis by anion binding. *J. Am. Chem. Soc.* 129: 13404–13405.

21 Mergott, D.J., Zuend, S.J., and Jacobsen, E.N. (2008). Catalytic asymmetric total synthesis of (+)-yohimbine. *Org. Lett.* 10: 745–748.

22 Peterson, E.A. and Jacobsen, E.N. (2009). Enantioselective, thiourea-catalyzed intermolecular addition of indoles to cyclic *N*-acyl iminium ions. *Angew. Chem. Int. Ed.* 48: 6328–6331.

23 Raheem, I.T., Thiara, P.S., and Jacobsen, E.N. (2008). Regio- and enantioselective catalytic cyclization of pyrroles onto *N*-acyliminium ions. *Org. Lett.* 10: 1577–1580.

24 Metz, A.E., Ramalingam, K., and Kozlowski, M.C. (2015). Xanthene-4,5-diamine derivatives: a study of anion-binding catalysts. *Tetrahedron Lett.* 56: 5180–5184.

25 Knowles, R.R., Lin, S., and Jacobsen, E.N. (2010). Enantioselective thiourea-catalyzed cationic polycyclizations. *J. Am. Chem. Soc.* 132: 5030–5032.

26 Knowles, R.R. and Jacobsen, E.N. (2010). Attractive noncovalent interactions in asymmetric catalysis: links between enzymes and small molecule catalysts. *Proc. Natl. Acad. Sci. USA*: 20678–20685.

27 Dunitz, J.D. (1995). Win some, lose some: enthalpy–entropy compensation in weak intermolecular interactions. *Chem. Biol.* 2: 709–712.

28 Calderone, C.T. and Williams, D.H. (2001). An enthalpic component in cooperativity: the relationship between enthalpy, entropy, and noncovalent structure in weak associations. *J. Am. Chem. Soc.* 123: 6262–6267.

29 Leung, D.H., Bergman, R.G., and Raymond, K.N. (2008). Enthalpy–entropy compensation reveals solvent reorganization as a driving force for supramolecular encapsulation in water. *J. Am. Chem. Soc.* 130: 2798–2805.

30 Park, Y., Schindler, C.S., and Jacobsen, E.N. (2016). Enantioselective aza-Sakurai cyclizations: dual role of thiourea as H-bond donor and Lewis base. *J. Am. Chem. Soc.* 138: 14848–14851.

31 Bendelsmith, A.J., Kim, S.C., Wasa, M. *et al.* (2019). Enantioselective synthesis of α-allyl amino esters via hydrogen-bond-donor catalysis. *J. Am. Chem. Soc.* 141: 11414–11419.

32 Banik, S.M., Levina, A., Hyde, A.M., and Jacobsen, E.N. (2017). Lewis acid enhancement by hydrogen-bond donors for asymmetric catalysis. *Science* 358: 761–764.

33 Wendlandt, A.E., Vangal, P., and Jacobsen, E.N. (2018). Quaternary stereocenters via an enantioconvergent catalytic S_N1 reaction. *Nature* 556: 447–451.

34 Taylor, M.S., Tokunaga, N., and Jacobsen, E.N. (2005). Enantioselective thiourea-catalyzed acyl-Mannich reactions of isoquinolines. *Angew. Chem. Int. Ed.* 44: 6700–6704.

35 Bergonzini, G., Schindler, C.S., Wallentin, C.-J. *et al.* (2014). Photoredox activation and anion binding catalysis in the dual catalytic enantioselective synthesis of β-amino esters. *Chem. Sci.* 5: 112–116.

36 Wasa, M., Liu, R.Y., Roche, S.P., and Jacobsen, E.N. (2014). Asymmetric Mannich synthesis of α-amino esters by anion-binding catalysis. *J. Am. Chem. Soc.* 136: 12872–12875.

37 Samanta, S.S. and Roche, S.P. (2019). Synthesis and reactivity of α-haloglycine esters: hyperconjugation in action. *Eur. J. Org. Chem.* (39): 6597–6605.

38 Yamaoka, Y., Miyabe, H., and Takemoto, Y. (2007). Catalytic enantioselective Petasis-type reaction of quinolines catalyzed by a newly designed thiourea catalyst. *J. Am. Chem. Soc.* 129: 6686–6687.

39 Reisman, S.E., Doyle, A.G., and Jacobsen, E.N. (2008). Enantioselective thiourea-catalyzed additions to oxocarbenium ions. *J. Am. Chem. Soc.* 130: 7198–7199.

40 Ford, D.D., Lehnherr, D., Kennedy, C.R., and Jacobsen, E.N. (2016). On- and off-cycle catalyst cooperativity in anion-binding catalysis. *J. Am. Chem. Soc.* 138: 7860–7863.

41 Kennedy, C.R., Lehnherr, D., Rajapaksa, N.S. *et al.* (2016). Mechanism-guided development of a highly active bis-thiourea catalyst for anion-abstraction catalysis. *J. Am. Chem. Soc.* 138: 13525–13528.

42 Ford, D.D., Lehnherr, D., Kennedy, C.R., and Jacobsen, E.N. (2016). Anion-abstraction catalysis: the cooperative mechanism of α-chloroether activation by dual hydrogen-bond donors. *ACS Catalysis* 6: 4616–4620.

43 Lehnherr, D., Ford, D.D., Bendelsmith, A.J. *et al.* (2016). Conformational control of chiral amido-thiourea catalysts enables improved activity and enantioselectivity. *Org. Lett.* 18: 3214–3217.

44 Yeung, C.S., Ziegler, R.E., Porco, J.A. Jr., and Jacobsen, E.N. (2014). Thiourea-catalyzed enantioselective addition of indoles to pyrones: alkaloid cores with quaternary carbons. *J. Am. Chem. Soc.* 136: 13614–13617.

45 Sun, L., Wu, X., Xiong, D.-C., and Ye, X.-S. (2016). Stereoselective Koenigs–Knorr glycosylation catalyzed by urea. *Angew. Chem. Int. Ed.* 55: 8041–8044.

46 Park, Y., Harper, K.C., Kuhl, N. *et al.* (2017). Macrocyclic bis-thioureas catalyze stereospecific glycosylation reactions. *Science* 355: 162–166.

47 Kwan, E.E., Park, Y., Besser, H.A. *et al.* (2017). Sensitive and accurate ^{13}C kinetic isotope effect measurements enabled by polarization transfer. *J. Am. Chem. Soc.* 139: 43–46.

48 Levi, S.M., Li, Q., Rötheli, A.R., and Jacobsen, E.N. (2019). Catalytic activation of glycosyl phosphates for stereoselective coupling reactions. *Proc. Natl. Acad. Sci. USA* 116: 35–39.

49 Mayfield, A.B., Metternich, J.B., Trotta, A.H., and Jacobsen, E.N. (2020). Stereospecific furanosylations catalyzed by bis-thiourea hydrogen-bond donors. *J. Am. Chem. Soc.* 142: 4061–4069.

50 Li, Q., Levi, S.M., and Jacobsen, E.N. (2020). Highly selective β-mannosylations and β-rhamnosylations catalyzed by bis-thiourea. *J. Am. Chem. Soc.* 142: 11865–11872.

51 Brown, A.R., Kuo, W.-H., and Jacobsen, E.N. (2010). Enantioselective catalytic α-alkylation of aldehydes via an S_N1 pathway. *J. Am. Chem. Soc.* 132: 9286–9288.

52 Kutateladze, D.A., Strassfeld, D.A., and Jacobsen, E.N. (2020). Enantioselective tail-to-head cyclizations catalyzed by dual-hydrogen-bond donors. *J. Am. Chem. Soc.* 142: 6951–6956.

53 Boiocchi, M., Del Boca, L., Gómez, D.E. *et al.* (2004). Nature of urea-fluoride interaction: incipient and definitive proton transfer. *J. Am. Chem. Soc.* 126: 16507–16514.

54 Engle, K.M., Pfeifer, L., Pidgeon, G.W. *et al.* (2015). Coordination diversity in hydrogen-bonded homoleptic fluoride-alcohol complexes modulates reactivity. *Chem. Sci.* 6: 5293–5302.

55 Pfeifer, L., Engle, K.M., Pidgeon, G.W. *et al.* (2016). Hydrogen-bonded homoleptic fluoride-diarylurea complexes: structure, reactivity, and coordinating power. *J. Am. Chem. Soc.* 138: 13314–13325.

56 Pupo, G., Ibba, F., Ascough, D.M.H. *et al.* (2018). Asymmetric nucleophilic fluorination under hydrogen bonding phase-transfer catalysis. *Science* 360: 638–642.

57 Ibba, F., Pupo, G., Thompson, A.L. *et al.* (2020). Impact of multiple hydrogen bonds with fluoride on catalysis: insight from NMR spectroscopy. *J. Am. Chem. Soc.* 142: 19731–19744.

58 Pupo, G., Vicini, A.C., Ascough, D.M.H. *et al.* (2019). Hydrogen bonding phase-transfer catalysis with potassium fluoride: enantioselective synthesis of β-fluoroamines. *J. Am. Chem. Soc.* 141: 2878–2883.

59 Roagna, G., Ascough, D.M.H., Ibba, F. *et al.* (2020). Hydrogen bonding phase-transfer catalysis with ionic reactants: enantioselective synthesis of γ-fluoroamines. *J. Am. Chem. Soc.* 142: 14045–14051.

60 Brindle, C.S., Yeung, C.S., and Jacobsen, E.N. (2013). Chiral β-iodoamines by urea-catalysed iodocyclization of trichloroacetimidates. *Chem. Sci.* 4: 2100–2104.

61 Craig, R., Litvajova, M., Cronin, S.A., and Connon, S.J. (2018). Enantioselective acyl-transfer catalysis by fluoride ions. *Chem. Commun.* 54: 10108–10111.

62 Trujillo, C., Litvajova, M., Cronin, S.A. *et al.* (2019). The Steglich rearrangement of 2-oxindole derivatives promoted by anion-based nucleophilic catalysis. *ChemCatChem* 11: 3776–3780.

63 Strassfeld, D.A., Wickens, Z.K., Picazo, E., and Jacobsen, E.N. (2020). Highly enantioselective, hydrogen-bond-donor catalyzed additions to oxetanes. *J. Am. Chem. Soc.* 142: 9175–9180.

4

Chiral Ureas, Thioureas, and Squaramides in Anion-Binding Catalysis with Co-catalytic Brønsted/Lewis Acids

Adam Trotta and Eric N. Jacobsen

Harvard University, Department of Chemistry and Chemical Biology, 12 Oxford Street, Cambridge, MA 02138, USA

4.1 Introduction

Chiral Brønsted acid and Lewis acid catalysis provides powerful strategies for promoting an extraordinary range of organic transformations proceeding through highly electrophilic, formally cationic intermediates [1–6]. In chiral Brønsted acid catalysis, stereocontrol is achieved via ion pairing between the protonated intermediate and the chiral conjugate base [6–10] (Scheme 4.1a). Traditional chiral Lewis-acid catalysis generally involves activation of a neutral substrate via complexation to a highly electrophilic, chiral-ligand-bound metal center [11–14] (Scheme 4.1b). A complementary strategy, which was only recently been implemented with success, involves cooperative action of an achiral Brønsted or Lewis acid with a chiral co-catalyst (Scheme 4.1c).

The chiral co-catalysts are typically H-bond donors capable of anion-binding the conjugate base or counterion of the achiral acid to form chiral ion pair intermediates. As outlined in this chapter, only a few examples of this strategy have been identified to date, but the strategy holds potential for broad application to the synthesis of important chiral compounds.

4.2 Carboxylic Acid Co-catalysts

The first reported example of a chiral H-bond-donor catalyst acting in concert with either a Brønsted or a Lewis acid co-catalyst was provided in the context of enantioselective Pictet–Spengler reactions (Scheme 4.2a) [15]. The combination of chiral thiourea catalyst **2** with benzoic acid was shown to catalyze the conversion of tryptamine derivatives **1** and aldehydes to unprotected tetrahydro-β-carbolines **3** in high yield and enantioselectivity. Unlike the acyl-Pictet–Spengler reaction that had been reported previously [16], the protio-Pictet–Spengler reaction displayed broad scope with respect to the tryptamine derivatives, enabling the synthesis

Anion-Binding Catalysis, First Edition. Edited by Olga García Mancheño.

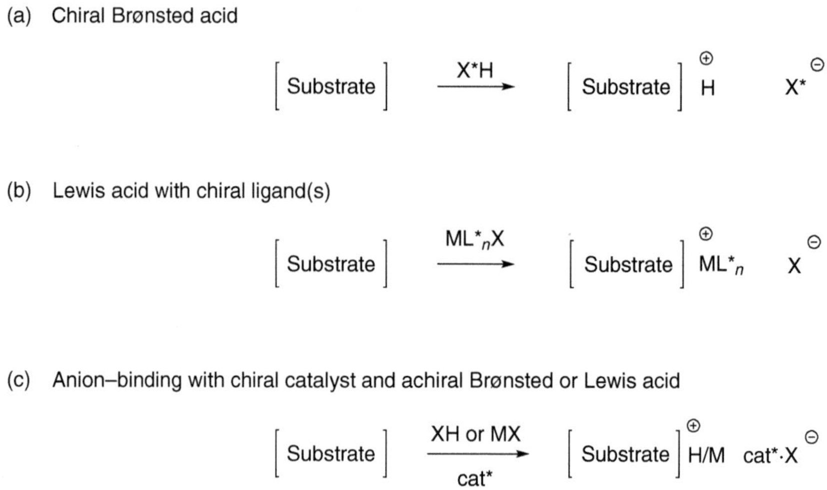

Scheme 4.1 Three strategies for generating carbocationic intermediates in chiral ion pairs: (a) activation by a chiral Brønsted acid (X*H), (b) activation by a Lewis acid bearing a chiral ligand (L*), and (c) activation using a chiral anion-binding catalyst. Source: Adapted from Refs. [6, 10–12].

of methoxy- and hydroxy-substituted tetrahydro-β-carbolines that are present in important natural products (e.g. reserpine, tubulosine, and the eudimistin family). The Brønsted acid co-catalyst was found to accelerate the reaction rate and was essential for benzaldehyde-derived substrates (Scheme 4.2a, R^2 = aryl), which were completely unreactive in the absence of benzoic acid. Aliphatic aldehydes (R^2 = alkyl) underwent reaction in the absence of benzoic acid co-catalyst, albeit very sluggishly.

In the original report on the protio-Pictet–Spengler, both the thiourea catalyst **2** and benzoic acid co-catalyst displayed poor catalytic turnover, requiring 20 mol% of each for a satisfactory reaction rate. Subsequent kinetic studies revealed that product inhibition is responsible for catalyst deactivation over the course of the reaction [17]. A modified protocol was devised involving *in situ* carbamate protection of the tetrahydro-β-carboline product, improving catalyst efficiency and enabling 10-fold lower loadings of both the thiourea and benzoic acid catalysts (2 mol%). Further mechanistic studies were conducted to elucidate the reaction mechanism with kinetic data and kinetic isotope effect studies providing evidence that rearomatization is the rate-limiting and enantioselectivity-determining step (Scheme 4.2b, III). A series of modified catalysts and co-catalysts were employed for structure–activity relationship studies, implicating the conjugate base of the carboxylic acid co-catalyst as the Brønsted base responsible for deprotonation during the critical rearomatization step **III**. Based on these studies, the catalytic cycle depicted in Scheme 4.2b was proposed, involving initial aldehyde condensation to yield imine **5**, followed by protonation by the thiourea–benzoic acid complex **4** to form the protio-iminium intermediate with a thiourea-stabilized benzoate counterion (**6**). Subsequent reversible alkylation generates cationic tricyclic intermediate **7**,

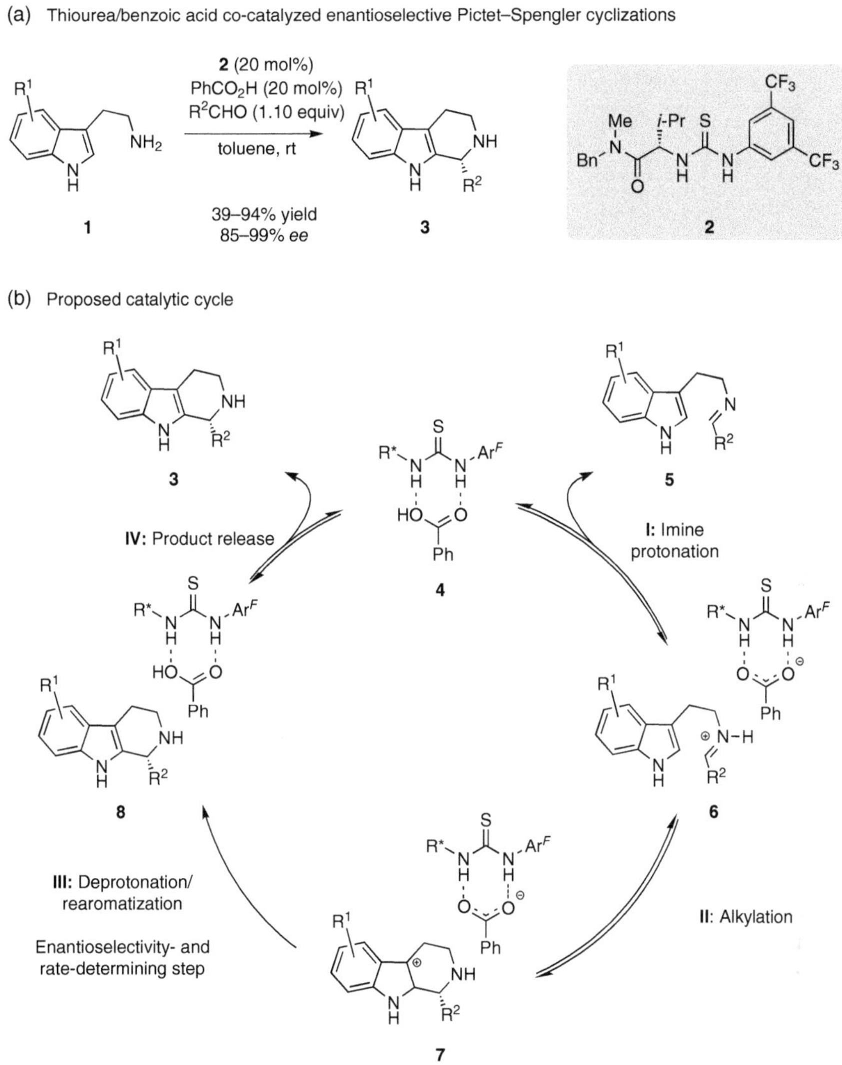

Scheme 4.2 (a) Enantioselective Pictet–Spengler reaction co-catalyzed by a chiral thiourea and benzoic acid and (b) proposed catalytic cycle.

which undergoes rate- and enantioselectivity-determining deprotonation by the catalyst-bound benzoate to generate the desired product **3**.

An enantioselective iso-Pictet–Spengler reaction was developed based on the optimized conditions for the protio-Pictet–Spengler reaction (Scheme 4.3) [18]. It was found that the concentration of the thiourea catalyst **10** must be equal to or greater than the concentration of the achiral carboxylic acid to prevent diminished enantioselectivities because of the acid-catalyzed racemic background reaction. Although only moderate enantioselectivities were achieved with certain substrates in the cyclization reaction (79–95% *ee*), a convenient protocol was devised for

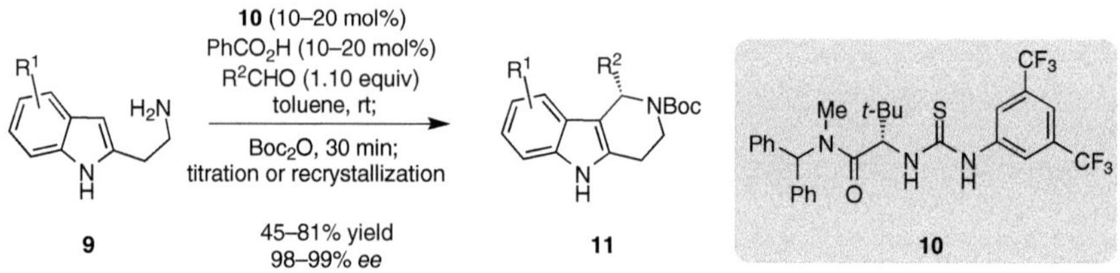

Scheme 4.3 Enantioselective iso-Pictet–Spengler reaction catalyzed by chiral thiourea **10** and co-catalytic benzoic acid. Source: Lee *et al.* [18].

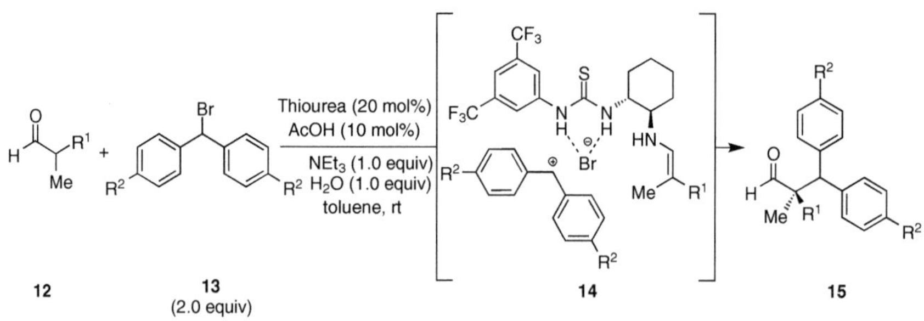

Scheme 4.4 Enantioselective α-alkylation of aldehydes catalyzed by a chiral thiourea and co-catalytic acetic acid. Source: Based on Brown *et al.* [19].

upgrading the *ee* of all of the tetrahydro-γ-carboline products. The crude products of the catalytic iso-Pictet–Spengler reaction were treated with Boc_2O, and the resulting carbamate-protected compounds could be readily upgraded to >99% *ee* by direct crystallization or trituration, providing a scalable and highly practical procedure for optically pure tetrahydro-γ-carboline derivatives **11**.

Brønsted acid/chiral H-bond donor co-catalysis was applied in a fundamentally different context in enantioselective α-alkylations of aldehydes (Scheme 4.4) [19]. It was found that primary amino thiourea derivatives promoted substitution of benzylic bromides with high enantioselectivity through the proposed intermediacy of a covalent catalyst–enamine derivative (e.g. **14**). Rate acceleration was observed upon addition of H_2O and co-catalytic AcOH, presumably through enablement of imine/enamine formation and imine hydrolysis [20]. Experiments were conducted to differentiate between an S_N1-like pathway proceeding through a discrete ion pair intermediate and an S_N2-like substitution mechanism. Kinetic isotope effect studies indicated a change of hybridization of the electrophilic carbon in the transition state, consistent with an S_N1-like pathway proceeding through a cationic sp^2-hybridized intermediate – further corroborated by a Hammett study consistent with the buildup of a high degree of positive charge in the rate-determining transition state. Thus, in this case, co-catalytic Brønsted acid does not appear to play a role in the enantiodetermining step but rather plays an indirect role in H-bond donor catalysis by accelerating imine/enamine formation and imine hydrolysis.

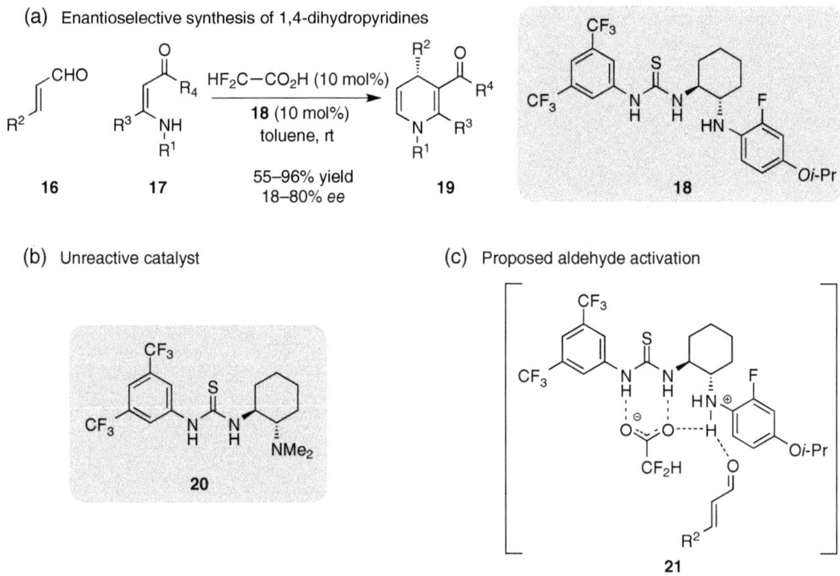

Scheme 4.5 (a) Enantioselective synthesis of 1,4-dihydropyridines with chiral thiourea catalyst **18** and co-catalytic difluoroacetic acid, (b) unreactive tertiary amine catalyst **20**, and (c) proposed mode of aldehyde activation.

Takemoto and coworkers applied the chiral 1,2-diaminocyclohexyl-derived thiourea **18** and co-catalytic Brønsted acid in an asymmetric synthesis of functionalized 1,4-dihydropyridines **19** from β-enamino esters **17** and α,β-unsaturated aldehydes **16** (Scheme 4.5a) [21, 22]. Difluoroacetic acid (DFA) alone was found to catalyze the desired transformation in a racemic manifold, but reactions carried out in the presence of tertiary-amino thiourea "Takemoto catalyst" **20** and DFA returned the starting materials. It was concluded that the conjugate acid of catalyst **20** was insufficiently acidic to activate the aldehyde substrate. The aniline-bearing catalyst **18** was prepared and shown to promote the enantioselective synthesis of the desired 1,4-dihydropyridine products. The Brønsted co-catalyst was shown to be essential for reactivity, as conducting the reaction with only the thiourea catalyst **18** returned the starting materials. The use of acetic acid in place of DFA resulted in improved enantioselectivity in the model reaction (39% to 78% *ee*) but with significantly decreased yield (72% to 11% yield). Taken together, these observations concluded to be consistent with aldehyde activation through thiourea–carboxylic acid complex **21** (Scheme 4.5c).

The 1,2-diaminocyclohexyl-derived amino thiourea **24** was utilized in conjunction with a second achiral thiourea **23** in enantioselective intramolecular [5+2] pyrylium-alkene cycloadditions (Scheme 4.6a) [23]. Acetic acid induced yield enhancements without altering the product *ee* when utilized as a co-catalyst, consistent with acceleration of imine/enamine formation and imine hydrolysis in an analogous manner to the α-alkylation of aldehydes [19]. Further evidence for enamine catalysis was found when the chiral thiourea catalyst **24** was replaced by

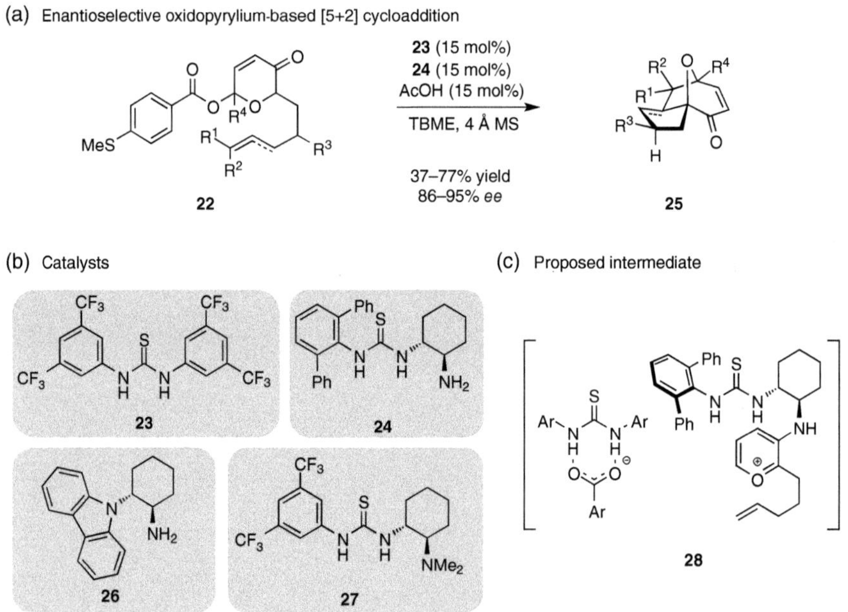

Scheme 4.6 (a) Enantioselective intramolecular [5+2] cycloaddition catalyzed by the combination of chiral thiourea **24** and achiral thiourea **24** and co-catalytic acetic acid, (b) H-bond-donor and primary amine catalysts, and (c) proposed pyrylium complex. Source: (a) Based on Burns *et al.* [23].

primary aminocarbazole **26**, producing the desired product in good yield and *ee* (58% yield, −85% *ee*). Thus, while the thiourea component of the optimal catalyst does influence the reaction enantioselectivity, it is not necessary for obtaining reactivity or high *ee*. Consistent with the hypothesis that an H-bond-donor catalyst is required for ionization to the pyrylium ion, primary aminocarbazole is unreactive in the absence of achiral thiourea **23**. Tertiary aminothiourea **27** is also unreactive with or without **23**, highlighting the necessity of a primary amine. Taken together, these observations are consistent with Brønsted-acid-accelerated enamine catalysis, in which condensation between the catalyst amine and the ketone yields a dienamine. Subsequent H-bond-donor-mediated benzoate abstraction would generate catalyst–pyrylium intermediate **28**, which could then undergo the desired cycloaddition (Scheme 4.6c).

The thiourea-catalyzed asymmetric [5+2] pyrylium-alkene cycloaddition was extended to intermolecular variants through subtle modifications to the reaction partners (Scheme 4.7) [24]. In particular, the electronic properties of the 2π component **30** were found to have a strong effect on reactivity. Electron-deficient olefins were completely inert, while electron-rich alkenes underwent cycloaddition enantioselectively, an observation consistent with the intermediacy of a cationic, electron-poor aminopyrylium (see Scheme 4.6c, **28**) rather than a zwitterionic oxidopyrylium intermediate. It was found that substitution at C6 of the pyranone led to significant improvements in the enantioselectivity. The reaction was further

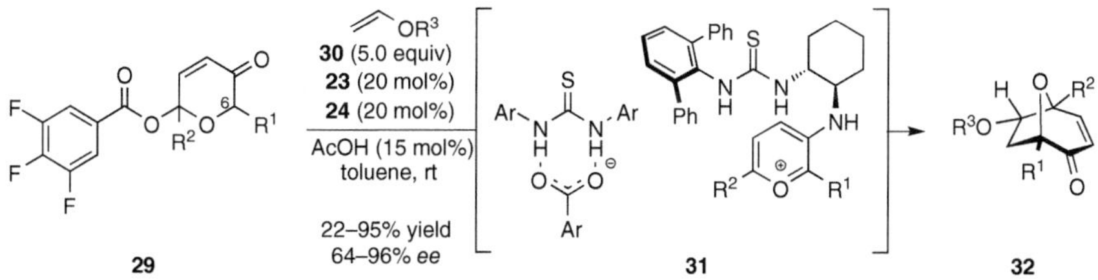

Scheme 4.7 Enantioselective intramolecular [5+2] cycloaddition catalyzed by the combination of a chiral thiourea and an achiral thiourea H-bond donor and co-catalytic acetic acid. Source: Based on Witten *et al.* [24].

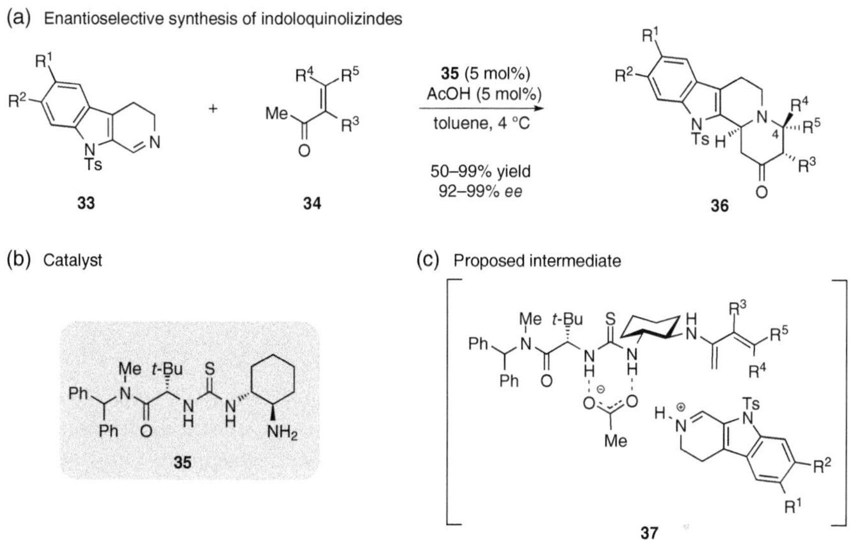

Scheme 4.8 (a) Enantioselective formal aza-Diels–Alder reaction catalyzed by a chiral thiourea and co-catalytic acetic acid, (b) optimal thiourea catalyst, and (c) proposed reactive intermediate. Source: (a) Based on Lalonde *et al.* [25].

improved through electronic tuning of the leaving group on the pyranone with the electron-deficient 3,4,5-trifluorobenzoate derivative **29** proving optimal.

A significant weak-Brønsted-acid co-catalyst effect was observed in enantioselective formal aza-Diels–Alder reactions between enones **34** and cyclic imines **33** (Scheme 4.8a) [25]. This effect was ascribed to acceleration of the condensation and/or hydrolysis steps integral to the proposed enamine catalysis cycle – analogous to other reactions thought to proceed through catalyst–enamine intermediates [19, 23]. The identity of the carboxylic acid had a measurable influence on the diastereoselectivity and enantioselectivity of the transformation, consistent with its presence in the diastereo- and enantiodetermining transition state. A mechanism was proposed involving enantiodeterming cyclization through the enamine–iminium ion pair complex **37** (Scheme 4.8c).

The formal aza-Diels–Alder protocol was subsequently utilized as a key step in the synthesis of (+)-reserpine (Scheme 4.9) [26]. This approach to the C ring of

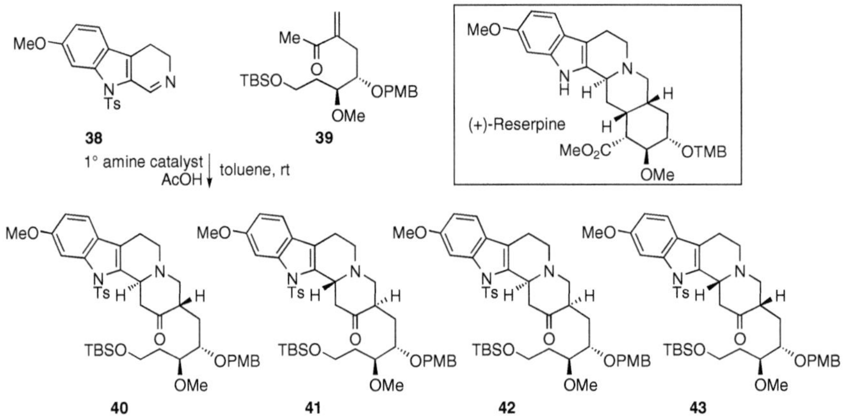

Entry	Amine (mol%)	AcOH (mol%)	Time (d)	Conv (%)	dr (**40**:**41**:**42**:**43**)
1	*n*-Hexylamine (100)	100	9	90	1.0 : 1.0 : 0 : 0
2	**35** (20)	20	6	>99	11.5 : 1.0 : 1.8 : 0
3	*ent*-**35** (20)	20	6	>99	1.0 : 11.9 : 0 : 2.1

Scheme 4.9 Application of the enantioselective formal aza-Diels–Alder to the synthesis of (+)-reserpine (OTMB = 3,4,5-trimethoxybenzoyl). Source: Based on Rajapaksa *et al.* [26].

reserpine is distinct from the previous successful syntheses, which have relied on a late-stage dehydrative C-ring synthesis from a 2,3-*seco*-derivative. The key coupling event proceeded with achiral *n*-hexylamine as the promoter to yield equal amounts of products **40** and **41**, revealing a lack of intrinsic substrate-induced diastereocontrol (entry 1). In contrast, high levels of chiral-catalyst-controlled diastereoselectivity were observed in the presence of aminothiourea **35** (entries 2 and 3), providing the desired diastereomer **40** in 76% isolated yield.

A collaborative effort between the Jacobsen lab and the Porco lab culminated in the discovery of a thiourea-catalyzed enantioselective addition of indoles (**44**) to pyrones (**45**), leveraging benzoic acid as a Brønsted acid co-catalyst (Scheme 4.10a) [27]. It was found that varying the identity or the amount of the Brønsted acid co-catalyst had little effect on the reaction outcome, consistent with the absence of the acid in the enantiodetermining step. In contrast, variation of the leaving group on the pyrone precursor had a notable effect on the enantioselectivity, with bromide proving optimal. These observations are consistent with an enantiodetermining step involving interactions between the catalyst and the pyrone leaving group, leading to the proposed intermediate **48** (Scheme 4.10c). The authors propose that the co-catalytic benzoic acid promotes desilylation of substrate **45** prior to abstraction of the leaving group by the H-bond-donor catalyst.

4.3 Sulfonic Acid Co-catalysts

The first report of co-catalysis between sulfonic acids and chiral H-bond donor catalysts was in the context of an asymmetric Povarov reaction catalyzed by chiral

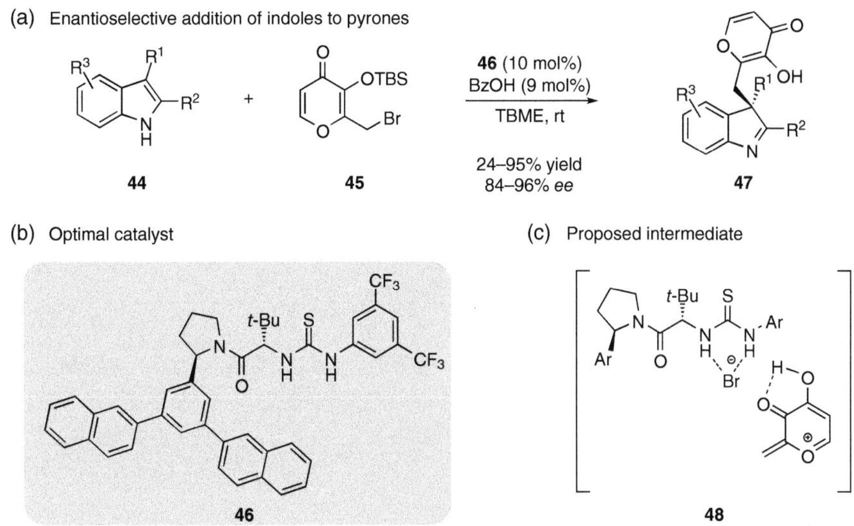

Scheme 4.10 (a) Enantioselective addition of indoles to pyrones catalyzed by a chiral thiourea and co-catalytic benzoic acid, (b) optimal thiourea catalyst, and (c) proposed reactive intermediate. Source: (a) Based on Yeung *et al.* [27].

sulfinamido urea derivative **51** and co-catalyzed by *o*-nitrobenzenesulfonic acid (2-NBSA) (Scheme 4.11) [28].

This formal [4+2] cycloaddition between *N*-aryl imines **49** and electron-rich olefins **50/53** affords tetrahydroquinoline derivatives **52/54** in high yield and *ee*. Best results were obtained when the ratio of urea to sulfonic acid was 2 : 1, suppressing the racemic pathway catalyzed by 2-NBSA alone. It was found that several sulfonic acid co-catalysts behaved similarly, and triflic acid (TfOH) was selected as a model for mechanistic analysis. Kinetic analysis of the reaction in the absence of H-bond donor catalyst was consistent with quantitative protonation of the imine and generation of an imine•TfOH resting state complex. When this complex was mixed with the bifunctional sulfinamidourea catalyst **51**, the ^{1}H NMR shift of the formyl proton exhibited a significant down-field shift, consistent with decreased charge separation [29]. Computational analysis using density functional theory suggested the basis for this effect, with a network of attractive interactions provided by the catalyst, including hydrogen bonding between the urea and the triflate and between the sulfinamide and the formyl proton (see Scheme 4.11c) [30]. These stabilizing interactions are consistent with the observed decreased rate of the urea-catalyzed reaction relative to the reaction catalyzed by TfOH alone. The enantioselective pathway is thus an example of "negative catalysis," wherein the reaction is channeled through a slower yet dominant co-catalytic pathway.

A modified version of the asymmetric Povarov reaction was applied in the preparative scale synthesis of **57** (Scheme 4.12) [31]. *p*-Toluenesolfonic acid (PTSA) was used as a practical alternative to PTSA and catalyst loadings were reduced with only slight reduction in product *ee* [28]. In this manner, the reaction was conducted

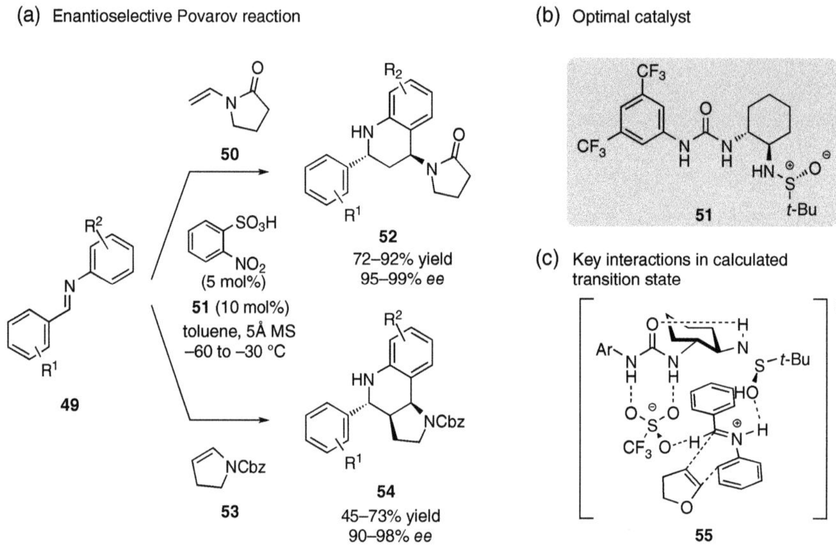

Scheme 4.11 (a) Enantioselective Povarov reaction catalyzed by a chiral urea and co-catalytic 2-nitrobenzenesulfonic acid, (b) optimal sulfinamide-bearing urea catalyst, and (c) key interactions in the computational model of the enantiodetermining transition state (B3LYP/6-31G(d)). Source: Based on Xu *et al.* [28].

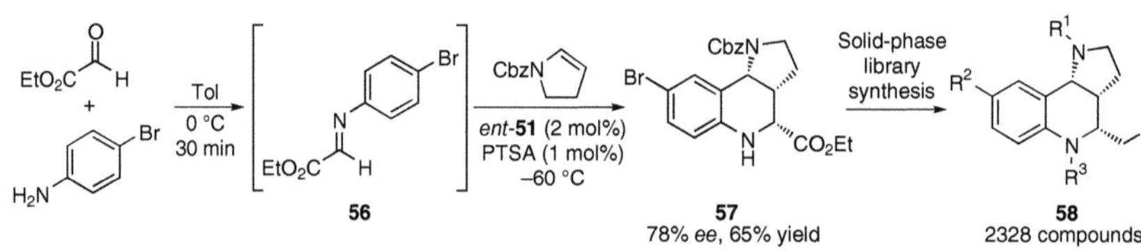

Scheme 4.12 Application of the enantioselective Povarov reaction to the synthesis of a library of 2,3,4-trisubstituted tetrahydroquinolines. Source: Based on Gerard *et al.* [31].

successfully with up to 66 mmol (17 g) of the imine substrate. The resultant product was elaborated via solid-phase synthesis to generate a strategically designed 2328-membered library of 2,3,4-trisubstituted tetrahydroquinolines.

A chiral thiourea-*p*-nitrobenzenesulfonic acid (4-NBSA) dual catalyst system was applied to the enantioselective ring opening of episulfonium ions with indole derivatives **66** (Scheme 4.13) [32]. The identity of the Brønsted acid co-catalyst proved to be highly consequential. Mineral acids with a nucleophilic counterion, such as HCl, produced only trace amounts of the desired indole-addition product, as the corresponding counterion-addition product predominated. In contrast, sulfonic acids afforded the desired indole adduct in good yield, and with enantioselectivities dependent on the structure of the acid. Reaction outcomes were insensitive to the identity of the acetimidate leaving group, consistent with the absence of that component and indicating instead that the sulfonate is associated with the H-bond donor in the enantiodetermining step. Reaction progress kinetic analysis [33] revealed a zeroth-order rate dependence on substrate **59** and a first-order rate dependence on

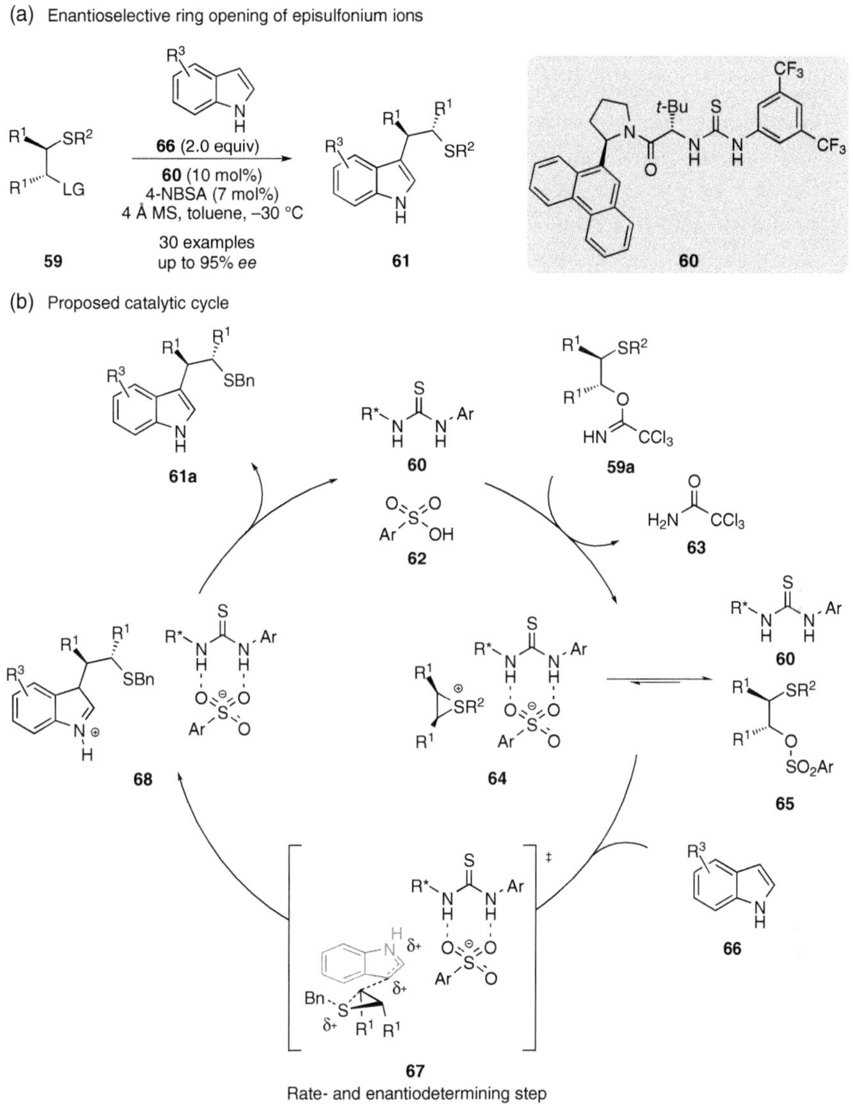

Scheme 4.13 (a) Enantioselective ring opening of episulfonium ions catalyzed by chiral thiourea **60** and co-catalytic 4-nitrobenzenesulfonic acid and (b) proposed catalytic cycle. Source: Lin and Jacobsen [32].

4-NBSA, consistent with a resting-state episulfonium sulfonate catalyst complex and supported by *in situ* IR analyses. A first-order dependence on indole was also observed, consistent with the presence of indole in the rate-determining transition structure (Scheme 4.13b, **67**). Moreover, studies of linear free-energy relationships and kinetic isotope effects were consistent with a rate-determining and enantiodetermining indole addition. Taken together, these observations led to the proposed catalytic cycle depicted in Scheme 4.13b.

4.4 Mineral Acid Co-catalysts

Mineral acids have been employed effectively as catalytic activators of electrophiles in H-bond-donor-catalyzed enantioselective anion-binding pathways. Co-catalysis with HCl was utilized in an enantioselective, thiourea-catalyzed intermolecular addition of indoles **66** to cyclic *N*-acyl iminium ions (Scheme 4.14a) [34].

Hydrochloric acid was generated *in situ* from the combination of catalytic H_2O and stoichiometric TMSCl, forming the reactive chlorolactam precursor **72** to the key *N*-acyl iminium ion intermediate from the acetoxylactam substrate **69**. Regeneration of HCl was postulated to occur through reaction with the acetic acid eliminated from substrate **69** (Scheme 4.14b). The *in-situ*-generated, co-catalytic Brønsted acid had a pronounced beneficial effect on reactivity, significantly improving the isolated yield of the desired product **71**. Strictly anhydrous conditions led to very poor reactivity, while elevated levels of water (>20 mol%) led to diminished enantioselectivity,

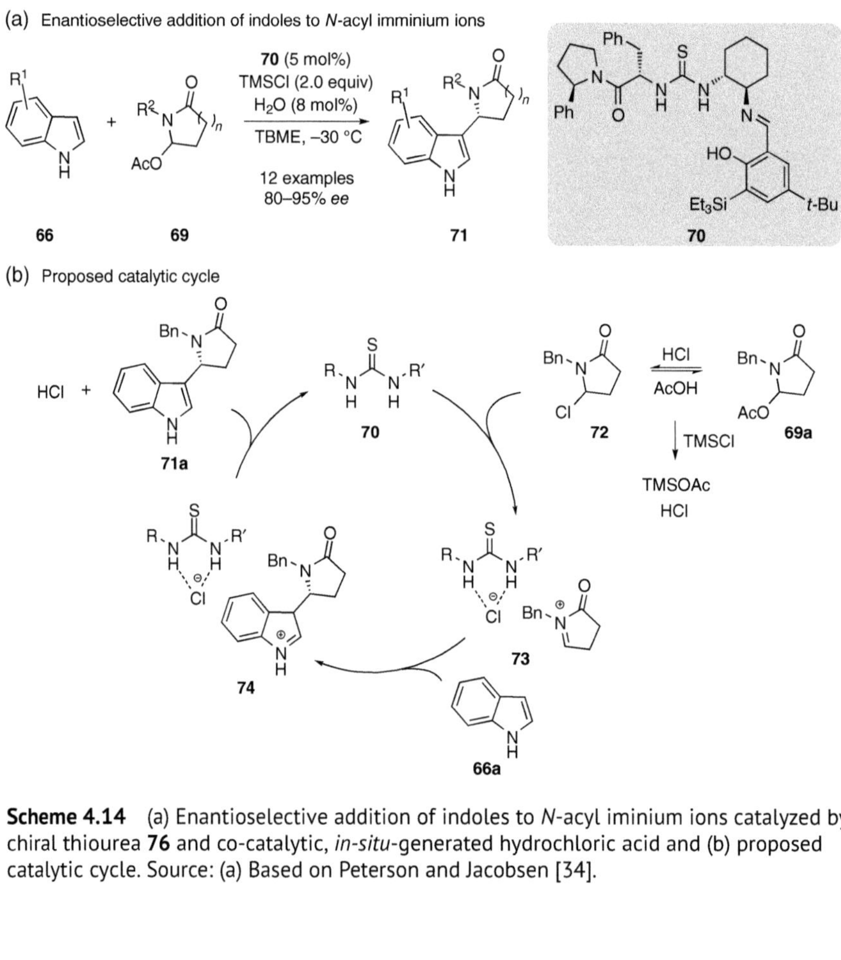

Scheme 4.14 (a) Enantioselective addition of indoles to *N*-acyl iminium ions catalyzed by chiral thiourea **76** and co-catalytic, *in-situ*-generated hydrochloric acid and (b) proposed catalytic cycle. Source: (a) Based on Peterson and Jacobsen [34].

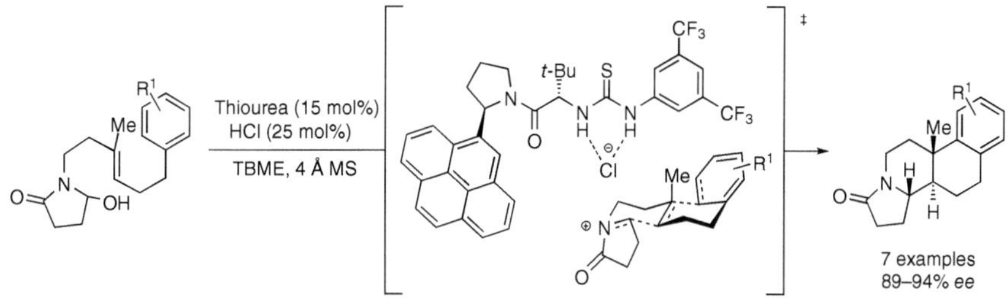

75 76 77

Scheme 4.15 Enantioselective polycyclization of hydroxylactams catalyzed by a chiral thiourea and co-catalytic HCl. Source: Based on Knowles *et al.* [35].

implying that a precise balance between the HCl and the catalyst is required for optimal results.

Co-catalytic HCl was also engaged in the promotion of *N*-acyl iminium ion pathways in thiourea-catalyzed cationic polycyclization reactions (Scheme 4.15) [35]. In the presence of thiourea catalysts and HCl, hydroxylactam **75** underwent a cascade cyclization to tetracycle **77** through the proposed intermediacy of the corresponding chlorolactam. The expanse of the arene on the arylpyrrolidino component of the catalyst was found to correlate directly with the enantioselectivity of the reaction. Linear correlations were observed between enantioselectivity and both the polarizability and quadrupole moment of the arene, consistent with selective stabilization of the major pathway through stabilizing cation–π interaction(s) [36–38].

Co-catalysis by HCl was shown to play an essential role in enantioselective selenocyclization reactions catalyzed by a chiral squaramide catalyst (Scheme 4.16a) [39]. In this system, cooperative Brønsted acid and Lewis base activation of the electrophilic selenium reagent *N*-phenylselenyl succinimide (NPSS, **79**) was proposed to generate a reactive ion pair **83** (Scheme 4.16b) in a step that may or may not be promoted by the squaramide catalyst [40]. Group transfer to olefin **74** generates rapidly equilibrating stereoisomeric seleniranium chloride ion pairs. The subsequent cyclization of **84** occurs enantioselectively because of the association with chiral squaramide.

A buffered mineral acid – in the form of amine•HCl **86** – was used by Seidel and coworkers in a catalytic enantioselective oxa-Pictet–Spengler reaction catalyzed by bis-thiourea **87** (Scheme 4.17a) [41]. The identity of the ammonium salt had a significant impact on the outcome of the reaction, while the HCl salts of a variety of secondary amines failed to promote a reaction when combined with bis-thiourea catalyst **87**, (*S*)-indoline-2-carboxylic acid methyl ester (**86**) facilitated rapid formation of the desired product in excellent yield and modest *ee*. The chloride, bromide, iodide, and triflate salts of catalyst **86** all provided the expected product in good yield but with the chloride salt furnishing the highest *ee*. These results are consistent with the ion-pairing mechanism depicted in Scheme 4.17b analogous to the thiourea/benzoic acid co-catalyzed Pictet–Spengler reaction discussed in Section 4.2.

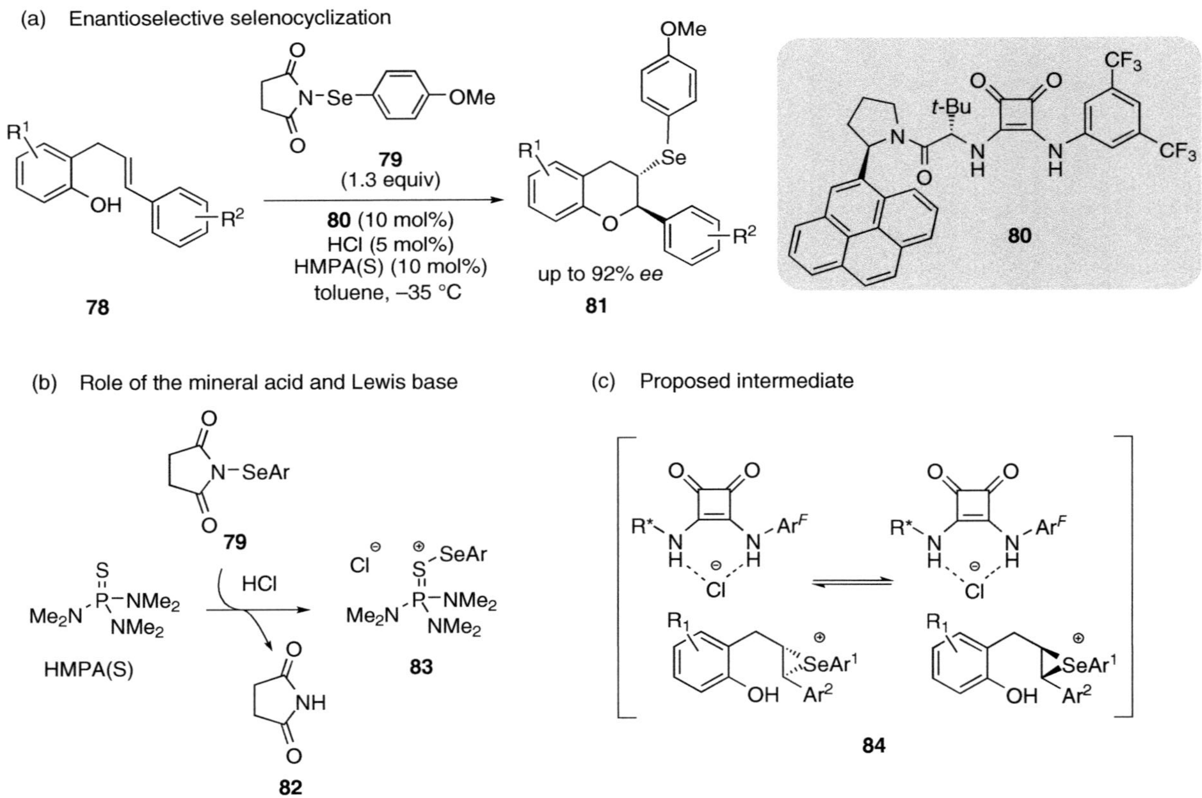

Scheme 4.16 (a) Enantioselective selenocyclization catalyzed by a chiral squaramide and co-catalytic hydrochloric acid, (b) proposed activation of *N*-arylselenyl succinimide with hydrochloric acid and tris-(dimethylamino)phosphorous sulfide, and (c) proposed seleniranium intermediate. Source: (a) Based on Zhang *et al.* [39].

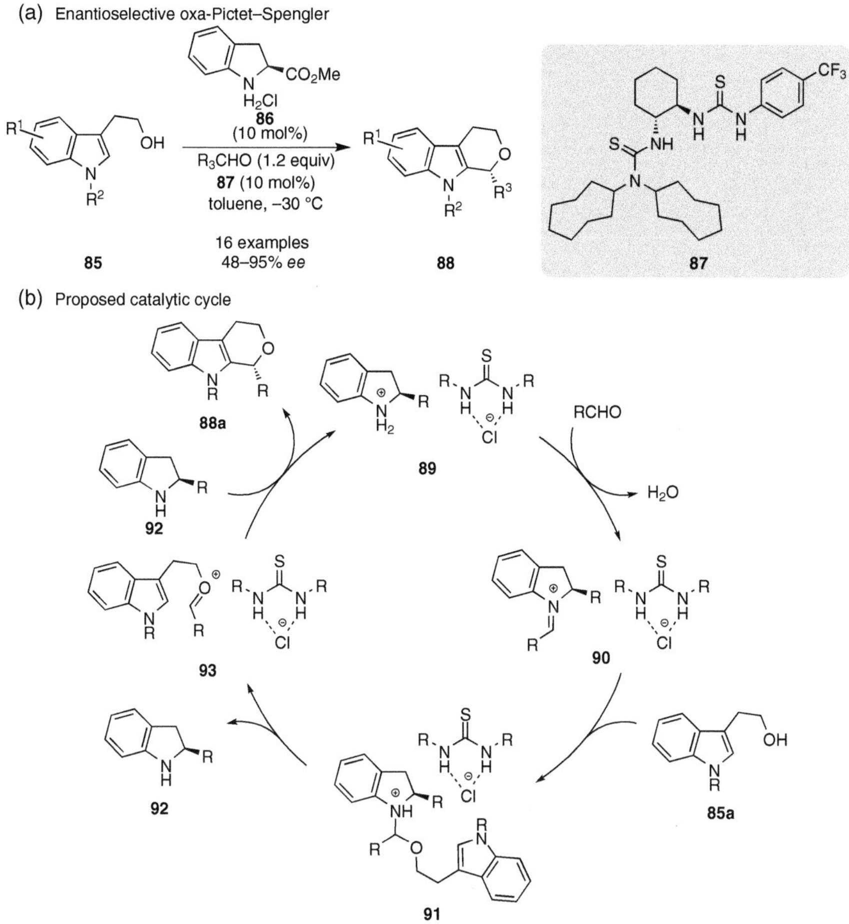

Scheme 4.17 (a) Enantioselective oxa-Pictet–Spengler catalyzed by chiral bis-thiourea **87** and co-catalytic amine•HCl salt **86**, and (b) proposed catalytic cycle. Source: (a) Based on Zhao *et al.* [41].

4.5 Lewis Acid Co-catalysts

Lewis acids have also proven to be compatible co-catalysts with anion-binding H-bond-donor catalysts. The cooperative action of Lewis acid and chiral H-bond-donor catalysts was applied to the asymmetric Nazarov cyclization of unactivated dienones (Scheme 4.18a) [42].

It was proposed that TMSOTf is activated by the squaramide catalyst **95** via anion binding (**97**) in a manner analogous to that documented in prior studies using chiral squaramides in conjunction with stoichiometric TMSOTf in anion-binding pathways [43, 44]. Activation of the dienone substrate **94** as catalyst-complexed ion pair **98** serves to promote 4π-electrocyclization, with subsequent proton transfer to form the observed product **96**. Kinetic isotope effect studies revealed that the

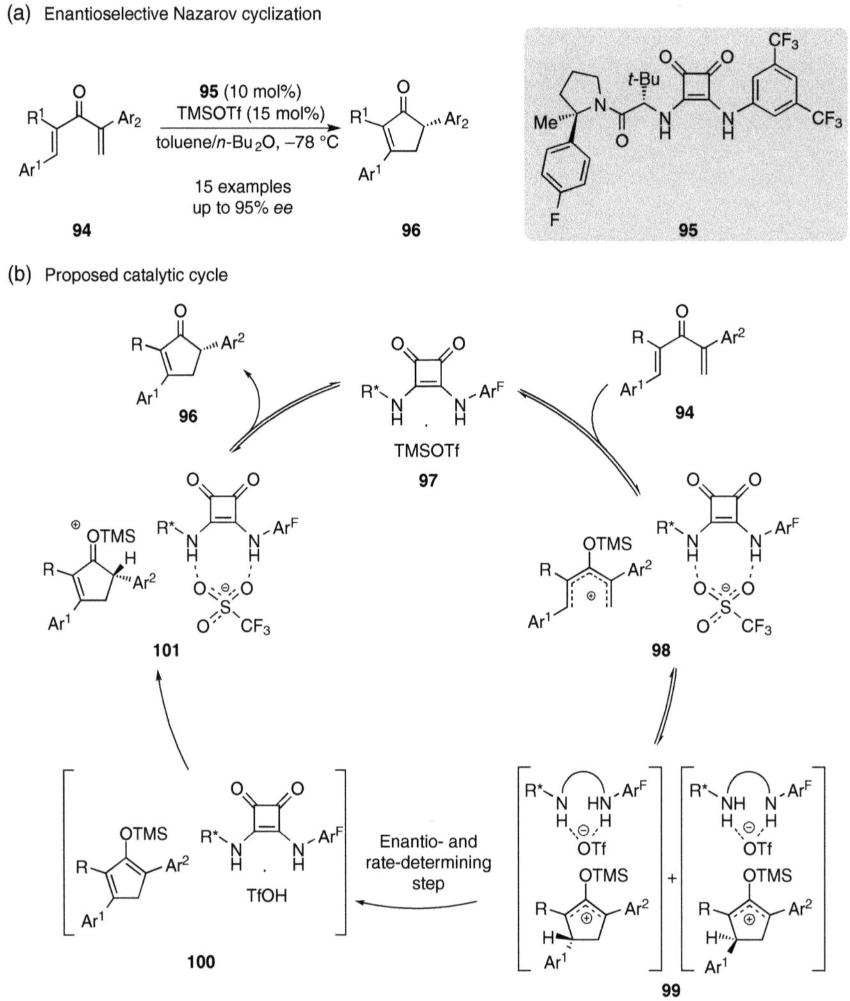

Scheme 4.18 (a) Enantioselective Nazarov electrocyclization catalyzed by chiral squaramide **95** and co-catalytic trimethylsilyl trifluoromethanesulfonate and (b) proposed catalytic cycle. Source: Metternich *et al.* [42].

4π-electrocyclization was reversible and that the proton transfer sequence includes the rate- and enantiodetermining step. The proton transfer was further shown through crossover studies to proceed intramolecularly. Given that a concerted [1,4]-shift from **99** would be forbidden suprafacially and sterically inaccessible antarafacially, a stepwise proton transfer sequence was invoked. Additional experiments were undertaken to determine the nature of the enantiodetermining step, with the results consistent with an HBD-promoted stereospecific proton transfer and catalyst-controlled enantioselective deprotonation of one of the rapidly interconverting stereoisomers of **99**. The Lewis acid co-catalyst is thus proposed to serve the dual role of electrophile activation and source of the anionic counterion that mediates the critical proton transfer step.

4.6 Conclusions and Outlook

The studies outlined above reveal a variety of different modes of co-catalysis between Brønsted/Lewis acids and chiral H-bond donors that ultimately promote and control the stereochemical outcomes of pathways involving cationic intermediates. Given that the only fundamental limitation to this approach is the requirement for a counterion on the Brønsted or Lewis acid capable of engaging in anion-binding with the chiral H-bond donor catalyst, it is anticipated that a wide variety of new applications of Brønsted/Lewis acid–chiral H-bond donor co-catalysis will emerge in the coming years.

References

1 Eigen, M. (1964). Proton transfer, acid-base catalysis, and enzymatic hydrolysis. Part I: Elementary processes. *Angew. Chem. Int. Ed.* 3: 1–19.

2 Santelli, M. and Pons, J.-M. (1995). *Lewis Acids and Selectivity in Organic Synthesis*. New York: CRC Press.

3 Yamamoto, H. (2000). *Lewis Acids in Organic Synthesis*. Weinheim: Wiley-VCH Verlag GmbH.

4 Cheon, C.H. and Yamamoto, H. (2011). Super Brønsted acid catalysis. *Chem. Commun.* 47: 3043–3056.

5 Schinzer, D. (2012). *Selectivities in Lewis acid promoted reactions*. Boston: Springer Science & Business Media.

6 Akiyama, T. (2007). Stronger Brønsted acids. *Chem. Rev.* 107: 5744–5758.

7 Yamamoto, H. and Futatsugi, K. (2005). "Designer Acids": combined acid catalysis for asymmetric synthesis. *Angew. Chem. Int. Ed.* 44: 1924–1942.

8 Terada, M. (2010). Chiral phosphoric acids as versatile catalysts for enantioselective transformations. *Synthesis*: 1929–1982.

9 Parmar, D., Sugiono, E., Raja, S., and Rueping, M. (2014). Complete field guide to asymmetric BINOL-phosphate derived Brønsted acid and metal catalysis: history and classification by mode of activation; Brønsted acidity, hydrogen bonding, ion pairing, and metal phosphates. *Chem. Rev.* 114: 9047–9153.

10 Akiyama, T. and Mori, K. (2015). Stronger Brønsted acids: recent progress. *Chem. Rev.* 115: 9277–9306.

11 Narasaka, K. (1991). Chiral Lewis acids in catalytic asymmetric reactions. *Synthesis*: 1–11.

12 Howell, G.P. (2012). Asymmetric and diastereoselective conjugate addition reactions: C—C bond formation at large scale. *Org. Process Res. Dev.* 16: 1258–1272.

13 Jørgensen, K.A., Johannsen, M., Yao, S. *et al.* (1999). Catalytic asymmetric addition reactions of carbonyls. A common catalytic approach. *Acc. Chem. Res.* 32: 605–613.

14 Christoffers, J., Koripelly, G., Rosiak, A., and Rössle, M. (2007). Recent advances in metal-catalyzed asymmetric conjugate additions. *Synthesis*: 1279–1300.

15 Klausen, R.S. and Jacobsen, E.N. (2009). Weak Brønsted acid–thiourea co-catalysis: enantioselective, catalytic protio-Pictet–Spengler reactions. *Org. Lett.* 11: 887–890.

16 Taylor, M.S. and Jacobsen, E.N. (2004). Highly enantioselective catalytic acyl-Pictet–Spengler reactions. *J. Am. Chem. Soc.* 126: 10558–10559.

17 Klausen, R.S., Kennedy, C.R., Hyde, A.M., and Jacobsen, E.N. (2017). Chiral thioureas promote enantioselective Pictet–Spengler cyclization by stabilizing every intermediate and transition state in the carboxylic acid-catalyzed reaction. *J. Am. Chem. Soc.* 139: 12299–12309.

18 Lee, Y., Klausen, R.S., and Jacobsen, E.N. (2011). Thiourea-catalyzed enantioselective iso-Pictet–Spengler reactions. *Org. Lett.* 13: 5564–5567.

19 Brown, A.R., Kuo, W.-H., and Jacobsen, E.N. (2010). Enantioselective catalytic α-alkylation of aldehydes via an S_N1 pathway. *J. Am. Chem. Soc.* 132: 9286–9288.

20 Huang, H. and Jacobsen, E.N. (2006). Highly enantioselective direct conjugate addition of ketones to nitroalkenes promoted by a chiral primary amine–thiourea catalyst. *J. Am. Chem. Soc.* 128: 7170–7171.

21 Yoshida, K., Inokuma, T., Takasu, K., and Takemoto, Y. (2010). Brønsted acid-thiourea co-catalysis: asymmetric synthesis of function-alized 1,4-dihydropyridines from β-enamino esters and α,β-unsaturated aldehydes. *Synlett*: 1865–1869.

22 Yoshida, K., Inokuma, T., Takasu, K., and Takemoto, Y. (2010). Catalytic asymmetric synthesis of both enantiomers of 4-substituted 1,4-dihydropyridines with the use of bifunctional thiourea-ammonium salts bearing different counterions. *Molecules* 15: 8305–8326.

23 Burns, N.Z., Witten, M.R., and Jacobsen, E.N. (2011). Dual catalysis in enantioselective oxidopyrylium-based [5+2] cycloadditions. *J. Am. Chem. Soc.* 133: 14578–14581.

24 Witten, M.R. and Jacobsen, E.N. (2014). Catalytic asymmetric synthesis of 8-oxabicyclooctanes by intermolecular [5+2] pyrylium cycloadditions. *Angew. Chem. Int. Ed.* 53: 5912–5916.

25 Lalonde, M.P., McGowan, M.A., Rajapaksa, N.S., and Jacobsen, E.N. (2013). Enantioselective formal aza-Diels–Alder reactions of enones with cyclic imines catalyzed by primary aminothioureas. *J. Am. Chem. Soc.* 135: 1891–1894.

26 Rajapaksa, N.S., McGowan, M.A., Rienzo, M., and Jacobsen, E.N. (2013). Enantioselective total synthesis of (+)-reserpine. *Org. Lett.* 15: 706–709.

27 Yeung, C.S., Ziegler, R.E., Porco, J.A., and Jacobsen, E.N. (2014). Thiourea-catalyzed enantioselective addition of indoles to pyrones: alkaloid cores with quaternary carbons. *J. Am. Chem. Soc.* 136: 13614–13617.

28 Xu, H., Zuend, S.J., Woll, M.G. *et al.* (2010). Asymmetric cooperative catalysis of strong Brønsted acid-promoted reactions using chiral ureas. *Science* 327: 986–990.

29 Mayr, H., Ofial, A.R., Würthwein, E.-U., and Aust, N.C. (1997). NMR spectroscopic evidence for the structure of iminium ion pairs. *J. Am. Chem. Soc.* 119: 12727–12733.

30 Elia, G.R., Childs, R.F., Britten, J.F., and Yang, D.S.C. (1996). Structure and wavelength modification in retinylidene iminium salts. *Can. J. Chem.* 74: 591–601.

31 Gerard, B., O'Shea, M.W., Donckele, E. *et al.* (2012). Application of a catalytic asymmetric Povarov reaction using chiral ureas to the synthesis of a tetrahydroquinoline library. *ACS Comb. Sci.* 14: 621–630.

32 Lin, S. and Jacobsen, E.N. (2012). Thiourea-catalysed ring opening of episulfonium ions with indole derivatives by means of stabilizing non-covalent interactions. *Nat. Chem.* 4: 817–824.

33 Blackmond, D.G. (2005). Reaction progress kinetic analysis: a powerful methodology for mechanistic studies of complex catalytic reactions. *Angew. Chem. Int. Ed.* 44: 4302–4320.

34 Peterson, E.A. and Jacobsen, E.N. (2009). Enantioselective, thiourea-catalyzed intermolecular addition of indoles to cyclic *N*-acyl iminium ions. *Angew. Chem. Int. Ed.* 48: 6328–6331.

35 Knowles, R.R., Lin, S., and Jacobsen, E.N. (2010). Enantioselective thiourea-catalyzed cationic polycyclizations. *J. Am. Chem. Soc.* 132: 5030–5032.

36 Gal, J.-F., Maria, P.-C., Decouzon, M. *et al.* (2003). Lithium-cation/π complexes of aromatic systems. The effect of increasing the number of fused rings. *J. Am. Chem. Soc.* 125: 10394–10401.

37 Ng, K.M., Ma, N.L., and Tsang, C.W. (1998). Cation–aromatic π interaction in the gas phase: an experimental study on relative silver (I) ion affinities of polyaromatic hydrocarbons. *Rapid Commun. Mass Spectrom.* 12: 1679–1684.

38 Vijay, D. and Sastry, G.N. (2008). Exploring the size dependence of cyclic and acyclic π-systems on cation–π binding. *Phys. Chem. Chem. Phys.* 10: 582–590.

39 Zhang, H., Lin, S., and Jacobsen, E.N. (2014). Enantioselective selenocyclization via dynamic kinetic resolution of seleniranium ions by hydrogen-bond donor catalysts. *J. Am. Chem. Soc.* 136: 16485–16488.

40 Denmark, S.E. and Collins, W.R. (2007). Lewis base activation of Lewis acids: development of a Lewis base catalyzed selenolactonization. *Org. Lett.* 9: 3801–3804.

41 Zhao, C., Chen, S.B., and Seidel, D. (2016). Direct formation of oxocarbenium ions under weakly acidic conditions: catalytic enantioselective Oxa-Pictet–Spengler reactions. *J. Am. Chem. Soc.* 138: 9053–9056.

42 Metternich, J.B., Reiterer, M., and Jacobsen, E.N. (2020). Asymmetric Nazarov cyclizations of unactivated dienones by hydrogen-bond-donor/Lewis acid co-catalyzed, enantioselective proton-transfer. *Adv. Synth. Catal.* 362: 4092–4097.

43 Banik, S.M., Levina, A., Hyde, A.M., and Jacobsen, E.N. (2017). Lewis acid enhancement by hydrogen-bond donors for asymmetric catalysis. *Science* 358: 761–764.

44 Wendlandt, A.E., Vangal, P., and Jacobsen, E.N. (2018). Quaternary stereocenters via an enantioconvergent catalytic S_N1 reaction. *Nature* 556: 447–451.

5

Anion-Binding Catalysis with Other Anions

Sankash Mitra and Santanu Mukherjee

Indian Institute of Science, Department of Organic Chemistry, C. V. Raman Road, Bangalore 560012, India

5.1 Introduction

Despite the advent of H-bonding [1] as a prominent force for substrate activation decades before [2] the discovery of Lewis acid catalysis [3], H-bonding catalysis remained in oblivion until nearly the end of the previous century except for a few sporadic reports. The lack of immediate attention of chemists toward H-bonding catalysis may be attributed to the loosely defined and less directional interactions compared to the well-understood geometry of Lewis acidic metal complexes [4]. Notwithstanding the late emergence, the past two decades have witnessed an explosive growth in the application of H-bond donor catalysts, especially in enantioselective catalysis [4, 5]. The initial efforts were primarily confined to the LUMO-lowering activation of neutral (e.g. aldehyde, ketone, imine, and epoxide) and partially charged (e.g. nitro and ester) functionalities using H-bond donors as catalysts [4, 5]. A combination of H-bonding with other covalent or non-covalent interactions in the form of bifunctional catalysts added another dimension by promoting dual-substrate activation [5e, 6]. In addition to the ever-inspiring Nature, the early works of Etter established the ability of H-bond donors to act as synthetic receptors for various small molecules (e.g. ketones, ethers, nitroarenes, sulfoxides, and phosphine oxides) [7], which laid the foundation for these discoveries.

Besides neutral guests, H-bond donor-based receptors were also shown to be efficient in selective recognition of anionic guests [8]. Depending on the shape (topology) of anions, H-bond donor hosts were designed. The structure of anions dictates not only their shape but also the number (and nature) of sites available for H-bonding. For example, halides and oxyanions can be considered spherical and monodentate, although each of them can form more than one H-bond simultaneously. On the other hand, sulfate (SO_4^{2-}), sulfonates (RSO_3^-), and phosphate (PO_4^{3-}) are tetrahedral and contain multiple H-bond acceptor sites. Carboxylates

Anion-Binding Catalysis, First Edition. Edited by Olga García Mancheño.

(RCO_2^-) and nitrate (NO_3^-) are planar and bidentate. Cyanide (CN^-) is unique as it is linear but can engage in H-bond formation with either carbon or nitrogen (ambidentate). Although less directional compared to the coordination sites in metal complexes, these studies clearly point out the underappreciated directionality of H-bonding and form the basis of anion-binding catalysis [9]. Therefore, catalyst design is perhaps the most important aspect of these studies because the distance between the H-bond donor sites is also important in addition to the number of such sites.

As the name implies, the concept of anion-binding catalysis is applicable to reactions proceeding via cationic intermediates [9a]. The binding of counteranions by H-bonding from the catalyst (i.e. anion recognition) not only helps in enhancing the electrophilicity of the cationic intermediates but also assist the ionization process itself. (Thio)ureas and squaramides are some of the most common H-bond donor catalysts used for such transformations. Undoubtedly, the highlight of anion-binding catalysis is in rendering the reactions enantioselective, particularly those reactions that are difficult to carry out enantioselectively using conventional modes of activation. With chiral H-bond donors, the complex formed with the anionic species can act as a chiral anion and gives rise to a chiral ion pair with cationic intermediates [9a]. The catalyst-bound anion can be a bystander, a reactive nucleophile, and even a reactive electrophile. During the bond formation within this ion pair or with an external (neutral) electrophile or nucleophile, effective transfer of stereochemical information from the chiral H-bond donor to the product takes place.

Some of these aspects are covered in the preceding and forthcoming chapters of this book. In this chapter, the catalytic reactions proceeding through the binding of less common anions such as cyanide, oxyanions, enolates, and carboxylates are discussed. The bifunctional H-bond donor catalysts bearing a Brønsted or Lewis basic functionality and the reactions promoted by such catalysts are not covered even though many of those reactions proceed through the stabilization of anionic intermediate (e.g. enolate) by one or more H-bonding and resulting in the formation of ion pairs. The coverage of this chapter is confined to reactions where the catalyst is primarily involved in anion binding. Other weak (and not-so-weak) non-covalent interactions (e.g. π-stacking and cation–π interaction) between the catalysts and the substrates/reagents could not be ignored.

5.2 Cyanide Anion

Compared to halides or oxyanions, binding of cyanide anion to H-bond donors is expected to be relatively weak. However, as in most of the reactions involving cyanide it acts as a nucleophile, stabilization of a cyanide anion through H-bonding (from the catalyst) can have significant implications on the reaction outcome. For example, such stabilization would enhance the acidity of otherwise weakly acidic HCN and lead to its facile ionization. The use of cyanide as a nucleophile is well known for reactions with imines and carbonyl compounds, some of which are discussed in this section.

5.2.1 Strecker Reaction

Enantioselective hydrocyanation of imines (the Strecker reaction) by Schiff base-(thio)urea organocatalyst, discovered by the Sigman and Jacobsen in 1998 [10], is not only among the first asymmetric versions of this reaction to be developed [11] but it also brought H-bonding catalysis into the limelight. In fact, this report may be credited for initiating the field of enantioselective H-bonding catalysis. This serendipitous yet remarkable discovery is also hailed as one of the rare early success stories of combinatorial chemistry. With the original objective of identifying a chiral tridentate Schiff base ligand for metal ions, the solid-supported "ligand" **C1** itself displayed the highest enantioselectivity (19% *ee*) when used as the catalyst for the addition of TBSCN to *N*-allylbenzaldimine **1a** (Scheme 5.1).

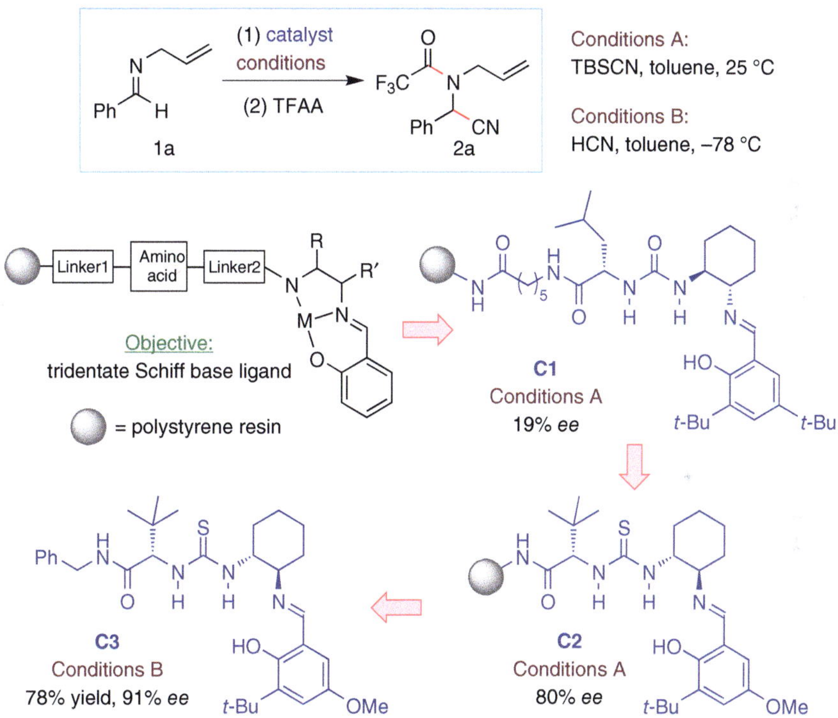

Scheme 5.1 Discovery of the Schiff base-(thio)urea organocatalyst for the enantioselective Strecker reaction by Sigman and Jacobsen [10]. Source: Based on Sigman and Jacobsen [10].

Triggered by this serendipitous lead, a systematic parallel library approach on solid-support eventually led to the discovery of a homogeneous thiourea derivative **C3**, which proved to be an efficient and general catalyst for the addition of HCN to *N*-allylaldimines **1**. Remarkably, aliphatic aldimines afforded the hydrocyanation products with >80% *ee* – the highest at this point of time.

Further catalyst optimization studies conducted by the Jacobsen group based on their earlier findings [10] revealed the importance of substituents on the salicylimine moiety besides the α-substituent on the amino acid residue [12]. Through the screening of a 70-membered library, Jacobsen and coworkers identified the urea derivative **C4** as a general catalyst for enantioselective hydrocyanation of a wide range of aromatic and aliphatic aldimines (**1**) with HCN (Scheme 5.2a). In fact, cyclic (*Z*)-imine such as 3,4-dihydroisoquinoline is a competent substrate for this reaction and afforded the product with 91% *ee*.

The practical potential of the protocol was demonstrated by a resin-bound thiourea derivative **C5**, which displayed comparable efficiency as the soluble urea derivative **C4** and could be recycled and reused by simple filtration [12]. To showcase the application, one of the Strecker products (**3**) was converted to D-*tert*-leucine **4** in essentially enantiopure form (Scheme 5.2b). The easy-to-prepare [14] soluble Schiff

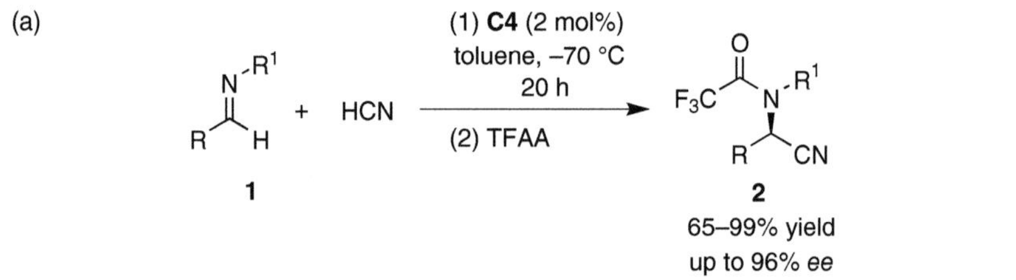

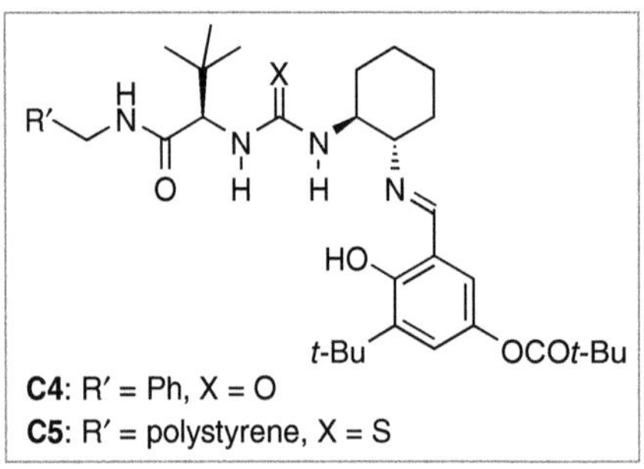

R	R^1	Yield (%)	*ee* (%)
Ph	Allyl	74	95
4-OMeC_6H_4	Allyl	98	95
3-OMeC_6H_4	Allyl	99	93
2-OMeC_6H_4	Allyl	93	77
t-Bu	Allyl	75	95
t-Bu	Bn	88	96
c-Hex	Bn	85	87
n-Pent	Bn	69	78

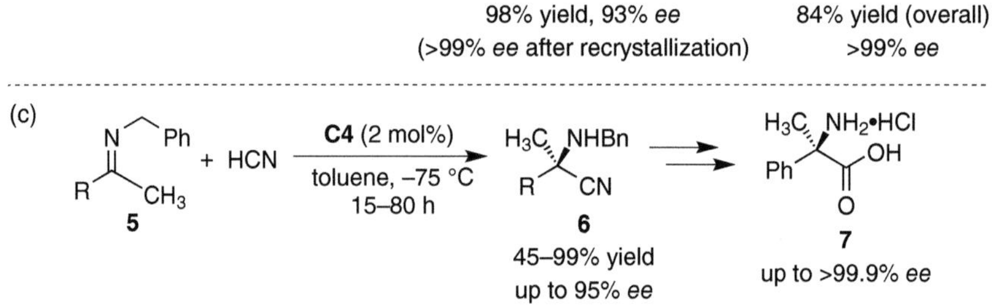

Scheme 5.2 Identification of a general catalyst for the enantioselective Strecker reaction by Jacobsen and coworkers [12] and Vachal and Jacobsen [13]. Source: Based on Vachal and Jacobsen [13].

base urea derivative **C4** was also shown to be effective for the hydrocyanation of ketimines **5** (Scheme 5.2c) [13]. Although the level of enantioselectivity remained somewhat less compared to aldimines, this protocol provides a practical route to α-quaternary α-amino acids **7**. Moreover, some of the hydrocyanation products could be isolated in enantiopure form after a single recrystallization.

The studies to improve the catalyst and reaction conditions were also accompanied by nearly a decade-long mechanistic investigation by Jacobsen and coworkers. Initial efforts based on the structure–activity relationship, NMR spectroscopy, kinetic studies, and density functional theory (DFT) calculations indicated that the (thio)urea functionality of the catalyst is the key to catalytic activity. Accordingly, a LUMO-lowering activation of imine by dual hydrogen bonding from (thio)urea NHs was proposed to take place through the (*Z*)-configuration of the imine (Scheme 5.3) [15]. In this catalyst–substrate complex, the large substituents on the imine carbon (e.g. Ph) and nitrogen (e.g. allyl) are directed away from the catalyst, whereas the small substituent R on the imine carbon is pointed directly into the catalyst. Addition of HCN to the Si-face of the imine over the less hindered diaminocyclohexane portion of the catalyst accounts for the observed stereochemical outcome. However, these studies failed to conclude whether this ground-state interaction is persevered in the rate- or enantioselectivity-determining transition state.

Nonetheless, these systematic investigations resulted in further improvement of the catalyst structure. The thiourea derivative **C6** thus identified turned out to be more enantioselective compared to its predecessors even for the hydrocyanation of

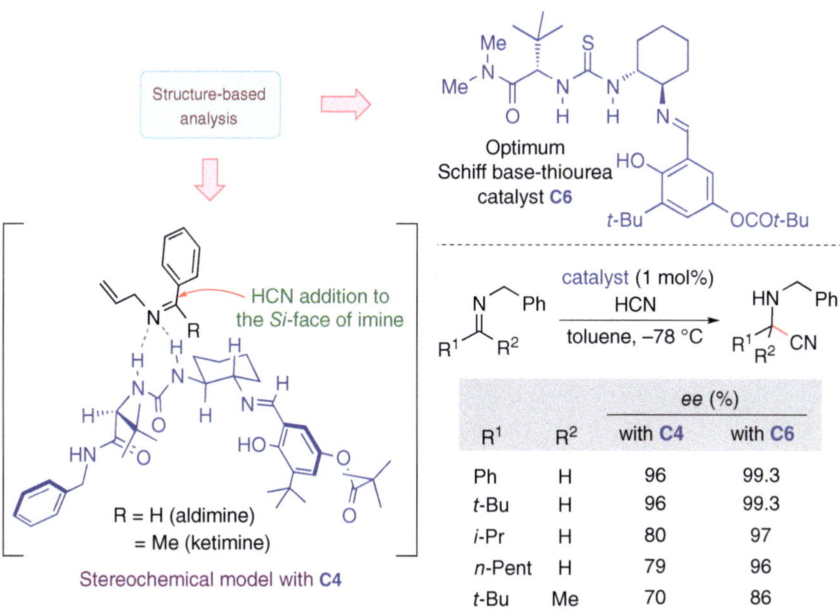

R¹	R²	*ee* (%) with C4	*ee* (%) with C6
Ph	H	96	99.3
t-Bu	H	96	99.3
i-Pr	H	80	97
n-Pent	H	79	96
t-Bu	Me	70	86

Scheme 5.3 Initially proposed stereochemical model and optimum Schiff base-thiourea catalyst for the enantioselective Strecker reaction by Vachal and Jacobsen [15]. Source: Based on Vachal and Jacobsen [15].

ketimines (Scheme 5.3) [15]. More importantly, this structure-based analyses open up the possibility of simplifying the catalyst even more as the Schiff base-(thio)urea derivatives contain sensitive function groups and require multi-step synthesis. In addition, the preparative scale application of these reactions is restricted by the indispensable use of hazardous cyanide source (either TMSCN/MeOH or HCN) and cryogenic conditions.

To overcome these drawbacks, Jacobsen and coworkers subsequently developed a simple, easy-to-prepare, and robust amido-thiourea catalyst **C7** bearing only one stereogenic center and devoid of any sensitive functionality (Scheme 5.4a) [16]. This remarkably efficient catalyst was found to effect the hydrocyanation of *N*-benzhydryl-protected imines (**8**) derived from a variety of aliphatic and (hetero)aromatic aldehydes. The corresponding α-amino nitriles (**9**) were obtained in near-quantitative yield and in most cases with excellent enantioselectivity. A fortuitous feature of this protocol is that using the catalyst (**C7**) derived from relatively inexpensive (*S*)-*tert*-leucine, (*R*)-α-amino nitriles are generated, which are precursors of unnatural amino acids.

(a)

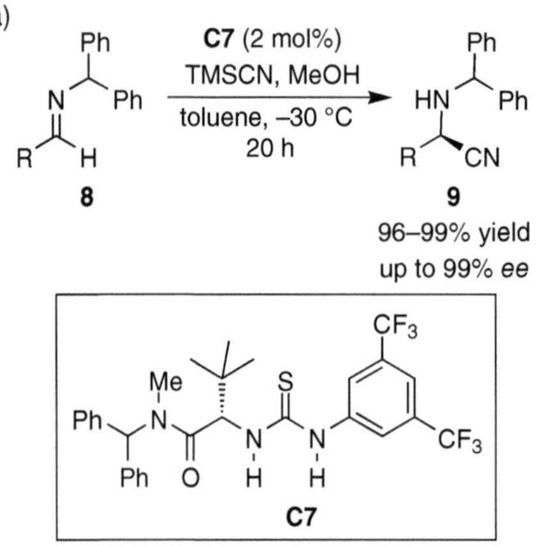

R	Yield (%)	*ee* (%)
Ph	98	98
4-OMeC$_6$H$_4$	99	99
4-CF$_3$C$_6$H$_4$	98[a]	96
3-BrC$_6$H$_4$	97[a]	92
2-OMeC$_6$H$_4$	97	88
3-Furyl	98	97
2-Thienyl	97[a]	95
t-Bu	99	93
c-Hex	99	74
1-Cyclohexenyl	99	95
(*E*)-Cinnamyl	99	95

[a] Reaction at 0 °C

(b)

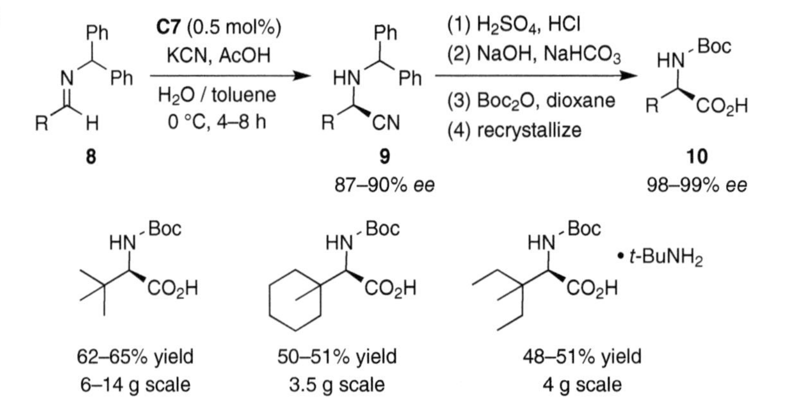

Scheme 5.4 A simplified amido-thiourea catalyst for scalable enantioselective Strecker reaction by Jacobsen and coworkers [16]. Source: Based on Zuend *et al.* [16].

Most importantly, the catalyst **C7**, devoid of any sensitive functionality, is compatible with biphasic (toluene/water) conditions and allows for the use of inexpensive and safer cyanide source KCN for reactions on a large scale. Thus, using only 0.5 mol% **C7**, this reaction can be scaled up to 25–100 mmol scale under more practical conditions (0 °C) with enantioselectivities similar to those observed in the small-scale homogeneous, TMSCN/MeOH-mediated reaction. The unpurified α-amino nitriles (**9**) obtained from this reaction could be converted in a two-step sequence to the synthetically relevant Boc-protected (*R*)-α-amino acids **10**, which were isolated in highly enantiomerically enriched form after recrystallization (Scheme 5.4b) [16]. The ultimate practical aspect of this protocol lies in its chromatographic purification-free conditions.

A fresh look at the reaction under this new catalyst system aided by experimental tools (e.g. kinetic studies, catalyst structure activity evaluation, Hammett analysis, and isotope labeling) and computational studies revealed a mechanistic scenario (Scheme 5.5) [17], which is completely different from the initially proposed stereochemical model (see Scheme 5.3). Instead of the direct general Brønsted acid activation of the free (*Z*)-imine by thiourea NHs, the reaction appears to be initiated by amido-thiourea-induced protonation of the (*E*)-imine with HNC – the tautomer of HCN (**A** → **B**, Scheme 5.5) [17]. The resulting iminium/cyanide ion pair (**B**) undergoes structural reorganization to form a ternary complex (**D**) with the catalyst, where H-bonding of the iminium N–H proton to the amide carbonyl oxygen leads to its stabilization and the cyanide anion is engaged in dual hydrogen bonding with thiourea NHs through its nitrogen. Although the dual H-bonding-mediated stabilization of cyanide anion is important for catalysis, the stabilization of the iminium cation is primarily responsible for controlling the enantioselectivity of the reaction. However, the catalyst-bound ion pair reorganization (**B** → **D**) appears to

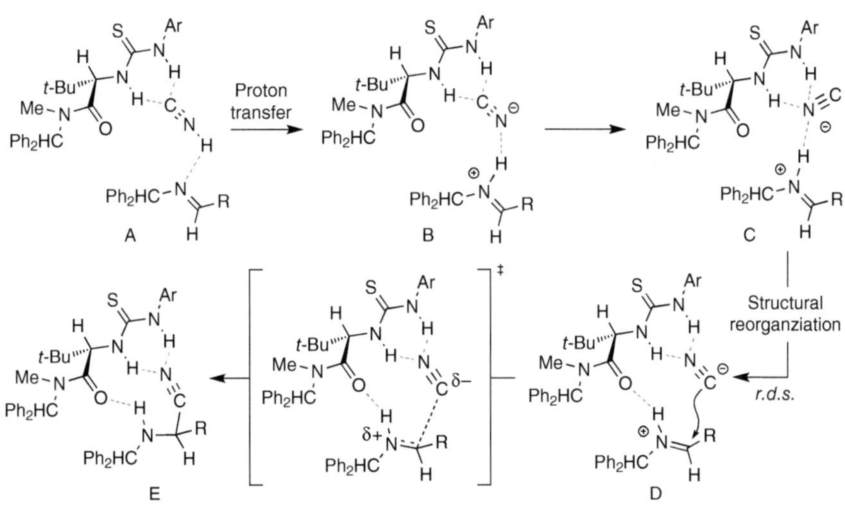

Scheme 5.5 Proposed anion-binding mechanism for the enantioselective Strecker reaction by Zuend and Jacobsen [17]. Source: Based on Zuend and Jacobsen [17].

be the rate-determining step of the reaction as the barrier for C—C bond formation (**D** → **E**) is rather low.

As of now, the reaction described in Scheme 5.4b remains the best general method for the enantioselective synthesis of α-amino acids in terms of operational simplicity, scalability, and the level of enantioselectivity.

5.2.2 Acylcyanation of Imines

With the objective of replacing highly toxic/poisonous cyanide sources in Strecker reaction, List and coworkers developed a novel variant, namely, an acylcyanation of aldimines (the acyl-Strecker reaction) using less toxic, easy-to-handle, and commercially available acetyl cyanide [18]. Under the influence of Jacobsen's Schiff base-thiourea catalyst **C6** [15], an assortment of *N*-benzyl aldimines (**11**) derived from (hetero)aromatic and aliphatic aldehydes was shown to furnish *N*-acetyl α-amino nitriles **13** in good to excellent yield and in most cases with high enantioselectivity (Scheme 5.6a). As in the case of Strecker reaction (see Scheme 5.2b), the *N*-acetyl α-amino nitrile products (**13**) could be readily converted to the corresponding free α-amino acids as demonstrated for L-*tert*-leucine *ent*-**4** [18].

Pan and List utilized the same catalyst and the similar reaction conditions to develop the first organocatalytic asymmetric three-component acyl-Strecker

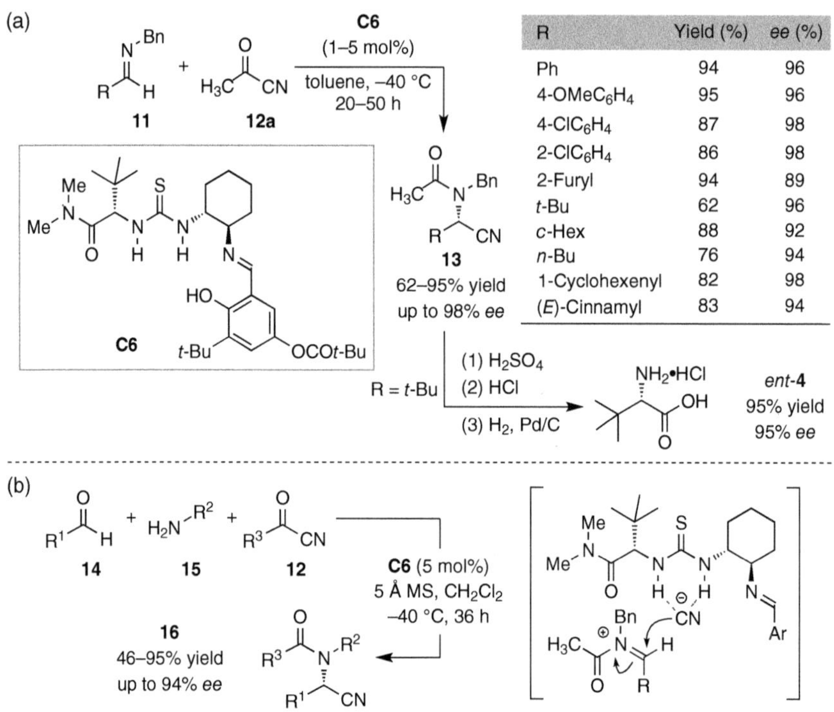

R	Yield (%)	*ee* (%)
Ph	94	96
4-OMeC$_6$H$_4$	95	96
4-ClC$_6$H$_4$	87	98
2-ClC$_6$H$_4$	86	98
2-Furyl	94	89
t-Bu	62	96
c-Hex	88	92
n-Bu	76	94
1-Cyclohexenyl	82	98
(*E*)-Cinnamyl	83	94

Scheme 5.6 Catalytic enantioselective acyl-Strecker reaction by List and coworkers [18–20]. Source: Adapted from Pan *et al.* [19, 20].

reaction (Scheme 5.6b) [19]. This operationally simple protocol relies on the *in situ* imine formation in the presence of 5 Å MS before the catalyst and acyl cyanide are introduced. While the products were isolated in similar yields, the enantioselectivity of the three-component reaction is slightly lower compared to that with pre-formed imine. Nevertheless, this protocol could find potential applications in diversity-oriented synthesis of α-amino acid derivatives.

The reaction most likely proceeds through the intermediacy of *N*-acetyl iminium ion formed by the reaction of imine **11** and acetyl cyanide **12a** (Scheme 5.6b) [20]. The thiourea-bound cyanide then acts as a chiral counteranion to form a tight ion pair. The addition of the chiral counteranion nucleophile to the *N*-acyl iminium cation within the tight ion pair is expected to be enantioselective. The same sense of stereoinduction as that of Jacobsen's Strecker reaction under the same catalyst **C6** (see Scheme 5.3) points to an analogous mechanism in both these reactions.

Unlike Strecker reaction discussed before [13], the acyl-Strecker protocol is only limited to aldimines as the reaction with ketimines under similar reaction conditions produced an amidine derivative as the major product [20].

5.3 Oxygen-Based Anions

The stabilization of tetrahedral alkoxide intermediate in oxyanion hole of serine protease provides the necessary impetus for the development of catalysis concepts involving the binding (and stabilization) of oxygen-based anions by H-bonding. Oxygen-based anions are common in organic synthesis, with the most frequently encountered ones being the enolates. A wide range of catalytic enantioselective transformations have been developed using bifunctional H-bonding catalysts, which involve enolates as the reactive intermediates. We will confine the discussion to only those reactions, where the H-bond donor catalysts are primarily involved in stabilizing the oxygen-based anions and do not directly participate in activating any other reactant.

5.3.1 Alkoxides and Enolates

The first examples of an oxyanion binding by a H-bond donor and its application in catalysis was somewhat unexpectedly encountered by Schreiner and Kotke in 2006 [21]. With the objective of developing an acid-free acetalization, Kotke and Schreiner realized the first catalytic acetalization of various aromatic as well as aliphatic aldehydes and ketones **17** using an electron-deficient thiourea derivative **C8** (Scheme 5.7a). This protocol, using alcohol as the solvent and the corresponding orthoformate as the stoichiometric dehydrating agent, requires very low catalyst loading of 0.01–1 mol% at room temperature to furnish the desired acetals or ketals **18** in moderate to excellent yield and a high turnover frequency (TOF) of ~600 h^{-1}. The authors have successfully expanded this mild and highly efficient protocol to convert acid-labile TBDMS-protected as well as unsaturated aldehydes to the corresponding acetals.

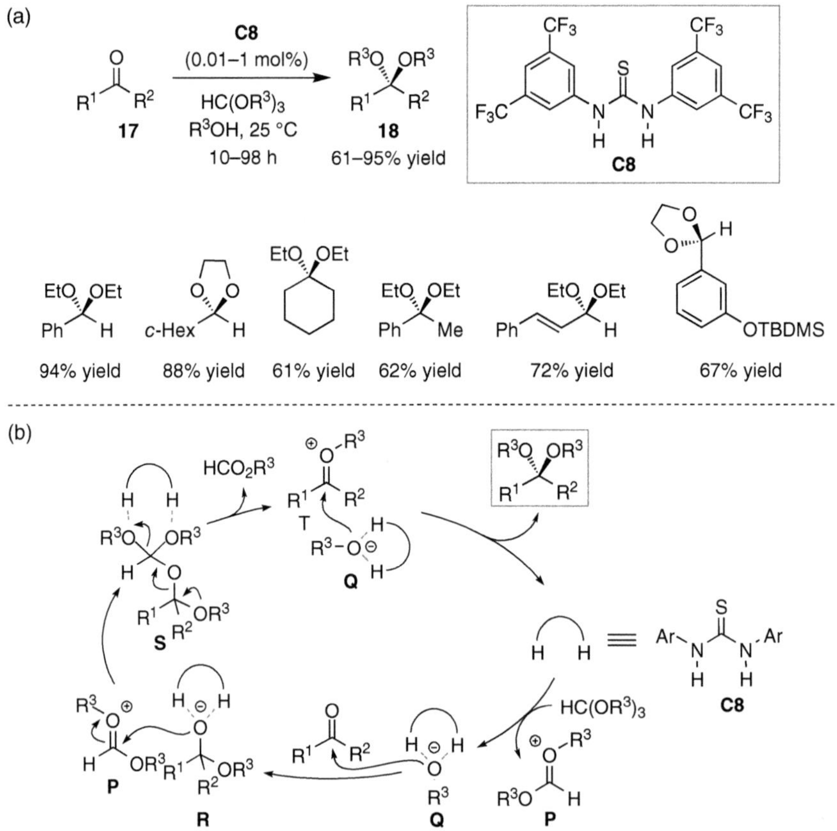

Scheme 5.7 Alkoxide-binding in thiourea-catalyzed acetalization of aldehydes and ketones by Kotke and Schreiner [21]. Source: Based on Kotke and Schreiner [21].

The original hypothesis of the direct LUMO-lowering activation of carbonyls and subsequent stabilization of the initially formed tetrahedral alkoxide (anionic hemiacetal) intermediate by dual H-bonding could not explain the failure of thioacetalization under the similar reaction conditions, despite thiols being more nucleophilic than aldehydes [21]. This observation led the authors to suggest an alternative mechanistic proposal involving thiourea-assisted heterolysis of orthoester to form an oxocarbenium ion **P** and a thiourea-bound alkoxide **Q** (Scheme 5.7b). The subsequent attack of the thiourea-bound alkoxide (**Q**) to the carbonyl generates the oxyanion (**R**) stabilized by dual hydrogen bonding from thiourea. This oxyanion further adds to the oxocarbenium ion **P** to produce a tetrahedral intermediate **S**, which upon thiourea-assisted fragmentation releases formate ester and generates a similar thiourea-bound alkoxide (**Q**) along with another oxocarbenium ion **T**. Recombination of these charged intermediates (**Q** and **T**) furnishes the desired acetal and regenerates the catalyst **C8**.

A similar catalysis concept was subsequently applied by Schreiner and Kotke for tetrahydropyranylation of a wide range of alcohols and phenols (Scheme 5.8) [22].

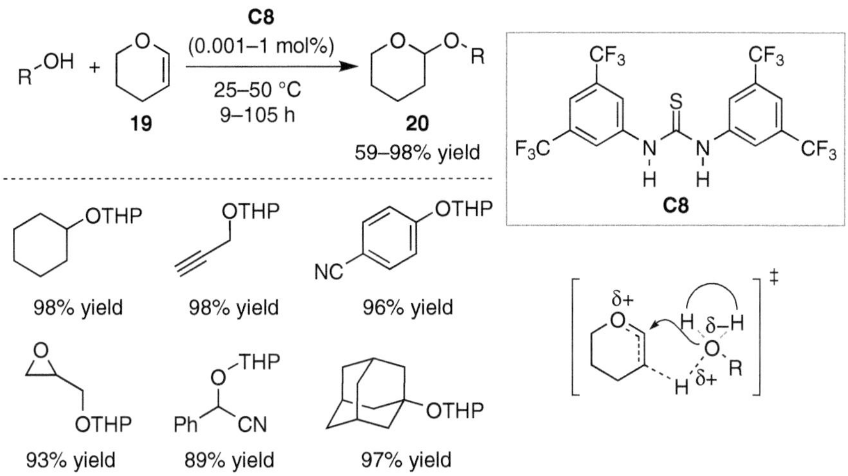

Scheme 5.8 Thiourea-catalyzed tetrahydropyranylation of alcohols and phenols by Kotke and Schreiner [22]. Source: Based on Kotke and Schreiner [22].

As in the case of acetalization, Schreiner's thiourea **C8** turned out to be a remarkably efficient catalyst for this acid-free hydroxy protection reaction. The catalyst loading could be reduced down to 0.001 mol%, which resulted in a TOF of 5700 h^{-1}. Not only dihydropyran (DHP, **19**) but other enol ethers such as 2-methoxypropene could also be employed under very similar reaction conditions. Mechanistic studies with the help of DFT calculations point to the enhancement of acidity of the alcohols through complexation with **C8**. Partial protonation of DHP with complexed alcohol, followed by addition of the H-bonded alkoxide to the incipient oxocarbenium ion results in the protected alcohol **20**.

A computational study by Varga, Pápai, and coworkers suggests an alternative mechanism, where thiourea **C8** acts as a proton donor to DHP to form the oxocarbenium ion [23]. The alcohol then adds to the oxocarbenium ion and returns the proton to the conjugate base of the catalyst. The authors provided sufficient experimental evidence in support of this pathway. Nearly equal efficacy of thiourea derivatives incapable of forming two simultaneous H-bonds (thiouracil and mono-*N*-methylated **C8**) showed the redundancy of the dual H-bonding from the catalyst (**C8**) in this reaction.

Considering the easy accessibility of the catalyst, simplicity of the protocol, and the relevance of the dihydropyran core in the field of carbohydrates, this acetal formation strategy is naturally suited for the application in glycosylation reactions. Indeed, Galan, McGarrigle, and coworkers applied this reaction for a direct α-selective glycosylation of galactals for the synthesis of 2-deoxyglycosides [24]. Using only 1 mol% of Schreiner's thiourea **C8**, a variety of glycosyl acceptors (**22**) underwent smooth reaction with diversely protected galactals (**21**) to furnish disaccharides (**23**) in excellent yield exclusively as α-anomers (Scheme 5.9). Only a slight excess (1.2 equiv) of galactal donors (**21**) is sufficient for this reaction, which was carried out in refluxing dichloromethane – conditions somewhat different from Schreiner's original report

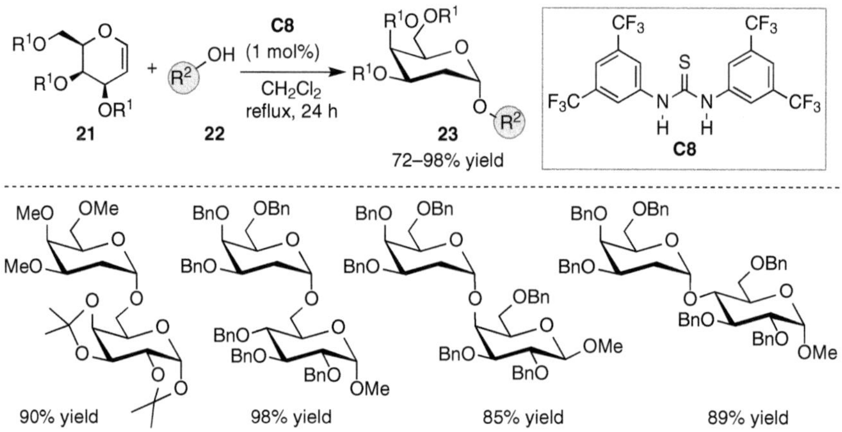

Scheme 5.9 Thiourea-catalyzed α-selective glycosylation by McGarrigle and coworkers [24]. Source: Based on Balmond *et al.* [24].

(see Scheme 5.8) [22]. A mechanism similar to the one shown in Scheme 5.8 is proposed where the diastereoselective C—H and C—O bond formations are hypothesized to take place from the less hindered α-face of galactal **21** [24].

The examples discussed so far involve the binding of anionic nucleophilic components with the H-bond donor catalyst. However, the anion-binding catalysis strategy is also amenable to electrophilic oxyanion intermediates. This apparently counterintuitive concept (oxyanion as electrophile!) is in fact completely rational considering the well-established LUMO-lowering activation of electrophiles under H-bonding catalysis.

Oxy-allyl cation is such a rare electrophilic species, whose intermediacy was first proposed in the Favorskii rearrangement way back in 1894 [25] and later on in various other transformations including Nazarov cyclization [26] and [4+3]-cycloaddition reactions [27]. While oxy-allyl cations were shown to be trapped by nucleophiles to form α-substituted carbonyls [28], development of an enantioselective variant of these reactions proved to be considerably challenging.

In 2016, MacMillan and coworkers unveiled an elegant strategy for the enantioselective substitution of α-tosyloxy ketones with π-nucleophiles [29]. The dual H-bond donor catalyst was not only hypothesized to facilitate the initial enolization through α-deprotonation (soft enolization) and subsequent fragmentation but the partial negative charge developed on the oxygen atom in oxy-allyl cation can also be stabilized through dual H-bonding. The latter interaction was thought to activate the oxy-allyl cation toward nucleophilic addition via LUMO lowering and with a suitable catalyst, in an enantioselective fashion by discriminating the enantiotopic faces of the oxy-allyl cation. Using the amino alcohol **C9** as an unusual hydrogen bond donor catalyst, a number of *N*-alkyl indole derivatives (**25**) were found to react with α-tosyloxy cyclopentanones (**24**) to form α-indolic cyclopentanones **26** in high yield with good to high enantioselectivity (Scheme 5.10).

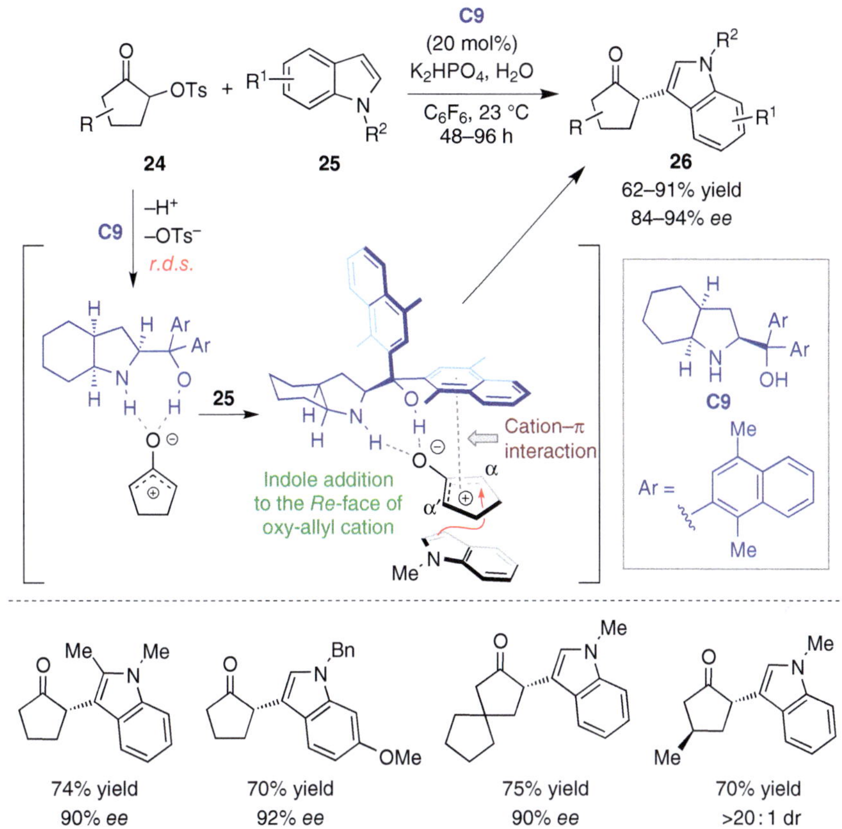

Scheme 5.10 Enantioselective nucleophilic addition to oxy-allyl cations under anion-binding catalysis by MacMillan and coworkers and He *et al.* [29, 30]. Source: Based on Liu *et al.* [29].

Initial rate measurements coupled with KIE studies and a number of control experiments revealed the deprotonation during the initial enolization to be rate determining [29]. However, the nature of tosyl ionization (stepwise or in concert with deprotonation) could not be concluded. Prefatory observations during catalyst screening indicated a cation–π interaction between the oxy-allyl cation and one of the naphthyl rings of the catalyst. This hypothesis was further corroborated by improved enantioselectivity in perfluorobenzene, which impedes competing interaction between the solvent and the oxy-allyl cation. This cation–π interaction effectively shields the Si-face (above face) of the oxy-allyl cation and facilitates the addition of indole **25** from the less sterically encumbered Re-face to deliver the enantioenriched α-indolic cyclopentanones **26** (Scheme 5.10). However, the reason for the preferred addition of indole to one of the two available α-carbons (α- vs. α′-) of oxy-allyl cation was not clear from this stereochemical model.

Houk and coworkers later revisited this reaction with the help of DFT calculations to explain the observed stereochemical outcome [30]. This study revealed the presence of electrostatic interactions between the *N*-methyl group of indole **25** and the cyclohexane ring of the catalyst **C9**. Although the energetically optimized major and minor transition state structures displayed a similar arrangement of the oxy-allyl moiety and the catalyst, they differ in distances between the hydrogen atom on the methyl group of indole and the cyclohexane ring of the catalyst. This accounts for the destabilization of minor TS through an unfavorable electrostatic interaction between the partially positive *N*-methyl group and partially positive hydrogens of the cyclohexane moiety. These theoretical findings were supported by the low enantioselectivity obtained experimentally with N-unprotected indole (**25**, $R^1 = R^2 = H$) and also when diarylprolinol was used as the catalyst.

5.3.2 Enolates of Lactones, Cyclic Anhydrides, and Imides

Cyclic enolates derived from lactones or lactams are excellent nucleophiles, whose reactions with electrophiles constitute a class of synthetically useful transformations. Owing to their inherent structural rigidity and the presence of at least two H-bond acceptor sites in close proximity, such cyclic enolates can easily form H-bonded complex with the likes of (thio)urea or squaramides. Consequently, chiral anion-binding catalysis appears as a perfect strategy for the enantioselective reactions of these nucleophiles, especially with cationic electrophiles through ion pairing.

Acyl activation of isoquinolines followed by nucleophilic addition, commonly known as the Reissert-type reaction [31], provides a unique approach for the construction of dihydroisoquinoline moiety. These reactions proceed through the formation of ion pairs consisting of *N*-acyl-isoquinolinium cation and the counteranion derived from the acyl donor. Capture of this anion through a chiral H-bond donor and subsequent attack to the *N*-acyl-isoquinolinium cation with an external nucleophile has emerged as an effective strategy for enantioselective dearomatization of isoquinolines [32]. As an alternative possibility with a suitable acyl donor, which can give rise to a nucleophilic counteranion, a completely atom-economic transformation can be accomplished.

In 2011, Seidel group documented an enantioselective addition of *O*-acylated azlactones (**27**) to isoquinolines (**28**) using this latter strategy (Scheme 5.11) [33]. Using 10 mol% thiourea **C10** as the catalyst, functionalized α,β-diamino acid derivatives **29** are formed in excellent yield with good diastereo- and enantioselectivity. Interestingly, a highly non-polar mesitylene/pentane (1 : 2) mixture turned out to be the optimum reaction medium, where a heterogeneous mixture is formed during the reaction. This heterogeneity is believed to be crucial in propelling the reaction in forward direction through precipitation of the product.

Initial addition of isoquinoline to *O*-acylated azlactone in the presence of thiourea **C10** results in the transfer of the acyl group to generate *N*-acyl-isoquinolinium cation along with azlactone-enolate stabilized by H-bonding from **C10** (Scheme 5.11) [33]. Nucleophilic addition of thiourea-bound enolate to the 1-position of *N*-acyl-isoquinolinium cation within tight ion pair gives rise to the observed face selectivity from both the reaction partners. Greater electrophilicity of the 1-position of

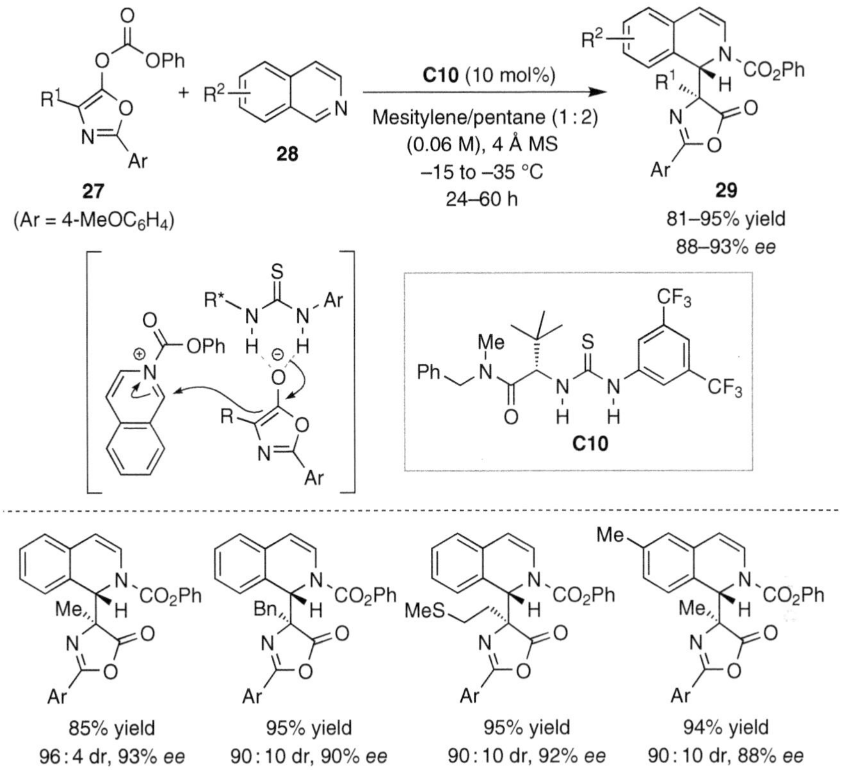

Scheme 5.11 Anion-binding catalysis in enantioselective addition of *O*-acylated azlactones to isoquinolines by Seidel and coworkers [33]. Source: Based on De *et al.* [33].

N-acyl-isoquinolinium cation compared to its acyl carbon is presumably responsible for the suppression of the Steglich rearrangement pathway.

Despite the electron-deficient character of α,β-double bond of allenoates, Lu group in 1995 have disclosed a phosphine-catalyzed addition of nucleophiles to the relatively electron-rich β,γ-double bond [34]. This transformation, also known as "*umpolung*" addition, enthralled several research groups including those of Zhang, Fu, and others to explore the enantioselective variant of this allylation reaction using chiral phosphine catalysts [35].

In 2013, Mukherjee's group devised an ingenious strategy of combining chiral squaramide as a H-bond donor with achiral phosphine [36]. This synergistic combination culminated in the first vinylogous umpolung addition to allenoates. An exhaustive screening of chiral H-bond donors, achiral phosphines, and other reaction parameters led to the identification of *tert*-leucine-derived chiral squaramide **C11** along with triphenylphosphine as the optimum catalyst combination for the vinylogous addition of deconjugated butenolides **30** to allenoates **31** (Scheme 5.12). The resulting γ-allyl butenolides (**32**) were obtained exclusively as the *E*-isomer in reasonable yield with moderate to good enantioselectivity. Notable rate enhancement was observed using a catalytic amount of 2,6-di-*tert*-butyl phenol

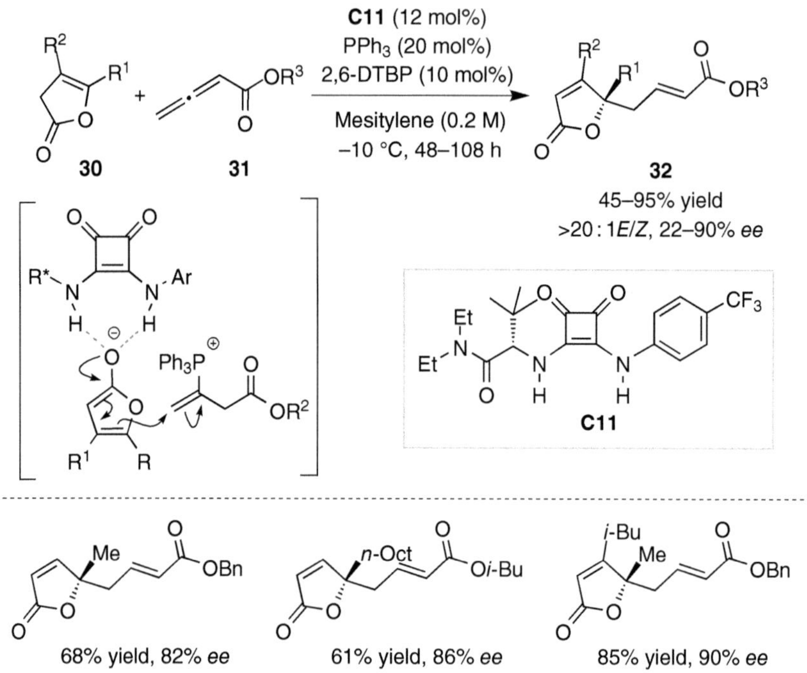

Scheme 5.12 Enantioselective vinylogous umpolung addition of deconjugated butenolides to allenoates under synergistic Lewis base and anion-binding catalysis by Kumar and Mukherjee [36]. Source: Based on Kumar and Mukherjee [36].

(2,6-DTBP), possibly by assisting proton transfer. This protocol is suitable for various γ-substituted and β,γ-disubstituted deconjugated butenolides. However, γ-phenyl-substituted butenolide provided the corresponding adduct with lower enantioselectivity, which marks a shortcoming of this protocol.

The conceptual uniqueness of this endeavor lies in the use of easily accessible achiral phosphine with chiral H-bond donor in lieu of chiral phosphine catalysts, which are generally expensive and vulnerable toward oxidation. This reaction is also the first example of a catalytic enantioselective Cγ—Cγ bond formation between two different carbonyl partners and emphasizes noteworthy utilization of chiral squaramide as an anion-binding catalyst for the first time.

Suitably substituted cyclic anhydrides can be enolized under mild basic conditions and the resulting cyclic enolates can participate in various C—C bond forming transformations. The condensation of enolizable cyclic anhydrides with imines, also known as the Castagnoli–Cushman reaction, is one such method and is particularly effective for the synthesis of lactam carboxylic acids [37]. Considering the commonly accepted mechanism of this reaction, which involves the formation of an ion pair consisting of the anhydride enolate and the iminium cation, Seidel, Vetticatt, and coworkers recognized the opportunity of utilizing the anion-binding strategy for developing an enantioselective Castagnoli–Cushman reaction [38]. In the presence of the amido-thiourea derivative **C12**, homophthalic anhydride (**33**) undergoes a facile reaction with diversely substituted aldimines (**1**) to generate δ-lactam carboxylic acids, which were isolated as the corresponding methyl ester **34** in good to high yield with high dr and *ee* in most cases (Scheme 5.13a). Interestingly, in the case of cinnamaldehyde-derived imine, an initial conjugate addition (leading to Tamura cycloaddition, *vide infra*) was found to compete with the desired

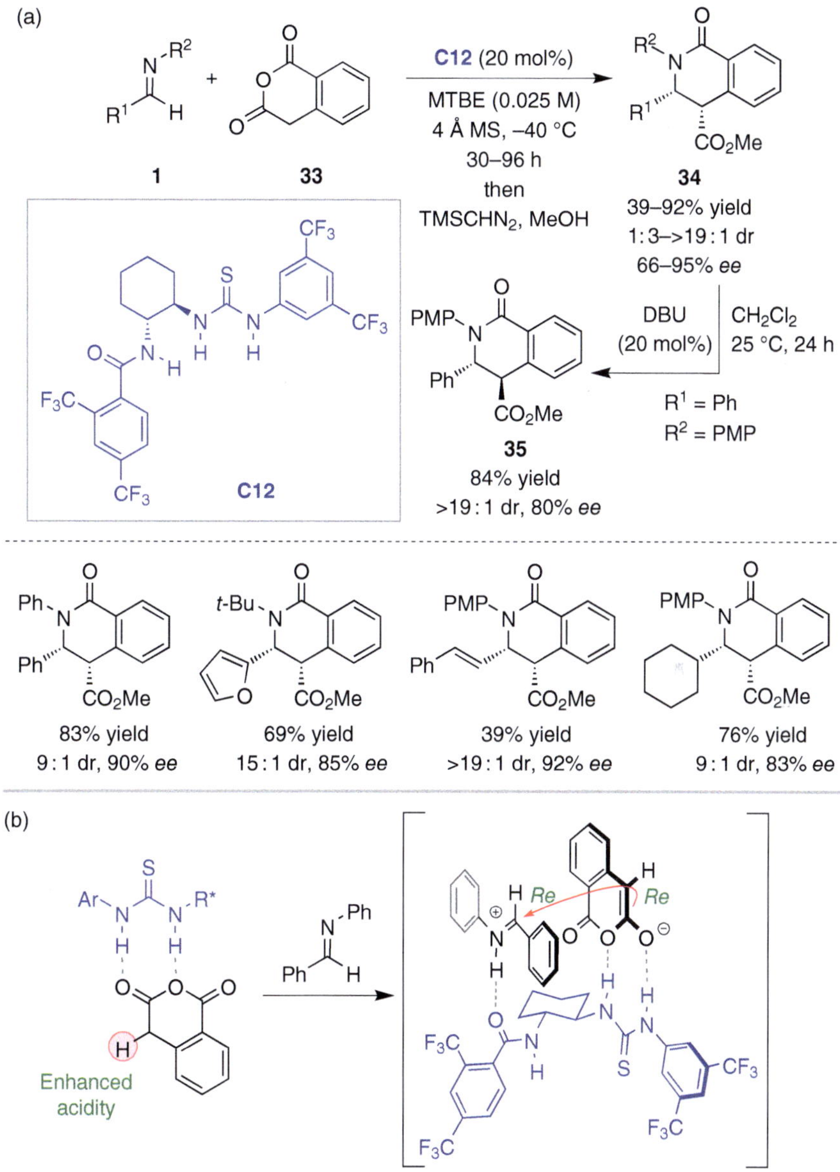

Scheme 5.13 Enantioselective Castagnoli–Cushman reaction of homophthalic anhydride under anion-binding catalysis by Vetticatt and coworkers [38]. Source: Based on Jarvis *et al.* [38].

pathway (via 1,2-addition or Mannich-type addition). While the products were predominantly formed as the kinetically favored *cis*-isomer (**34**), epimerization to the more stable *trans*-isomer (**35**) is possible upon treatment with DBU, albeit with a slight erosion of enantiopurity.

The catalytic hypothesis is based on the dual role of **C12**: (i) acidity enhancement of homophthalic anhydride (**33**) by forming H-bonded complex and (ii) creation of a chiral ion pair by binding with anhydride enolate (Scheme 5.13b). DFT calculations supported this hypothesis and rationalized the formation of (*S*,*S*)-**34** by nucleophilic addition from the Re-face of the catalyst-bound enolate to the Re-face of protonated imine [38]. The orientation of the protonated imine is benefited from a strong H-bonding interaction with the amide oxygen of the catalyst. Overall, the neutral catalyst brings the two reactive ionic intermediates by concurrently interacting with both of them.

Compared to homophthalic anhydride, most other anhydrides (e.g. succinic anhydride) are not sufficiently acidic to allow enolization with imine as the base. Therefore, the scope of Castagnoli–Cushman and related reactions with respect to anhydrides has been rather limited with the exception of α-aryl-substituted succinic anhydrides [37d].

Following the trail of the enantioselective Castagnoli–Cushman reaction of homophthalic anhydride developed by Seidel group [38], Connon and coworkers applied the same strategy for an analogous reaction with α-aryl-substituted succinic anhydrides (**36**) using bisurea **C13** as the catalyst (Scheme 5.14) [39]. The presence of an electron-deficient aryl substituent at the α-position of succinic anhydrides is essential for the enolate formation. Similarly, aldimines derived from electron-rich aromatic aldehydes proved to be superior to electron-deficient aromatic or aliphatic aldehydes. The resulting stereochemically rich γ-lactam carboxylic acids were formed exclusively as a single diastereoisomer and isolated as methyl ester **37** in excellent yield with good to excellent *ee*. DFT calculations point to a mechanism similar to the one proposed by Seidel and Vetticatt for homophthalic anhydride [38]. The binding of the imine **1** in the hydrophobic pocket created between the two bis(trifluoromethyl)phenyl rings of the catalyst **C13** is facilitated through multiple π–π interactions with its aryl residues and explains the requirement of electron-rich aldimines.

The Castagnoli–Cushman reaction proceeds through an initial 1,2-addition of anhydride enolates to imines, which is a Mannich-type reaction. An analogous pathway with α,β-unsaturated imines proceeding through a conjugate addition as the initial C—C bond formation step followed by cyclization could lead to a carbocyclic skeleton. This was found to be a competing pathway in the Castagnoli–Cushman reaction with cinnamaldehyde-derived imine, reported by Seidel, Vetticatt, and coworkers (see Scheme 5.13) [38]. The possibility of regiocontrol through the use of bulky N-substituent on α,β-unsaturated imine was established back in 1990s [40]. However, an enantioselective version of this 3,4-cycloaddition, also known as Tamura cycloaddition [41], was unknown until recently. In 2019, Connon and coworkers developed an enantioselective Tamura cycloaddition between homophthalic anhydrides (**33**) and *N*-trityl α,β-unsaturated imines (**38**) using the

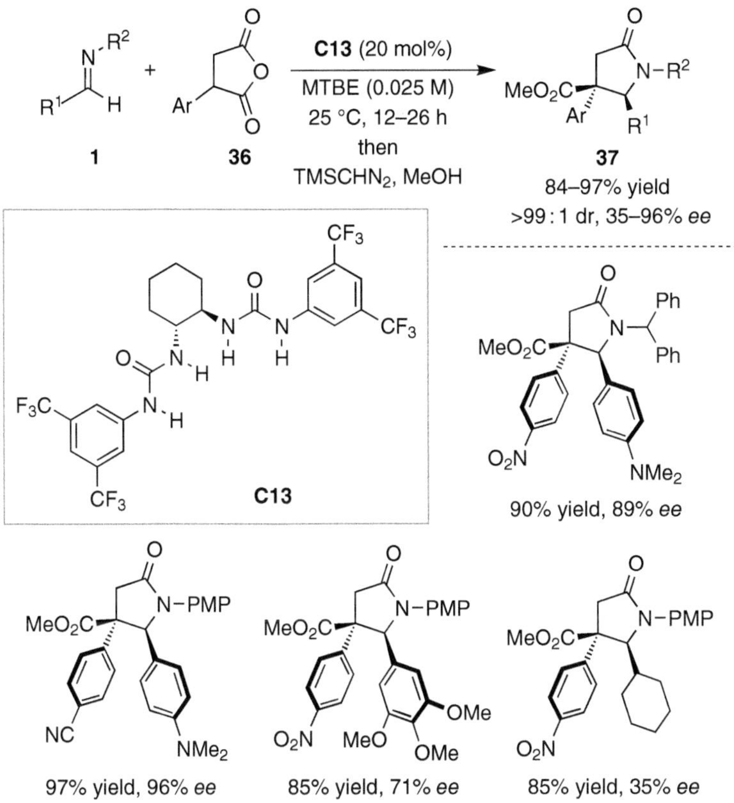

Scheme 5.14 Enantioselective Castagnoli–Cushman reaction of α-aryl succinic anhydrides under anion-binding catalysis by Connon and coworkers [39]. Source: Based on Collar *et al.* [39].

amido-thiourea derivative **C14** (Scheme 5.15) [42]. In the presence of only 5 mol% of **C14**, this energetically facile transformation proceeds efficiently with a wide variety of β-substituted imines at −40 °C to furnish highly substituted α-tetralone derivatives (**39**) in very high yield with moderate to excellent dr and mostly with excellent *ee*. Interestingly, the products (**39**) contain a 1,3-dicarbonyl-derived enamine functionality, which can be reacted with suitable electrophiles. This was demonstrated with a α-chlorination reaction. Hydrolysis of the resulting *N*-trityl imine **40** followed by oxidation and decarboxylation led to α-chlorotetralone derivative **41** as a single diastereomer.

DFT calculations revealed the stepwise nature of this reaction involving a 1,4-adduct as the intermediate and the irreversible cyclization step [42]. The binding of the enolate and the iminium ion to the catalyst is reminiscent to the one described above (see Scheme 5.13), where the bulky *tert*-butyl group on the catalyst **C14** is essential to ensure high enantioselectivity. It is rather curious that the rate-determining steps are different for the pathways leading to the major and minor enantiomers.

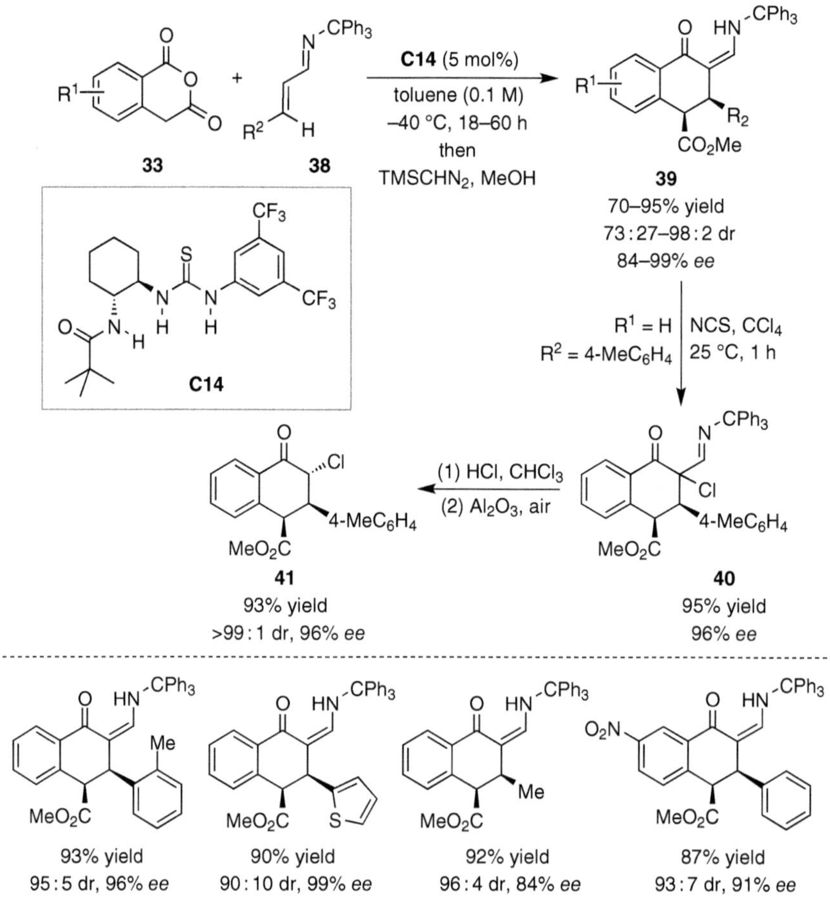

Scheme 5.15 Enantioselective Tamura cycloaddition under anion-binding catalysis by Connon and coworkers [42]. Source: Based on Collar *et al.* [42].

The reactions discussed so far in this section rely on the stabilization of the anionic nucleophiles by H-bonding from the catalysts. Apart from this commonly encountered scenario, the H-bond donor catalyst can also activate a cationic electrophile by stabilizing its counteranion. Anion binding of this type has the potential to orient the reaction partners to enable an enantioface-selective bond formation.

Such an activation mode was demonstrated by Jacobsen and coworkers for an enantioselective chlorination reaction of α-tertiary silyl ketene acetals (**42**) with *N*-chlorosuccinimide (NCS) **43** (Scheme 5.16) [43]. The *tert*-leucine-arylpyrrolidine squaramide catalyst **C15** appears to play multiple roles in this reaction, which proceeds through a rate-determining transfer of Cl^+ from NCS to the silyl ketene acetals **42** [44]. The polarized transition state leading to the ion pair intermediate is proposed to be stabilized by H-bonding to the ensuing succinimide anion and a concomitant cation–π interaction of the developing positive charge on the silyl ketene acetal with the phenanthrylpyrrolidine moiety on the catalyst **C15** [43].

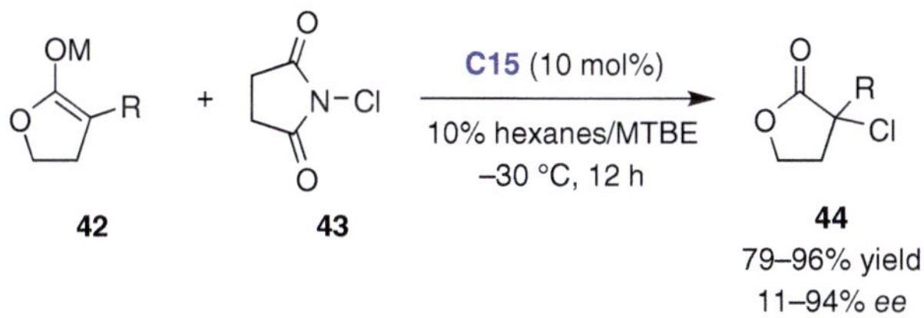

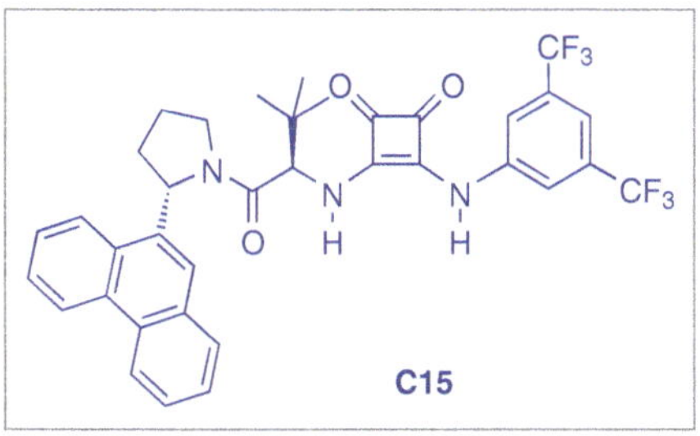

R	M	Yield (%)	*ee* (%)
Ph	TMS	91	90
2-Naphthyl	TMS	90	92
4-ClC$_6$H$_4$	TMS	96	94
3-OMeC$_6$H$_4$	TMS	84	80
3-Thienyl	TMS	88	58
4-OMeC$_6$H$_4$	TMS	95	11
Ph	TES	90	86
Ph	TBS	79	83

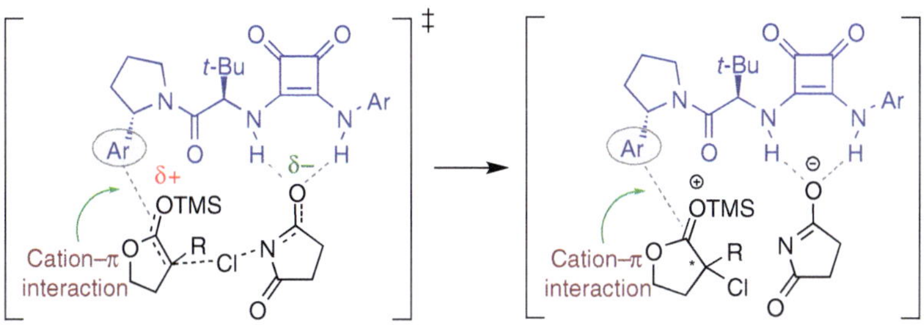

Scheme 5.16 Enantioselective α-chlorination of silyl ketene acetals under anion-binding catalysis by Jacobsen and coworkers [43]. Source: Based on Liu *et al.* [43].

Highly non-polar reaction medium, where NCS is sparingly soluble, is essential for obtaining a good level of enantioselectivity. Therefore, the squaramide catalyst may also help in solubilizing NCS through H-bonding interaction. The optimized conditions were found to be tolerated by electron-neutral or mildly electron-deficient aryl substituted cyclic silyl ketene acetals (**42**) to afford α-chlorolactones **44**, bearing a quaternary stereocenter, in high yield with good enantioselectivity. However, enantiocontrol was dramatically reduced for **42** bearing electron-rich aryl substituent, thus leaving room for improvement. The involvement of neutral silyl ketene acetals **42** (as opposed to the corresponding enolate) in the enantiodetermining C—Cl bond formation step is clearly evident from the influence of the silyl group on the enantioselectivity of this α-chlorination reaction.

In addition to the reactions discussed above, there are a number of examples where enolates derived from cyclic lactones or imides participate in H-bonding interaction with the catalyst. Perhaps the most prominent examples are the electrophilic halogen-induced difunctionalizations of olefins [45]. The counteranions derived from the electrophilic halogen source (e.g. succinimide anion from *N*-halosuccinimide) are often H-bonded with the catalyst [46]. Such anion-binding stabilization has been shown to have visible influence not only on the reactivity but also on the stereoselectivity of these reactions. However, in a majority of these reactions, the catalysts also harbor Brønsted/Lewis basic sites and proceed through bi- or even multi-functional mode of activation. Therefore, these reactions are kept outside the scope of our discussion.

5.3.3 Carboxylates

Carboxylic acids are possibly the most commonly used organic acids in organic synthesis. A wide variety of chiral carboxylic acids in enantiomerically pure form are also available from natural resources. Nevertheless, the application of chiral carboxylic acids as a catalyst in enantioselective transformations remained underdeveloped until rather recently [47]. If one of the reasons for this paucity is the weaker Brønsted acidity, especially when compared to the more widely used phosphoric acids, difficulty in creating a suitable chiral environment around the carboxylic acid functionality is certainly a key contributing factor. Moreover, the tendency of carboxylate anion to participate in H-bonding with protonated substrates markedly attenuates the extent of LUMO-lowering substrate activation, although this interaction is often essential for high enantioselectivity. It is only in 2007 the Maruoka group developed the axially chiral dicarboxylic acids based on binaphthyl backbone [48] and successfully applied them for a series of catalytic enantioselective transformations [49]. Despite being recognized as the most widely used chiral carboxylic acid catalysts, these axially chiral dicarboxylic acid derivatives suffer from their multi-step synthesis [50].

Yamamoto pioneered the concept of "Brønsted acid-assisted Brønsted acid catalysis (BBA)", where the H-bonding interaction with a second (similar or different) Brønsted acid leads to the enhancement of Brønsted acidity of the parent Brønsted acid [51]. The foundation of this concept is the additional stabilization of the conjugate base of the acid of interest. Such cooperative augmentation of reactivity is a common phenomenon in enzymatic reactions. In fact, intramolecular H-bonding between the two carboxylic acid groups is believed to be responsible for increased acidity of Maruoka's axially chiral dicarboxylic acids [49].

Schreiner and coworkers reported a regioselective alcoholysis of styrene oxides **45** with a variety of alcohols using a combination of an equimolar mixture (1 mol% each) of mandelic acid (**C16**) and thiourea **C8** as the catalyst (Scheme 5.17a) [52]. Supported by DFT calculations, mandelic acid is proposed to coordinate with thiourea **C8** through double H-bonding and forms an additional intramolecular H-bond with the α-hydroxy group in a *cis*-hydroxy conformation. Such an array of H-bonding enhances the acidity of the α-hydroxy group, which then activates the epoxides by a single-point H-bond toward regioselective nucleophilic attack by alcohols at the benzylic position (Scheme 5.17b).

In 2009, Jacobsen and Klausen applied the BBA concept to combine an achiral carboxylic acid ($PhCO_2H$) with a chiral thiourea derivative **C17** to carry out an enantioselective Pictet–Spengler reaction (Scheme 5.18a) [53]. This operationally simple protocol involves the treatment of tryptamines **47** and a number of aromatic

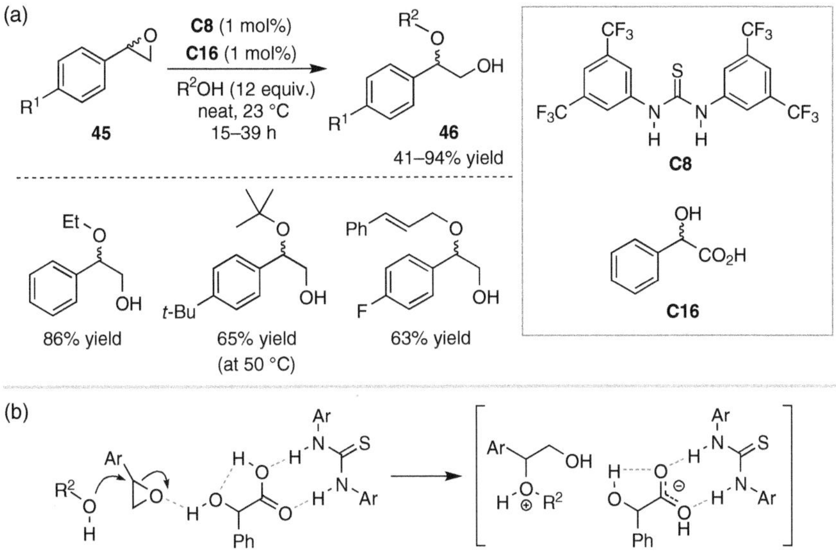

Scheme 5.17 Alcoholysis of styrene oxides under thiourea-carboxylic acid dual catalysis by Schreiner and coworkers [52]. Source: Based on Weil *et al.* [52].

or aliphatic aldehydes (**14**) with the catalyst combinations ($PhCO_2H$ and **C17**) either at room temperature or a slightly elevated temperature (35 °C). Under these conditions, the *in situ* formed imines underwent smooth cyclization to generate tetrahydro-β-carbolines **48** in moderate to good yield with good to excellent enantioselectivity.

The reaction is believed to take place through the protonation of the *in situ* formed imine with benzoic acid, whose acidity is enhanced by association with the thiourea derivative **C17** (Scheme 5.18b) [53]. The resulting thiourea-bound carboxylate forms a chiral ion pair with the iminium cation and triggers the enantioselective cyclization. Deprotonative rearomatization produces tetrahydro-β-carboline **48** and regenerate the thiourea co-catalyst **C17**. The same strategy was later on extended to 2-substituted indolylethylamines (isotryptamines) for an iso-Pictet–Spengler reaction [54].

In 2013, Seidel and coworkers introduced the concept of conjugate-base-stabilized Brønsted acid catalysts, where both the carboxylic acid and the thiourea functionalities are connected through a linker within the same molecule [55]. This elegant design offers stabilization of the corresponding conjugate base (carboxylate) by means of intramolecular H-bonding with the tethered anion receptor thiourea moiety (Figure 5.1). Apart from enhancing the acidity of the carboxylic acid, which was found to be the case in a later study [56], this internal anion binding weakens the ability of carboxylate anion to form H-bond with the protonated substrates, thereby resulting in strict ion pairing (*ion pair type-I*). In addition to the direct activation of substrates through protonation, such conjugate-base-stabilized carboxylic acid catalysts could also enable condensation of two substrates to produce a cationic intermediate, which can form an ion pair with the internally stabilized carboxylate

(a)

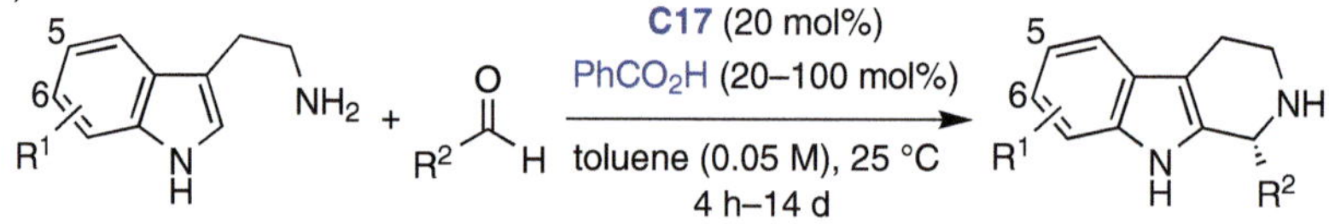

47 **14** **48**

39–94% yield
85–99% *ee*

R^1	R^2	Yield (%)	*ee* (%)
HH	2-BrC_6H_4	45[a]	95
6-OMe	Ph	94	86
6-OMe	4-ClC_6H_4	78	94
6-OMe	3-BrC_6H_4	87	94
6-OMe	*i*-Pr	60	88
5-OMe	2-BrC_6H_4	82[a]	99
5-OMe	*i*-Pr	39[a]	88

[a] Reaction at 35 °C

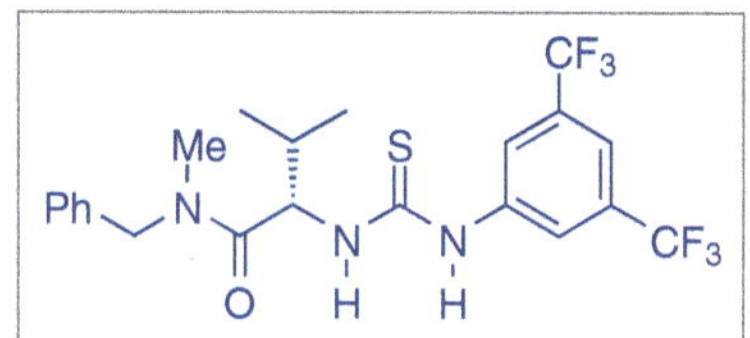

C17

(b)

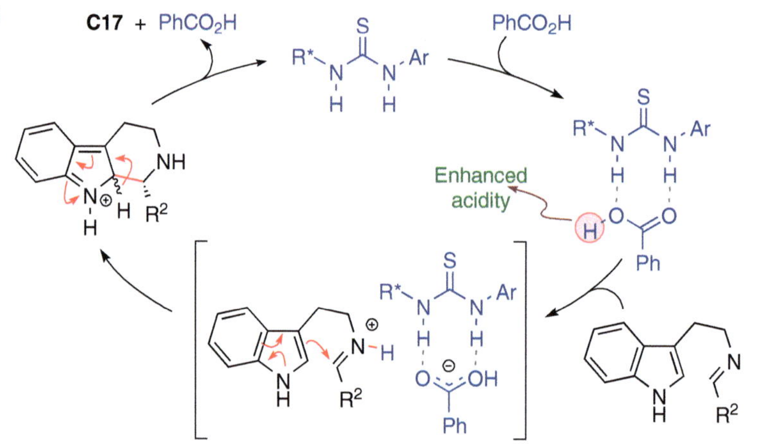

Scheme 5.18 Enantioselective Pictet–Spengler reaction under thiourea-carboxylic acid dual catalysis by Jacobsen and coworkers [53]. Source: Based on Klausen *et al.* [53].

anion (*ion pair type-II*). Because of the presence of the rigid chiral carboxylate anion, efficient transfer of stereochemical information can be expected from these ion pairs.

Besides entropic advantages associated with intramolecular chelation, the covalent attachment of these two functionalities offers an additional benefit over the dual catalysis approach discussed above with respect to the potential "background" reaction by the achiral carboxylic acid co-catalyst.

Seidel group successfully implemented this catalysis concept to achieve the first enantioselective Povarov reaction of secondary aromatic amines (Scheme 5.19) [55]. Using 20 mol% of an easy-to-prepare thiourea-carboxylic acid derivative **C18** as the catalyst, an array of cyclic secondary amines (**49**) and diversely substituted aldehydes (**14**) were assembled with 1-vinyl-pyrrolidine-2-one (**50**) to furnish the tricyclic

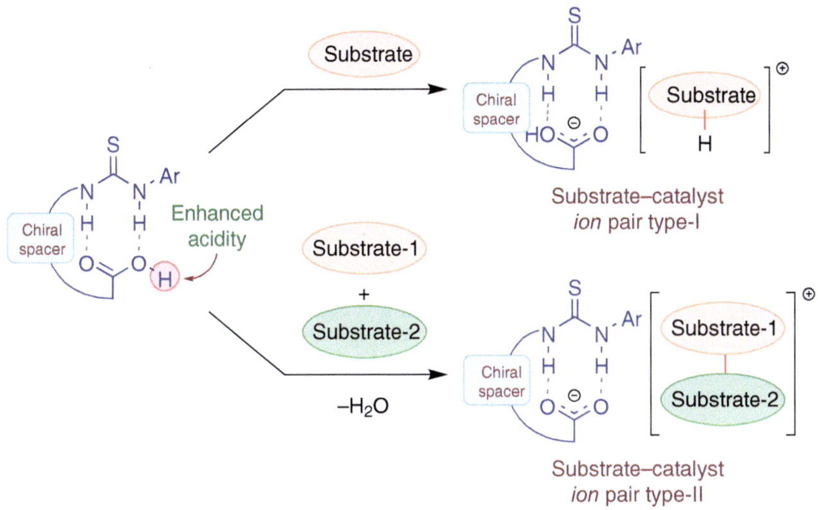

Figure 5.1 Substrate activation by conjugate-base-stabilized carboxylic acid catalysts [55].

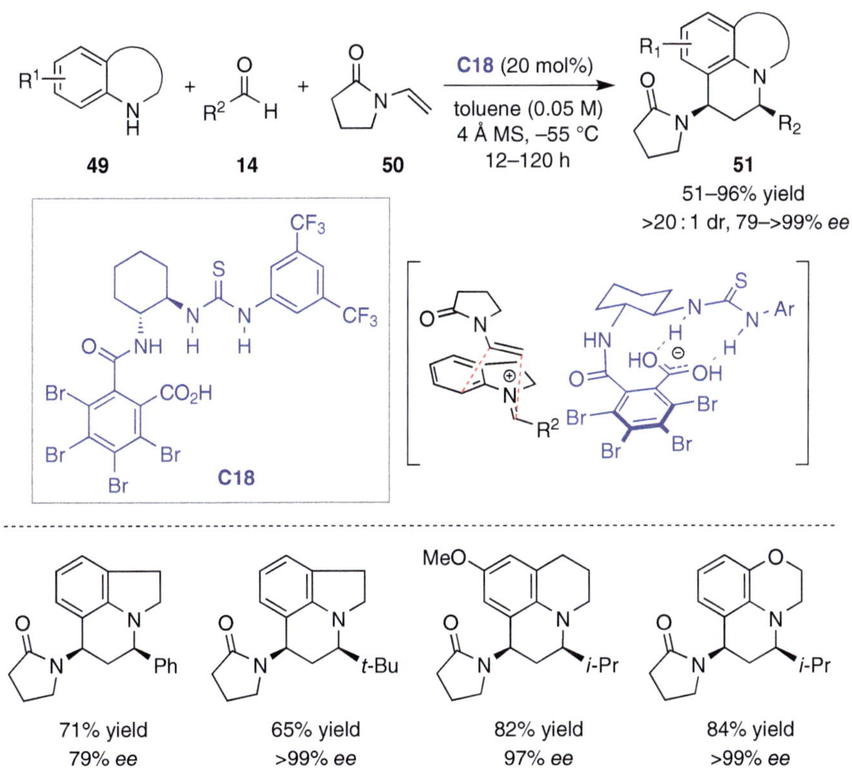

Scheme 5.19 Enantioselective Povarov reaction under conjugate-base-stabilized carboxylic acid catalysis by Seidel and coworkers [55]. Source: Based on Min *et al.* [55].

products **51** in good to excellent yields with good to high enantioselectivity. It must be noted that aliphatic aldehydes, which are known to be less reactive because of their propensity to form enamines, performed better than aromatic aldehydes under the optimized reaction conditions. The reaction is likely to proceed through the formation of an ion pair consisting of the *in situ* generated iminium cation (azadiene) and the internally stabilized carboxylate anion (*ion pair type-II*). An inverse electron demand [4+2]-cycloaddition reaction of 1-vinyl-pyrrolidine-2-one (**50**) with the cationic azadiene within the chiral ion pair leads to the transfer of stereochemical information from the catalyst to the product.

Preliminary structural analysis of the catalyst through X-ray diffraction revealed the presence of an internal H-bonding between the amide N–H proton and the sulfur atom of thiourea [55]. This H-bonding is likely to enhance the acidity of thiourea NHs and result in better anion binding.

The same catalyst and the similar reaction conditions were extended by Seidel and coworkers to accomplish an intramolecular Povarov reaction [57]. Thus, in the presence of 20 mol% of the thiourea-carboxylic acid catalyst **C18**, secondary aromatic amines (**49**) react with aldehyde containing dienophiles (**52**) to afford tetra- and pentacyclic products (**53**), having three contiguous stereogenic centers, in good to excellent yield with excellent diastereo- and enantioselectivity (Scheme 5.20). This reaction represents the first example of a catalytic enantioselective intramolecular aza-Diels–Alder reaction. Surprisingly, *rac*-**C18** was found to be virtually ineffective in catalyzing the formation of racemic product [57]. Moreover, a non-linear effect study showed the formation of the product with almost identical (high) level of enantioselectivity, even with a catalyst (**C18**) sample having 10% *ee*. However, both the reaction rate and diastereoselectivity steadily deteriorated with the erosion of catalyst *ee*. This observation points to a possible aggregation between (*R*,*R*)- and (*S*,*S*)-enantiomers of the catalyst. In fact, aggregation of enantiomerically pure catalyst, as observed previously in the X-ray structure [55], might also be relevant to the catalytic pathway. This catalyst system is rather sensitive to the substitution pattern on the secondary amine substrate as considerably diminished enantioselectivity was observed with 2-phenyl indoline [58].

With the objective of overcoming the low reactivity observed in the Pictet–Spengler reaction of unmodified tryptamine (**47**) reported by Jacobsen *et al.* (see Scheme 5.18), Seidel and coworkers applied internally conjugate-base-stabilized thiourea-carboxylic acid to this reaction [59]. Owing to the attenuation of H-bonding between the protonated imine and the thiourea-bound carboxylate within the ion pair (*ion pair type-I*), the iminium ion was expected to display enhanced electrophilicity. This was indeed found to be the case when **C18** was used as the catalyst (Scheme 5.21). However, severe product inhibition was encountered because of the higher basicity of the tetrahydro-β-carboline product compared to the *in situ* formed imine. The use of a stoichiometric amount of poorly soluble achiral carboxylic acid (e.g. malonic acid) proved effective in mitigating the product inhibition while maintaining high enantioselectivity. Yet, because of the lack of generality of this strategy, the authors opted for *in situ* Boc protection of the products. Under these conditions, electron-poor aromatic aldehydes (**14**) furnished

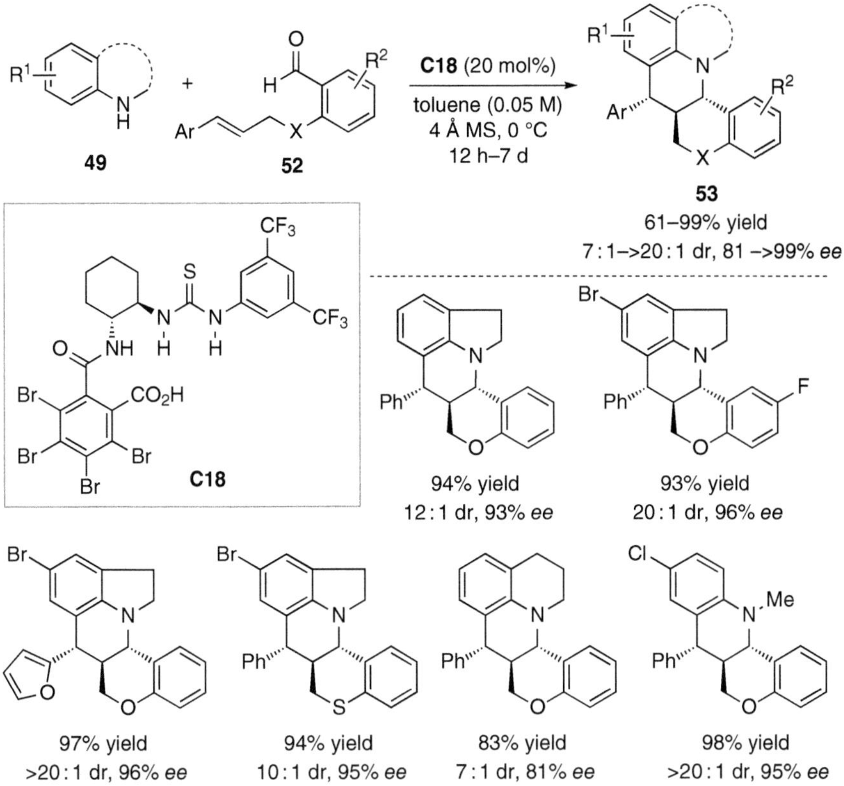

Scheme 5.20 Enantioselective intramolecular Povarov reaction under conjugate-base-stabilized carboxylic acid catalysis by Seidel and coworkers [57]. Source: Based on Min *et al.* [57].

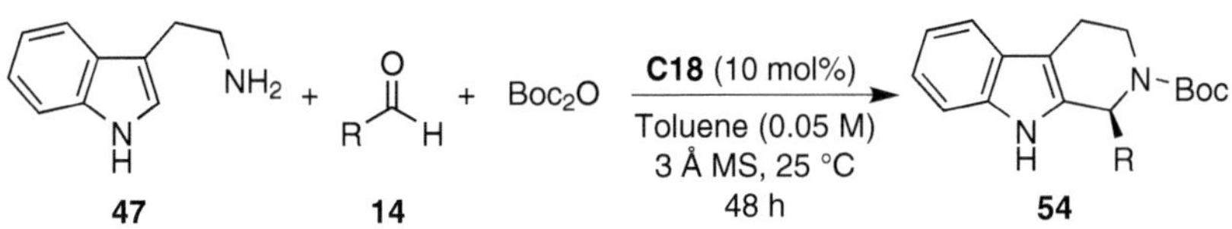

R	Yield (%)	*ee* (%)
Ph	86	87
4-ClC_6H_4	95	92
4-CNC_6H_4	95	92
4-MeC_6H_4	28[a]	64
4-$OMeC_6H_4$	12[a]	32
3-$OMeC_6H_4$	73[a]	84
2-ClC_6H_4	83	81
2-MeC_6H_4	62[a]	54
t-Bu	23[a]	62

[a] Reaction over 96 h.

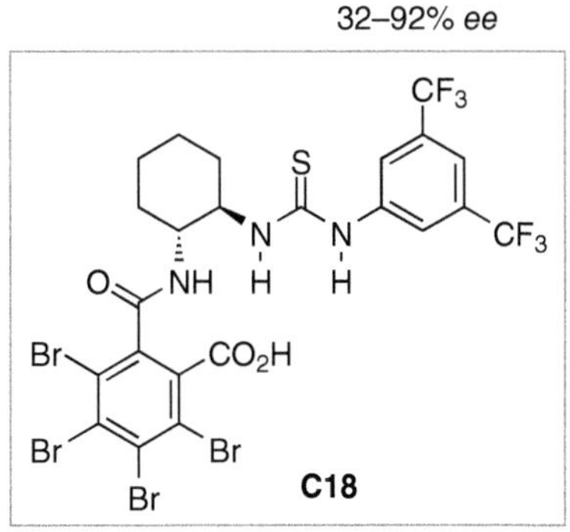

Scheme 5.21 Enantioselective Pictet–Spengler reaction of unmodified tryptamine under conjugate-base-stabilized carboxylic acid catalysis by Seidel and coworkers [59]. Source: Based on Mittal *et al.* [59].

N-Boc tetrahydro-β-carbolines (**54**) in good to high yield and good enantioselectivity. However, the products derived from aliphatic and electron-rich aromatic aldehydes were obtained with poor yield and enantioselectivity. In a way, this protocol complements Jacobsen's acyl-Pictet–Spengler reaction [60], which is better suited for aliphatic rather than aromatic aldehydes.

A recent study by Seidel, Yamanaka, and coworkers was directed to reveal the modus operandi of the catalyst **C18** in asymmetric Pictet–Spengler reaction [61]. Thorough X-ray structure analysis of the tetrabutyl ammonium salt of **C18** shows that the catalyst anion in the solid state exists as a 1D polymer chain through intermolecular N–H· · ·O interactions of both the thiourea NHs with both the carboxylate oxygens of a neighboring molecule. However, structure–activity relationship and computational studies point to the involvement of a discrete bowl-shaped internally H-bond-stabilized carboxylate anion in the rate- and enantiodetermining deprotonation/aromatization step. In this well-organized transition state, the catalyst anion interacts with the substrate through a number of non-covalent interactions including H-bonding, halogen bonding, and π–π interaction. In the TS leading to the major product enantiomer, the cationic intermediate fits more favorably with this chiral anion by maximizing these interactions and accounts for the large difference in interaction energy between the two diastereomeric TSs.

Considering the efficacy of the thiourea-carboxylic acid derivatives in enantioselectively catalyzing the reactions proceeding through iminium ion intermediates, Seidel and coworkers employed this class of compounds as a co-catalyst for the copper-catalyzed addition of terminal alkynes to *in situ* generated iminium ions [62]. This three-component reaction involving an aldehyde, an amine, and a terminal alkyne, commonly known as a A^3 reaction, is possibly the most straightforward method for the enantioselective synthesis of propargylic amines. However, an enantioselective A^3 reaction with secondary amines presents a considerable challenge because of the inherent difficulty associated with the activation of the intermediate iminium ions, i.e. the lack of available lone pair of electrons on nitrogen unlike imine, which can be activated via H-bonding, protonation, or metal coordination toward enantioselective addition. This constraint provides an ideal opportunity to test the potential of the conjugate-base-stabilized carboxylic acid catalysts in rendering the A^3 reaction enantioselective through the formation of a rigid ion pair with iminium ion.

Using the combination of 4 mol% of CuI and 3 mol% thiourea-carboxylic acid **C19**, cyclic secondary amines (**55**), an array of aldehydes (**14**), and a variety of terminal alkynes (**56**) were efficiently assembled to furnish propargylamines (**57**) in good to excellent yield and in most cases with high level of enantioselectivity (Scheme 5.22) [62].

This catalyst system is particularly suitable for aromatic aldehydes as considerably lower enantioselectivity was obtained with aliphatic aldehyde. However, alkynes with aromatic, aliphatic, as well as alkenyl substituents were tolerated. While a similar level of outcome was observed with pyrrolidine and azepane, six-membered cyclic amines were found to be less reactive and furnished the products with less *ee*. The authors could showcase the utility of the propargylamine products by converting

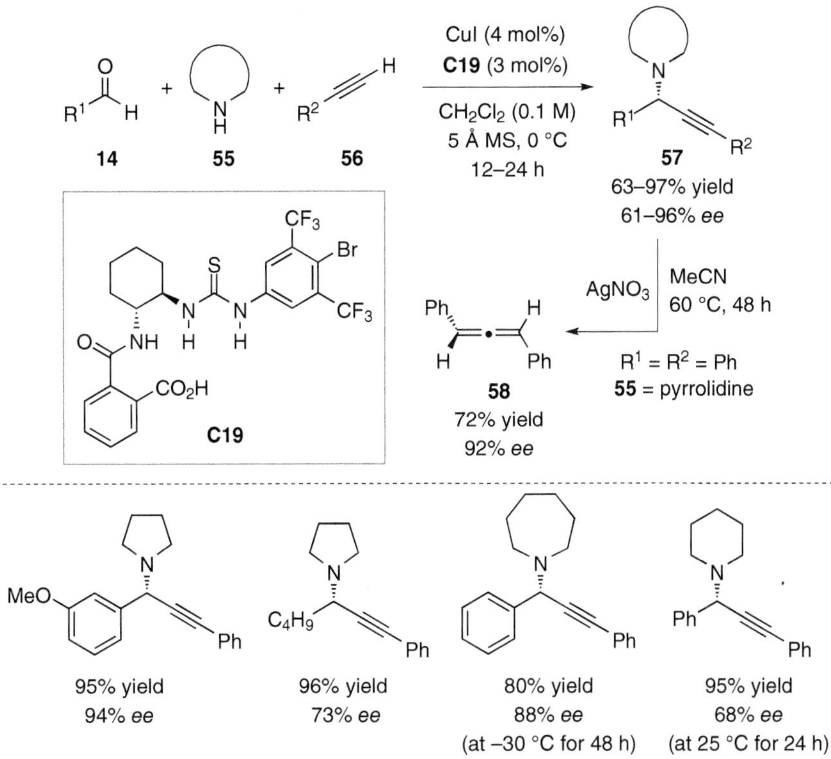

Scheme 5.22 Enantioselective A^3 reaction of secondary amines with cooperative Cu(I)/thiourea-carboxylic acid catalysis by Zhao and Seidel [62]. Source: Zhao and Seidel [62].

them to chiral allenes (e.g. **58**) in a single step with complete transfer of point-to-axial chirality. Although the exact role of the catalyst **C19** in determining enantioselectivity remained unclear, a number of possibilities exist including the involvement of thiourea sulfur center as a ligand for copper(I), and the formation of a cuprate complex with the carboxylate besides the formation of the originally conceived rigid ion pair with iminium ion.

The compatibility of these internally H-bonded chiral carboxylate anions with more electrophilic and much less developed oxocarbenium ions was recently demonstrated by Seidel and coworkers for an analogous oxa-Pictet–Spengler reaction [56a]. Thus, when tryptophols **59** was combined with dimethyl ketals **18** in the presence of only 10 mol% of thiourea-carboxylic acid catalyst **C20**, tetrahydropyranoindoles **60** were obtained in excellent yield with high enantioselectivity (Scheme 5.23a). The catalyst **C20** was found to be remarkably efficient for this reaction and tolerated diversely substituted tryptophols (**59**) and a wide variety of dimethyl ketals (**18**). In fact, spirocyclic product could also be prepared with good enantioselectivity.

A few control experiments established the formation of a mixed ketal **M** by transketalization of dimethyl ketal **18** with tryptophol **59** (Scheme 5.23b) [56a]. Acid-catalyzed elimination of methanol from **M** can give rise to two stereoisomeric

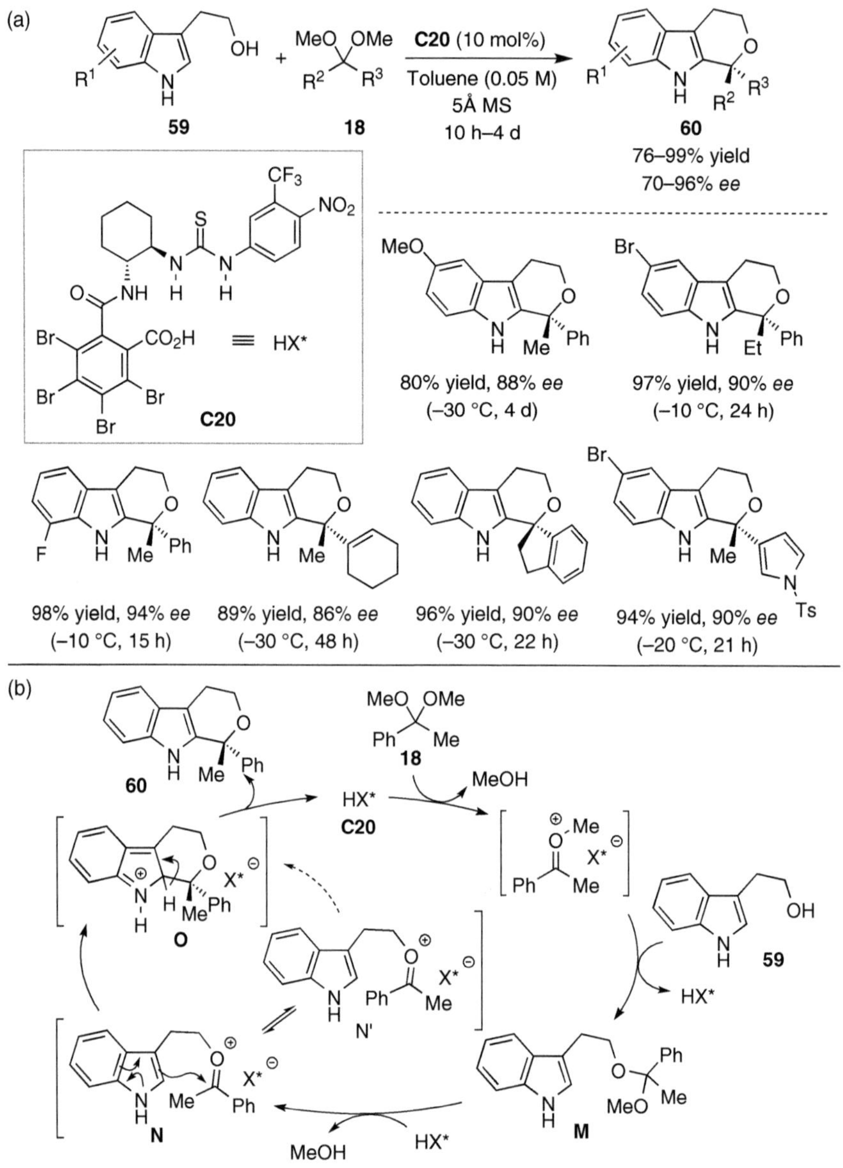

Scheme 5.23 Enantioselective oxa-Pictet–Spengler reaction under conjugate-base-stabilized carboxylic acid catalysis by Seidel and coworkers [56a]. Source: Based on Zhu *et al.* [56a].

oxocarbenium ions **N** and **N′**, one of which participates in the subsequent cyclization under the influence of the catalyst-derived chiral carboxylate anion. Deprotonative rearomatization of indole in **O** then liberates the product **60** along with the concomitant regeneration of the catalyst **C20**. The indole NH is turned out to be the key for rendering this reaction enantioselective as this NH in one

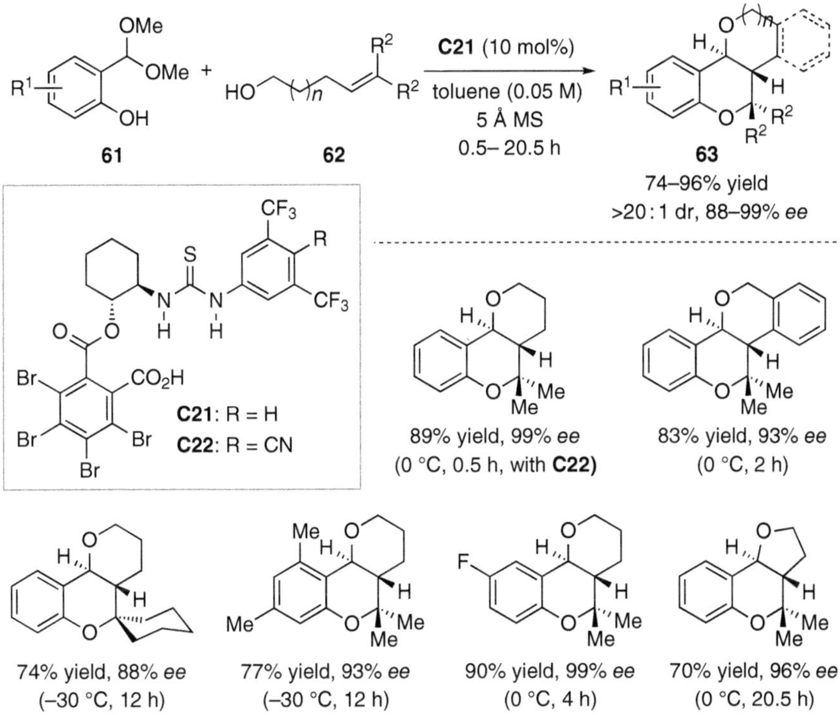

Scheme 5.24 Enantioselective [4+2]-cycloaddition of acetals under conjugate-base-stabilized carboxylic acid catalysis by Seidel and coworkers [56b]. Source: Zhu *et al.* [56b].

of the critical intermediates interact with the catalyst in the enantiodetermining step [56a].

The ability of the conjugate-base-stabilized chiral carboxylic acids in catalyzing reactions proceeding through oxocarbenium ions prompted the development of a highly efficient [4+2]-cycloaddition of salicylaldehyde-derived acetals (**61**) with homoallylic or bishomoallylic alcohols (**62**) by the Seidel group (Scheme 5.24) [56b]. During the course of this study, it was revealed that the acidity (pK_a) of the catalysts does not correlate with their catalytic activity. Therefore, changing the linker of the thiourea-carboxylic acid derivative from amide to ester, although led to somewhat reduced acidity [pK_a (MeCN) of **C18** and **C21** are 12.4 and 13.6, respectively], **C21** turned out to be a remarkably more active and selective catalyst compared to the previously described **C18** (*vide supra*). Only 10 mol% of **C21** is sufficient to catalyze the reaction between **61** and **62** and furnished pyranochromanes and furanochromane (**63**) exclusively as the *trans*-fused diastereomers generally in high yield with excellent enantioselectivity. In fact, with a more active ester-linked thiourea-carboxylic acid derivative **C22**, the catalyst loading could be reduced to 0.5 mol% for a larger scale reaction at higher concentration without compromising the stereoselectivities of the reaction.

The reaction was found to be stereospecific with respect to the double bond geometry of the bishomoallylic alcohol [56b]. While this observation does point to a concerted [4+2]-cycloaddition with the *in situ* formed *ortho*-quinone methide, a Prins-type pathway involving a tertiary carbocation intermediate appears to be more likely based on a study by Spivey and coworkers [63]. Interestingly, the catalyst **C21** proved to be more active and selective for this reaction compared to chiral phosphoric acids with similar acidity or even more acidic imidodiphosphoric acid catalysts (IDP) [64].

Within a few years since their emergence [55], these conjugate-base-stabilized chiral carboxylic acids proved their potential in catalyzing some of the challenging transformations enantioselectively. Equipped with the prospect of acidity modulation and structural variation, this class of catalysts has the potential to effect transformations other than those proceeding under Brønsted acid activation. One such example, namely, the sulfenylation of indoles and pyrroles, was reported by Gustafson and coworkers using a *cis*-1,2-diaminocyclohexane-derived thiourea-carboxylic acid as the catalyst [65]. More exciting developments, especially in the realm of enantioselective catalysis, under this class of thiourea-carboxylic acid should follow.

5.4 Conclusions and Outlook

In this chapter, we attempted to cover the reactions, where H-bonding interactions of cyanide, oxyanions, enolates, and carboxylates with the catalyst constitute the primary activation mode in dictating the reactivity and/or enantioselectivity. Apart from the examples discussed above, there are plenty of reactions involving the stabilization of anionic reactive intermediates by H-bonding with the catalyst and the influence of such anion binding in determining the enantioselectivity. This phenomenon is particularly common for reactions catalyzed by bifunctional H-bond donors such as amino(thio)urea or aminosquaramide derivatives. In fact, any conjugate addition reaction to unsaturated carbonyl compounds catalyzed by such bifunctional catalysts invariably comprises the stabilization of the enolate intermediate by H-bonding.

The field of anion-binding catalysis is merely one and half decade old and may therefore be considered at its infancy. Nevertheless, the range of applications already established is remarkable and at times surpasses the traditional modes (i.e. Lewis/Brønsted acid/base) of catalysis. Undoubtedly, the most notable feature of anion-binding catalysis lies in rendering those reactions enantioselective, where the cationic intermediate is devoid of any apparent binding site and therefore difficult to activate with the traditional modes. The easy accessibility and the air and moisture stability of the catalysts used for anion-binding catalysis allows for the reactions to be carried out with a fairly simple reaction setup without the need for rigorous exclusion of air and moisture. The latter factor ensures scalability and reproducibility of these reactions.

As the reactions proceeding via anion-binding activation necessarily involves cationic intermediates, it should be possible to carry out at least some of the transition metal-catalyzed transformations in an enantioselective fashion under anion-binding catalysis. If successful, this strategy would obviate the need of relatively expensive chiral ligands. Efforts in this direction can be expected in the near future.

References

1 Desiraju, G.R. (2011). A bond by any other name. *Angew. Chem. Int. Ed.* 50: 52–59.

2 Wassermann, A. (1942). Homogeneous catalysis of diene syntheses. A new type of third-order reaction. *J. Chem. Soc.*: 618–621.

3 Yates, P. and Eaton, P. (1960). Acceleration of the Diels–Alder reaction by aluminum chloride. *J. Am. Chem. Soc.* 82: 4436–4437.

4 Taylor, M.S. and Jacobsen, E.N. (2006). Asymmetric catalysis by chiral hydrogen-bond donors. *Angew. Chem. Int. Ed.* 45: 1520–1543.

5 (a) Doyle, A.G. and Jacobsen, E.N. (2007). Small-molecule H-bond donors in asymmetric catalysis. *Chem. Rev.* 107: 5713–5743.
(b) Schreiner, P.R. (2003). Metal-free organocatalysis through explicit hydrogen bonding interactions. *Chem. Soc. Rev.* 32: 289–296.
(c) Connon, S.J. (2006). Organocatalysis mediated by (thio)urea derivatives. *Chem.–Eur. J.* 12: 5418–5427.
(d) Albrecht, Ł., Jiang, H., and Jørgensen, K.A. (2014). Hydrogen-bonding in aminocatalysis: from proline and beyond. *Chem.–Eur. J.* 20: 358–368.
(e) Alemán, J., Parra, A., Jiang, H., and Jørgensen, K.A. (2011). Squaramides: bridging from molecular recognition to bifunctional organocatalysis. *Chem.–Eur. J.* 17: 6890–6899.

6 (a) Connon, S.J. (2008). Asymmetric catalysis with bifunctional cinchona alkaloid-based urea and thiourea organocatalysts. *Chem. Commun.*: 2499–2510.
(b) Chauhan, P., Mahajan, S., Kaya, U. *et al.* (2015). Bifunctional amine-squaramides: powerful hydrogen-bonding organocatalysts for asymmetric domino/cascade reactions. *Adv. Synth. Catal.* 357: 253–281.
(c) Fang, X. and Wang, C.-J. (2015). Recent advances in asymmetric organocatalysis mediated by bifunctional amine–thioureas bearing multiple hydrogen-bonding donors. *Chem. Commun.* 51: 1185–1197.
(d) Wei, Y.-L.S.Y. and Shi, M. (2017). Applications of chiral thiourea-amine/phosphine organocatalysts in catalytic asymmetric reactions. *ChemCatChem* 9: 718–727.

7 (a) Etter, M.C., Urbañczyk-Lipkowska, Z., Zia-Ebrahimi, M., and Panunto, T.W. (1990). Hydrogen bond directed cocrystallization and molecular recognition properties of diarylureas. *J. Am. Chem. Soc.* 112: 8415–8426.
(b) Etter, M.C. (1990). Encoding and decoding hydrogen-bond patterns of organic compounds. *Acc. Chem. Res.* 23: 120–126.

8 (a) Fan, E., Van Arman, S.A., Kincaid, S., and Hamilton, A.D. (1993). Molecular recognition: hydrogen-bonding receptors that function in highly competitive solvents. *J. Am. Chem. Soc.* 115: 369–370.

(b) Albert, J.S. and Hamilton, A.D. (1993). Synthetic analogs of the ristocetin binding site: neutral, multidentate receptors for carboxylate recognition. *Tetrahedron Lett.* 34: 7363–7366.
(c) Smith, P.J., Reddington, M.V., and Wilcox, C.S. (1992). Ion pair binding by a urea in chloroform solution. *Tetrahedron Lett.* 33: 6085–6088.
(d) Zhang, Z. and Schreiner, P.R. (2009). (Thio)urea organocatalysis – what can be learnt from anion recognition? *Chem. Soc. Rev.* 38: 1187–1198.

9 (a) Brak, K. and Jacobsen, E.N. (2013). Asymmetric ion-pairing catalysis. *Angew. Chem. Int. Ed.* 52: 534–561.
(b) Lacour, J. and Moraleda, D. (2009). Chiral anion-mediated asymmetric ion pairing chemistry. *Chem. Commun.*: 7073–7089.
(c) Brière, J.-F., Oudeyer, S., Dalla, V., and Levacher, V. (2012). Recent advances in cooperative ion pairing in asymmetric organocatalysis. *Chem. Soc. Rev.* 41: 1696–1707.
(d) Mahlau, M. and List, B. (2013). Asymmetric counteranion-directed catalysis: concept, definition, and applications. *Angew. Chem. Int. Ed.* 52: 518–533.
(e) Phipps, R.J., Hamilton, G.L., and Toste, F.D. (2012). The progression of chiral anions from concepts to applications in asymmetric catalysis. *Nat. Chem.* 4: 603–614.
(f) Beckendorf, S., Asmus, S., and García Mancheño, O. (2012). H-donor anion acceptor organocatalysis – the ionic electrophile activation approach. *ChemCatChem* 4: 926–936.
(g) Visco, M.D., Attard, J., Guan, Y., and Mattson, A.E. (2017). Anion-binding catalyst designs for enantioselective synthesis. *Tetrahedron Lett.* 58: 2623–2628.

10 Sigman, M.S. and Jacobsen, E.N. (1998). Schiff base catalysts for the asymmetric Strecker reaction identified and optimized from parallel synthetic libraries. *J. Am. Chem. Soc.* 120: 4901–4902.

11 Sigman, M.S. and Jacobsen, E.N. (1998). Enantioselective addition of hydrogen cyanide to imines catalyzed by a chiral (Salen)Al(III) complex. *J. Am. Chem. Soc.* 120: 5315–5316.

12 Sigman, M.S., Vachal, P., and Jacobsen, E.N. (2000). A general catalyst for the asymmetric Strecker reaction. *Angew. Chem. Int. Ed.* 39: 1279–1281.

13 Vachal, P. and Jacobsen, E.N. (2000). Enantioselective catalytic addition of HCN to ketoimines. Catalytic synthesis of quaternary amino acids. *Org. Lett.* 2: 867–870.

14 Su, J.T., Vachal, P., and Jacobsen, E.N. (2000). Practical synthesis of a soluble Schiff base catalyst for the asymmetric Strecker reaction. *Adv. Synth. Catal.* 343: 197–200.

15 Vachal, P. and Jacobsen, E.N. (2002). Structure-based analysis and optimization of a highly enantioselective catalyst for the Strecker reaction. *J. Am. Chem. Soc.* 124: 10012–10014.

16 Zuend, S.J., Coughlin, M.P., Lalonde, M.P., and Jacobsen, E.N. (2009). Scaleable catalytic asymmetric Strecker syntheses of unnatural α-amino acids. *Nature* 461: 968–971.

17 Zuend, S.J. and Jacobsen, E.N. (2009). Mechanism of amido-thiourea catalyzed enantioselective imine hydrocyanation: transition state stabilization via multiple non-covalent interactions. *J. Am. Chem. Soc.* 131: 15358–15374.
18 Pan, S.C., Zhou, J., and List, B. (2007). Catalytic asymmetric acylcyanation of imines. *Angew. Chem. Int. Ed.* 46: 612–614.
19 Pan, S.C. and List, B. (2007). Catalytic asymmetric three-component acyl-Strecker reaction. *Org. Lett.* 9: 1149–1151.
20 Pan, S.C. and List, B. (2008). The catalytic acylcyanation of imines. *Chem. Asian. J.* 3: 430–437.
21 Kotke, M. and Schreiner, P.R. (2006). Acid-free, organocatalytic acetalization. *Tetrahedron* 62: 434–439.
22 Kotke, M. and Schreiner, P.R. (2007). Generally applicable organocatalytic tetrahydropyranylation of hydroxy functionalities with very low catalyst loading. *Synthesis*: 779–790.
23 Madarász, Á., Dósa, Z., Varga, S. *et al.* (2016). Thiourea derivatives as Brønsted acid organocatalysts. *ACS Catal.* 6: 4379–4387.
24 Balmond, E.I., Coe, D.M., Galan, M.C., and McGarrigle, E.M. (2012). α-Selective organocatalytic synthesis of 2-deoxygalactosides. *Angew. Chem. Int. Ed.* 51: 9152–9155.
25 (a) Favorskii, A.E. (1894). *J. Russ. Phys.-Chem. Soc.* 26: 559.
(b) Loftfield, R.B. (1951). The alkaline rearrangement of α-haloketones. II. The mechanism of the Faworskii reaction. *J. Am. Chem. Soc.* 73: 4707–4714.
26 (a) Vaidya, T., Eisenberg, R., and Frontier, A.J. (2011). Catalytic Nazarov cyclization: the state of the art. *ChemCatChem* 3: 1531–1548.
(b) Wenz, D.R. and de Alaniz, J.R. (2015). The Nazarov cyclization: a valuable method to synthesize fully substituted carbon stereocenters. *Eur. J. Org. Chem.*: 23–37.
27 (a) Harmata, M. (2010). The (4+3)-cycloaddition reaction: simple allylic cations as dienophiles. *Chem. Commun.* 46: 8886–8903.
(b) Harmata, M. (2010). The (4+3)-cycloaddition reaction: heteroatom-substituted allylic cations as dienophiles. *Chem. Commun.* 46: 8904–8922.
28 (a) Tang, Q., Chen, X., Tiwari, B., and Chi, Y.R. (2012). Addition of indoles to oxyallyl cations for facile access to a-indole carbonyl compounds. *Org. Lett.* 14: 1922–1925.
(b) Vander Wal, M.N., Dilger, A.K., and MacMillan, D.W.C. (2013). Development of a generic activation mode: nucleophilic α-substitution of ketones via oxy-allyl cations. *Chem. Sci.* 4: 3075–3079.
29 Liu, C., Oblak, E.Z., Vander Wal, M.N. *et al.* (2016). Oxy-allyl cation catalysis: an enantioselective electrophilic activation mode. *J. Am. Chem. Soc.* 138: 2134–2137.
30 He, C.Q., Yu, P., Lam, Y., and Houk, K.N. (2017). Origins of stereoselectivity in chiral aminoalcohol catalysis of oxyallyl cation–indole reactions. *Org. Lett.* 19: 5685–5688.
31 (a) Reissert, A. (1905). Ueber die Einführung der Benzoyl-gruppe in tertiäre cyclische Basen. *Ber. Dtsch. Chem. Ges.* 38: 1603–1614.
(b) Ahamed, M. and Todd, M.H. (2010). Catalytic asymmetric additions of

carbon-centered nucleophiles to nitrogen-containing aromatic heterocycles. *Eur. J. Org. Chem.*: 5935–5942.

32 (a) Taylor, M.S., Tokunaga, N., and Jacobsen, E.N. (2005). Enantioselective thiourea-catalyzed acyl-Mannich reactions of isoquinolines. *Angew. Chem. Int. Ed.* 44: 6700–6704.
(b) Schafer, A.G., Wieting, J.M., Fisher, T.J., and Mattson, A.E. (2013). Chiral silanediols in anion-binding catalysis. *Angew. Chem. Int. Ed.* 52: 11321–11324.
(c) Wieting, J.M., Fisher, T.J., Schafer, A.G. *et al.* (2015). Preparation and catalytic activity of BINOL-derived silanediols. *Eur. J. Org. Chem.*: 525–533.
(d) Zurro, M., Asmus, S., Bamberger, J. *et al.* (2016). Chiral triazoles in anion-binding catalysis: new entry to enantioselective Reissert-type reactions. *Chem.-Eur. J.* 22: 3785–3793.
(e) Ray Choudhury, A. and Mukherjee, S. (2016). Enantioselective dearomatization of isoquinolines by anion-binding catalysis en route to cyclic α-aminophosphonates. *Chem. Sci.* 7: 6940–6945.

33 De, C.K., Mittal, N., and Seidel, D. (2011). A dual-catalysis approach to the asymmetric steglich rearrangement and catalytic enantioselective addition of *O*-acylated azlactones to isoquinolines. *J. Am. Chem. Soc.* 133: 16802–16805.

34 Zhang, C. and Lu, X. (1995). Umpolung addition reaction of nucelophiles to 2,3-butadienoates catalyzed by a phosphine. *Synlett*: 645–646.

35 (a) Chen, Z., Zhu, G., Jiang, Q. *et al.* (1998). Asymmetric formation of quaternary carbon centers catalyzed by novel chiral 2,5-dialkyl-7-phenyl-7-phosphabicyclo-[2.2.1]heptanes. *J. Org. Chem.* 63: 5631–5635.
(b) Smith, S.W. and Fu, G.C. (2009). Asymmetric carbon–carbon bond formation γ to a carbonyl group: phosphine-catalyzed addition of nitromethane to allenes. *J. Am. Chem. Soc.* 131: 14231–14233.
(c) Sinisi, R., Sun, J., and Fu, G.C. (2010). Phosphine-catalyzed asymmetric additions of malonate esters to γ-substituted allenoates and allenamides. *Proc. Natl. Acad. Sci. USA* 107: 20652–20654.
(d) Wang, T., Yao, W., Zhong, F. *et al.* (2014). Phosphine-catalyzed enantioselective γ-addition of 3-substituted oxindoles to 2,3-butadienoates and 2-butynoates: use of prochiral nucleophiles. *Angew. Chem. Int. Ed.* 53: 2964–2968.

36 Kumar, V. and Mukherjee, S. (2013). Synergistic Lewis base and anion-binding catalysis for the enantioselective vinylogous addition of deconjugated butenolides to allenoates. *Chem. Commun.* 49: 11203–11205.

37 (a) Castagnoli, N. (1969). Condensation of succinic anhydride with *N*-benzylidene-*N*-methylamine. Stereoselective synthesis of *trans*- and *cis*-1-methyl-4-carboxy-5-phenyl-2-pyrrolidinone. *J. Org. Chem.* 34: 3187–3189.
(b) Castagnoli, N. and Cushman, M. (1971). Condensation of succinic anhydrides with Schiff bases. Scope and mechanism. *J. Org. Chem.* 36: 3404–3406.
(c) Cushman, M., Gentry, J., and Dekow, F.W. (1977). Condensation of imines with homophthalic anhydrides. A convergent synthesis of *cis*- and *trans*-13-methyltetrahydroprotoberberines. *J. Org. Chem.* 42: 1111–1116.
(d) González-López, M. and Shaw, J.T. (2009). Cyclic anhydrides in formal cycloadditions and multicomponent reactions. *Chem. Rev.* 109: 164–189.

38 Jarvis, C.L., Hirschi, J.S., Vetticatt, M.J., and Seidel, D. (2017). Catalytic enantioselective synthesis of lactams through formal [4+2] cycloaddition of imines with homophthalic anhydride. *Angew. Chem. Int. Ed.* 56: 2670–2674.

39 Collar, A.G., Trujillo, C., Lockett-Walters, B. *et al.* (2019). Catalytic asymmetric γ-lactam synthesis from enolisable anhydrides and imines. *Chem.-Eur. J.* 25: 7275–7279.

40 (a) Georgieva, A., Stanoeva, E., Spassov, S. *et al.* (1995). Cyclocondensations of homophthalic anhydrides with 1-aza-1,3-dienes. *Tetrahedron* 51: 6099–6114.
(b) Georgieva, A., Spassov, S., Stanoeva, E. *et al.* (1997). Effect of the structure of 1-aza-1,3-dienes on 1,2- versus 3,4-selectivity in cycloaddition reactions with homophthalic anhydride. *J. Chem. Research (S)*: 148–149.

41 Tamura, Y., Wada, A., Sasho, M., and Kita, Y. (1981). Cycloaddition of homophthalic anhydride: a new and simple route to linearly condensed phenolic compoumds. *Tetrahedron Lett.* 22: 4283–4286.

42 Collar, A.G., Trujillo, C., and Connon, S.J. (2019). Highly enantio- and diastereoselective catalytic asymmetric Tamura cycloaddition reactions. *Chem.-Eur. J.* 25: 7270–7274.

43 Liu, R.Y., Wasa, M., and Jacobsen, E.N. (2015). Enantioselective synthesis of tertiary α-chloro esters by non-covalent catalysis. *Tetrahedron Lett.* 56: 3428–3430.

44 (a) Hambly, G.F. and Chan, T.H. (1986). Reactions of enol silyl ethers with *N*-halosuccinimide – a stepwise process. *Tetrahedron Lett.* 27: 2563–2566.
(b) Ohkata, K., Mase, M., and Akiba, K.-y. (1987). Reaction of silyl enol ethers with *N*-chlorosuccinimide: trapping of the siloxycarbinyl cation by an azide anion. *J. Chem. Soc., Chem. Commun.*: 1727–1728.

45 (a) Denmark, S.E., Kuester, W.E., and Burk, M.T. (2012). Catalytic, asymmetric halofunctionalization of alkenes – a critical perspective. *Angew. Chem. Int. Ed.* 51: 10938–10953.
(b) Cheng, Y.A., Yu, W.Z., and Yeung, Y.-Y. (2014). Recent advances in asymmetric intra- and intermolecular halofunctionalizations of alkenes. *Org. Biomol. Chem.* 12: 2333–2343.
(c) Tripathi, C.B. and Mukherjee, S. (2014). Catalytic enantioselective halocyclizations beyond lactones: emerging routes to enantioenriched nitrogenous heterocycles. *Synlett* 25: 163–169.

46 Veitch, G.E. and Jacobsen, E.N. (2010). Tertiary aminourea-catalyzed enantioselective iodolactonization. *Angew. Chem. Int. Ed.* 49: 7332–7335.

47 (a) Min, C. and Seidel, D. (2017). Asymmetric Brønsted acid catalysis with chiral carboxylic acids. *Chem. Soc. Rev.* 46: 5889–5902.
(b) Akiyama, T. and Mori, K. (2015). Stronger Brønsted acids: recent progress. *Chem. Rev.* 115: 9277–9306.

48 Hashimoto, T. and Maruoka, K. (2007). Design of axially chiral dicarboxylic acid for asymmetric Mannich reaction of arylaldehyde N-Boc imines and diazo compounds. *J. Am. Chem. Soc.* 129: 10054–10055.

49 Hashimoto, T. and Maruoka, K. (2013). Development of axially chiral dicarboxylic acid catalyzed asymmetric transformations. *J. Synth. Org. Chem. Jpn.* 71: 472–479.

50 Hashimoto, T. and Maruoka, K. (2008). Design of an axially chiral dicarboxylic acid and its application in syntheses of optically active β-amino acids and β-amino phosphonic acid derivatives. *Synthesis*: 3703–3706.

51 Yamamoto, H. and Futatsugi, K. (2005). "Designer Acids": combined acid catalysis for asymmetric synthesis. *Angew. Chem. Int. Ed.* 44: 1924–1942.

52 Weil, T., Kotke, M., Kleiner, C.M., and Schreiner, P.R. (2008). Cooperative Brønsted acid-type organocatalysis: alcoholysis of styrene oxides. *Org. Lett.* 10: 1513–1516.

53 Klausen, R.S. and Jacobsen, E.N. (2009). Weak Brønsted acid-thiourea co-catalysis: enantioselective, catalytic Protio–Pictet–Spengler reactions. *Org. Lett.* 11: 887–890.

54 Lee, Y., Klausen, R.S., and Jacobsen, E.N. (2011). Thiourea-catalyzed enantioselective iso-Pictet–Spengler reactions. *Org. Lett.* 13: 5564–5567.

55 Min, C., Mittal, N., Sun, D.X., and Seidel, D. (2013). Conjugate-base-stabilized Brønsted acids as asymmetric catalysts: enantioselective Povarov reactions with secondary aromatic amines. *Angew. Chem. Int. Ed.* 52: 14084–14088.

56 (a) Zhu, Z., Odagi, M., Zhao, C. *et al.* (2020). Highly acidic conjugate-base-stabilized carboxylic acids catalyze enantioselective oxa-Pictet–Spengler reactions with ketals. *Angew. Chem. Int. Ed.* 59: 2028–2032. (b) Zhu, Z., Odagi, M., Supantanapong, N. *et al.* (2020). Modular design of chiral conjugate-base-stabilized carboxylic acids: catalytic enantioselective [4+2] cycloadditions of acetals. *J. Am. Chem. Soc.* 142: 15252–15258.

57 Min, C., Lin, C.-T., and Seidel, D. (2015). Catalytic enantioselective intramolecular aza-Diels–Alder reactions. *Angew. Chem. Int. Ed.* 54: 6608–6612.

58 Min, C. and Seidel, D. (2016). Stereochemically rich polycyclic amines from the kinetic resolution of indolines through intramolecular Povarov reactions. *Chem.-Eur. J.* 22: 10817–10820.

59 Mittal, N., Sun, D.X., and Seidel, D. (2014). Conjugate-base-stabilized Brønsted acids: catalytic enantioselective Pictet–Spengler reactions with unmodified tryptamine. *Org. Lett.* 16: 1012–1015.

60 Taylor, M.S. and Jacobsen, E.N. (2004). Highly enantioselective catalytic acyl-Pictet–Spengler reactions. *J. Am. Chem. Soc.* 126: 10558–10559.

61 Odagi, M., Araki, H., Min, C. *et al.* (2019). Insights into the structure and function of a chiral conjugate-base-stabilized Brønsted acid catalyst. *Eur. J. Org. Chem.*: 486–492.

62 Zhao, C. and Seidel, D. (2015). Enantioselective A^3 reactions of secondary amines with a Cu(I)/acid–thiourea catalyst combination. *J. Am. Chem. Soc.* 137: 4650–4653.

63 Nielsen, C.D.-T., Mooij, W.J., Sale, D. *et al.* (2019). Reversibility and reactivity in an acid catalyzed cyclocondensation to give furanochromanes – a reaction at the 'Oxonium-Prins' vs. '*Ortho*-Quinonemethide Cycloaddition' mechanistic nexus. *Chem. Sci.* 10: 406–412.

64 Xie, Y. and List, B. (2017). Catalytic asymmetric intramolecular [4+2] cycloaddition of in situ generated *ortho*-quinone methides. *Angew. Chem. Int. Ed.* 56: 4936–4940.

65 Nalbandian, C.J., Miller, E.M., Toenjes, S.T., and Gustafson, J.L. (2017). A conjugate Lewis base-Brønsted acid catalyst for the sulfenylation of nitrogen containing heterocycles under mild conditions. *Chem. Commun.* 53: 1494–1497.

6

Silanediols, Phosphoramides, and Other OH- and NH-Based H-Donor Catalysts

Alexandria Leveille and Anita Mattson

Worcester Polytechnic Institute, Department of Chemistry and Biochemistry, 60 Prescott Street., Worcester, MA 01605, USA

6.1 Introduction

The expectations of an anion binding catalyst are high. The chemistry community prefers a readily accessible scaffold that can be rendered chiral with ease and affords excellent enantiomeric excesses at low loadings in a broad range of reactions. These demanding goals can be difficult to reach, in part due to the young age of the field of anion-binding catalysts and the giant gaps in knowledge that have yet to be filled. For instance, anion binding catalysts can suffer from high catalyst loadings and limited reaction scopes. Moreover, the design and optimization of the anion binding catalyst structure is largely achieved by trial and error as few predictive strategies for catalyst optimization currently exist. These gaps in knowledge become even more apparent as chemists branch out into the discoveries of new scaffolds of anion binding catalysts, such as the ones covered in this chapter.

This chapter covers several promising families of hydrogen bond donor catalysts that are in the early stages of development (Figure 6.1). The chapter begins by overviewing anion binding catalysts that are reliant on dual O—H bond donors, such as the silanediols and disiloxane-1,3-diols. The next sections cover scaffolds that bind through unique geometries, including the triphosphoramides and the cyclophosphazanes. The chapter ends with the introduction of several catalyst scaffolds that have just made their entry into the field, including diaminoxanthene derivatives, croconamides, and pyrrole-based anion binding catalysts.

6.2 Silanediols

6.2.1 Introduction

The modern applications of silanediols in the field of anion binding catalysis grew from the long roots that the silanol functional group has in hydrogen bonding and molecular recognition [1–5]. It is well established in the literature that silanols,

Anion-Binding Catalysis, First Edition. Edited by Olga García Mancheño.

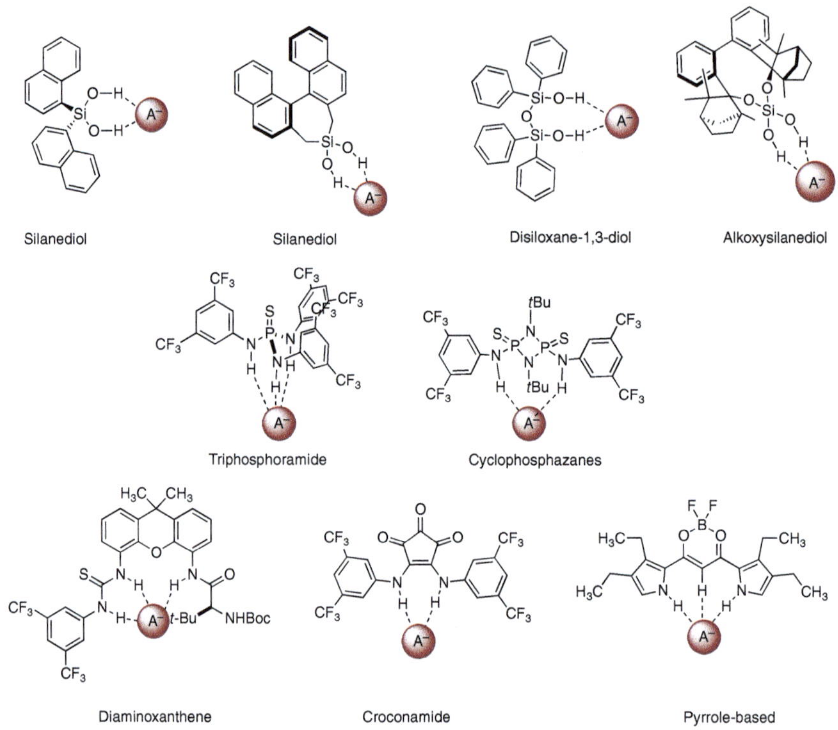

Figure 6.1 Examples of anion-binding catalysts covered in this chapter.

organic molecules with Si—OH bonds, can interact with both organic and inorganic species through hydrogen bonds [4, 5]. These non-covalent forces give rise to lucrative properties that can be taken advantage in multiple fields, including catalysis, sensing, and drug discovery [6]. The rich chemical history of silanols gives rise to a stimulating landscape that has been reviewed thoroughly and, therefore, we will not be covering the details of silanol hydrogen bonding [4–11]. Instead, this section focuses on the anion-binding and associated catalytic properties of silanediols, silanols with two Si—OH bonds, and their closely related structures.

6.2.2 Overview of Silanols in Anion Binding and Catalysis

In 2006, Kondo *et al.* introduced an attractive dinaphthyl silanediol and provided evidence, including data from UV–vis and ^{1}H nuclear magnetic resonance (NMR) titration experiments, that revealed the ability of silanediols to recognize anions (e.g. chloride, bromide, and acetate) through hydrogen bonding interactions (Figure 6.2) [12]. The association constant of di(1-naphthyl)silanediol with the chloride anion was found to be $1.44 \pm 0.11 \times 10^{2}$ M^{-1} in dimethyl sulfoxide (DMSO). Kondo *et al.* expanded the studies to include disiloxane-1,3-diols that bind anions through two Si—OH hydrogen bonds in 2007 [13]. Franz and coworkers reported on the self-association and anion binding properties of select silanediols in 2012 [14].

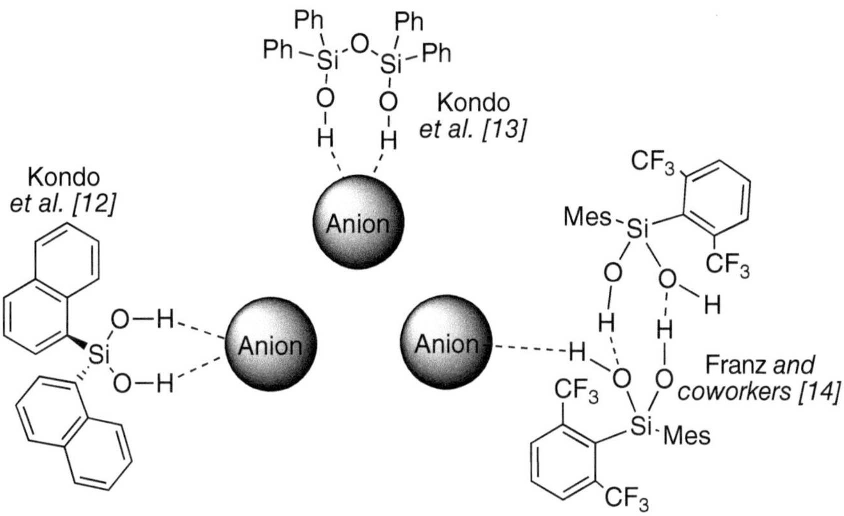

Figure 6.2 Examples of silanols in anion recognition. Source: Based on Kondo *et al.* [12].

These significant contributions to the field likely played substantial roles in the advancement of silanediol and, more recently, disiloxane-1,3-diol anion binding catalysis.

6.2.3 Silanediol Anion-Binding Catalysis

Inspired by their anion binding properties, the Mattson group reported the first example of silanediol anion binding catalysis in 2013 [15, 16]. Beyond demonstrating the anion binding catalyst abilities of the achiral dinaphthyl silanediol reported by Kondo *et al.* [12], this work also detailed the synthesis of the first chiral silanediol **1** built for anion binding catalysis: a C_2-symmetric silanediol derived from 1,1′-bi-2-naphthol (BINOL, **2**, Figure 6.3). Features of this catalyst design include the following: (i) it is stable to silica gel chromatography, (ii) both enantiomers are readily accessible from commercially available starting materials, and (iii) the backbone can be modified to achieve optimal levels of enantiocontrol. An association constant of **1** and chloride in DMSO was found to be $2.19 \pm 0.03 \times 10^2\ M^{-1}$.

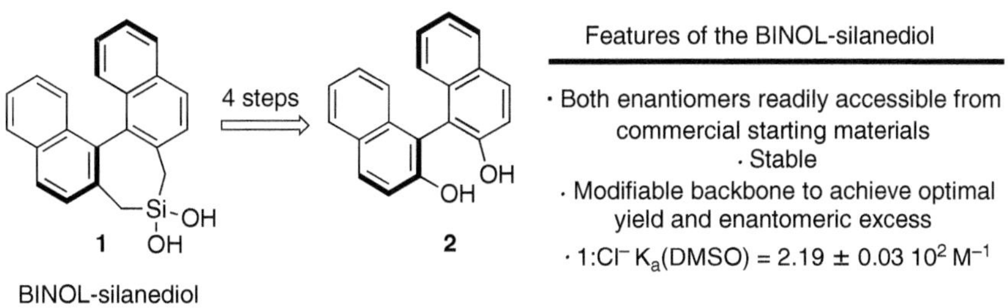

Figure 6.3 Features of chiral BINOL-silanediols.

The catalytic abilities of BINOL-silanediol **1a** were first demonstrated in the addition of silyl ketene acetals to *N*-acyl isoquinolinium ions, a reaction that is known to benefit from anion-binding thio(urea) catalysis (Scheme 6.1). Specifically, in the first report, silyl ketene acetal **5** was added to a variety of isoquinolinium ions to afford isoquinolines **6** in up to 80% yield and up to 75 : 25 enantiomeric ratio. The proposed reaction pathway involves silanediol-catalyzed ionization of **3** to generate ion pair **4**.

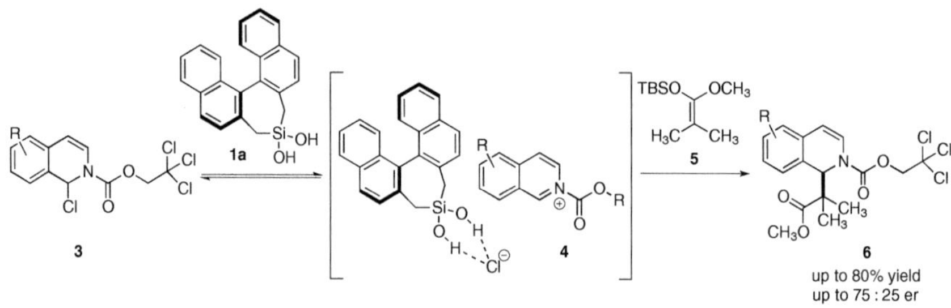

Scheme 6.1 BINOL-silanediol **1** as an anion binding catalyst in the preparation of **6**.

Inspired by the initial success of the silanediol-catalyzed addition of silyl ketene acetals to *N*-acyliminium ions, significant effort was invested in advancing a silanediol catalyst that would lead to improved levels of enantiocontrol. Gratifyingly, in 2015, a novel tetraphenylsilanediol (**1d**) catalyst was synthesized and gave rise to nearly 90 : 10 enantiomeric ratio in the addition of **5** to **3** (Scheme 6.2) [15]. An association constant of $3.10 \pm 0.10 \times 10^2\ M^{-1}$ in DMSO was observed for **1d**:Cl^-. Taken together, these data demonstrated the promise of silanediols in anion-binding catalysis, including the ability to strategically modify the silanediol to achieve highly enantioselective processes.

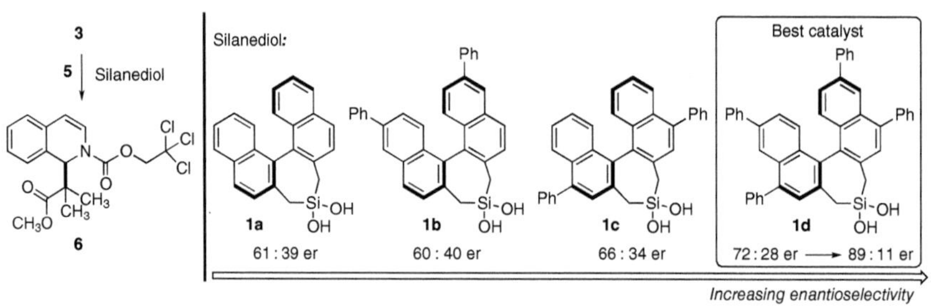

Scheme 6.2 Influence of substituted BINOL-silanediols on enantiomeric ration in the addition of **5** to **3**. Source: Based on Wieting *et al.* [15].

The impact of anion binding catalysis on the synthesis of 2-alkyl-chroman-4-ones, oxygen heterocycles with direct applications in bioactive target synthesis, was revealed in 2016 (Scheme 6.3) [17]. Specifically, Mattson and coworkers demonstrated that silanediol catalysts (**1**) enable the enantiocontrolled addition of silyl ketene acetals **5** to benzopyrylium triflates **8**, reactive species generated in situ, giving rise to **10** in high yield with decent levels of enantiocontrol. In this work, the

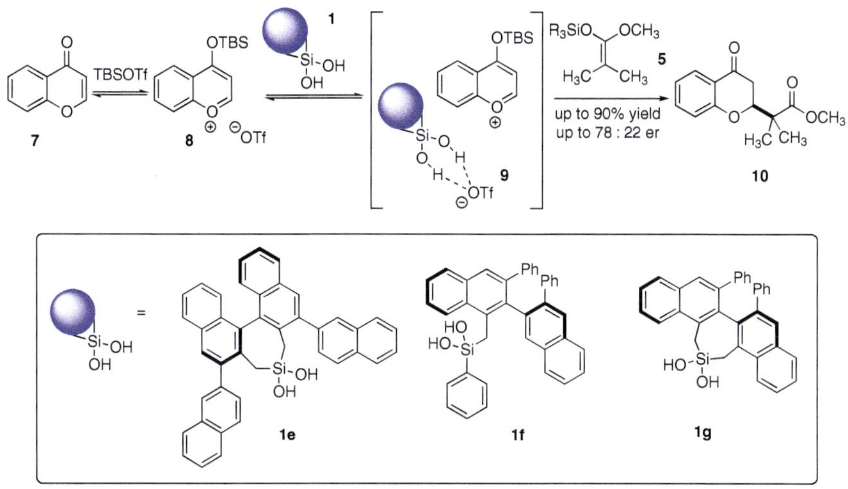

Scheme 6.3 Impact of silanediols on the synthesis of **10**. Source: Based on Hardman-Baldwin *et al.* [17].

family of silanediol catalysts was expanded to include, 3,3′-substituted BINOL-based silanediols (**1e**), acyclic 3,3′-diphenyl-2,2′-bi-1-naphthalol (VANOL)-based silanediols (**1f**), and cyclic VANOL-derived silanediols (**1g**). In this process, the 3,3′-substituted BINOL-silanediol **1e** and the cyclic VANOL-silanediol **1g** proved most enantioselective. The scope of the work included a collaboration between the Kondo and Mattson groups to execute UV–vis titration studies to collect evidence in support of association between triflate and silanediol **1a**. Indeed, the results suggest that association is plausible with an observed association constant of $2.31 \pm 0.52 \times 10^3\ M^{-1}$ in chloroform.

In 2018, Mattson and coworkers extended the use of silanediols beyond the typical substrate activation (Scheme 6.4a): silanediols were used to enhance the reactivity of Lewis acid catalysts (Scheme 6.4b) [18].

In this new application for anion-binding catalysts, the combination of BINOL-silanediol **1a** and $Cu(OTf)_2$ catalyzes the enantioselective addition of indoles to alkylidene malonates with high levels of enantiocontrol (up to 93 : 7 er). A hypothesized reaction pathway is depicted in Scheme 6.5. The silanediol interacts with the $Cu(OTf)_2$ to give rise to an enhanced Lewis acid. This new catalyst complex goes on to activate the alkylidene malonate **11** via ion pair **II** for reaction with the indole nucleophile (**12**). The subsequent release of the product **13** simultaneously regenerates the catalyst system for further reaction. Experimental support for enhanced Lewis acid complex was derived from UV–vis studies and control experiments.

Hardman-Baldwin and Mattson reported the application of silanediol catalysis in carbon dioxide fixation in 2014 (Scheme 6.6) [19]. Specifically, BINOL-silanediol **1a** was applied to influence the metal-free insertion of carbon dioxide into epoxides (e.g. **14**) to give rise to lactones (e.g. **15**) in the presence of a tetrabutylammonium iodide additive. Under the reaction conditions, the silanediol out-performed

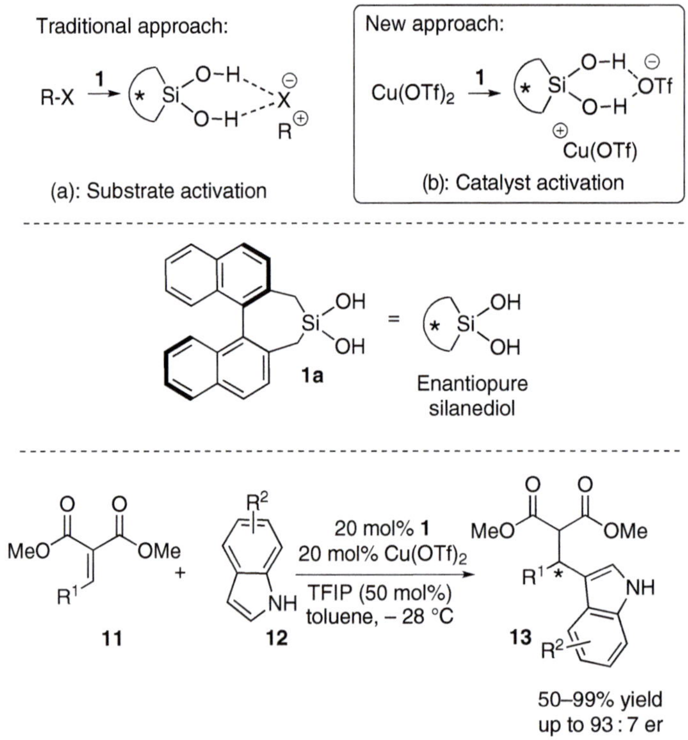

Scheme 6.4 Silanediol activation of Lewis acids. Source: Based on Guan *et al.* [18].

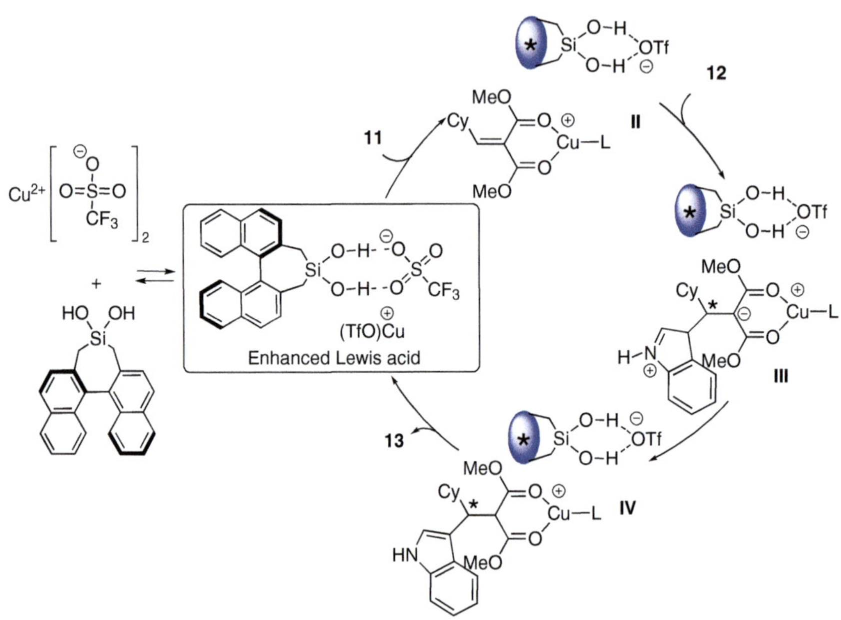

Scheme 6.5 Plausible catalytic cycle for silanediol and copper(II) triflate-catalyzed reactions of **11** and **12**.

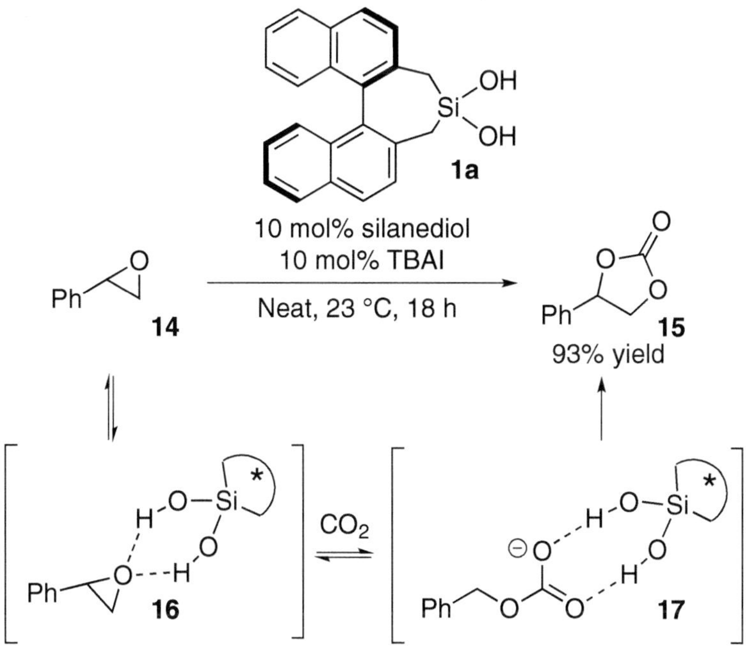

Scheme 6.6 Silanediol-catalyzed carbon dioxide fixation. Source: Based on Hardman-Baldwin *et al.* [19].

traditional urea and thiourea catalysts in the preparation of **15**, giving rise to the desired carbonate in 93% yield. In the plausible reaction pathway, it was hypothesized that the anion-binding abilities of the silanediol facilitate epoxide ring opening via intermediates such as **17**.

6.2.4 Alkoxysilanediol Anion Binding Catalysis

In 2019, Goldfuss and coworkers developed a new class of enantiopure biphenyl-2,2′-bisfenchol-based (BIFOL) silanediol catalysts (**18**, Figure 6.4) [20]. Having previously demonstrated that the sterically encumbered BIFOL backbone is unexpectedly stable against hydrolysis in the context of chlorophosphite ligands [21], Goldfuss and coworkers postulated that BIFOL-based silanediols may operate as a new family of anion binding catalysts. In addition to its stability, the straightforward preparation of the BIFOL-silanediol is attractive; it is readily prepared from BIFOL in just two steps. The chloride-binding abilities of BIFOL-silanediol **18**, possibly resembling an anion-bound complex like **18:A**$^-$, was studied with UV–vis titration experiment and a binding constant of 5274.9 M^{-1} was observed in chloroform (Scheme 6.7). This binding constant is similar to the binding constant observed for di(1-naphthyl)silanediol, which was 4688.0 M^{-1} in chloroform.

The abilities of BIFOL-silanediol **18** as an anion binding catalyst were put to the test in the addition of silyl ketene acetals to three different substrates: (i) *N*-acylisoquinolinium ions (Scheme 6.8a), (ii) oxocarbenium ions (Scheme 6.8b),

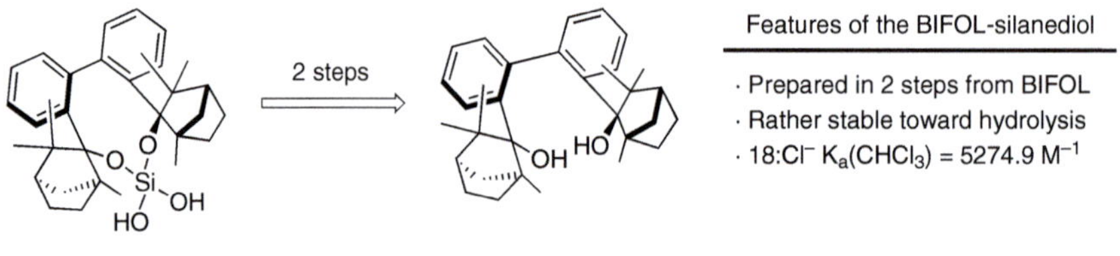

18 BIFOL-silanediol   **19** BIFOL

Figure 6.4 Features of chiral BIFOL-silanediol. Source: Fox *et al.* [20]. Licensed under CC BY 4.0.

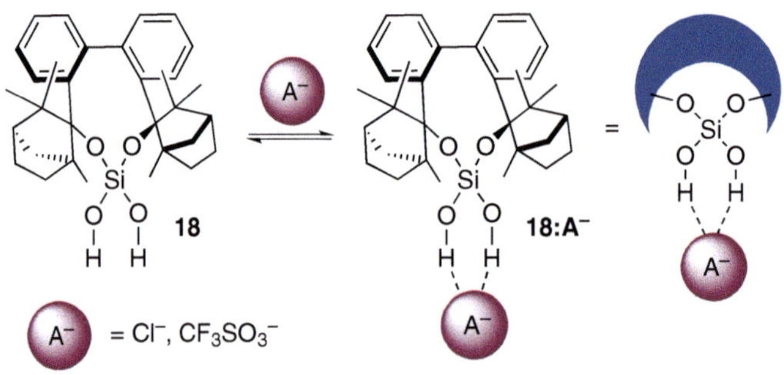

Scheme 6.7 Plausible anion binding of **18** with chloride and triflate.

and (iii) pyrylium ions (Scheme 6.8c). The addition of silyl ketene acetal (**5**) to **3**, plausibly via ion pair **20**, proceeded with high yields and up to 56 : 44 enantiomeric ratio. The reaction of **5** with likely ion pair **22** gave rise to product **23** in high yield but minimal enantiomeric ratios. Likewise, the addition of silyl ketene acetals to plausible ion pair **24** gave rise to product **10** in excellent yield, but the enantioselectivity was low. The observed stability of the BIFOL backbone against hydrolysis, along with the catalysts' ability to recognize a variety of anions, suggests potential for further development.

6.3 Siloxanes

Quickly following their initial 2006 studies focused on silanediols recognition of select anions through hydrogen bonding, Kondo *et al.* quantified the host:guest interactions of select disiloxane-1,3-diols and anions in both $CDCl_3$ and CD_3CN (Figure 6.5) [13]. Capitalizing on the work of Hossain and Hursthouse, Kondo and coworkers centered their studies on 1,1,3,3,-tetraphenyl-1,3-disiloxane-1,3-diol (**25**) [22]. Through ^1H NMR titration experiments, the binding constants of each chloride, bromide, iodide, and acetate were deduced. The association constant of **25** with chloride was found to be $670 \pm 44\ M^{-1}$ in $CH_3CN\text{-}d_5$.

1,1,3,3-Tetraferrocenyldisiloxane-1,3-diol (**26**) was advanced in 2016 by Cuadrado and coworkers as a redox-active anion receptor (Scheme 6.9) [23]. The authors found evidence of anion binding of siloxane **26** with chloride and acetate in the

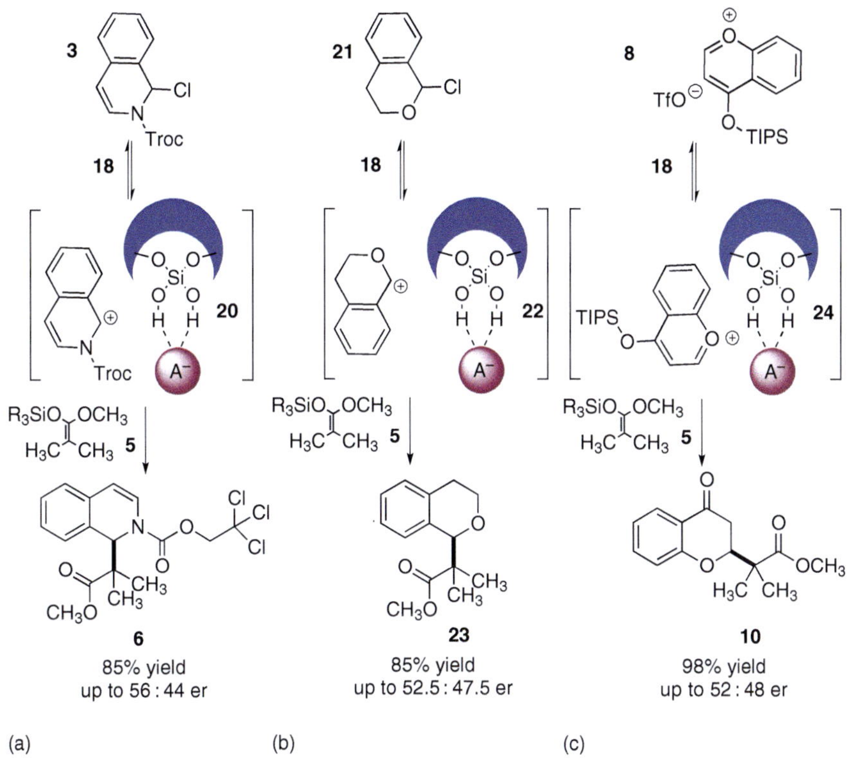

Scheme 6.8 Anion-binding catalysis with **18**.

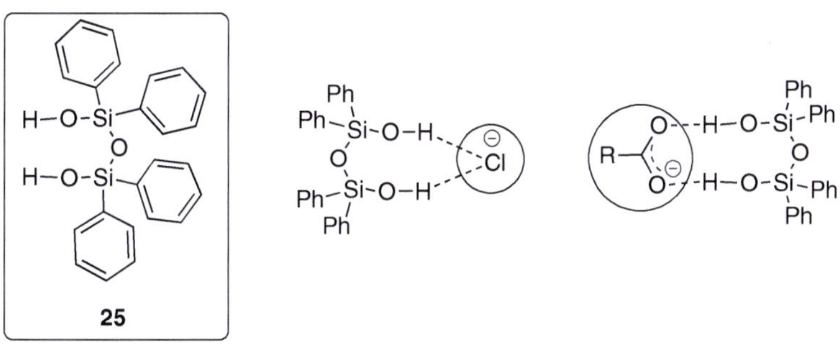

Figure 6.5 Examples of anion recognition of disiloxane-1,3-diols. Source: Based on Kondo *et al.* [13].

liquid phase through ^{1}H NMR experiments and electrochemical studies. That data suggested that **26** may bind to chloride in a 1 : 1 host guest interaction, whereas acetate may bind to **26** in both 2 : 1 and 1 : 1 ratios. The authors observed that diferrocenylsilanediol has a higher association constant for chloride than 1,1,3,3-tetraferrocenyldisiloxane-1,3-diol.

The catalytic abilities of a small family of achiral disiloxane-1,3-diols were demonstrated in 2016 in the racemic addition of silyl ketene acetals to isoquinolinium ions

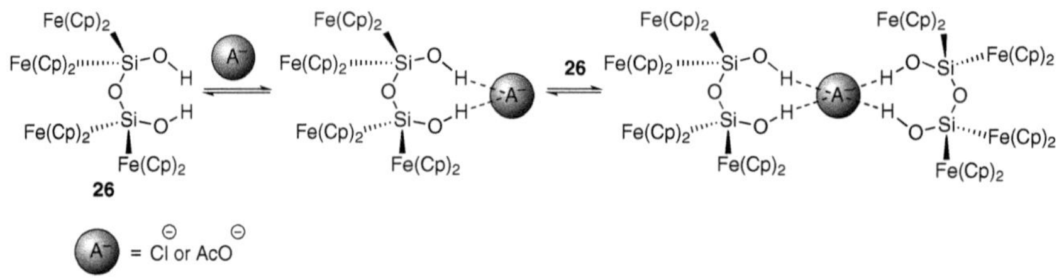

Scheme 6.9 Examples of anion binding of disiloxane-1,3-diol **26**. Source: Based on Bruna *et al.* [23].

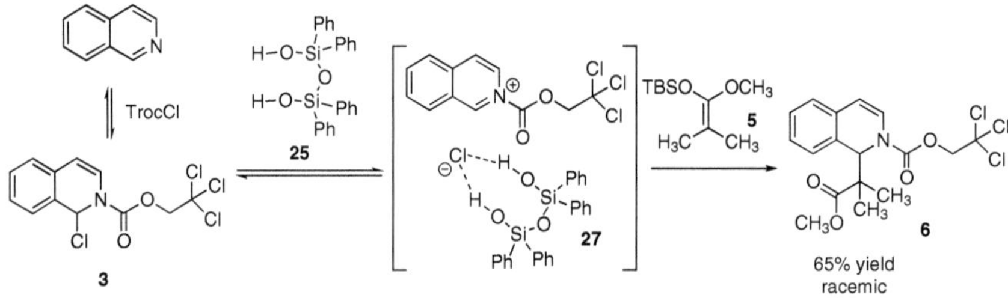

Scheme 6.10 Achiral disiloxane-1,3-diol as an anion-binding catalyst. Source: Based on Diemoz *et al.* [24].

(Scheme 6.10) [24]. It was proposed that the disiloxane-1,3-diol **25** operates through anion binding to the chloride ion to facilitate the formation of the ion pair **27**, eventually giving rise to product **6** in 65% yield. Control experiments confirmed the importance of both silanol groups in obtaining the highest yields.

6.4 Thiophosphoramides

(Thio)phosphoramides were introduced to the field of hydrogen bonding catalysis in 2009 by Shea and coworkers as three-point binding species that may afford improvements in catalysis over traditional two-point binding urea and thiourea catalysts (Scheme 6.11) [25]. Traditional hydrogen bond donor catalysts, such as (thio)ureas, often require high catalyst loading and long reaction times and can be unstable at higher temperatures. Shea *et al.* reasoned that three N—H

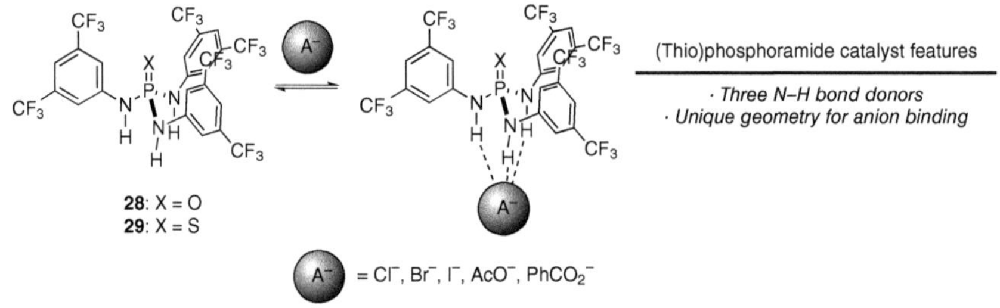

Scheme 6.11 (Thio)phosphoramide anion binding. Source: Based on Rodriguez *et al.* [25].

bonds supplied by phosphoric triamides and thiophosphoric triamides may render these species more catalytically active than their (thio)urea counterparts. The phosphoric triamide **28** and thiophosphoric triamide **29** were prepared by the reaction of 3,5-((bis)trifluoromethyl)aniline (**30**) to phosphoryl oxychloride and thiophosphoryl chloride to give rise to **28** and **29** in 49% and 72% yield, respectively (Scheme 6.12). Shea and coworkers found that **28** and **29** offer modest improvements in catalytic activity over conventional thioureas when compared directly as hydrogen bond donor catalysts in Friedel–Crafts and Baylis–Hillman reactions.

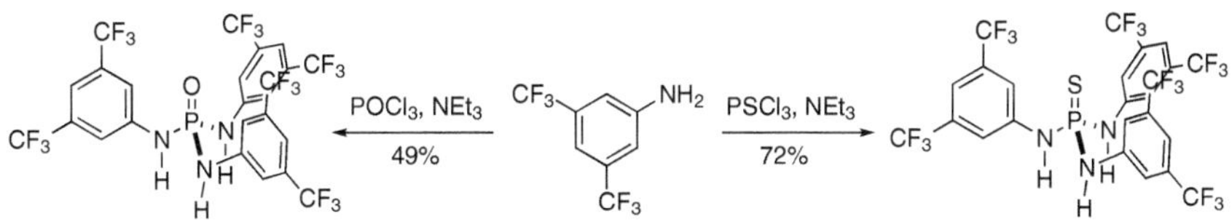

Scheme 6.12 The synthesis of phosphoric triamide **28** and thiophosphoric triamide **29**.

Inspired by the demonstrated hydrogen bonding abilities of (thio)phosphoric triamides **28** and **29**, Gale and coworkers executed studies to probe the anion-binding abilities of these new hydrogen bond donors in 2013 [26]. Through X-ray crystallographic analysis and NMR titration studies, the Gale team collected evidence of the anion-binding abilities of **28** and **29**. A number of anions, including halides, acetate, nitrate, and sulfates, were observed to be associated with **28** and/or **29**. Notably, phosphoramide **28** was found to form a strong 1 : 1 complex with sulfate (K_a = (DMSO-d_6/0.5% H_2O) > 10^4 M^{-1}). It was hypothesized that this strong 1 : 1 complex may be the result of the sulfate and phosphoramide interaction through three hydrogen bonds (Figure 6.6).

In 2013, Nagorny and coworkers demonstrated that the anion-binding abilities of thiophosphoramides can be uniquely advantageous in the catalysis of ionic Diels–Alder reactions [27]. Specifically, the [2+4] reaction of commercially available acrolein acetal **31** and cyclopentadiene **32** is catalyzed by a combination of *p*-toluene sulfonic acid and thiophosphoramide **28** to give rise to the cycloadduct **33** in 98% conversion and 3 : 1 *endo*/*exo* (Scheme 6.13). In a comparison of catalysts, the thiophosphoramide **28** outperformed more conventional thiourea and squaramide

Figure 6.6 Phosphoramide participating in plausible three-point binding with SO_4^{2-}.

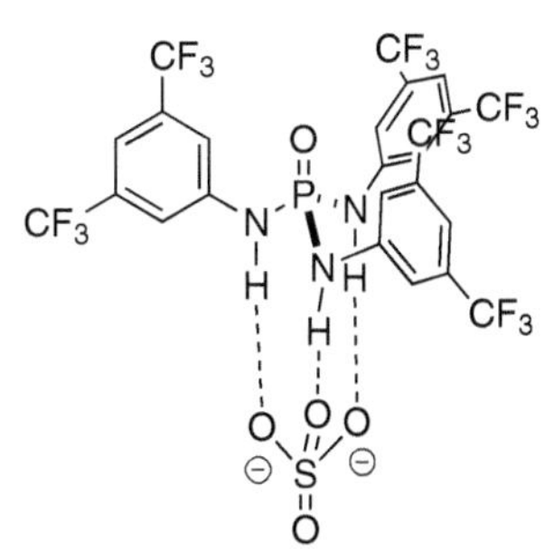

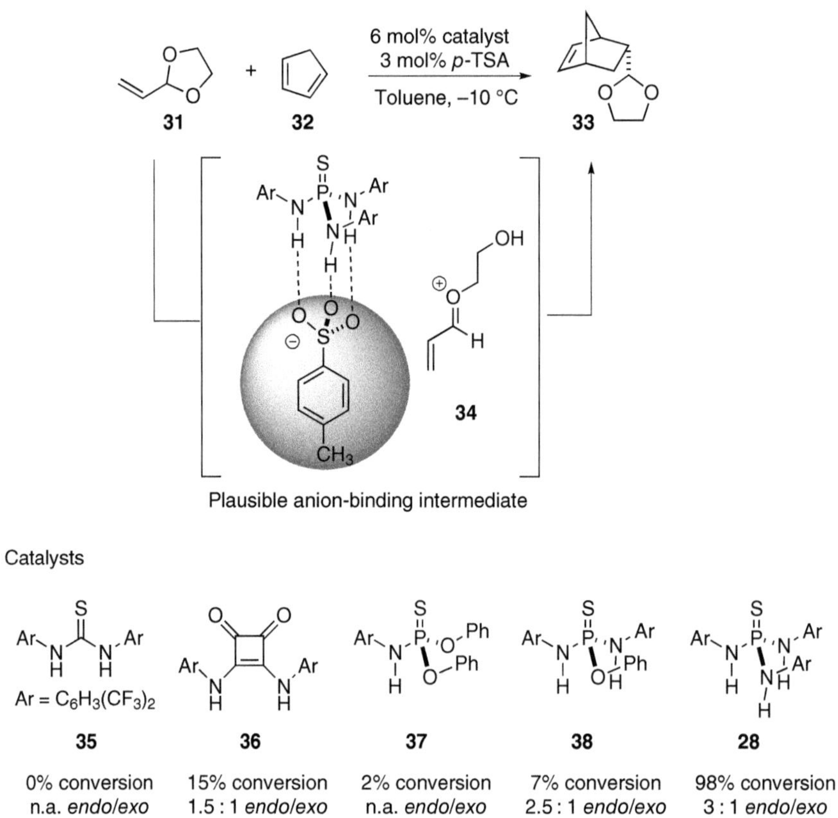

Scheme 6.13 Thiophosphoramide-catalyzed Diels–Alder.

anion-binding catalysts **35** and **36**. The thiophosphoramide **28** also afforded significantly higher conversions than thiophosphoramides **37** or **38**. The authors suggest that the superior catalyst abilities of **28**, in comparison to **35** and **36**, may be a result of the geometry of the three N—H bonds being best-suited for interaction with the anion through hydrogen bonding, such as in plausible intermediate **34**. A variety of acetals and dienes proved to be tolerant of the reaction conditions and the resultant cycloadducts were isolated in 57–91% yields.

Nagorny and coworkers continued their investigations with thiophosphoramides and in 2015 described their ability to perform as a cocatalyst in $Cu(OTf)_2$-catalyzed arylation reactions of potassium carboxylates with diaryliodonium salts (Scheme 6.14) [28]. Control experiments proved that both catalysts were necessary for the enhanced reactivity of the diaryliodonium salts, where no reaction is observed in the absence of thiophosphoramide. Other hydrogen bond donor (HBD) catalysts were screened, including a thiourea and a squaramide, which resulted in little to no activity and thereby suggesting that thiophosphoramide is able to bind to the counterion of the Cu(II) catalyst, as well as the counterion of the diaryliodonium salt (e.g. **41**), in a unique manner.

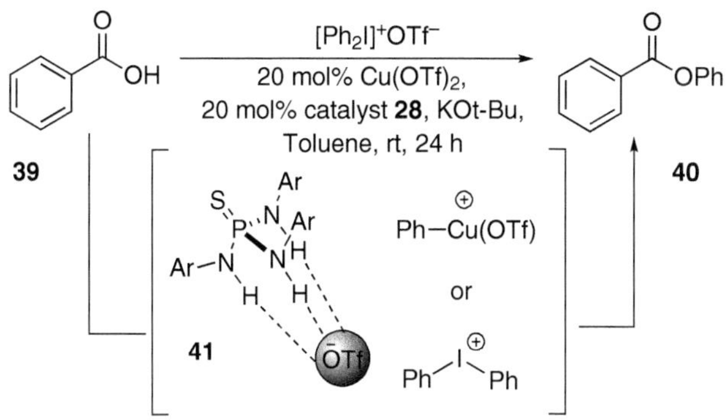

Scheme 6.14 (Thio)phosphoramide cocatalyst in $Cu(OTf)_2$-catalyzed arylation reactions of potassium carboxylates with diaryliodonium salts. Source: Based on Bhattarai *et al.* [28].

6.5 Cyclodiphosphazanes

Cyclodiphosphazanes are four-membered heterocycles containing alternating phosphorous and nitrogen atoms (**42–43**, Scheme 6.15) [29]. They are easily synthesized from amines and PCl_3, thus enabling access to large families of cyclodiphosphazanes with a variety of substituents. Originally, they were studied as ligands in organometallic catalysis, but their applications in supramolecular chemistry and catalysis are gaining traction. In the context of hydrogen bond donor and/or anion binding catalysis, cyclodiphosphazanes may benefit from increased hydrogen bond strengths.

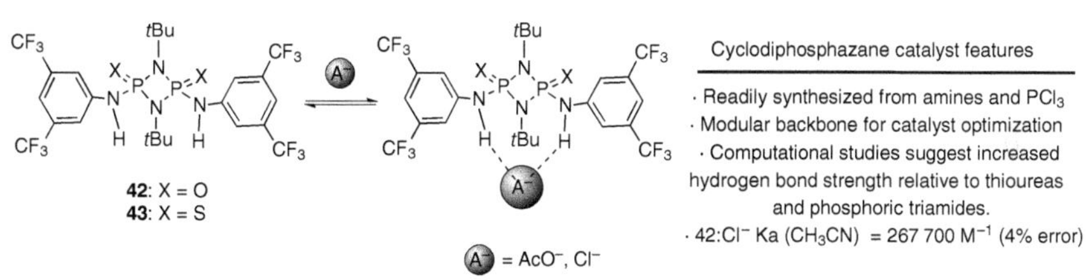

Scheme 6.15 Cyclodiphosphazanes in anion binding and catalysis. Source: Based on Balakrishna [29].

In 2014, Goldfuss and coworkers developed a new class of cyclodiphosphazane catalysts and demonstrated the first successful application of this type of hydrogen bond donor in enantioselective organocatalysis [30, 31]. Inspiration for the development of these catalysts stemmed from the hypothesis that by improving the rigidity of literature-known phosphoric triamide catalysts while maintaining steric bulk should result in improved catalytic performance. This hypothesis was further supported by computational studies of various HBDs with nitrobenzene, which suggested that the cyclodiphosphazane will have shorter bond lengths and stronger hydrogen bonding abilities similar to that of the squaramide compared to the phosphoric triamide catalysts (Figure 6.7).

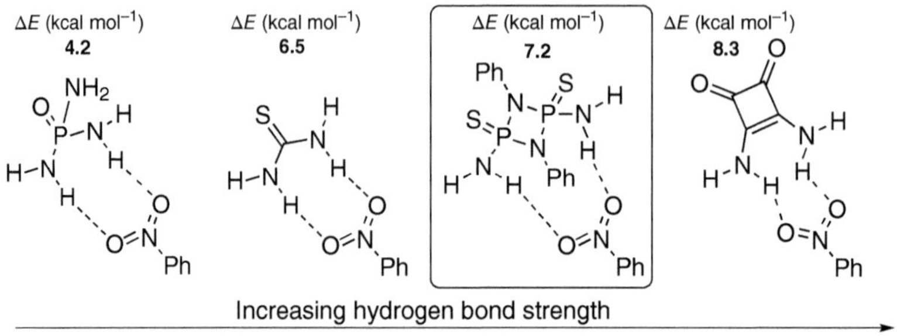

Figure 6.7 Comparison of cyclodiphosphazane hydrogen bond strength to nitrobenzene with other popular hydrogen bond donors.

With a small library of chiral cyclodiphosphazanes in hand, Goldfuss and coworkers tested their catalytic abilities in an asymmetric Michael addition of β-nitrostyrene to 2-hydroxy-1,4-naphthoquinone (**44**). An enantiomeric ratio of up to 87.5 : 12.5 was achieved using a cyclohexanediamine-derived cyclodiphosphazane catalyst (Scheme 6.16). The reaction may proceed through plausible transition state **45**. Having demonstrated their ability as organocatalysts, the Goldfuss team proceeded to examine the cyclodiphosphaxane catalyst scaffold in anion recognition catalysis.

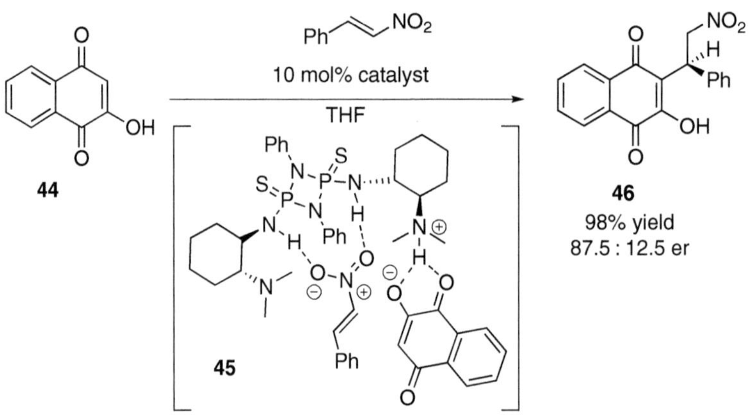

Scheme 6.16 Chiral cyclodiphosphazane as an enantioselective hydrogen bond donor catalyst.

In 2018, Goldfuss and coworkers published their investigations on the synthesis of several new cyclodiphosphazanes as well as their abilities to catalyze *N*-acyl Mannich reactions through proposed anion binding (Scheme 6.17) [32]. Although several substitution patterns were analyzed on the cyclodiphosphazane backbone, the O(=P)cyclodiphosphazane **42** demonstrated superior binding to chloride (log $K_a = 5.91$), while S(=P) cyclodiphosphazane exhibited stronger binding with acetate (log K_a = 6.45) in UV–vis titration experiments. Thus, the addition of silyl ketene acetal **5** to **3** under the influence of 20 mol% **42** gave rise to a desired product **6** in 45% yield. The anion-binding abilities of **42** were invoked in the plausible ion-pair

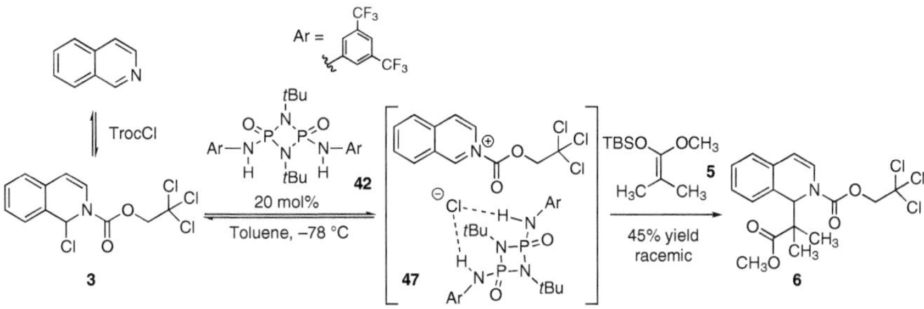

Scheme 6.17 Cyclophosphazane-catalyzed *N*-acyl Mannich reaction. Source: Based on Wolf *et al.* [32].

intermediate **47**. Although the reaction was performed only with an achiral catalyst, thus generating the racemic product, these studies highlight the potential for this new family of catalysts in future catalyst design and anion recognition research.

6.6 Other Examples

6.6.1 Xanthene–Diamine Scaffold

The 4,5-diaminoxanthene core is a promising scaffold in anion binding catalysis as multiple hydrogen bond donors can be installed for the purpose of molecular recognition. Indeed, the molecular recognition properties of the 4,5-diaminoxanthene core has been taken advantage of in the design of both molecular receptors and hydrogen bond donor catalysts. In 2015, Kozlowski and coworkers demonstrated that 4,5-diaminoxanthene derivatives can operate as anion-binding catalysts, such as **48–50** (Figure 6.8) [33]. In this study, Kozlowski and coworkers identified the equilibrium constants of several of their derivatives with chloride ions: **48**:Cl^- = 56 in pyridine-d_5; **49**:Cl^- = 159 in DMSO-d_6; and **50**:Cl^- = 127 in DMSO-d_6.

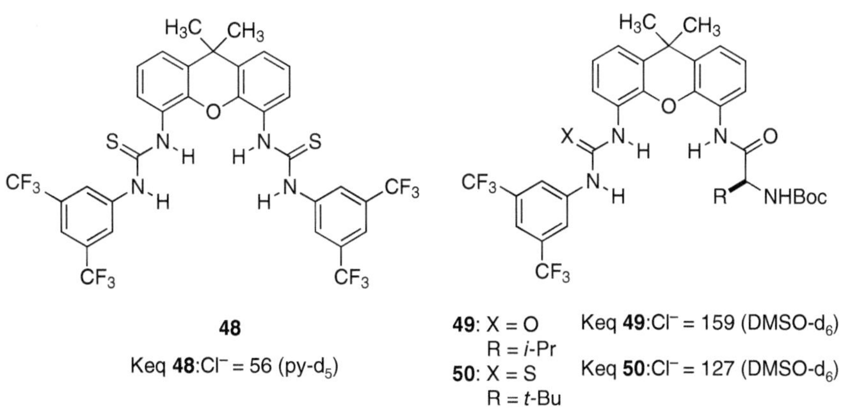

Figure 6.8 Examples of xanthene diamine scaffolds explored in anion-binding catalysis. Source: Based on Metz *et al.* [33].

With evidence of chloride recognition, the authors explored the anion-binding abilities in intramolecular Pictet–Spengler reactions inspired by Jacobsen and coworkers [34]. For instance, the intramolecular cyclization of three-substituted furan **51** was catalyzed to give rise to **52** in 42% conversion and 24% enantiomeric excess under the influence of 30 mol% **49** (Scheme 6.18). This process may take advantage of chloride binding, such as plausible species **53**.

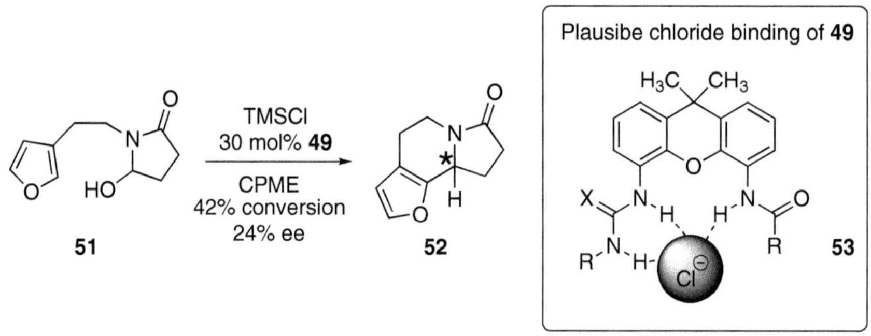

Scheme 6.18 Example of xanthene diamine catalyst in Pictet–Spengler reaction.

6.6.2 Croconamides

Introduced as organocatalysts in 2017 by Pittelkow and coworkers, croconamides are dual N—H bond donors derived from croconic acid (**54**, Scheme 6.19) [35]. Pittelkow *et al.* hypothesized that, due to the analog's nature of croconamides to squaramides, species such as **54** may both recognize anions and operate as hydrogen bond donor catalysts. The anion binding properties of **54a** and **54b** were studied by NMR and UV/vis titration experiments. From these studies, it was determined that **54** associates with several anions, including chloride, bromide, iodide, nitrate, and hydrogen sulfate. The association constant (K_a) for **54a**:Cl^- = $1.7 \times 10^6 \pm 3.5 \times 10^5\ M^{-1}$ in methylene chloride.

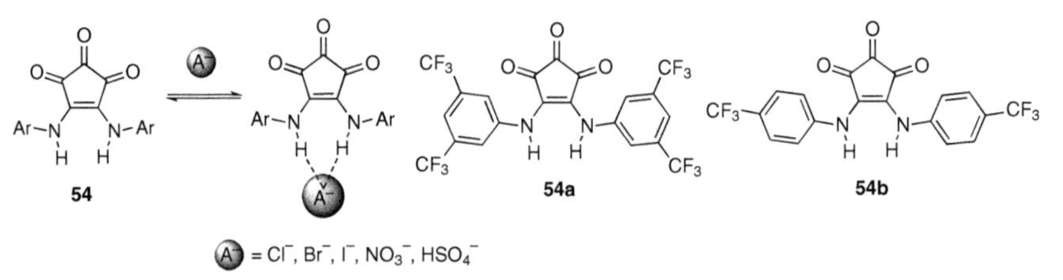

Scheme 6.19 Anion recognition abilities of croconamides. Source: Based on Jeppesen *et al.* [35].

The catalytic abilities of croconamides **54a** and **54b** were studied in the tetrahydropyranylation reaction of substituted phenols (Scheme 6.20). Both **54a** and **54b** were found to accelerate the rate of the reaction. The authors postulate that the mechanism may involve an organized complexation of the catalyst, phenol, and dihydropyran.

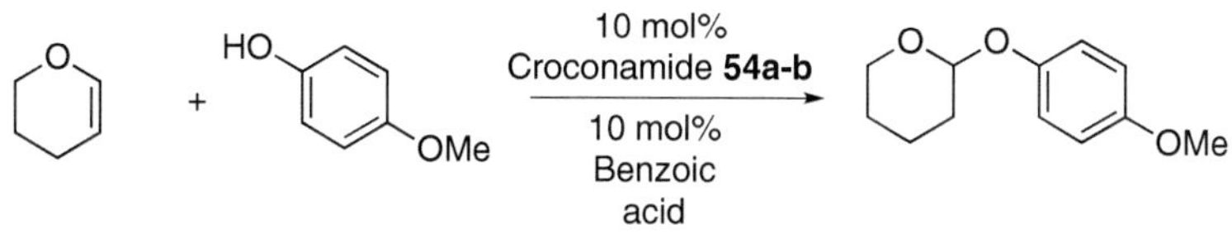

Scheme 6.20 Croconamides as catalysts in tetrahydropyranylation.

6.6.3 Pyrrole-Based Anion-Binding Catalyst

Hirata and Maeda capitalized on the anion recognition abilities of dipyrrolyldiketone boron complexes for anion-binding catalysis [36]. Dipyrrolyldiketone boron complexes, such as **55**, are effective in anion binding, giving rise to planar anion complexes (Scheme 6.21). Stemming from this backdrop, Hirata and Maeda investigated the ability of **55** to operate as a catalyst in an *N*-acyl Mannich reaction likely proceeding through anion-binding catalysis (Scheme 6.22). Specifically, under the influence of 5 mol% **55**, the addition of silyl ketene acetal **5** to plausible ion pair **56** gave rise to **57** and **58** in 84% yield and in a 90/10 ratio.

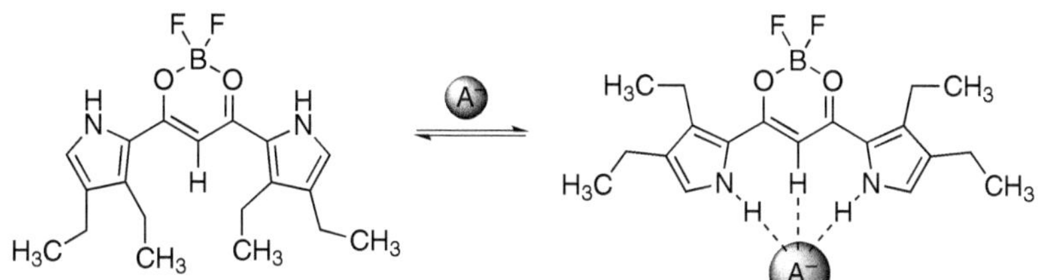

55

Scheme 6.21 Anion recognition of dipyrrolyldiketone boron complexes.

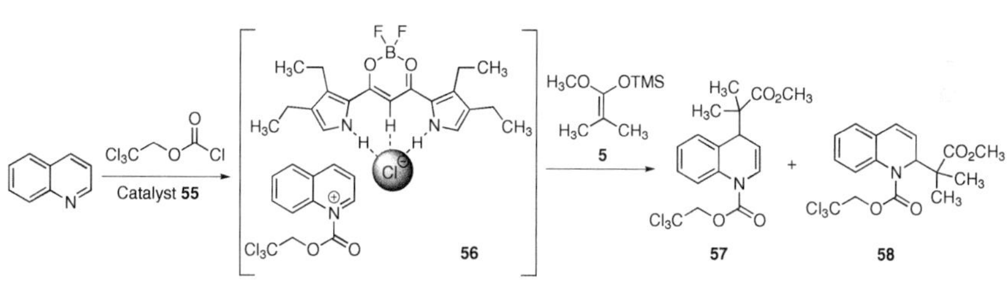

Scheme 6.22 Dipyrrolyldiketone boron complex **55** as a catalyst in an *N*-acyl Mannich reaction.

6.6.4 Bisamidine Catalysts

Göbel and coworkers reported the use of chiral C_2-symmetric bisamidines **60** as Brønsted bases for catalyzing the Diels–Alder reaction between anthrones **59** and maleimides **61** [37] (Scheme 6.23). The deprotonation of **59** led to the formation of the proposed ion pair **63**, which underwent the cycloaddition reaction with **61** to give rise the cycloadducts **62** in moderate to good yields and, generally, low to moderate enantioselectivities (up to 76% ee).

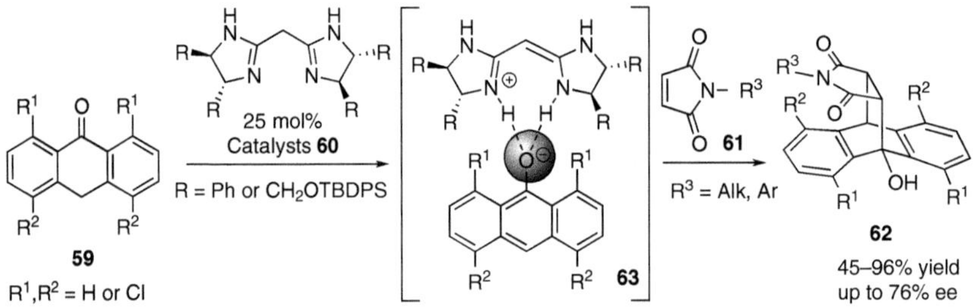

Scheme 6.23 C_2-symmetric bisamidines **60** as catalysts in a Diels–Alder reaction with anthrones. Source: Based on Akalay *et al.* [37].

6.7 Conclusions

A great deal of inspiration for the development of new anion binding catalysts can be found on the interface of supramolecular chemistry and anion recognition. This chapter reviewed several new O—H and N—H hydrogen bond donor scaffolds for anion binding catalysis that started as platforms for anion recognition. Certainly, there is a requirement that an anion-binding catalyst participates in anion recognition; however, that is not the only requirement that makes a good anion binding catalyst. As the field moves forward, features in the anion-binding catalyst that should be considered include the fit of the anion into the hydrogen bond donor functional group and design elements that can lead to enantiocontrol. As anion binding catalysis continues to grow, it is likely that the field will continue to see the development of new families of anion-binding catalysts that can address limitations, such as high catalyst loadings and limited reaction scopes, that are currently facing the field.

References

1 Dilthey, W. and Eduardoff, F. (1904). Über die Darstellung von Phenylsiliciumverbindungen. *Ber. Dtsch. Chem. Ges.* 37: 1139–1149.

2 Martin, G. and Kipping, F.S. (1909). Benzyl and ethyl derivatives of silicon tetrachloride. *J. Chem. Soc.* 95: 302–314.

3 Raabe, G. and Michl, J. (1989). *The Chemistry of Organic Silicon Compounds.* Part 2. Chichester, UK: Wiley.

4 (a) Lickiss, P.D. (1995). The synthesis and structure of organosilanols. *Adv. Inorg. Chem.* 1995 (*42*): 147–262.
(b) Voigt, A., Murugavel, R., Ritter, U., and Roesky, H.W. (1996). Infrared and ^{29}Si NMR spectroscopic investigations on metallasiloxanes derived from organosilanetriols. *J. Organomet. Chem.* 521: 279–286.

5 Chandrasekhar, V., Boomishankar, R., and Nagendran, S. (2004). Recent developments in the synthesis and structure of organosilanols. *Chem. Rev.* 104: 5847–5910.

6 Wieting, J.M., Hardman-Baldwin, A.M., Visco, M.D., and Mattson, A.E. (2016). Silanediol recognition in catalysis, sensing, and drug discovery. *Aldrichim. Acta* 49: 15–20.

7 Min, G.K., Hernandez, D., and Skrydstrup, T. (2013). Efficient routes to carbon–silicon bond formation for the synthesis of silicon-containing peptides and azasilaheterocycles. *Acc. Chem. Res.* 46: 457–470.

8 Sieburth, S.M. and Chen, C.A. (2006, 2006). Silanediol protease inhibitors: from conception to validation. *Eur. J. Org. Chem.*: 311–322.

9 Sieburth, S.M. (2014). *Bio-inspired Silicon-based Materials*, vol. 5. Dordrecht, The Netherlands: Springer.

10 Franz, A.K. and Wilson, S.O. (2013). Organosilicon molecules with medicinal applications. *J. Med. Chem.* 56: 388–405.

11 Auvil, T.J., Schafer, A.G., and Mattson, A.E. (2014, 2014). Design strategies for enhanced hydrogen-bond donor catalysts. *Eur. J. Org. Chem.*: 2633–2646.

12 Kondo, S., Harada, T., Tanaka, R., and Unno, M. (2006). Anion recognition by a silanediol-based receptor. *Org. Lett.* 8: 4621–4624.

13 Kondo, S.I., Fukuda, A., Yamamura, T. *et al.* (2007). Anion recognition by a disiloxane-1,3-diol in organic solvents. *Tetrahedron Lett.* 48: 7946–7949.

14 Wilson, S.O., Tran, N.T., and Franz, A.K. (2012). NMR and X-ray studies of hydrogen bonding for amide-containing silanediols. *Organometallics* 31: 6715–6718.

15 Wieting, J.M., Fisher, T.J., Schafer, A.G. *et al.* (2015, 2015). Preparation and catalytic activity of BINOL-derived silanediols. *Eur. J. Org. Chem.*: 525–533.

16 Schafer, A.G., Wieting, J.M., Fisher, T.J., and Mattson, A.E. (2013). Chiral silanediols in anion-binding catalysis. *Angew. Chem. Int. Ed.* 52: 11321–11324.

17 Hardman-Baldwin, A.M., Visco, M.D., Wieting, J.M. *et al.* (2016). Silanediol-catalyzed chromenone functionalization. *Org. Lett.* 18: 3766–3769.

18 Guan, Y., Attard, J.W., Visco, M.D. *et al.* (2018). Enantioselective catalyst systems from copper(II) triflate and BINOL-silanediol. *Chem. Eur. J.* 24: 7123–7127.

19 Hardman-Baldwin, A.M. and Mattson, A.E. (2014). Silanediol-catalyzed carbon dioxide fixation. *ChemSusChem* 7: 3275–3278.

20 Fox, F., Neudorfl, J.M., and Goldfuss, B. (2019). Silanediol versus chlorosilanol: hydrolyses and hydrogen-bonding catalyses with fenchole-based silanes. *Beilstein J. Org. Chem.* 15: 167–186.

21 Kop-Weiershausen, T., Lex, J., Neudorfl, J.M., and Goldfuss, B. (2005). An exceptional P–H phosphonite: biphenyl-2,2′-bisfenchylchlorophosphite and derived ligands (BIFOPs) in enantioselective copper-catalyzed 1,4-additions. *Beilstein J. Org. Chem.* 1 (6).

22 Hossain, M.A. and Hursthouse, M.B. (1988). Organosilanols – the structure of 1,1,3,3-tetraphenyldisiloxane-1,3-diol. *J. Cryst. Spectrosc.* 18: 227–234.

23 Bruna, S., Garrido-Castro, A.F., Perles, J. *et al.* (2016). Multi-ferrocene-containing silanols as redox-active anion receptors. *Organometallics* 35: 3507–3519.

24 Diemoz, K.M., Wilson, S.O., and Franz, A.K. (2016). Synthesis of structurally varied 1,3-disiloxanediols and their activity as anion-binding catalysts. *Chem. Eur. J.* 22: 18349–18353.

25 Rodriguez, A.A., Yoo, H., Ziller, J.W., and Shea, K.J. (2009). New architectures in hydrogen bond catalysis. *Tetrahedron Lett.* 50: 6830–6833.

26 Cranwell, P.B., Hiscock, J.R., Haynes, C.J.E. *et al.* (2013). Anion recognition and transport properties of sulfamide-, phosphoric triamide- and thiophosphoric triamide-based receptors. *Chem. Commun.* 49: 874–876.

27 Borovika, A., Tang, P.I., Klapman, S., and Nagorny, P. (2013). Thiophosphoramide-based cooperative catalysts for Bronsted acid promoted ionic Diels–Alder reactions. *Angew. Chem. Int. Ed.* 52: 13424–13428.

28 Bhattarai, B., Tay, J.H., and Nagorny, P. (2015). Thiophosphoramides as cooperative catalysts for copper-catalyzed arylation of carboxylates with diaryliodonium salts. *Chem. Commun.* 51: 5398–5401.

29 Balakrishna, M.S. (2016). Cyclodiphosphazanes: options are endless. *Dalton Trans.* 45: 12252–12282.

30 Klare, H., Hanft, S., Neudorfl, J.M. *et al.* (2014). Anion recognition with hydrogen-bonding cyclodiphosphazanes. *Chem. Eur. J.* 20: 11847–11855.

31 Klare, H., Neudorfl, J.M., and Goldfuss, B. (2014). New hydrogen-bonding organocatalysts: chiral cyclophosphazanes and phosphorus amides as catalysts for asymmetric Michael additions. *Beilstein J. Org. Chem.* 10: 224–236.

32 Wolf, F.F., Neudorfl, J.M., and Goldfuss, B. (2018). Hydrogen-bonding cyclodiphosphazanes: superior effects of 3,5-$(CF_3)_2$-substitution in anion-recognition and counter-ion catalysis. *New J. Chem.* 42: 4854–4870.

33 Metz, A.E., Ramalingam, K., and Kozlowski, M.C. (2015). Xanthene-4,5-diamine derivatives: a study of anion-binding catalysis. *Tetrahedron Lett.* 56: 5180–5184.

34 Raheem, I.T., Thiara, P.S., Peterson, E.A., and Jacobsen, E.N. (2007). Enantioselective Pictet–Spengler-type cyclizations of hydroxylactams: H-bond donor catalysis by anion binding. *J. Am. Chem. Soc.* 129: 13404–13405.

35 Jeppesen, A., Nielsen, B.E., Larsen, D. *et al.* (2017). Croconamides: a new dual hydrogen bond donating motif for anion recognition and organocatalysis. *Org. Biomol. Chem.* 15: 2784–2790.

36 Hirata, G. and Maeda, H. (2018). Pyrrole-based anion-responsive π-electronic molecules as hydrogen-bonding catalysts. *Org. Lett.* 20: 2853–2856.

37 Akalay, D., Dürner, G., Bats, J.W., and Göbel, M.W. (2008). C_2-symmetric bisamidines: chiral Brønsted bases catalysing the Diels–Alder reaction of anthrones. *Beilstein J. Org. Chem.* **4**: 28.

7

1,2,3-Triazoles and 1,2,3-Triazolium Ions as Catalysts

Kohsuke Ohmatsu and Takashi Ooi

Nagoya University, Institute of Transformative Bio-Molecules, Graduate School of Engineering, Department of Molecular and Macromolecular Chemistry, D2-1 Furo-cho, Chikusa, Nagoya 464-8601, Japan

7.1 Introduction

1,2,3-Triazole is a unique heterocycle with phenomenal stability compared to other organic compounds having contiguous three nitrogen atoms. The history of triazole synthesis can be traced back to late nineteenth century when 1,3-dipolar cycloaddition between phenyl azide and dimethyl acetylenedicarboxylate was reported [1]. While this synthetic method attracted less attention for a long period, the concept of click chemistry, introduced by Sharpless in 2001, triggered its extensive applications. Click chemistry refers to the reactions that are high yielding, atom-economical, chemoselective, wide in scope, operationally simple, effectively proceed in a range of solvents including aqueous media, and can be conducted under mild conditions [2]. Copper(I)-catalyzed cycloaddition of organic azides with terminal alkynes, named Huisgen cycloaddition, to produce 1,4-disubstituted 1,2,3-triazoles is now regarded as one of the prime examples of click chemistry [3]. Particular attention was initially focused on the triazole formation as an orthogonal cross-linking of two organic molecules because the rigidity of the disubstituted aromatic ring system can prevent interactions between the linked components. This feature made the triazole unit a versatile linker for small molecules [4], biohybrid materials [5], and polymers [6]. In conjunction with enormous opportunities for the development of triazole-containing molecules, the intriguing properties of 1,2,3-triazole as a functional group has also garnered increasing attention in various fields of chemical science.

The three nitrogen atoms of 1,2,3-triazoles cause strong polarization within their aromatic π system, where N(2) and N(3) atoms have partial negative charges [7]. Together with the lone electron pairs, these two nitrogen atoms of triazole rings have the ability pertinent to coordinate with metals and also to act as hydrogen bond acceptors (Figure 7.1) [8]. In addition, because of the polarization of triazole ring as well as the σ-accepting ability of the pyrrole-type N(1) atom, the adjacent C(5) proton exhibits notable C–H acidity. The pK_a value in DMSO for C(5) proton of

Anion-Binding Catalysis, First Edition. Edited by Olga García Mancheño.

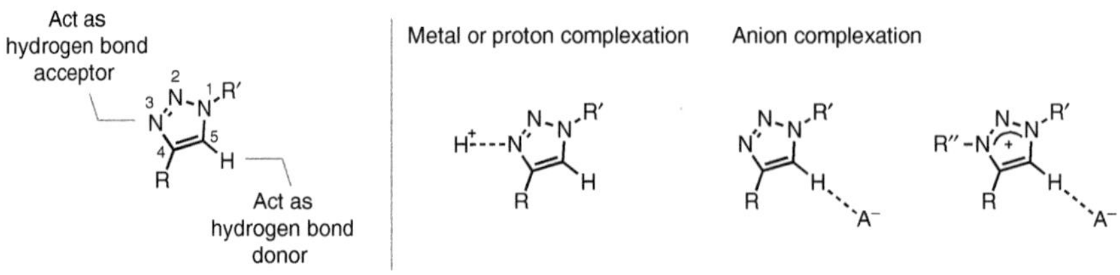

Figure 7.1 1,2,3-Triazole as a multiple binding donor for metal ions, proton, and anions. Source: Based on Bryant and Bunz [8].

1-methyl-1*H*-1,2,3-triazole is 27.8, which is much lower than that for the C(2) proton of 1-methylimidazole (34.1) and the C(5) proton of 1-methylpyrazole (33.8) [9]. Such attribute provides 1,2,3-triazole with the capability of acting as an anion-binding motif through the C(5)—H· · ·anion hydrogen bonding interaction. Furthermore, this C–H acidity as well as anion-binding ability can be enhanced by converting a 1,2,3-triazole unit to a 1,2,3-triazolium ion [10].

The unique hydrogen-bonding ability of 1,2,3-trizoles has been initially exploited in the field of supramolecular chemistry. In 2008, Flood reported the utilization of the anion-binding property of triazole motif in the design of macrocyclic anion receptors **1** (Figure 7.2) [11]. Macrocyclic triazolophane **1** consists of four triazoles and four phenyl rings and the eight C—H bonds in total converge to the central cavity. The internal diameter of **1** is about 3.8 Å, and this cavity size is matched with chloride and bromide ions, of which ionic radii are 1.81 and 1.96 Å, respectively. The actual associating affinity was quantified through the titration experiments of **1** with halide ions: exceptionally large association constants were obtained for chloride ($K_a = 11 \times 10^6\ M^{-1}$) and bromide ($K_a = 7.5 \times 10^6\ M^{-1}$). In contrast, smaller fluoride ion and larger iodide ion showed much smaller association constants because of the mismatch between the size of cavity and the ionic radii of these anions.

As a proof of the importance of preorganized macrocyclic structure of **1**, Flood and coworkers also synthesized an acyclic analog **2** having four triazole units, three *t*-butyl-substituted phenyl groups, and two terminal phenyl rings [11b]. While **2** can form the folded complex with halide ions through the same type and almost the same number of hydrogen bonds with **1**, the association constants for the halide binding with **2** are negligibly small compared to those with **1**. These large differences in association constants clearly indicate that preorganized structure of triazole-containing molecular frameworks is crucial for exerting high anion-binding ability [12].

In parallel with Flood's initial report, Craig introduced a very similar acyclic tetrakistriazole **3** possessing ester-substituted phenylene motifs and revealed that **3** exhibited relatively large binding strengths toward chloride ions ($K_a = 1.7 \times 10^4\ M^{-1}$) [13]. Since the ester substituents can induce strong polarization effect owing to the electron-withdrawing nature, the introduction of such functionalities is a judicious strategy for improving the anion-binding affinity. In addition, Craig confirmed that ester-substituted phenylene-linked bistriazole **4** acted as a weaker chloride receptor than **3**, corroborating the importance of the number of triazole motifs for acquiring the ability as an anion receptor.

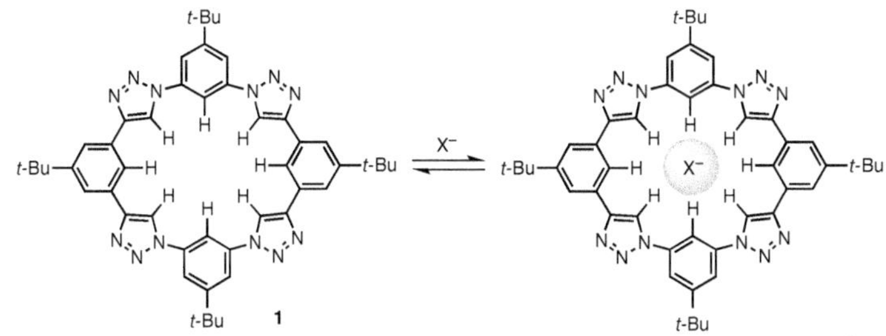

1

X	K_a (M^{-1})
F	280 000
Cl	11 000 000
Br	7 500 000
I	17 000

in CD_2Cl_2

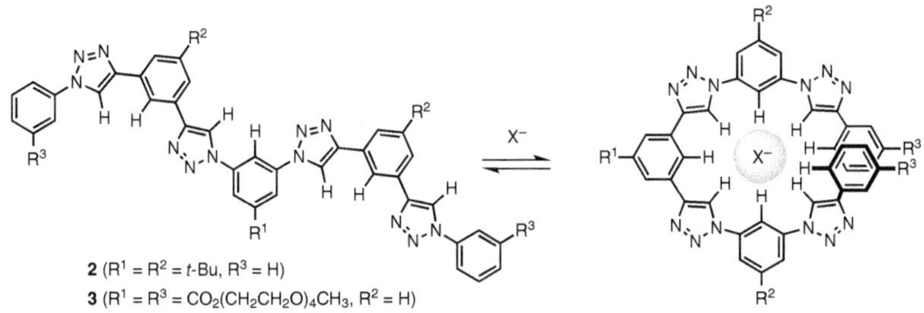

2 ($R^1 = R^2 =$ t-Bu, R^3 = H)
3 ($R^1 = R^3 = CO_2(CH_2CH_2O)_4CH_3$, R^2 = H)

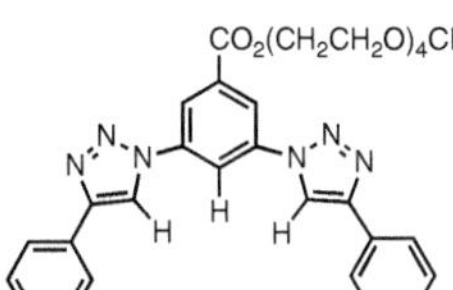

4

Oligotriazole	X	K_a (M^{-1})	Solvent
2	Cl	6.9	CD_2Cl_2
2	Br	268	CD_2Cl_2
3	Cl	17,000	Acetone-d_6
4	Cl	1,300	Acetone-d_6

Figure 7.2 1,2,3-Triazole-containing macrocyclic anion receptor and its acyclic analogs. Source: Adapted from Li and Flood. [11], and Juwarker et al. [13].

As already described, 1,2,3-triazolium ion has much stronger anion-binding ability than the neutral 1,2,3-triazole ring. For instance, Pandey demonstrated that steroid-based cyclic bis-1,2,3-triazoliums of type **5** exhibited remarkable ability to recognize anions through C(5)—H···anion hydrogen bonding interaction (Figure 7.3) [10a]. Bistriazolium ion **5a** showed the highest binding affinity for fluoride ion, and the NMR titrations indicated the order of fluoride > chloride > bromide > iodide > acetate ion (AcO^-) in regard to the association constants. Furthermore, no detectable binding was observed with dihydrogen phosphate ion ($H_2PO_4^-$) probably because of the mismatch between the size of the cavity of **5a** and the anion. In contrast, bistriazolium ion **5b** having slightly larger cavity than **5a** showed the largest association constants with $H_2PO_4^-$. Furthermore, acyclic analog **6** exerted much stronger ability in association with $H_2PO_4^-$ as compared to the preorganized cyclic bistriazolium ions, **5a** and **5b**. The higher binding affinity of acyclic bistriazolium **6** might be attributed to its flexible structure beneficial for adapting the favorable geometry for the binding of an anion. In addition, the relative aptitude of acyclic triazoliums for anion binding was clearly different from the tendency observed with non-cationic oligotriazoles, featuring the difference between 1,2,3-triazolium ions and neutral 1,2,3-triazoles in terms of the structural requirements for exerting strong anion-binding ability.

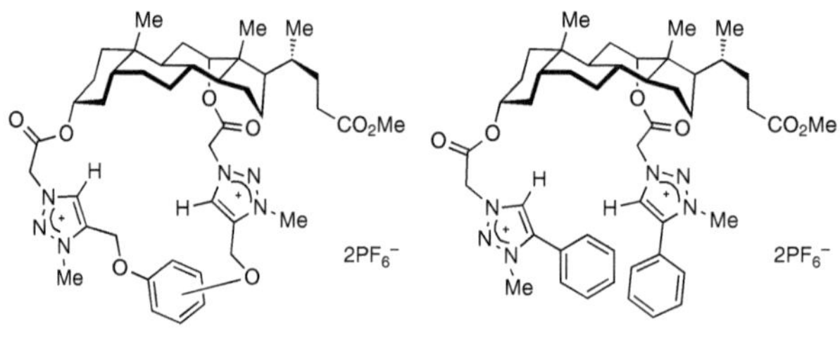

meta **5a**
para **5b**

6

5 or **6** $\xrightleftharpoons{X^-}$ **5·X** or **6·X**

Triazolium	X	K_a (M^{-1})
5a	Cl	270
5a	H_2PO_4	No binding
5b	Cl	690
5b	H_2PO_4	1100
6	Cl	390
6	H_2PO_4	1920

in $CDCl_3$

Figure 7.3 1,2,3-Triazolium-containing anion receptors. Source: Based on Kumar and Pandey [10a].

7.2 Triazole-Based Anion-Binding Molecular Catalysts

The concept of neutral anion-binding organic molecular catalysts consists in the activation of the cationic electrophile through the association of its counteranion with hydrogen bond donor catalyst, thus facilitating and/or controlling the attack of the nucleophile. While this mode of catalysis is operative mainly with ureas and thioureas [14, 15], in recent years, 1,2,3-triazole-based molecules have also emerged as powerful anion-binding catalysts.

As a leading contribution, Mancheño developed a series of bistriazoles **7a–7e** featuring a *meta*-phenylene motif and two side substituents [16]. The NMR titration experiments revealed moderate to good binding affinity of bistriazoles **7** with chloride ion, and **7e** bearing 3,5-bis(trifluoromethyl)phenyl groups showed the largest association constants (Figure 7.4). The choice of solvent was also important for increasing the association constants, and tetrahydrofuran (THF) was found to be optimal. The catalytic performance of **7** was evaluated in the N-alkylation of benzylamine with 4-dimethylamino-*N*-triphenylmethyl-pyridinium chloride **8**, which was a precursor of the electrophilic trityl ion. No product formation was observed in THF at 45 °C in the absence of the catalysts, whereas *N*-tritylated product **9** was obtained in moderate to high yields in the presence of bistriazoles **7**. Noteworthy is that there was an evident correlation between the anion-binding ability of the catalysts and the efficiency of the N-alkylation reaction. Thus, the highest yield of **9** was attained when the reaction was conducted with **7e**.

As an application of enantiospecific C–H azidation of Tröger bases, Lacour and coworkers synthesized bistriazoles **10** [17]. Compared to meta-phenylene derivative **10b**, pyridine-tethered **10a** showed a weaker binding ability toward chloride ion in the titration experiments in acetone-d_6 probably because the lone pair on nitrogen induced a repulsion with the anion (Figure 7.5). Nevertheless, these two derivatives exhibited comparable catalytic activity in the reactions of benzyl alcohol or primary amines with **8** in THF, where tritylated products **11–13** were formed in uniformly good yields.

Chiral variant of triazole-based anion-binding catalysts were also introduced by Mancheño and coworkers. For instance, they developed 1,2-cyclohexanediamine-

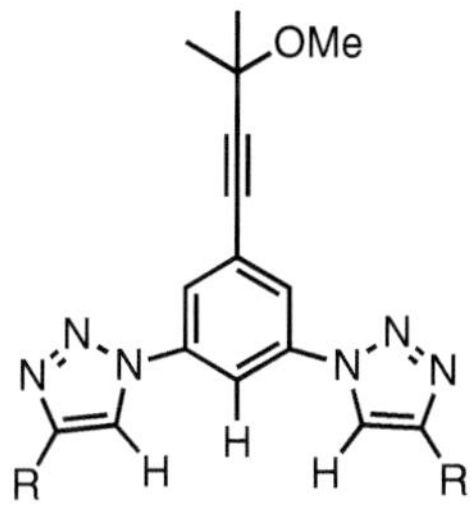

7a: R = *i*-Pr
7b: R = C_6F_5
7c: R = 4-$Me_2NC_6H_4$
7d: R = Ph
7e: R = 3,5-$(CF_3)_2C_6H_3$

7 $\xrightleftharpoons{Cl^-}$ **7·Cl**

Bistriazole	Solvent	K_a (M^{-1})
7a	Acetone	34
7b	Acetone	111
7c	Acetone	152
7d	Acetone	342
7e	Acetone	458
7e	CH_2Cl_2	108
7e	THF	1417

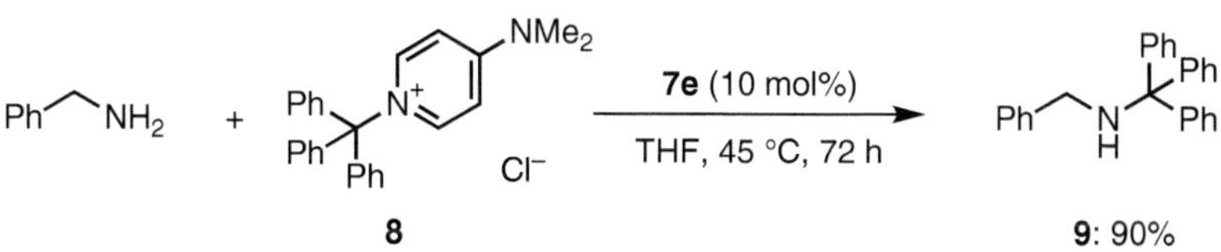

Order of catalytic activity: **7e** > **7d** > **7c** > **7b** > **7a**

Figure 7.4 Structure of bistriazoles **7** and application for alkylation of amine.

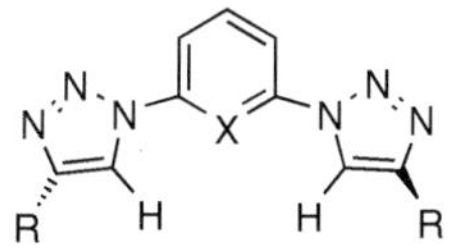

10a (X = N), **10b** (X = CH)

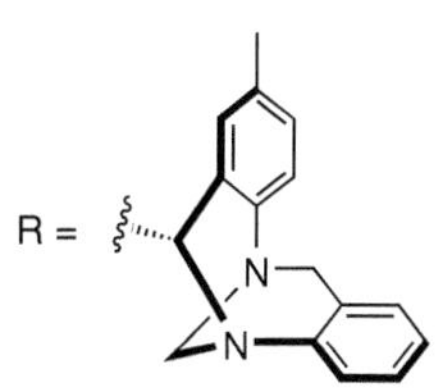

10 $\xrightleftharpoons{Cl^-}$ **10·Cl**

Bistriazole	Solvent	K_a (M^{-1})
10a	Acetone	11.5
10b	Acetone	47.9

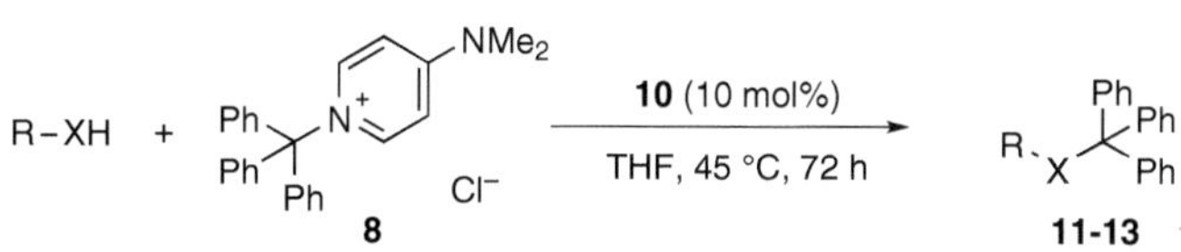

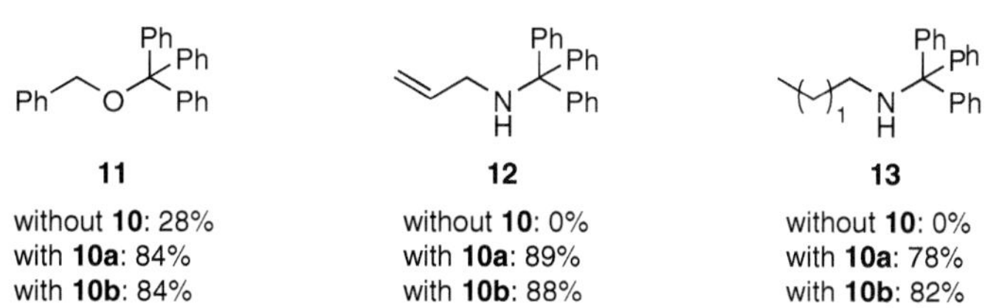

Figure 7.5 Tröger base-derived bistriazoles **10** and application for tritylation reactions.

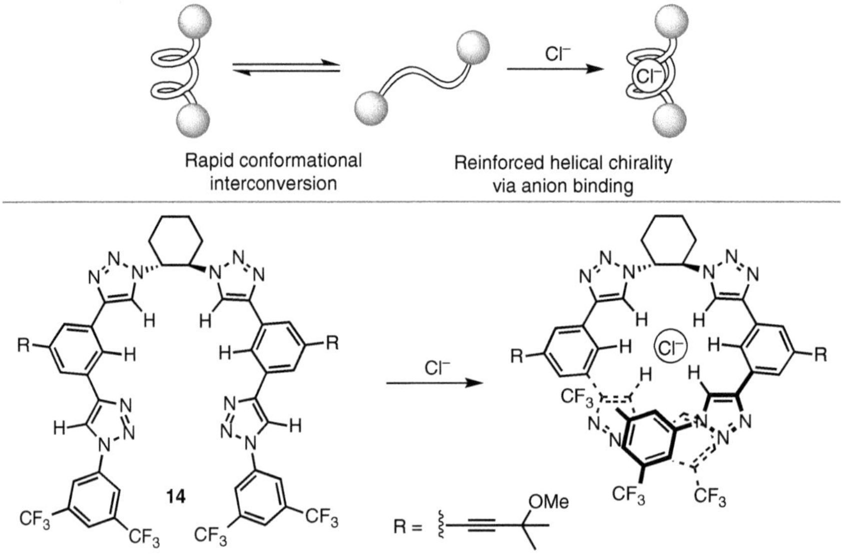

Figure 7.6 Structure of tetrakistriazole **14** and formation of chiral helical anion complex.

derived chiral tetrakistriazole **14** and evaluated its binding ability as well as catalytic performance [18]. The circular dichroism (CD) titration of **14** with tetrabutylammonium chloride clearly showed a strong CD effect along with the increase in the amount of chloride ions in the solution. This result suggested that **14** existed as various flexible open-chain conformers in the absence of anions and the rigid chiral helical complex was formed upon addition and binding of chloride ions (Figure 7.6). Computational analysis also suggested that the helical 1 : 1 **14** – chloride ion complex was the most stable among other possible complexes [19].

To assess the viability of employing **14** as a chiral anion-binding catalyst, Mancheño applied it to acyl-Mannich reaction of quinolines with ketene silyl acetals, of which the related transformations had been established by using thiourea-based chiral catalysts [20, 21]. The reaction of quinoline with 2,2,2-trichloroethyl chloroformate (TrocCl) and ketene silyl acetal **15a** in the presence of 10 mol% of **14** in methyl *t*-butyl ether (MTBE) afforded the corresponding Mannich adduct **16** in good yield with high enantioselectivity (Figure 7.7). In this reaction, *N*-Troc quinolinium chloride initially formed from quinoline and TrocCl and its chloride ion was captured by **14**. Thus, the generated chiral triazole-bound ion pair intermediate underwent bond formation with **15a** in an enantioselective manner. The high level of enantiocontrol was attributed to the induced helical chirality based on the observation that chiral bistriazole **17**, which would be unable to display helical chirality upon binding to chloride ion, was not enantioselective. The enantioselectivity also depended on the structure of nucleophiles and acylating reagents to generate quinolinium chloride. While ketene silyl acetals with bulky isopropyl or *t*-butyl substituent (**15a** or **15b**) gave high enantioselectivity, the methyl ester derivative **15c** provided much lower selectivity. Regarding the acylating reagents, TrocCl was

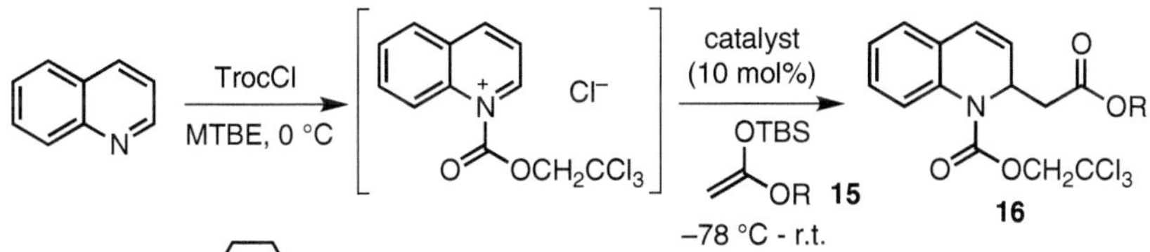

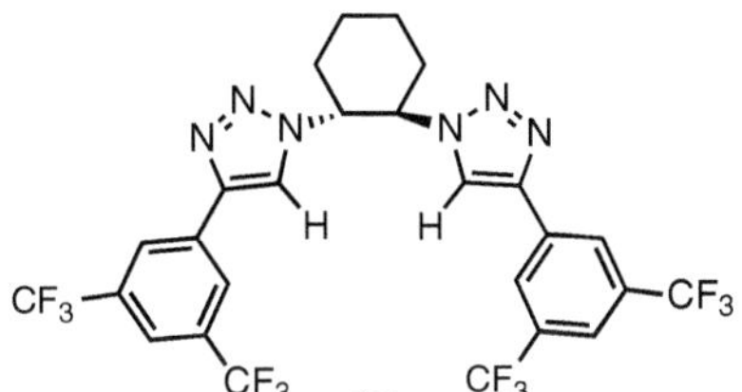

17

Catalyst	**15** (R)	Yield (%)	e.r.
14	**15a** (*i*-Pr)	76	96 : 4
17	**15a** (*i*-Pr)	91	41 : 59
14	**15b** (*t*-Bu)	58	98 : 2
14	**15c** (Me)	56	64 : 36

Figure 7.7 Asymmetric acyl Mannich reaction of quinoline with ketene silyl acetals.

found to be particularly effective: the use of other analogs, such as acetyl chloride and benzyl chloroformate (CbzCl), led to poor asymmetric induction.

Tetrakistriazoles of type **14** also acted as effective catalysts for the reactions of isoquinolines with ketene silyl acetals and the reaction of quinolines with ketene silyl thioacetals [22] and silyl alkylphosphites [23]. In addition, the development of **14**-catalyzed asymmetric acyl Mannich reaction of pyridines was especially notable [24]. The nucleophilic attack to pyridine derivatives leads to the loss of their aromaticity, whereas quinolines and isoquinolines retain the aromaticity in the fused benzene ring in similar transformations. Therefore, acyl Mannich reaction of pyridines is energetically more demanding. Furthermore, the control of regioselectivity is a difficult issue because the nucleophilic addition can proceed at both C4 and C2 positions of the pyridine ring. During the course of application of the oligotriazoles, Mancheño found that chiral tetrakistriazole **14** proved uniquely effective for acyl Mannich reaction of pyridines with ketene silyl acetals. Specifically, 5 mol% of **14** catalyzed the addition of **15a** to pyridinium ion, which was generated *in situ* from 2-methylpyridine and TrocCl to give the corresponding dihydropyridine **18** in high yield with excellent enantioselectivity (Figure 7.8). This process proceeded with a high level of regiocontrol, where C2 alkylation was more productive than C4 alkylation. The distinct catalytic performance of the chiral tetrakistriazole was unambiguously demonstrated by the reactions with the other representative chiral anion-binding catalysts, such as chiral thioureas **19**, **20**, and squaramide **21**, which led to the preferential C4 alkylation and afforded nearly racemic C2 and C4 adducts, respectively. Superior stereocontrolling ability of chiral tetrakistriazoles was also highlighted in asymmetric acyl Mannich reaction of quinazoline [25].

The remarkable capability of chiral tetrakistriazole **14** as an anion-binding catalyst was further demonstrated in the asymmetric dearomatization of pyrylium derivatives [26]. The nucleophilic addition of ketene silyl acetal **15a** to pyrylium ion, which was generated *in situ* from 4-chromenone and *tert*-butyldimethylsilyl triflate (TBSOTf), proceeded smoothly in the presence of 10 mol% of **14** to afford

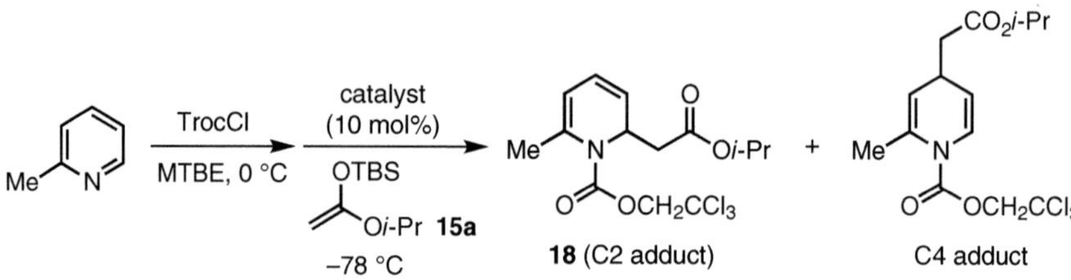

catalyst	Yield (%)	C2 : C4	e.r. of **18**
14	87	92 : 8	97 : 3
19	66	11 : 89	57 : 43
20	21	17 : 83	52 : 48
21	26	14 : 86	54 : 46

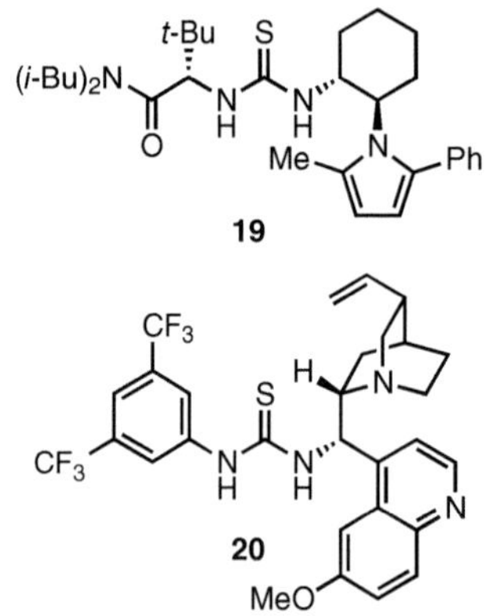

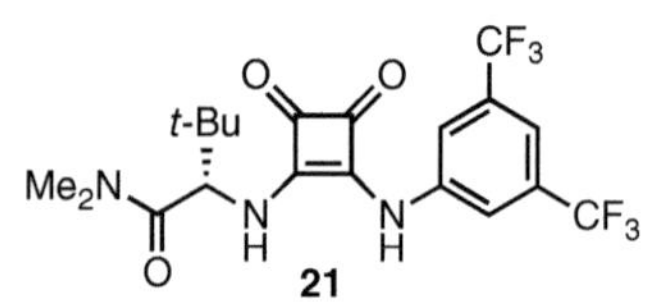

Figure 7.8 Asymmetric acyl Mannich reaction of 2-methylpyridine using representative chiral anion-binding catalysts.

the product **22** in a highly enantioenriched form (Figure 7.9). In contrast, chiral thiourea **23** and squaramide **21** were fare less effective, particularly in terms of stereocontrol. An additional salient aspect is that catalyst loading of **14** could be reduced to 0.05 mol%: **22** was obtained in high yield without a significant loss of enantioselectivity. While the NMR titration of **14** with tetrabutylammonium triflate (TBAOTf) in acetone-d_6 showed almost nonexistent binding of **14** toward triflate, theoretical calculation indicated that a 1 : 1 complex of **14** and triflate ion could form to a considerable extent in toluene at −78 °C [19].

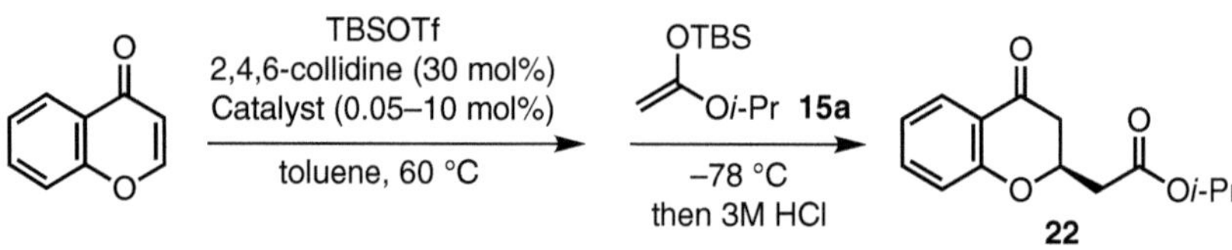

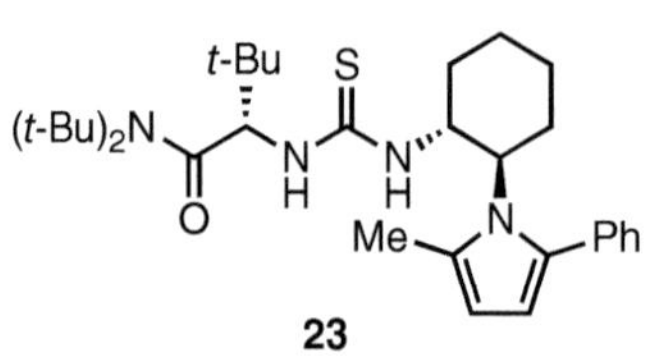

Catalyst (mol%)	Yield (%)	e.r. of **14**
14 (10)	92	91 : 9
23 (10)	82	57 : 43
21 (10)	59	51 : 49
14 (0.05)	82	88 : 12

Figure 7.9 Enantioselective dearomatization of pyrylium derivatives.

Based on the importance of helical chirality of oligotriazole–anion complexes for asymmetric induction, Feringa and coworker developed a tetrakistriazole-functionalized chiral switchable catalyst that enabled the enantiodivergent synthesis [27] in the reactions with a single set of substrates under identical conditions [28]. Their tactics for the design of switchable catalysts was the use of molecular motors, which can be interconverted between different stereoisomers with the aid of light and heat [29]. Concretely, the triazole-functionalized molecular motor **24** comprising two branches, each of them having phenylene-tethered bistriazole motif, was developed (Figure 7.10). While the two bistriazoles are far from each other in *trans*-isomer of **24**, these units come in close proximity through the light-induced interconversion to *cis*-isomer. Specifically, UV (312 nm) light irradiation of (*R*,*R*)-(*P*,*P*)-*trans*-**24** facilitated isomerization to (*R*,*R*)-(*M*,*M*)-*cis*-**24**, and this process was confirmed to be reversed by the irradiation with 365 nm light. In contrast, heating of (*R*,*R*)-(*M*,*M*)-*cis*-**24** in THF at 60 °C promoted thermal helix inversion to give (*R*,*R*)-(*P*,*P*)-*cis*-**24**. The behavior of these three interconvertible isomers of **24** as anion-binding catalysts was evaluated in the asymmetric reaction

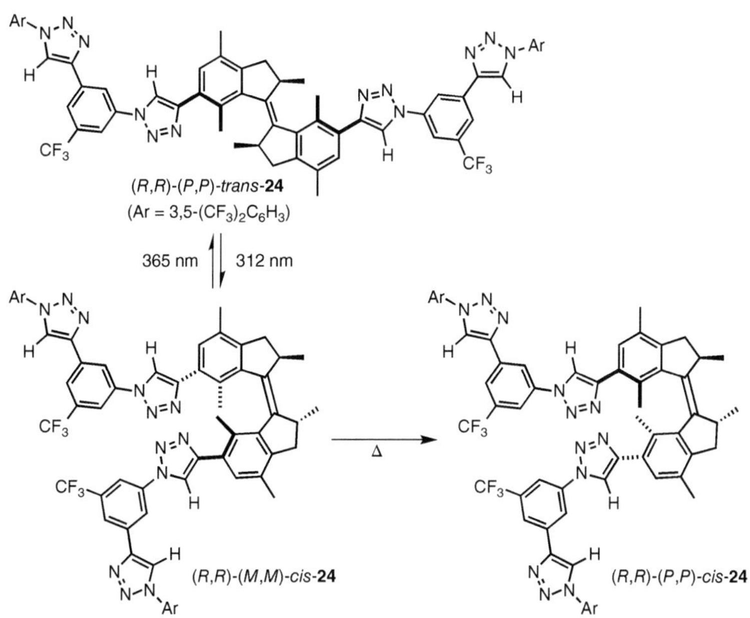

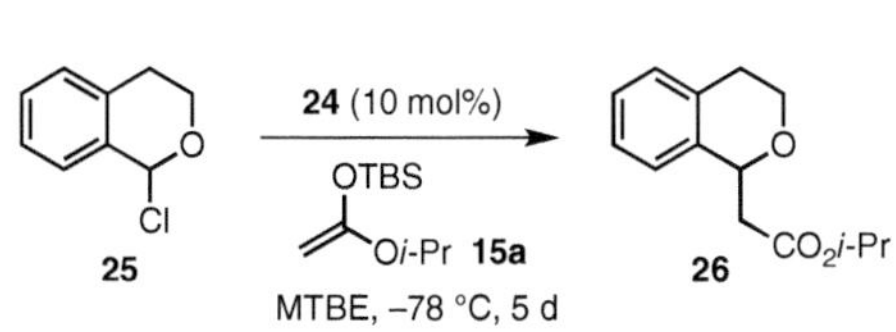

24	Yield (%)	e.r.
(*P*,*P*)-*trans*-**24**	48	51 : 49
(*M*,*M*)-*cis*-**24**	47	74 : 26 (*R*)
(*P*,*P*)-*cis*-**24**	90	10 : 90 (*S*)

Figure 7.10 Triazole-functionalized molecular motor as a switchable chiral anion-binding catalyst.

of 1-chloroisochroman **25** with ketene silyl acetal **15a**. The treatment of **25** with **15a** and a catalytic amount (10 mol%) of (*R*,*R*)-(*P*,*P*)-*trans*-**24** in MTBE at −80 °C for five days resulted in the formation of the product **26** in moderate yield and nearly racemic form. This negligible enantioselectivity was likely attributed to the *trans* conformation of **24** unable to form helical chiral structure upon chloride binding. In fact, the reaction with (*R*,*R*)-(*M*,*M*)-*cis*-**24**, whose formation of supramolecular helical structure was indicated by CD spectroscopy, afforded (*R*)-**26** in moderate yield with moderate enantioselectivity under similar conditions. In addition, (*R*,*R*)-(*P*,*P*)-*cis*-**24** having helical chirality opposite to (*R*,*R*)-(*M*,*M*)-*cis*-**24** preferentially gave the opposite enantiomer product (*S*)-**26** in high yield with good enantioselectivity. These results indicated that helical chirality of the oligotriazole–chloride complex is a dominant factor to determine the stereochemistry of the reaction products. The different catalytic activities between (*M*,*M*)-*cis*-**24** and (*P*,*P*)-*cis*-**24** would stem from less favorable conformation of (*M*,*M*)-*cis*-**24** for chloride binding.

Besides the introduction of electron-withdrawing groups and the design of oligotriazoles with a preorganized structure, protonation of 1,2,3-triazole, that is the formation of N(3)-protonated triazolium ion, can also enhance its anion-binding ability. For instance, ^{1}H NMR titration experiments with chiral bistriazole **27a**, which was developed by our group (Figure 7.11) and tetrabutylammonium bromide (TBAB) or HBr revealed a clear difference in the degree of anion binding [30]. While the peaks of both triazole C(5) proton and amide proton of **27a** underwent only slight downfield shifts upon treating with TBAB, drastic downfield shifts of these signals were observed in response to the addition of HBr. In addition, a continuous variation plot (Job plot) revealed that **27a** and HBr could form a 1 : 2 complex. The molar concentration of a 1 : 1 **27a**–HBr complex was estimated to be maximum with the addition of 1 equiv of HBr, and the concentration of the 1 : 2 complex seemed to exceed that of the 1 : 1 complex around 1.5 equiv of HBr added. The first and second association constants (K^1 and K^2) were determined to be approximately 7.8×10^5 M^{-1} and 1.8×10^4 M^{-1}, respectively. Based on this intriguing characteristic, our group developed a chiral supramolecular Brønsted acid catalyst generated from **27** and simple achiral Brønsted acid.

The reaction of indole with *N*-sulfonyl imine **28** in the presence of **27a** and trifluoromethanesulfonic acid (TfOH) afforded a mixture of Friedel–Crafts adduct **29** and undesired side product **30**. The side product **30** was formed through the acid-promoted elimination of sulfonamide from **29**, followed by the addition of indole. This undesired overreaction could be suppressed by attenuating the acidity of the assembled catalyst. Indeed, the use of weaker Brønsted acids such as methanesulfonic acid and HCl instead of TfOH led to the improvement in chemoselectivity. The change of achiral acids also had an impact on the stereocontrolling ability of the supramolecular catalyst and HCl was optimal both in terms of chemoselectivity and enantioselectivity. As the increased amount of achiral acid toward **27a** induced difference in the structure of the supramolecular complex, the ratio of **27a** and HCl was also an important parameter that affects the catalytic performance: the enantioselectivity was gradually decreased increase in the amount of HCl to 1.5, 1.8, and 2 equiv toward **21a**. Eventually, the structurally optimized

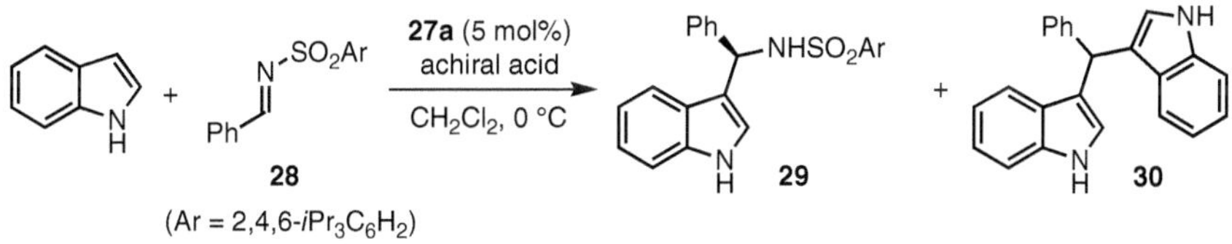

Achiral acid (mol%)	**29**/**30**	Yield (%) of **29**	*ee* (%)
TfOH (5)	3 : 1	59	48
HBr (5)	16 : 1	48	60
$MeSO_3H$ (5)	>20 : 1	36	40
HCl (5)	**29** only	33	59
HCl (7.5)	**29** only	34	55
HCl (9)	**29** only	41	53
HCl (10)	**29** only	39	51

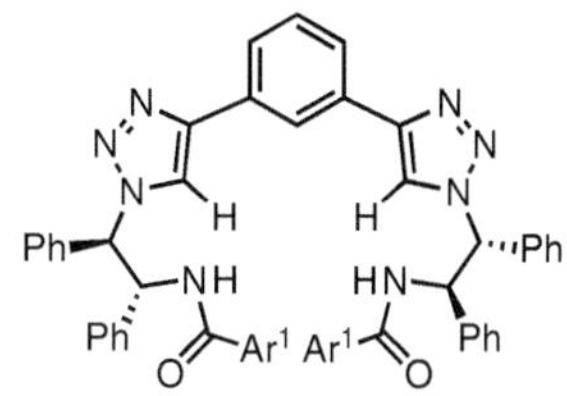

27a (Ar^1 = Ph)
27b (Ar^1 = 4-tBuC_6H_4)

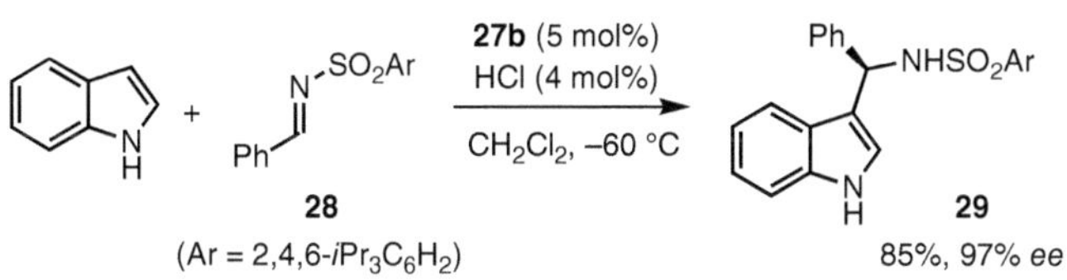

Figure 7.11 Protonated chiral bistriazolium ion as a supramolecular chiral Brønsted acid catalyst.

bistriazole **27b** and slightly reduced amount of HCl allowed the efficient and highly enantioselective Friedel–Craft reaction to produce **29**.

7.3 Triazolium Ions as Organic Molecular Catalysts with Anion-Binding Ability

Chiral 1,2,3-triazolium ions were also developed as organic molecular catalysts and successfully applied for a variety of asymmetric transformations. In 2011, our group introduced a chiral 1,2,3-triazolium ion of type **31** as a new class of anion-binding catalyst (Figure 7.12) [31]. The molecular structure of **31** features an acyclic chiral framework possessing a triazolium core and amide functionality. One of the advantages of this structure is the ready accessibility from terminal alkynes and α-amino acid-derived chiral azides. In addition, structural modification of **31** can be easily implemented in the course of synthesis using different reagents and conditions, thus providing sufficiently large catalyst library.

The flexible acyclic structure of chiral catalysts is generally considered to be disadvantageous for stereocontrol of asymmetric transformations because of the numerous possibilities of competing transition-state structures. While acyclic

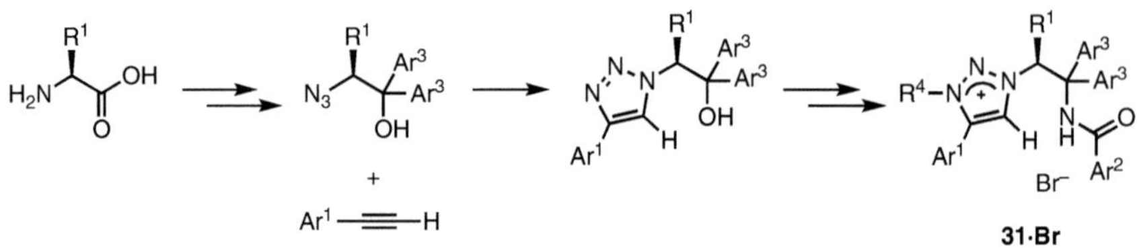

Figure 7.12 Structure and synthesis of chiral 1,2,3-triazolium bromide **31**·Br. Source: Based on Ohmatsu *et al.* [31].

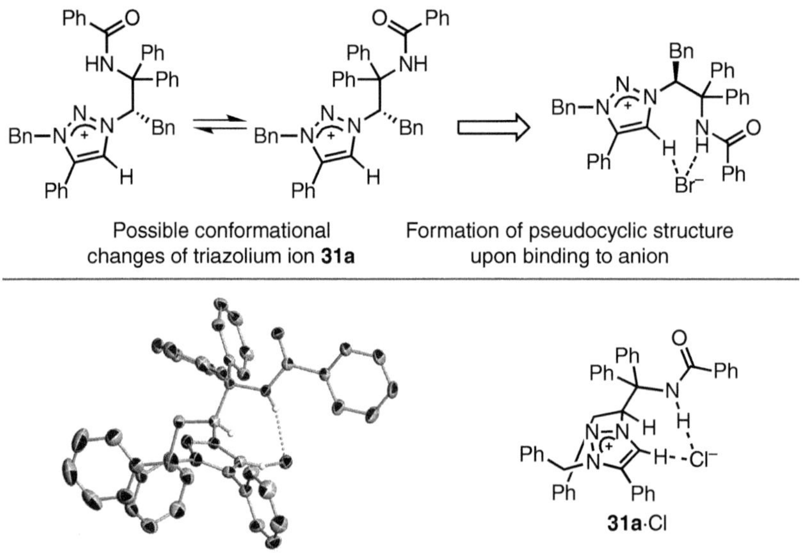

Figure 7.13 Flexibility of 1,2,3-triazolium ion **31** and formation of pseudocyclic structure.

triazolium ion **31** itself would be flexible, it can form pseudocyclic structure upon binding to an anion through two hydrogen bonds, i.e. a C(5)—H· · ·anion and an amide N—H· · ····anion (Figure 7.13). In addition, flexibility of **31** would be advantageous for the formation of a structured ion pair complex with various anions of different sizes. The three-dimensional molecular structure of chiral triazolium chloride **31a**·Cl was unambiguously determined by single-crystal X-ray diffraction analysis. The chloride ion was located in proximity to the triazolium C(5) proton and the amide proton, clearly indicating the existence of two hydrogen bonds. The evidence also came from the ^{1}H NMR spectra of **31a** paired with a series of counterions, such as tetrafluoroborate, bromide, chloride, and acetate. With the increase in Brønsted basicity of counterions, significant and simultaneous downfield shifts were observed for the triazolium C(5) proton and amide proton, verifying that both of the protons were strongly bound to the anion regardless of its size.

Cationic triazoliums can be regarded as analogs of quaternary onium ions, and thus, this cationic anion-binding catalyst can be utilized for asymmetric phase-transfer reactions involving a variety of anionic reactive species [32]. The catalytic performance of triazolium ion **31** was initially evaluated in the asymmetric

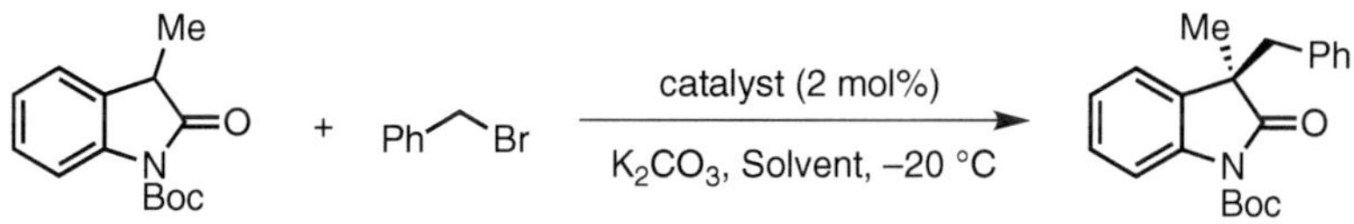

32 **33**

31·Br **34·Br**

31a ($Ar^1 = Ar^2$ = Ph)
31b (Ar^1 = *o*-Tol, Ar^2 = Ph)
31c (Ar^1 = *o*-PhC_6H_4, Ar^2 = Ph)
31d (Ar^1 = *o*-PhC_6H_4, Ar^2 = 3,5-$Cl_2C_6H_3$)

34a (R^1 = Me, R^2 = H)
34b (R^1 = H, R^2 = Me)

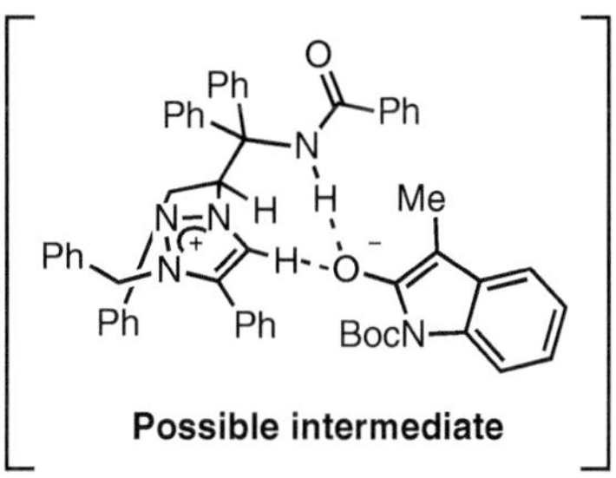

catalyst	Solvent	Yield (%)	*ee* (%)
31a·Br	CH_2Cl_2	69	58
34a·Br	CH_2Cl_2	65	–4
34b·Br	CH_2Cl_2	72	9
31a·Br	EtOAc	85	56
31b·Br	EtOAc	99	84
31c·Br	EtOAc	99	93
31d·Br	EtOAc	99	97

Figure 7.14 Application of 1,2,3-triazolium salts **31**·Br for asymmetric alkylation of oxindole.

phase-transfer alkylation of oxindoles, of which the framework is often found in natural alkaloids and pharmaceuticals [33]. The use of 2 mol% of triazolium bromide **31a**·Br for the reaction of *N*-Boc-protected 3-methyloxindole **32** with benzyl bromide in the presence of potassium carbonate resulted in the formation of the corresponding benzylated oxindole **33** with moderate enantioselectivity (Figure 7.14). Noteworthy is that C(5)-methyltriazolium bromide **34a**·Br and *N*-methylbenzamido triazolium bromide **34b**·Br afforded nearly racemic **33**, suggesting that the formation of pseudocyclic ion pair between **31a** and intermediary enolate ion via two hydrogen bonds was essential for asymmetric induction. As the above-described three-dimensional structure of **31a**·Cl revealed that the phenyl groups of **31a** at C-4 of triazolium and at the amide carbonyl were located closely to the paring anion, the modification of these substituents was expected to increase the stereocontrolling ability of **31**. Indeed, the reactions with triazolium bromides **31b**·Br and **31c**·Br having bulky *ortho*-substituted phenyl appendage at triazolium C(4) position proceeded with the significant improvement in enantioselectivity.

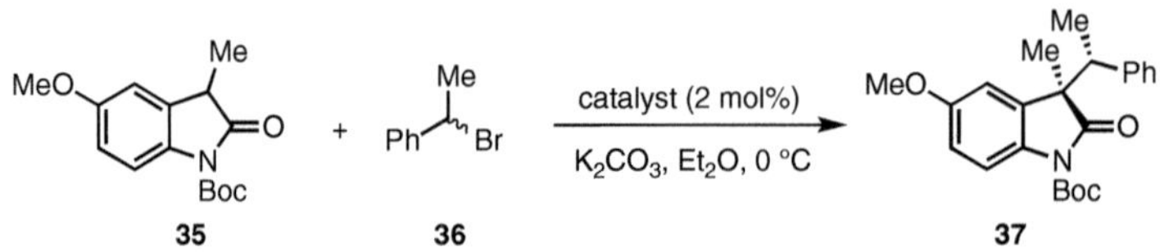

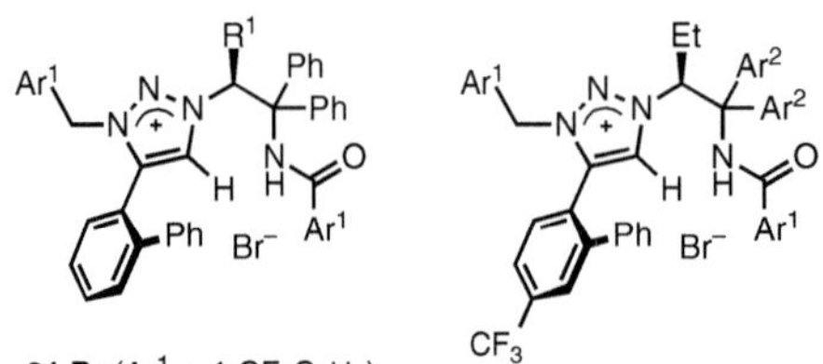

31·Br (Ar^1 = 4-$CF_3C_6H_4$)
31e (R^1 = CH_2Ph)
31f (R^1 = CH_2OH)
31g (R^1 = Et)

31h·Br
(Ar^1 = 4-$CF_3C_6H_4$)
(Ar^2 = 4-ClC_6H_4)

catalyst	Yield (%)	dr	*ee* (%)
31e·Br	67	1.6 : 1	87/94
31f·Br	71	6.3 : 1	97/89
31g·Br	56	9.4 : 1	98/91
31h·Br	99	>20 : 1	99/58

Figure 7.15 Asymmetric alkylation of oxindole with racemic secondary alkyl bromide. Source: Based on Ohmatsu *et al.* [34].

Further tuning of the amide substituent of **31** allowed to identify the optimal catalyst **31d**·Br that exerted excellent activity and stereoselectivity.

The preeminent stereocontrolling ability of chiral 1,2,3-triazolium ion **31** was further demonstrated through the realization of challenging asymmetric alkylation-type reactions. For example, the structurally modified chiral triazolium ion enabled a highly enantio- and diastereoselective alkylation of oxindoles with racemic secondary alkyl bromides to construct vicinal all-carbon quaternary and tertiary stereocenters (Figure 7.15) [34], where the catalyst was required to achieve the enantiofacial discrimination of the prochiral enolates and the kinetic resolution of the chiral alkyl halides. The reaction of *N*-Boc oxindole **35** with 2 equiv of racemic (1-bromoethyl)benzene (**36**) in the presence of potassium carbonate and 2 mol% of phenylalanine-derived triazolium bromide **31e**·Br gave the corresponding alkylated product **37** in moderate yield with low diastereo- yet high enantioselectivity. The alkyl substituent of the amino acid origin (R^1) was found to have a notable impact on the ability of **31** to discriminate the central chirality of **36** for attaining a satisfactory level of diastereocontrol: serine- and ethylglycine-derived catalysts **31f**·Br and **31g**·Br allowed the reaction to proceed with much higher diastereoselectivities. Further structural modification of **31** reached identification of **31h**·Br as an optimal catalyst, which efficiently promoted the alkylation to furnish **37** quantitatively with almost complete diastereo- and enantiocontrol.

The stereoselective nucleophilic substitution at the tetrasubstituted chiral carbon of tertiary (pseudo)halides [35] provides a solid platform for the construction of contiguous all-carbon quaternary stereocenters, while it remains one of the most challenging objectives in the field of asymmetric synthesis [36]. Our group addressed this issue by exploiting the ability of triazolium ion **31** to discriminate the central chirality of non-ionic electrophiles, such as chiral alkyl halides. Specifically, catalytic ring-opening alkylation of racemic 2,2-disubstituted aziridines with 3-substituted oxindoles was developed [37].

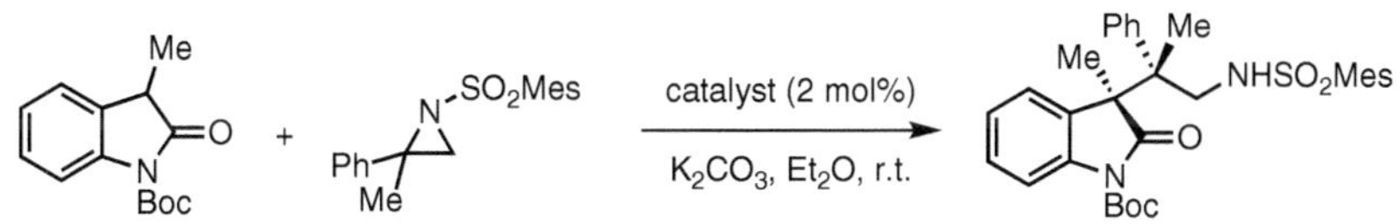

32 **38** (2 equiv) **39**

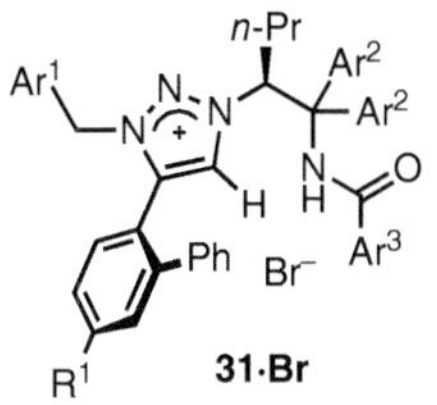

31·Br

catalyst (R^1, Ar^1, Ar^2, Ar^3)	Yield (%)	dr	*ee* (%)
31i·Br (H, Ph, Ph, Ph)	56	4.4 : 1	93/86
31j·Br (CF_3, 4-$CF_3C_6H_4$, Ph, Ph)	95	16 : 1	99/nd
31k·Br (CF_3, 4-$CF_3C_6H_4$, 4-ClC_6H_4, 4-Tol)	99	>20 : 1	99/nd

Figure 7.16 Asymmetric ring-opening substitution of 2,2-disubstituted aziridine.

When oxindole **32** and 2 equiv of racemic *N*-sulfonyl 2-methyl-2-phenylaziridine (**38**) were treated with a catalytic amount of triazolium bromide **31i**·Br and excess potassium carbonate, the ring-opening substitution occurred exclusively at the fully substituted carbon to afford a diastereomeric mixture of enantiomerically enriched aminoalkylated product **39** in high yield (Figure 7.16). It should be noted that the reaction did not take place at all in the absence of **31**·Br. Thus, the triazolium ion would be essential not only for the stereocontrol of bond-forming event but also for the promotion of ring-opening substitution. In addition, similar to the asymmetric alkylation with secondary alkyl halides, the diastereoselectivity heavily depended on the structure of **31**. In particular, the introduction of electron-withdrawing trifluoromethyl groups on the aromatic rings of **31** significantly improved the stereocontrolling ability and also the catalytic activity: the use of **31j**·Br as a catalyst led to the nearly quantitative production of **39** with high diastereo- and complete enantioselectivity, and further selectivity improvement was achieved by using **31k**·Br. In this ring-opening alkylation, the starting aziridines were recovered in an optically active form. The absolute configurations of the recovered aziridine and alkylated product, assigned by single-crystal X-ray diffraction analysis, confirmed that (*S*)-aziridine was preferentially consumed and that the ring-opening substitution of aziridine mainly proceeded in a stereoinvertive manner.

Homo- and heterodialkylation of bisoxindoles with primary alkyl bromides was also accomplished by using a different type of chiral triazolium catalyst. 1,2-Dialkylation would be rather difficult because of the increased steric repulsion between vicinal two carbons in the second alkylation [38]. For executing this mode of carbon–carbon bond formations, Tian and coworkers developed a chiral spirocyclic triazolium ion **40** possessing a lactam moiety as an additional potential hydrogen-bonding functionality (Figure 7.17) [39]. X-ray diffraction analysis of **40a**·Br showed that the triazolium C(5)-proton of **40a** dominantly coordinated with bromide ion and a hydrogen on the lactam nitrogen was disposed to the opposite direction. This structural information suggested that **40** would associate with a nucleophilic anion via electrostatic interaction and a single hydrogen bond

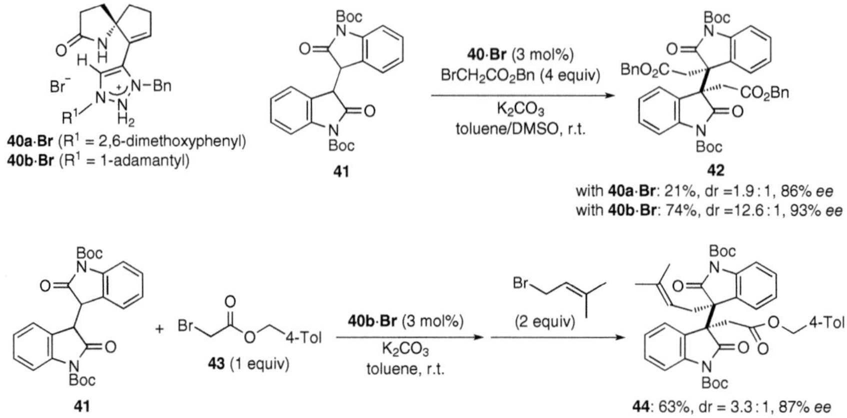

Figure 7.17 Asymmetric dialkylation of bisoxindole. Source: Based on Chen *et al.* [39].

in the transition state. The initial attempt by the use of **40a**·Br as a catalyst for dialkylation of bisoxindole **41** with 4 equivalents of benzyl bromoacetate resulted in low yield of dialkylated product **42** with low diastereoselectivity; however, 1-adamantyl-substituted catalyst **40b**·Br exerted much higher catalytic activity and improved the stereocontrolling ability. This catalytic system also enabled one-pot heterodialkylation with two different alkyl bromides. For instance, the treatment of **41** with 1 equiv of bromoacetate **43** under the catalysis of **40b**·Br for 24 hours and the subsequent addition of 3 equiv of 3-methylcrotyl bromide led to the formation of heterodialkylated oxindole **44** in moderate yield with good enantioselectivity.

Apart from the asymmetric reactions of enolates with (pseudo)alkyl halides, our triazolium ion **31** was proven to be applicable to stereocontrol in the transformations with other electrophiles. For instance, L-leucine-derived catalyst **31l**·Br acted as an efficient catalyst for the asymmetric Mannich-type reaction of α-cyano α-sulfonyl carbanions (Figure 7.18) [40]. Concretely, the reaction of *N*-Boc imine **45** with α-cyanosulfone **46** under the influence of a catalytic amount of cesium carbonate and **31l**·Br proceeded smoothly to give β-amino α-cyanosulfone **47** quantitatively with good diastereo- and high enantioselectivity. A range of substituted *N*-Boc

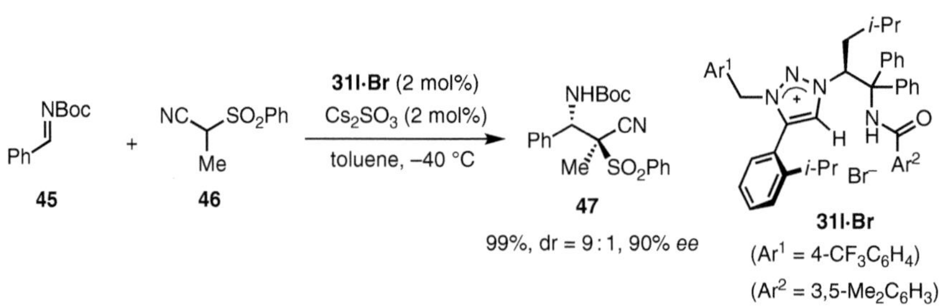

Figure 7.18 Asymmetric Mannich-type reaction of α-cyanosulfone. Source: Ohmatsu *et al.* [40].

imines including aromatic, heteroaromatic, and aliphatic aldehyde-derived imines were well tolerated with high stereoselectivities.

The other example highlighting the unique catalytic function of triazolium ion **31** is the development of asymmetric direct α-amination of carbonyl compounds. Chiral α-aminocarbonyl compounds represent a ubiquitous structural motif in a large family of natural products and pharmaceuticals [41]. Especially in the context of medicinal chemistry, not only the structure of the organic framework around the carbonyl group but also the nitrogen substituents are of critical importance because of the binding properties of the amine components, whose steric and electronic features have a profound effect on the biological activity. Accordingly, the development of general methods for the stereoselective construction of chiral aminocarbonyls with sufficient structural variation of the nitrogen substituents poses an important and challenging task for chemical synthesis. Among the reaction manifolds available for the synthesis of chiral α-aminocarbonyls, the catalytic asymmetric α-amination of carbonyl compounds is one of the most efficient and straightforward methods, and a large number of stereoselective C—N bond-forming reactions have been reported [42]. However, the direct enantioselective installation of a diverse range of non-protected amines onto carbonyl α-carbons had remained unrealized.

Under these circumstances, our group designed a strategy for modular direct α-amination of carbonyl compounds based on the use of hydroxylamines as an amine source. While hydroxylamine is nucleophilic, its electronic nature could be inversed via the addition to trichloroacetonitrile to form *O*-trichloroacetimino hydroxylamine. For the actual promotion of the direct amination of carbonyl compounds with hydroxylamines and trichloroacetonitrile, the use of triazolium ion as a catalyst turned out to be crucial. For instance, representative phase-transfer catalysts, such as tetrabutylammonium and benzyltrimethylammonium bromides, were inactive for the reaction of *N*-Boc 3-phenyloxindole **48** with *N*-cyclohexyl hydroxylamine (**49**) and trichloroacetonitrile, and none of the product formation was detected (Figure 7.19). In contrast, 3-aminooxindole **50** was obtained, albeit in low yield, when employing simple achiral triazolium bromide **51a**·Br as a catalyst. It was also worthy to note that structurally similar yet C(5)-methylated catalyst **51b**·Br did not promote amination at all. Thus, the hydrogen-bonding ability of triazolium ion would be essential for this reaction. Chiral triazolium ion **31** possessing an amide functionality as an additional hydrogen-bonding donor site exhibited high catalytic activity, and amination product **50** was formed in high yield with complete enantioselectivity by the use of adequately modified catalyst **31m**·Br. This amination reaction would proceed through the substitution of intermediary generated *O*-trichloroacetimino hydroxylamine by the oxindole enolate, the former of which was observed by ^{1}H NMR spectroscopic analysis of a solution of **49** and trichloroacetonitrile in toluene-d_4. Using this protocol, a wide range of non-protected aliphatic and aromatic amino groups could be directly introduced at C3 carbon of oxindoles. With respect to carbonyl nucleophiles, β-keto ester **52** and α-cyano ester **54** were smoothly aminated to produce highly enantioenriched **53** and **55** under similar reaction conditions. Furthermore, oxindole **56**

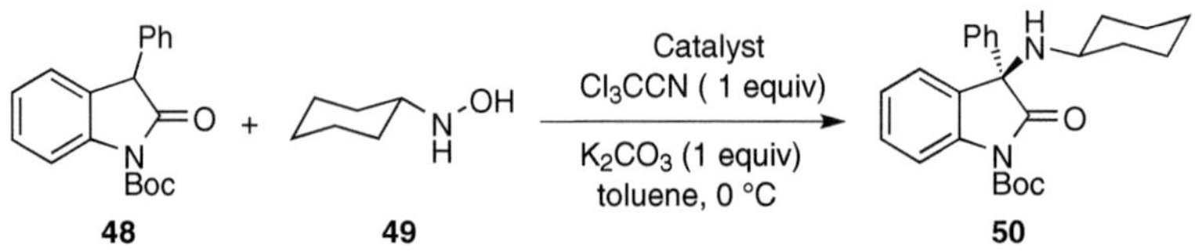

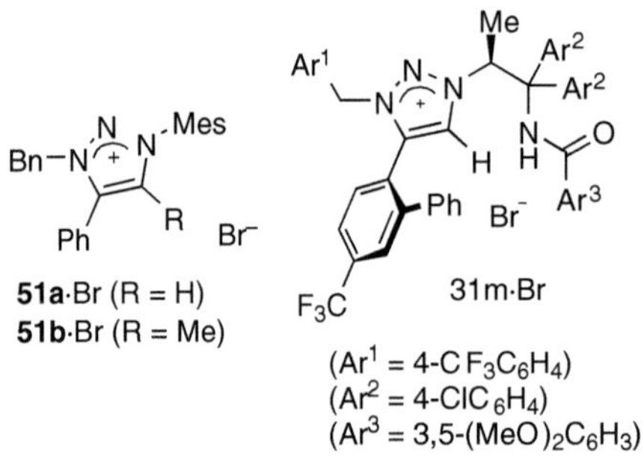

Catalyst	Yield (%)	*ee* (%)
n-Bu_4N·Br (10 mol%)	0	-
$BnNMe_3$·Br (10 mol%)	0	-
51a·Br (10 mol%)	10	-
51b·Br (10 mol%)	0	-
31m·Br (10 mol%)	80	99

Formation of *O*-trichloroacetimino hydroxylamine

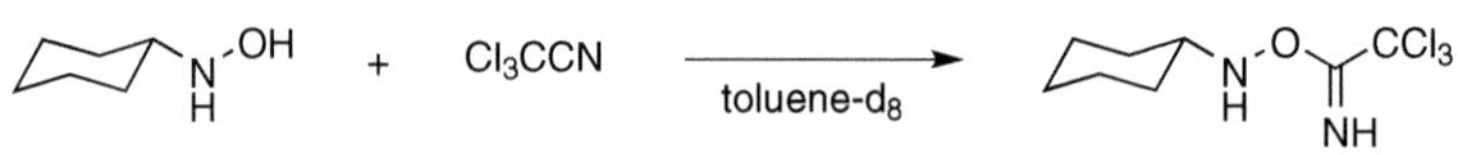

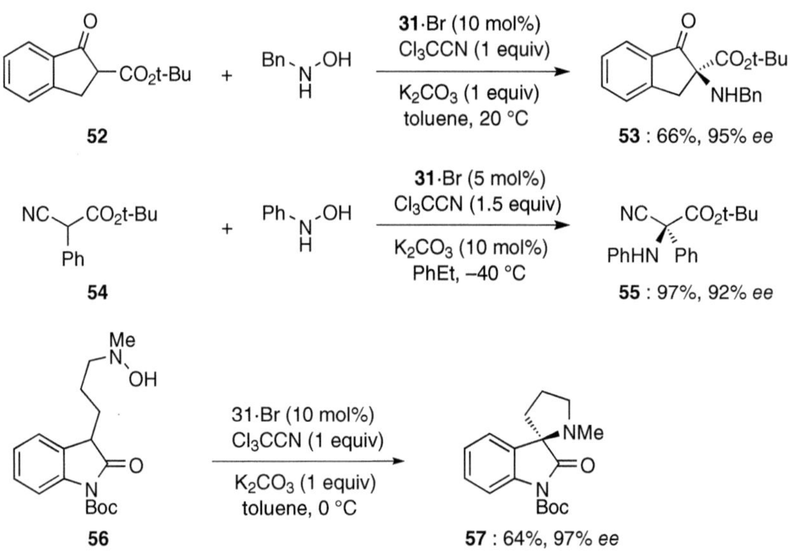

Figure 7.19 Asymmetric α-amination of carbonyl compounds with hydroxylamines.

bearing *N*-hydroxyl-*N*-methylamino group underwent intramolecular ring-closing amination to afford pyrrolidinyl spirooxindole **57**.

The strategy for the electrophilic activation of hydroxylamine by trichloroacetonitrile can also be utilized for the activation of hydrogen peroxide. Trichloroacetonitrile is well known to react with hydrogen peroxide under basic conditions to form

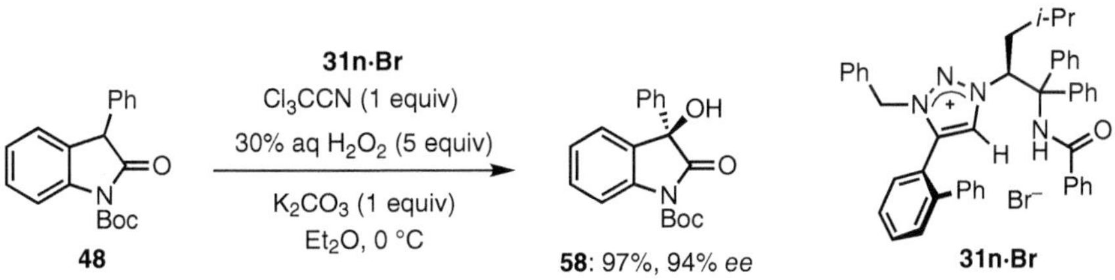

Figure 7.20 Asymmetric α-hydroxylation of oxindole with aqueous hydrogen peroxide. Source: Ohmatsu *et al.* [44].

peroxy trichloroacetimidic acid, which is an active intermediate in Payne oxidations of alkenes and imines to produce epoxides and oxaziridines, respectively [43]. The catalysis of triazolium ion **31** made it feasible to apply this intermediate for the direct α-hydroxylation of carbonyl compounds. The use of **31n**·Br as a phase-transfer catalyst enabled the reaction of *N*-Boc oxindole **48** with aqueous hydrogen peroxide to produce 3-hydroxy oxindole **58** with high efficiency and rigorous absolute stereocontrol (Figure 7.20) [44].

Chloride ion appears frequently in anion-binding catalysis, where the association of chloride ion with chiral catalysts is an important elementary step for the generation of cationic intermediates. Because of its weak basicity and nucleophilicity, however, the use of chloride ion as a nucleophile for the reactions with carbon electrophiles has been scarcely reported [45]. Our group developed a unique asymmetric chloride ring-opening reaction, where the combination of chiral triazolium chloride **31o**·Cl and trimethylsilyl chloride allowed the facile transfer of chloride ion to *N*-sulfonyl meso aziridine **59** to give **60** with an excellent level of enantiocontrol (Figure 7.21) [46]. Because the individual use of a stoichiometric amount of triazolium chloride or trimethylsilyl chloride led to quantitative recovery of **59**,

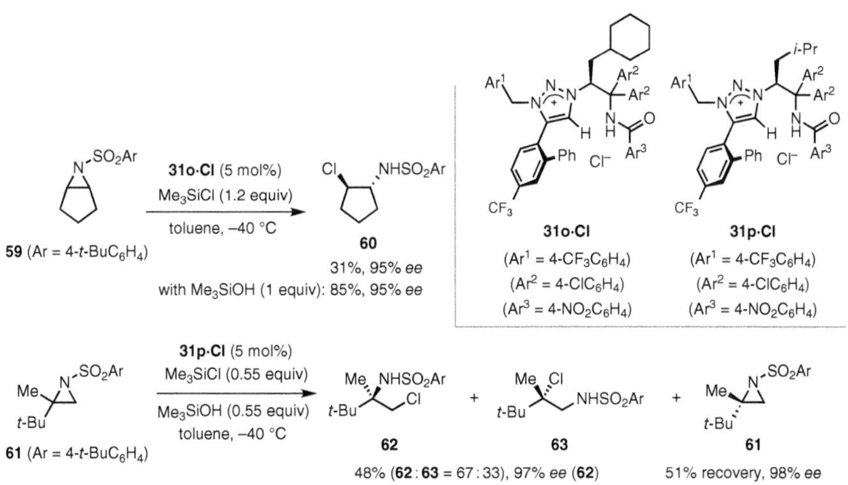

Figure 7.21 Asymmetric chloride ring opening of meso and terminal aziridines. Source: Based on Ohmatsu *et al.* [46].

this ring-opening reaction would proceed via hypervalent chlorosilicate as a reactive intermediate [47]. The intervention of silicate was also suggested by the ^{29}Si NMR spectroscopic analysis of a 1 : 1 mixture of triazolium chloride and trimethylsilyl chloride in tolune-d_8; this showed a single peak at $\delta = 6.6$ ppm, which was considerably shifted upfield from the signal of Me_3SiCl (29.5 ppm). While the yields of **60** in the reactions with a series of catalysts **31**·Cl and trimethylsilyl chloride were uniformly insufficient, the addition of 1 equiv of trimethylsilanol significantly improved the conversion of **59** to **60** without the loss of enantioselectivity. Meanwhile, the combined use of **31p**·Cl, trimethylsilyl chloride, and trimethylsilanol enabled the kinetic resolution of terminal aziridines. For instance, exposure of racemic 2,2-disubtituted aziridine **61** to this condition gave rise to a regioisomeric mixture of the chlorinated products (**62**:**63** = 67 : 33) in 48% yield, and the enantiomeric excess of the major β-chloro-*tert*-amine **63** was 97% *ee*. Importantly, aziridine **61** was recovered in 51% yield in a nearly enantiopure form.

7.4 Triazolium Ions in Dual Functional Catalysts

Design of multifunctional catalysts is a powerful strategy for achieving high levels of stereoselectivities in difficult-to-control asymmetric transformation or the realization of tandem reactions that require multiple activation of several reactants. Peters developed chiral multifunctional Lewis acid catalyst bearing two cobalt(III) centers and two triazolium ion components for the asymmetric conjugate addition of oxindoles to maleimides [48]. The 1,1′-binaphthol-tethered biscobalt/bistriazolium complex **64** promoted the reaction of *N*-Boc oxindole **32** with maleimide **65** to furnish the corresponding adduct **66** in high yield with complete stereoselectivity (Figure 7.22). It was confirmed that the monometallic/monotriazolium complex **67** gave moderate yield and slightly lower diastereo- and enantioselectivity. In sharp contrast, poor reaction efficiency and much lower stereoselectivities were observed when the reaction was performed with monocobalt complex **68** possessing an imidazolium ion. These results clearly indicated that the presence of the triazolium ion in the multifunctional Lewis acid catalysts was crucial for high catalytic activity and stereocontrolling ability. The crucial importance of the triazolium units was further supported by conducting the reaction with biscobalt complex **69** having non-cationic two triazoles, which afforded **66** in low yield with negligible diastereo- and enantioselectivity.

A unique dual functional catalyst enabling the switch of anion-binding catalysis and aminocatalysis was developed by Leigh and coworkers [49]. They designed a rotaxane **70** consisting of amine-tethered bistriazolium ion and dibenzo-24-crown-8 (Figure 7.23). Because macrocyclic ether is preferentially associated with triazolium ion, anion-binding catalysis of **70** is in the "off" state and aminocatalysis is in the "on" state under basic or neutral conditions. On the other hand, when dibenzylamine moiety of **70** is protonated, macrocyclic ether is locked onto the resulting ammonium ion. Thus, aminocatalysis of **70**·H^+ is in the "off" state and anion-binding catalysis is in the "on" state. The switching of these two different

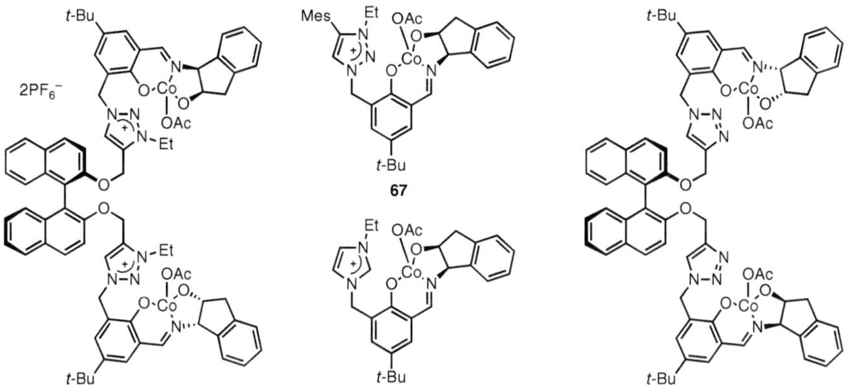

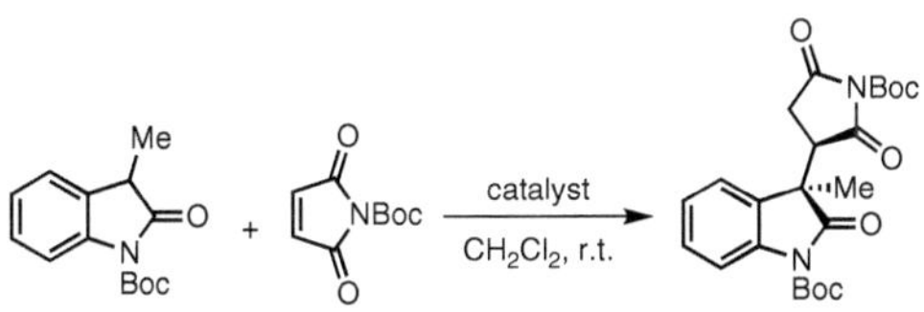

catalyst (mol%)	Yield (%)	d.r.	*ee* (%)
64 (5)	91	99 : 1	99
67 (10)	54	95 : 5	−95
68 (10)	29	33 : 67	−8
69 (5)	32	44 : 56	−15

Figure 7.22 Multifunctional Lewis acid/triazolium complexes for asymmetric conjugate addition.

catalyses was explicitly demonstrated through the application for the tandem reaction of 1-chloroisochroman (**25**) with enol silyl ether **71** and bis-sulfone **72**. The treatment of **25** with **70**·H$^+$ (trifluoroacetic acid salt of **70**, anion-binding catalysis "on") facilitated the generation of the intermediary oxonium ion via chloride binding, thus allowing the bond formation with **71** to afford the corresponding adduct **73** in high yield: this reaction did not proceed at all by using **70** (anion-binding catalysis "off") instead of **70**·H$^+$. Subsequent one-pot treatment with sodium methoxide to convert **70**·H$^+$ to **70** (aminocatalysis "on"), followed by the addition of **72**, led to the formation of **74** through enamine activation of the aldehyde **73** by **70**.

7.5 Conclusion

As illustrated and summarized in this chapter, the development of 1,2,3-triazole- and 1,2,3-triazolium-based organic molecular catalysts has made steady growth in the last decade. The advantages of triazole derivatives are the ease of their synthesis via click chemistry and inherent modularity originating from the facile structural modification by using different combinations of organic azides and terminal alkynes. In addition, their anion-binding ability can be readily tuned through the electronic and/or steric manipulation of the molecular structure, thus allowing for the rapid discovery of effective catalysts for target transformations. Despite these attractive features, the examples of newly devised catalysts and their successful applications

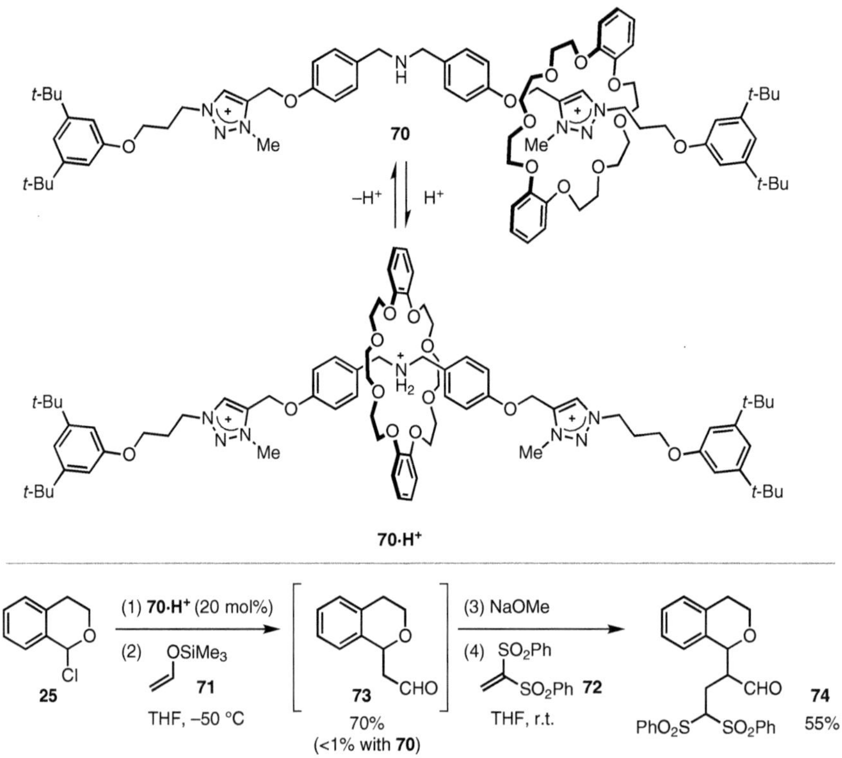

Figure 7.23 Triazolium-functionalized rotaxane catalyst **70** for tandem three-component reaction.

for reaction development are still limited in number. Therefore, more and more different approaches toward the catalyst design are being awaited, and precise and deep understanding of the relationship between the catalyst structures and their reactivity and/or selectivity in the applications to a wide variety of molecular transformations is a key to open a new avenue of anion-binding catalysis.

References

1 Michael, A. (1893). The reaction of diazobenzolimide with acetylenedicarboxylic acid methylester. *J. Prakt. Chem.* 48: 94–95.

2 Kolb, H., Finn, M.G., and Sharpless, K.B. (2001). Click chemistry: diverse chemical function from a few good reactions. *Angew. Chem. Int. Ed.* 40: 2004–2021.

3 (a) Huisgen, R. (1963). 1,3-Dipolar cycloadditions. Past and future. *Angew. Chem. Int. Ed.* 2: 565–598. (b) Huisgen, R. (1963). Kinetics and mechanism of 1,3-dipolar cycloadditions. *Angew. Chem. Int. Ed.* 2: 633–645.

4 (a) Kolb, H.C. and Sharpless, K.B. (2003). The growing impact of click chemistry on drug discovery. *Drug Discovery Today* 8: 1128–1137. (b) Le Droumaguet, C.,

Wang, C., and Wang, Q. (2010). Fluorogenic click reaction. *Chem. Soc. Rev.* 39: 1233–1239.

5 Iha, R.K., Wooley, K.L., Nyström, A.M. *et al.* (2009). Applications of orthogonal "Click" chemistries in the synthesis of functional soft materials. *Chem. Rev.* 109: 5620–5686.

6 (a) Fournier, D., Hoogenboom, R., and Schubert, U.S. (2007). Clicking polymers: a straightforward approach to novel macromolecular architectures. *Chem. Soc. Rev.* 36: 1369–1380. (b) Binder, W.H. and Sachsenhofer, R. (2007). 'Click' chemistry in polymer and materials science. *Macromol. Rapid Commun.* 28: 15–54. (c) Binder, W.H. and Sachsenhofer, R. (2008). 'Click' chemistry in polymer and material science: an update. *Macromol. Rapid Commun.* 29: 952–981. (d) Golas, P.L. and Matyjaszewski, K. (2010). Marrying click chemistry with polymerization: expanding the scope of polymeric materials. *Chem. Soc. Rev.* 39: 1338–1354.

7 Weigert, F.J. and Roberts, J.D. (1968). Nuclear magnetic resonance spectroscopy. Carbon-13 spectra of five-membered aromatic heterocycles. *J. Am. Chem. Soc.* 90: 3543–3549.

8 Bryant, J.J. and Bunz, U.H.F. (2013). Click to bind: metal sensors. *Chem. Asian J.* 8: 1354–1367.

9 Matulis, V.E., Halauko, Y.S., Ivashkevich, O.A., and Gaponik, P.N. (2009). CH acidity of five-membered nitrogen-containing heterocycles: DFT investigation. *THEOCHEM* 909: 19–24.

10 (a) Kumar, A. and Pandey, P.S. (2008). Anion recognition by 1,2,3-triazolium receptors: application of click chemistry in anion recognition. *Org. Lett.* 10: 165–168. (b) Mullen, K.M., Mercurio, J., Serpell, C.J., and Beer, P.D. (2009). Exploiting the 1,2,3-triazolium motif in anion-templated formation of a bromide-selective rotaxane host assembly. *Angew. Chem. Int. Ed.* 48: 4781–4784. (c) Schulze, B., Friebe, C., Hager, M.D. *et al.* (2010). Anion complexation by triazolium "Ligands": mono- and bis-tridentate complexes of sulfate. *Org. Lett.* 12: 2710–2713.

11 (a) Li, Y. and Flood, A.H. (2008). Pure C–H hydrogen bonding to chloride ions: a preorganized and rigid macrocyclic receptor. *Angew. Chem. Int. Ed.* 47: 2649–2652. (b) Li, Y. and Flood, A.H. (2008). Strong, size-selective, and electronically tunable C–H···halide binding with steric control over aggregation from synthetically modular, shape-persistent [3_4]triazolophanes. *J. Am. Chem. Soc.* 130: 12111–12122.

12 Lee, S., Hua, Y., Park, H., and Flood, A.H. (2010). Intramolecular hydrogen bonds preorganize an aryl-triazole receptor into a crescent for chloride binding. *Org. Lett.* 12: 2100–2102.

13 Juwarker, H., Lenhardt, J.M., Pham, D.M., and Craig, S.L. (2008). 1,2,3-Triazole CH···Cl^- contacts guide anion binding and concomitant folding in 1,4-diaryl triazole oligomers. *Angew. Chem. Int. Ed.* 47: 3740–3743.

14 Taylor, M.S. and Jacobsen, E.N. (2004). Highly enantioselective catalytic acyl-Pictet–Spengler reactions. *J. Am. Chem. Soc.* 126: 10558–10559.

15 (a) Brak, K. and Jacobsen, E.N. (2013). Asymmetric ion-pairing catalysis. *Angew. Chem. Int. Ed.* 52: 534–561. (b) Visco, M.D., Attard, J., Guan, Y., and Mattson,

A.E. (2017). Anion-binding catalyst designs for enantioselective synthesis. *Tetrahedron Lett.* 58: 2623–2628.

16 (a) Beckendorf, S., Asmus, S., Mück-Lichtenfeld, C., and García Mancheño, O. (2013). "Click" bis-triazoles as neutral C–H⋯anion-acceptor organocatalysts. *Chem. Eur. J.* 19: 1581–1585. (b) Asmus, S., Beckendorf, S., Zurro, M. *et al.* (2014). Influence of the substitution and conformation of C–H-bond-based bis-triazole acceptors in anion-binding catalysis. *Chem. Asian J.* 9: 2178–2186.

17 Bosmani, A., Pujari, S.A., Besnard, C. *et al.* (2017). Stereoselective and enantiospecific mono- and bis-C–H azidation of tröger bases: insight on bridgehead iminium intermediates and application to anion-binding catalysis. *Chem. Eur. J.* 23: 8678–8684.

18 Zurro, M., Asmus, S., Beckendorf, S. *et al.* (2014). Chiral helical oligotriazoles: new class of anion-binding catalysts for the asymmetric dearomatization of electron-deficient *N*-heteroarenes. *J. Am. Chem. Soc.* 136: 13999–14002.

19 Piekarski, D.G., Steinforth, P., Gómez-Martínez, M. *et al.* (2020). Insight into the folding and cooperative multi-recognition mechanism in supramolecular anion-binding catalysis. *Chem. Eur. J.* 26: 17598–17603.

20 (a) Taylor, M.S., Tokunaga, N., and Jacobsen, E.N. (2005). Enantioselective thiourea-catalyzed acyl-Mannich reactions of isoquinolines. *Angew. Chem. Int. Ed.* 44: 6700–6704. (b) Yamaoka, Y., Miyabe, H., and Takemoto, Y. (2007). Catalytic enantioselective Petasis-type reaction of quinolines catalyzed by a newly designed thiourea catalyst. *J. Am. Chem. Soc.* 129: 6686–6687.

21 Ahamed, M. and Todd, M.H. (2010). Catalytic asymmetric additions of carbon-centered nucleophiles to nitrogen-containing aromatic heterocycles. *Eur. J. Org. Chem.*: 5935–5942.

22 (a) Zurro, M., Asmus, S., Bamberger, J. *et al.* (2016). Chiral triazoles in anion-binding catalysis: new entry to enantioselective Reissert-type reactions. *Chem. Eur. J.* 22: 3785–3793. (b) Duong, Q.-N., Schifferer, L., and García Mancheño, O. (2019). Nucleophile screening in anion-binding Reissert-type reactions of quinolines with chiral tetrakis(triazole) catalysts. *Eur. J. Org. Chem.*: 5452–5461.

23 Fischer, T., Duong, Q.-N., and García Mancheño, O. (2017). Triazole-based anion-binding catalysis for the enantioselective dearomatization of *N*-heteroarenes with P-nucleophiles. *Chem. Eur. J.* 23: 5983–5987.

24 García Mancheño, O., Asmus, S., Zurro, M., and Fischer, T. (2015). Highly enantioselective nucleophilic dearomatization of pyridines by anion-binding catalysis. *Angew. Chem. Int. Ed.* 54: 8823–8827.

25 Fischer, T., Bamberger, J., and García Mancheño, O. (2016). Asymmetric nucleophilic dearomatization of diazarenes by anion-binding catalysis. *Org. Biomol. Chem.* 14: 5794–5802.

26 Fischer, T., Bamberger, J., Gómez-Martínez, M. *et al.* (2019). Helical multi-coordination anion-binding catalysts enable highly enantioselective dearomatization of pyrylium derivatives. *Angew. Chem. Int. Ed.* 58: 3217–3221.

27 Beletskaya, I.P., Najera, C., and Yus, M. (2018). Stereodivergent catalysis. *Chem. Rev.* 118: 5080–5200.

28 Dorel, R. and Feringa, B.L. (2020). Stereodivergent anion binding catalysis with molecular motors. *Angew. Chem. Int. Ed.* 59: 785–789.
29 Koumura, N., Zijlstra, R.W.J., van Delden, R.A. *et al.* (1999). Light-driven monodirectional molecular rotor. *Nature* 401: 152–155.
30 Furukawa, Y., Suzuki, R., Nakashima, T. *et al.* (2018). Protonated bis-1,2,3-triazole as an anion-binding chiral Brønsted acid for catalytic asymmetric Friedel–Crafts reaction of indoles with imines. *Bull. Chem. Soc. Jpn.* 91: 1252–1257.
31 Ohmatsu, K., Kiyokawa, M., and Ooi, T. (2011). Chiral 1,2,3-triazoliums as new cationic organic catalysts with anion-recognition ability: application to asymmetric alkylation of oxindoles. *J. Am. Chem. Soc.* 133: 1307–1309.
32 (a) Ooi, T. and Maruoka, K. (2007). Recent advances in asymmetric phase-transfer catalysis. *Angew. Chem. Int. Ed.* 46: 4222–4266. (b) Shirakawa, S. and Maruoka, K. (2013). Recent developments in asymmetric phase-transfer reactions. *Angew. Chem. Int. Ed.* 52: 4312–4348. (c) Liu, S., Kumatabara, Y., and Shirakawa, S. (2016). Chiral quaternary phosphonium salts as phase-transfer catalysts for environmentally benign asymmetric transformations. *Green Chem.* 18: 331–341.
33 (a) Trost, B.M. and Brennan, M.K. (2009). Asymmetric syntheses of oxindole and indole spirocyclic alkaloid natural products. *Synthesis*: 3003–3025. (b) Zhou, F., Liu, Y.-L., and Zhou, J. (2010). Catalytic asymmetric synthesis of oxindoles bearing a tetrasubstituted stereocenter at the C-3 position. *Adv. Synth. Catal.* 352: 1381–1407.
34 Ohmatsu, K., Furukawa, Y., Kiyokawa, M., and Ooi, T. (2017). Diastereo- and enantioselective phase-transfer alkylation of 3-substituted oxindoles with racemic secondary alkyl halides. *Chem. Commun.* 53: 13113–13116.
35 (a) Trost, B.M. and Jiang, C. (2001). atom economic asymmetric creation of quaternary carbon: regio- and enantioselective reactions of a vinylepoxide with a carbon nucleophile. *J. Am. Chem. Soc.* 123: 12907–12908. (b) Trost, B.M. and Jiang, C. (2003). Pd-catalyzed asymmetric allylic alkylation. A short route to the cyclopentyl core of viridenomycin. *Org. Lett.* 5: 1563–1565. (c) Du, C., Li, L., Li, Y., and Xie, Z. (2009). Construction of two vicinal quaternary carbons by asymmetric allylic alkylation: total synthesis of hyperolactone C and (−)-Biyouyanagin A. *Angew. Chem. Int. Ed.* 48: 7853–7856. (d) Wu, W.-Q., Ding, C.-H., and Hou, X.-L. (2012). Pd-catalyzed diastereo- and enantioselective [3+2]-cycloaddition reaction of vinyl epoxide with nitroalkenes. *Synlett* 23: 1035–1038. (e) Pineschi, M., Del Moro, F., Crotti, P. *et al.* (2004). Catalytic regiodivergent kinetic resolution of allylic epoxides: a new entry to allylic and homoallylic alcohols with high optical purity. *J. Organomet. Chem.* 69: 2099–2105. (f) Liu, X., Wang, P., Bai, L. *et al.* (2018). Construction of vicinal all-carbon quaternary stereocenters enabled by a catalytic asymmetric dearomatization reaction of β-naphthols with 3-bromooxindoles. *ACS Catal.* 8: 10888–10894.
36 (a) Peterson, E.A. and Overman, L.E. (2004). Contiguous stereogenic quaternary carbons: a daunting challenge in natural products synthesis. *Proc. Natl. Acad. Sci. U.S.A.* 101: 11943–11948. (b) Trost, B.M. and Jiang, C. (2006). Catalytic

enantioselective construction of all-carbon quaternary stereocenters. *Synthesis*: 369–396. (c) Hawner, C. and Alexakis, A. (2010). Metal-catalyzed asymmetric conjugate addition reaction: formation of quaternary stereocenters. *Chem. Commun.* 46: 7295–7306. (d) Das, J.P. and Marek, I. (2011). Enantioselective synthesis of all-carbon quaternary stereogenic centers in acyclic systems. *Chem. Commun.* 47: 4593–4623. (e) Zhou, F., Zhu, L., Pan, B.-W. *et al.* (2020). Catalytic enantioselective construction of vicinal quaternary carbon stereocenters. *Chem. Sci.* 11: 9341–9365.

37 Ohmatsu, K., Ando, Y., and Ooi, T. (2013). Asymmetric substitution at the tetrasubstituted chiral carbon: catalytic ring-opening alkylation of racemic 2,2-disubstituted aziridines with 3-substituted oxindoles. *J. Am. Chem. Soc.* 135: 18706–18709.

38 (a) Mitsunuma, H., Shibasaki, M., Kanai, M., and Matsunaga, S. (2012). Catalytic asymmetric total synthesis of chimonanthine, folicanthine, and calycanthine through double Michael reaction of bisoxindole. *Angew. Chem. Int. Ed.* 51: 5217–5221. (b) Trost, B.M. and Osipov, M. (2013). Palladium-catalyzed asymmetric construction of vicinal all-carbon quaternary stereocenters and its application to the synthesis of cyclotryptamine alkaloids. *Angew. Chem. Int. Ed.* 52: 9176–9181. (c) Ghosh, S., Bhunia, S., Kakde, B.N. *et al.* (2014). Enantioselective construction of vicinal all-carbon quaternary centers via catalytic double asymmetric decarboxylative allylation. *Chem. Commun.* 50: 2434–2437.

39 Chen, S.-K., Ma, W.-Q., Yan, Z.-B. *et al.* (2018). Organo-cation catalyzed asymmetric homo/heterodialkylation of bisoxindoles: construction of vicinal all-carbon quaternary stereocenters and total synthesis of (−)-chimonanthidine. *J. Am. Chem. Soc.* 140: 10099–10103.

40 Ohmatsu, K., Goto, A., and Ooi, T. (2012). Catalytic asymmetric Mannich-type reactions of α-cyano α-sulfonyl carbanions. *Chem. Commun.* 48: 7913–7915.

41 Hili, R. and Yudin, A.K. (2006). Making carbon–nitrogen bonds in biological and chemical synthesis. *Nat. Chem. Biol.* 2: 284–287.

42 (a) Janey, J.M. (2005). Recent advances in catalytic, enantioselective α aminations and α oxygenations of carbonyl compounds. *Angew. Chem. Int. Ed.* 44: 4292–4300. (b) Smith, A.M.R. and Hii, K.K. (2011). Transition metal catalyzed enantioselective α-heterofunctionalization of carbonyl compounds. *Chem. Rev.* 111: 1637–1656. (c) Zhou, F., Liao, F.-M., Yu, J.-S., and Zhou, J. (2014). Catalytic asymmetric electrophilic amination reactions to form nitrogen-bearing tetrasubstituted carbon stereocenters. *Synthesis* 46: 2983–3003. (d) Maji, B. and Yamamoto, H. (2015). Use of in situ generated nitrosocarbonyl compounds in catalytic asymmetric α-hydroxylation and α-amination reactions. *Bull. Chem. Soc. Jpn.* 88: 753–762. (e) Tian, J.-S., Ng, K.W.J., Wong, J.-R., and Loh, T.-P. (2012). α-Amination of aldehydes catalyzed by in situ generated hypoiodite. *Angew. Chem. Int. Ed.* 51: 9105–9109.

43 (a) Payne, G.B., Deming, P.H., and Williams, P.H. (1961). Reactions of hydrogen peroxide. VII. Alkali-catalyzed epoxidation and oxidation using a nitrile as co-reactant. *J. Organomet. Chem.* 26: 659–663. (b) Bach, R.D. and Knight, J.W. (1981). Epoxidation of olefins by hydrogen peroxide–acetonitrile: cis-cyclooctene oxide. *Org. Synth.* 60: 63. (c) Arias, L.A., Adkins, S., Nagel, C.J., and Bach, R.D. (1983). Epoxidation of alkenes with trichloroacetonitrile/hydrogen peroxide in a neutral biphasic solvent system. *J. Organomet. Chem.* 48: 888–890. (d) Kraïem, J., Kacem, Y., Khiari, J., and Hassine, B.B. (2001). stereospecific synthesis of 2-alkyl-3-aryloxaziridines from prochiral and chiral *N*-arylidenealkylamines by use of the benzonitrile–hydrogen peroxide system. *Synth. Commun.* 31: 263–271. (e) Kraïem, J., Othman, R.B., and Hassine, B.B. (2004). Synthesis of oxaziridines by oxidation of imines with the trichloroacetonitrile–hydrogen peroxide system. *C.R. Chim.* 7: 1119–1126. (f) Tka, N., Kraïem, J., and Hassine, B.B. (2012). Enantio- and diastereoselective oxidation of *N*-alkylimines using chiral α-bromonitriles and hydrogen peroxide system. *Synth. Commun.* 42: 2994–3003.

44 Ohmatsu, K., Ando, Y., and Ooi, T. (2017). In situ electrophilic activation of hydrogen peroxide for catalytic asymmetric α-hydroxylation of 3-substituted oxindoles. *Synlett* 28: 1291–1294.

45 Mita, T. and Jacobsen, E.N. (2009). Bifunctional asymmetric catalysis with hydrogen chloride: enantioselective ring opening of aziridines catalyzed by a phosphinothiourea. *Synlett*: 1680–1684.

46 Ohmatsu, K., Hamajima, Y., and Ooi, T. (2012). Catalytic asymmetric ring openings of meso and terminal aziridines with halides mediated by chiral 1,2,3-triazolium silicates. *J. Am. Chem. Soc.* 134: 8794–8797.

47 (a) Chuit, C., Corriu, R.J.P., Reye, C., and Young, J.C. (1993). Reactivity of penta- and hexacoordinate silicon compounds and their role as reaction intermediates. *Chem. Rev.* 93: 1371–1448. (b) Rendler, S. and Oestreich, M. (2005). Hypervalent silicon as a reactive site in selective bond-forming processes. *Synthesis*: 1727–1747. (c) Orito, Y. and Nakajima, M. (2006). Lewis base catalyzed asymmetric reactions involving hypervalent silicate intermediates. *Synthesis*: 1391–1401. (d) Benaglia, M., Guizzetti, S., and Pignataro, L. (2008). Stereoselective reactions involving hypervalent silicate complexes. *Coord. Chem. Rev.* 252: 492–512.

48 Schmid, J., Junge, T., Lang, J. *et al.* (2019). Polyfunctional bis-Lewis-acid-/bis-triazolium catalysts for stereoselective 1,4-additions of 2-oxindoles to maleimides. *Angew. Chem. Int. Ed.* 58: 5447–5451.

49 Eichstaedt, K., Jaramillo-García, J., Leigh, D.A. *et al.* (2017). Switching between anion-binding catalysis and aminocatalysis with a rotaxane dual-function catalyst. *J. Am. Chem. Soc.* 139: 9376–9381.

8

Quaternary Ammonium, Phosphonium, and Tertiary Sulfonium Salts as Hydrogen-Bonding Catalysts

Seiji Shirakawa

Nagasaki University, Graduate School of Fisheries and Environmental Sciences, Department of Environmental Science, 1-14 Bunkyo-machi, Nagasaki 852-8521, Japan

8.1 Introduction

Onium salts are recognized as an important class of compounds not only in the field of chemical science but also in the fields of material and biological sciences [1]. The powerful synthetic utility of onium salts such as quaternary ammonium, phosphonium, and tertiary sulfonium salts is well established in the field of organic chemistry. These onium salt compounds are frequently utilized as useful organic reagents in the construction of important building blocks. The reactions using these reagents are listed in the textbooks of organic chemistry as very important, named reactions [2]. These named reactions mainly rely on the acidity of α-hydrogen atoms in the onium salts to generate reactive onium ylide intermediates (Figure 8.1).

Furthermore, quaternary ammonium and phosphonium salts are known as reliable organocatalysts in ion-pairing catalysis via the activation of an anionic nucleophile [3]. Chiral versions of quaternary ammonium and phosphonium salts have also been developed for asymmetric transformations [4]. A valuable example of chiral tertiary sulfonium salts as effective asymmetric organocatalysts has also been reported [5]. Although ion-pairing catalysis is a major direction for onium salt catalysis, hydrogen-bonding catalysis of onium salts has received attention in recent years as a new possibility for onium salt chemistry [6]. In the catalytic mode, the onium salt catalysts activate electrophiles via hydrogen-bonding interactions with the α-hydrogen atoms of the onium salts to promote organic reactions [7]. This means that the mechanism for onium salt catalysis is the reverse of ion-pairing catalysis, which activates nucleophiles.

8.2 Hydrogen-Bonding Ability of Quaternary Ammonium Salts

Quaternary ammonium salts such as tetrabutylammonium halides are recognized as representative organocatalysts and are mainly employed in phase transfer

Anion-Binding Catalysis, First Edition. Edited by Olga García Mancheño.

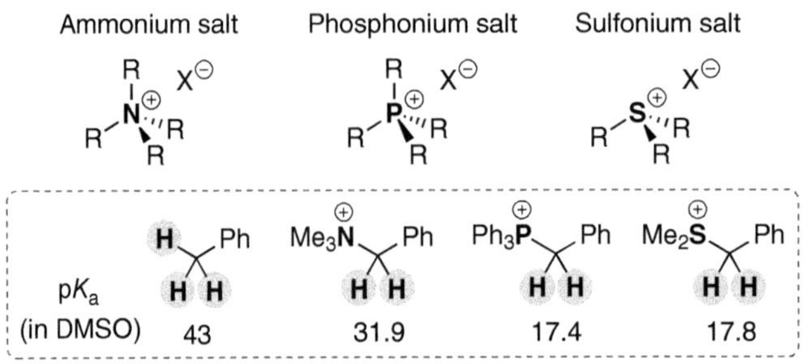

Figure 8.1 Acidity of α-hydrogen atoms in onium salts.

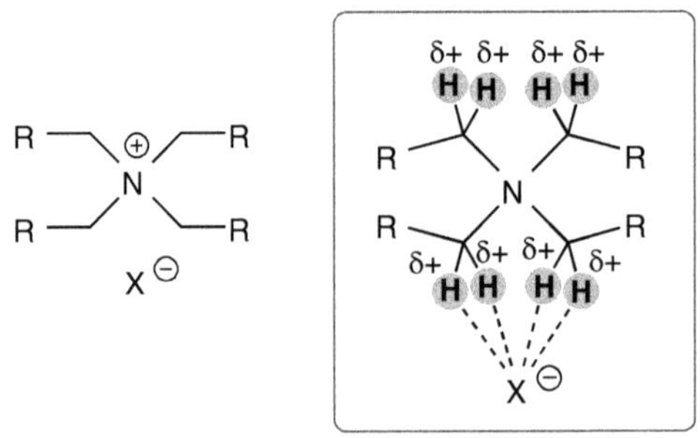

(a) (b)

Figure 8.2 Actual ionic structure of tetraalkylammonium salts. (a) Commonly expressed structure. (b) Actual structure.

catalysis and base catalysis to activate anionic nucleophiles through the formation of an ion pair with a quaternary ammonium cation [3, 4]. Although the structures of tetraalkylammonium salts are commonly expressed as example (a) in Figure 8.2, the actual ionic structures are discussed differently. The positive charges of ammonium salts are delocalized on the α-hydrogen atoms, which are known to interact with an anionic counterion through hydrogen-bonding interactions, as shown in example (b) (Figure 8.2).

In a pioneering report, Reetz *et al.* used X-ray crystal analysis of tetrabutylammonium salts to discuss delocalization of the positive charge in quaternary ammonium salts [8]. Clear hydrogen-bonding interactions between the α-hydrogen atoms of tetrabutylammonium cation and anionic enolate oxygens were observed in the crystal structure of tetrabutylammonium enolate (Figure 8.3).

Density functional theory (DFT) calculations also supported the delocalized structures of tetraalkylammonium salts [9], which included chiral quaternary ammonium salts for asymmetric phase transfer reactions [10]. For example, Shibasaki and coworkers discussed the importance of the interactions between α-hydrogen atoms on the chiral tetraalkylammonium salt catalyst and the enolate oxygen in the transition-state models of highly enantioselective phase transfer alkylations (Scheme 8.1) [10a]. Other research groups have focused on the similar types of hydrogen-bonding interactions in the asymmetric phase transfer reactions via the use of representative chiral quaternary ammonium salts such as cinchona

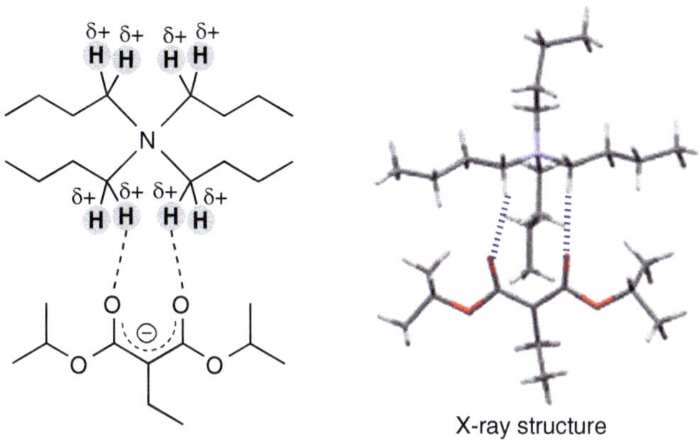

Figure 8.3 Evidence for the delocalized structure of tetraalkylammonium salts via X-ray crystal analysis.

alkaloid-derived catalysts, Maruoka-type 1,1′-bi-2-naphthol (BINOL)-derived catalysts, and diaminocyclohexane-derived bifunctional catalysts.

The hydrogen-bonding interactions of quaternary ammonium salts are also important in the field of biological chemistry. The importance of the hydrogen-bonding donor ability of biologically active ammonium salts has also been noted in the field of medicinal chemistry [11].

8.3 Hydrogen-Bonding Catalysis of Quaternary Ammonium Salts

Although the interesting hydrogen-bonding ability of the α-hydrogen atoms on quaternary ammonium salts is often discussed in the literature, hydrogen-bonding catalysis, which more directly utilizes the ability, had remained out of the spotlight until recently. Bibal and coworkers reported the potential ability of quaternary ammonium salts as hydrogen-bonding catalysts to activate epoxides in 2014 [12]. Alkyl-ammonium and pyridinium salt catalysts possessing a tetrakis[3,5-bis(trifluoromethyl)phenyl]borate ($BArF^-$; barfate) as the non-coordinating counter-anion promote the ring-opening reactions of epoxides with amines (Scheme 8.2).

We also became interested in the hydrogen-bonding catalysis of quaternary ammonium salts. To realize the efficient hydrogen-bonding catalysis of a quaternary ammonium salt, we designed a new ammonium salt catalyst **1** (Figure 8.4) [13]. The quaternary ammonium salt **1** possesses carboxylate groups at the α-carbon atoms, which enhance the acidity of the α-hydrogen atoms. Furthermore, the six-membered structure of the piperidine backbone fixes the acidic α-hydrogen atoms to a position that is appropriate for bidentate binding to an anionic group.

We performed X-ray diffraction analysis of our quaternary ammonium iodide **1a** to obtain the structural information of ammonium salts **1** (Figure 8.5). The crystal structure of ammonium iodide **1a** provided important structural information. We observed hydrogen-bonding interactions between the α-hydrogen atoms and the

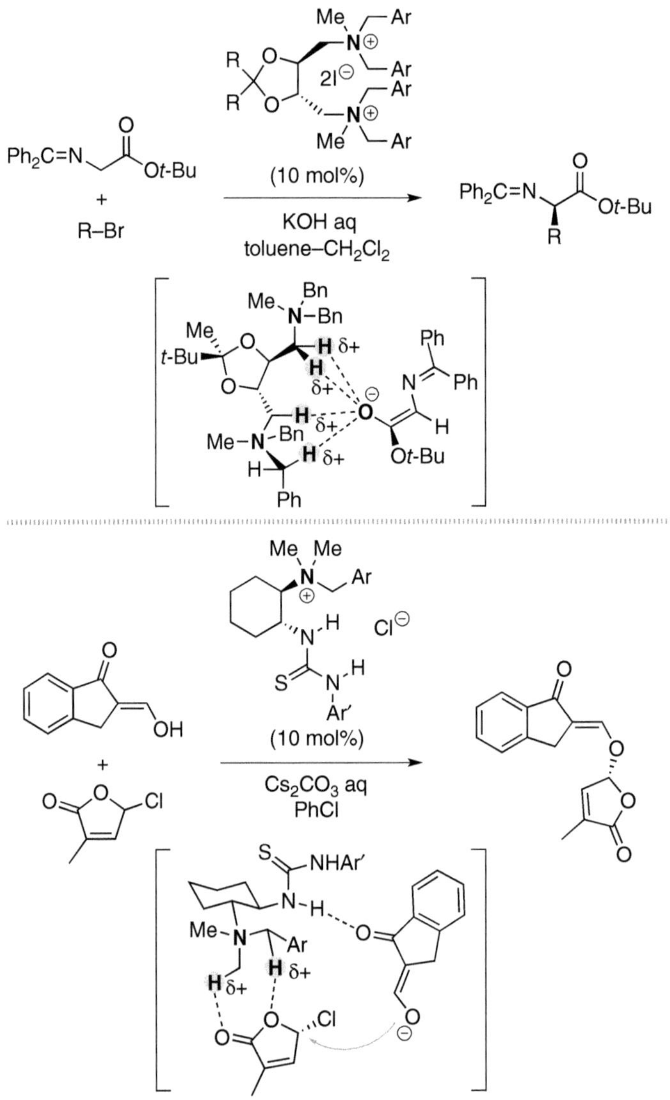

Scheme 8.1 Importance of hydrogen-bonding interactions in asymmetric phase transfer reactions. Source: Based on Ohshima *et al.* [10a].

iodide counterion and noted three α-hydrogen atoms including one hydrogen of the *N*-methyl group bound to the iodide anion. A similar binding mode was observed in the crystal structure of piperidine-derived ammonium iodide **2a**, but the distances between the α-hydrogen atoms and the iodide counterion were shorter in the newly designed ammonium iodide **1a** bearing carboxylate groups. These results indicate that the carboxylate moieties of quaternary ammonium salts **1** enhance the hydrogen-bonding ability of the α-hydrogen atoms.

With important structural information of the newly designed quaternary ammonium salts **1** in hand, the catalytic ability of **1** as a hydrogen-bonding

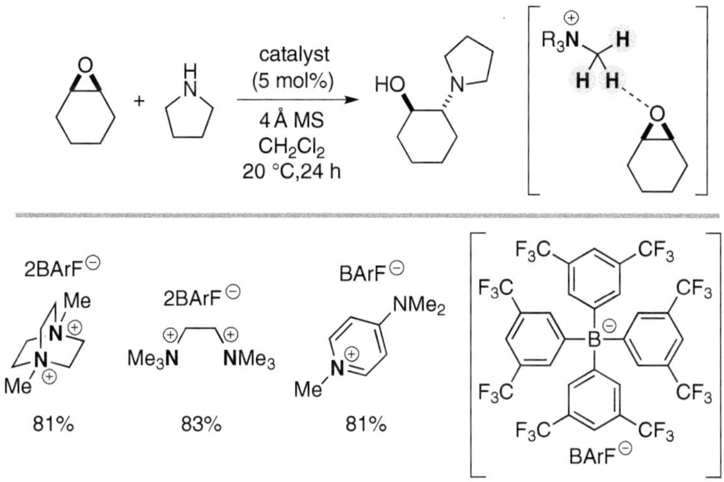

Scheme 8.2 *N*-methylammonium salts as hydrogen-bonding catalysts in ring-opening reactions with epoxide.

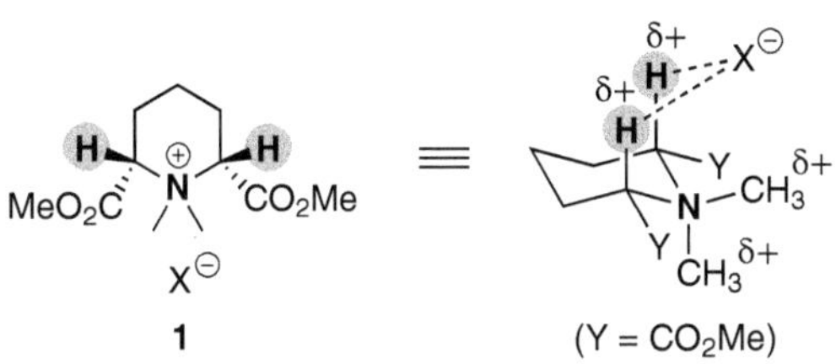

Figure 8.4 Newly designed quaternary ammonium salt catalyst **1**. Source: Shirakawa *et al.* [13].

catalyst was investigated in a Mannich-type reaction involving a *N*-acylisoquinoline derivative, which was generated *in situ* from 2,2,2-trichloroethyl chloroformate (TrocCl) and isoquinoline, as a benchmark reaction (Scheme 8.3). In the absence of a catalyst, the reaction with ketene silyl acetal proceeded slowly to give the corresponding Mannich product in a 7% yield at −78 °C for 3 hours. When our quaternary ammonium iodide **1a** was employed as a hydrogen-bonding catalyst, the reaction was moderately promoted to produce the Mannich product in a 38% yield. On the other hand, very low catalytic activity was observed when the reaction was performed with piperidine-derived simple quaternary ammonium iodide **2a**. Also, the corresponding tertiary amine **3** showed almost no catalytic activity. These results clearly suggest that both the quaternary ammonium moiety and the carboxylate groups in catalyst **1a** are important factors in promoting the reaction as a hydrogen-bonding catalyst. Exchanging the iodide counter-anion in ammonium iodide **1a** for $BArF^-$ as the non-coordinating counter-anion (**1b**) further improved the catalytic activity, and the reaction that used quaternary ammonium barfate catalyst **1b** gave the product in a 61% yield. The Mannich product could be obtained in a high yield with a prolonged reaction time. The hydrogen-bonding activation of the chloroisoquinoline-type substrate by ammonium salts **1** was demonstrated

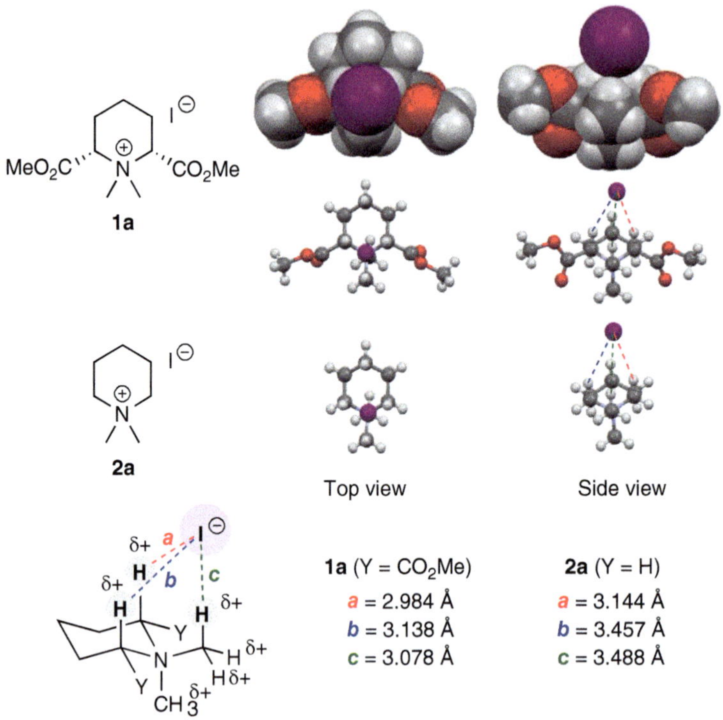

Figure 8.5 X-ray crystal structures of quaternary ammonium iodides.

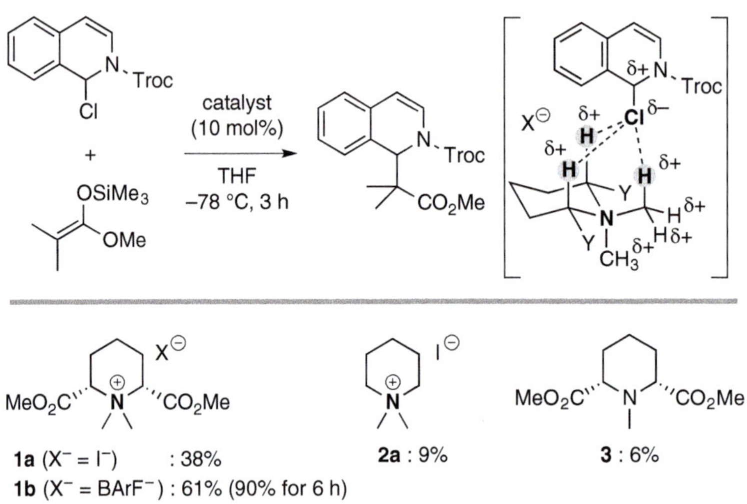

Scheme 8.3 Hydrogen-bonding catalysis of newly designed quaternary ammonium salts **1** in a Mannich-type reaction.

in several control experiments and nuclear magnetic resonance (NMR) titration studies.

Newly designed cyclic quaternary ammonium salts **1** worked as hydrogen-bonding catalysts in the Mannich-type reaction of *N*-acylisoquinoline derivatives. To expand the utility of quaternary ammonium salts as hydrogen-bonding catalysts, we were next interested in the activation of imines as electrophiles. In 2011, Park *et al.* reported that the tetraalkylammonium bromide and iodide catalysts could promote aza-Diels–Alder reactions of aldimines with a Danishefsky's diene [14]. In their report on the aza-Diels–Alder reactions of imines, several interesting observations were reported, but the activation mode by the quaternary ammonium salt catalysts has remained unclear. The report by Park combined with our findings led us to hypothesize that tetraalkylammonium salts could activate aldimines via hydrogen-bonding interactions with α-hydrogen atoms to promote the aza-Diels–Alder reaction with a Danishefsky's diene.

To prove our hypothesis, we investigated the structure/catalytic activity relationships of quaternary ammonium salts in an aza-Diels–Alder reaction of *N*-phenylbenzaldimine and a Danishefsky's diene (Scheme 8.4) [15]. In the absence of a catalyst, the reaction in dichloromethane proceeded very slowly at 0 °C to give an aza-Diels–Alder product in a 4% yield after 3 hours. When acyclic tetrabutylammonium iodide was used as a catalyst under these conditions, the reaction was only

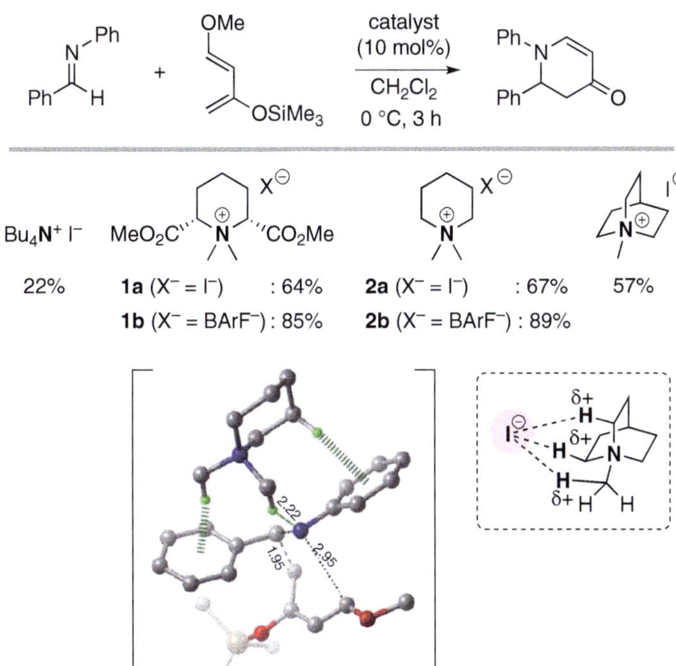

Scheme 8.4 Quaternary ammonium salt-catalyzed aza-Diels–Alder reactions and the transition-state structure. Source: Based on Kumatabara *et al.* [15].

promoted to a moderate extent (22% yield). Next, our originally designed quaternary ammonium iodide **1a** was employed as a hydrogen-bonding catalyst. As expected, the reaction was efficiently accelerated by ammonium iodide **1a** bearing carboxylate groups, which was an effective hydrogen-bonding catalyst for the Mannich-type reaction in Scheme 8.3 and produced an aza-Diels–Alder product in a 64% yield.

The simple piperidine-derived catalyst **2a** was also an effective catalyst for the aza-Diels–Alder reaction, and the product was obtained in a 67% yield. Although catalyst **1a** possessed more of the acidic α-hydrogen atoms that show superior activity to **2a** in our previous work on Mannich-type reactions, similar reactivities between **1a** and **2a** were observed in the present aza-Diels–Alder reaction. These different trends could be explained by the differences in the structures of the electrophiles because steric repulsion was present between the phenyl group of the imines and the carboxylate groups of catalyst **1a**. As a result, catalysts **1a** and **2a** exhibited similar reactivity for the aza-Diels–Alder reaction with imines. Furthermore, quinuclidine-derived ammonium iodide showed higher reactivity than tetrabutylammonium iodide (57% vs. 22% yields). These results clearly suggested that the arrangement of α-hydrogen atoms to appropriate positions on ammonium salt compounds, which were induced by the cyclic structure, is essential in the design of effective hydrogen-bonding catalysts based on tetraalkylammonium salts. In order to further improve the catalytic activity of quaternary ammonium salt catalysts for use in aza-Diels–Alder reactions, $BArF^-$ was employed as a non-coordinating counter-anion. Finally, we obtained the aza-Diels–Alder product in high yields via the use of ammonium barfate catalysts **1b** and **2b**. In independent investigations, both our group [16] and Houk group [17] used DFT calculations to examine the activation modes of piperidine-derived quaternary ammonium salts **2**.

Electron-deficient alkyl-pyridinium salt-catalyzed Mukaiyama aldol-type reactions using 1-chloroisochroman were reported by Berkessel *et al.* in 2014 (Scheme 8.5) [18]. Highly electron-deficient pyridinium tetraphenylborate **4a** worked as an efficient catalyst for the Mukaiyama aldol-type reaction with ketene silyl acetal to provide the target product in a high yield. On the other hand, neither the corresponding electron-deficient pyridinium bromide **4b** nor the pyridinium tetraphenylborates **4c** and **4d** promoted the Mukaiyama aldol-type reaction under the same reaction conditions.

Based on the X-ray crystal structures of **4a** and **4b**, Berkessel proposed that anion–π interactions are the key to promoting the initial reaction process, which is a chloride abstraction process involving a 1-chloroisochroman substrate (Figure 8.6). Another possibility for the activation mode of pyridinium salt catalysts was suggested by other researchers in a review article. Based on the X-ray crystal analysis of **4b**, Nagorny and Sun suggested that the pyridinium salt catalyst activates the 1-chloroisochroman substrate through hydrogen-bonding interactions with acidic hydrogen atoms on the pyridinium salt [19]. An X-ray crystal analysis of **4b** supported their suggestion.

Tse, Yeung, and coworkers prepared zwitterionic-type iminium and ammonium salt catalysts **5** for transesterification [20]. The catalytic activities of zwitterionic catalysts **5** have been discussed in the transesterification of triglyceride to fatty acid methyl ester (Scheme 8.6).

The transesterification was efficiently promoted by zwitterionic catalysts **5a** and **5b** based on an amide anion/iminium cation pair. Catalytic activities of

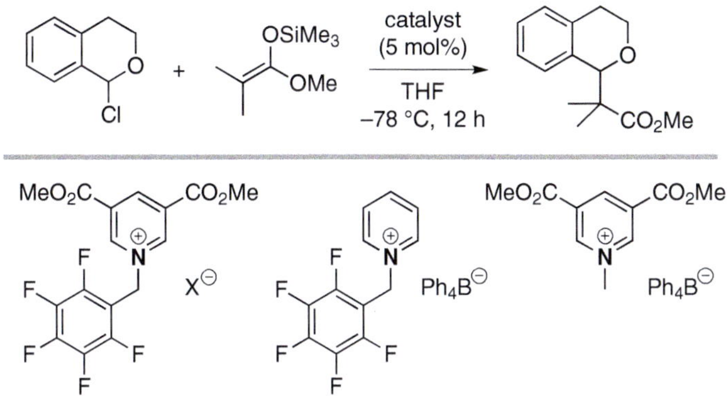

4a ($X^- = Ph_4B^-$): 90%
4b ($X^- = Br^-$): 0%
4c : 0%
4d : 0%

Scheme 8.5 Electron-deficient pyridinium salt catalysts for Mukaiyama aldol-type reactions. Source: Based on Berkessel *et al.* [18].

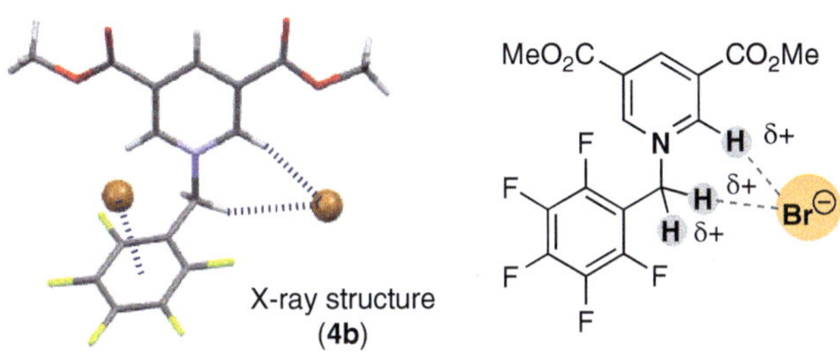

Figure 8.6 X-ray crystal structure of electron-deficient pyridinium bromide **4b**.

zwitterionic-type tetraalkylammonium salts **5c** and **5d** were also examined. Although catalyst **5c** with a pyrrolidine backbone moderately promoted the transesterification, almost no catalytic activity was observed with catalyst **5d**. The zwitterionic catalyst **5a** could also be applied to transesterification with simple substrates, and computational mechanistic studies were performed with catalyst **5a**. Based on the results of experimental and theoretical studies, they concluded that a catalyst can activate the carbonyl on esters via hydrogen-bonding interactions with acidic α-hydrogen atoms on the pyrrolidine backbone.

Related zwitterionic-type iminium salt catalyst **5e** could be applied to the deacylative dibromination of β-oxo amides (Scheme 8.7) [21]. The acidic α-hydrogen atoms of a dimethylamino moiety on the catalyst-activated carbonyl oxygen on the brominated intermediate in the deacylation step.

8.4 Hydrogen-Bonding Catalysis of Quaternary Phosphonium Salts

Acetonyltriphenylphosphonium bromide **6a** is a reactive compound because the acidity of its α-hydrogen atoms is higher than that of either phenol or acetic acid, and it is utilized as a Wittig reagent for the synthesis of α,β-unsaturated carbonyls

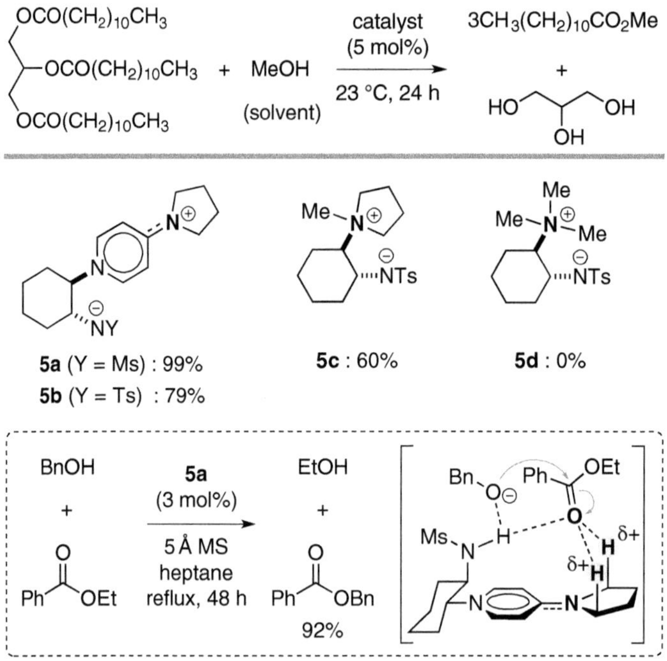

Scheme 8.6 Zwitterionic-type iminium and ammonium salt catalysts for transesterification.

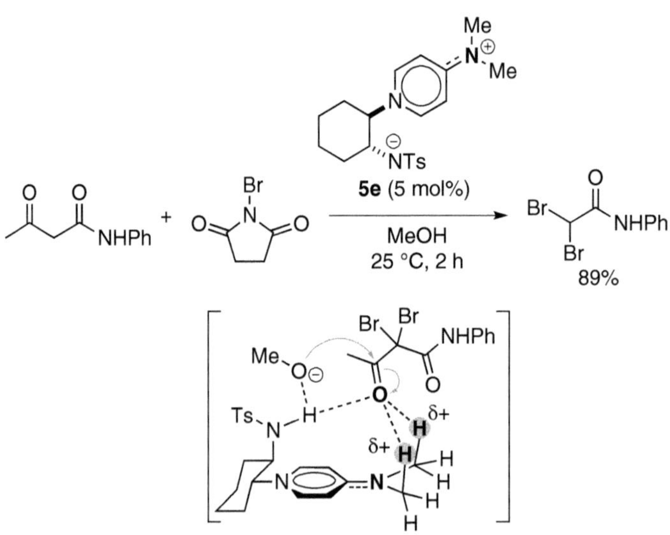

Scheme 8.7 Application of zwitterionic-type iminium salt catalyst to deacylative dibromination. Source: Based on Ke *et al.* [21].

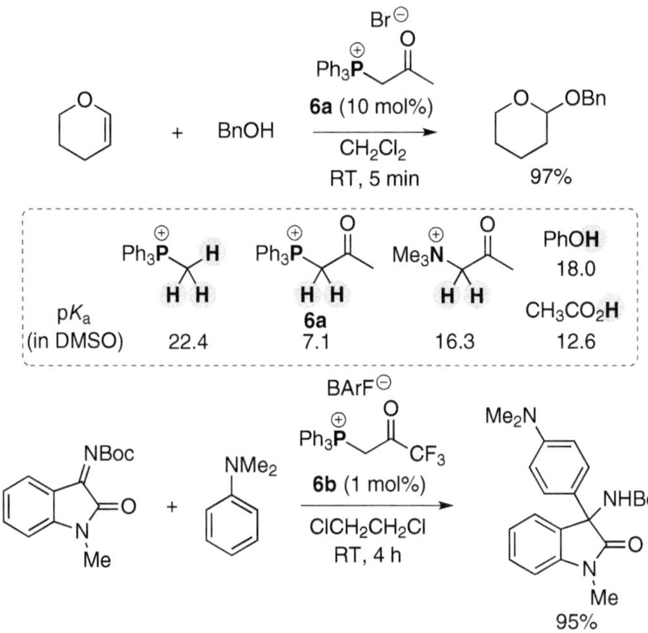

Scheme 8.8 Acetonyl- and trifluoroacetonyl-triphenylphosphonium salts as hydrogen-bonding catalysts.

[22]. Also, another potential of acetonyltriphenylphosphonium bromide **6a** was demonstrated by Hon and Lee in 1999 [23]. Hon employed the phosphonium bromide **6a** as a catalyst for acetalization (Scheme 8.8). Based on Hon's discovery, Chen examined the catalytic ability of trifluoroacetonyl-triphenylphosphonium barfate **6b** for Friedel–Crafts-type reactions with isatin-derived *N*-Boc ketoimines [24]. The phosphonium barfate **6b** possesses a non-coordinating counter-anion that has proven to be worked as an effective hydrogen-bonding catalyst in Friedel–Crafts-type reactions with isatin-derived *N*-Boc ketoimines.

In 1989, Mukaiyama *et al.* reported aldol and Mannich reactions catalyzed by bis-quaternary phosphonium triflate **7a** (Scheme 8.9) [25]. In these reactions, bis-quaternary phosphonium salts worked as catalysts that were more effective than mono-quaternary phosphonium salts. In 2018, Ingleson, Stephan, and coworkers examined the catalytic abilities of tris-quaternary phosphonium salts in the reduction of imines generated *in situ* [26]. The tris-phosphonium triflate **7b** could efficiently promote the reduction to provide target secondary amine products. Although Ingleson and Stephan mentioned that the tris-phosphonium salts worked as Lewis acid catalysts in this report, we suggested that **7b** might behave as a hydrogen-bonding catalyst [6]. Multiple hydrogen-bonding interactions between the acidic α-hydrogen atoms of tris-phosphonium salt and the anionic oxygen of triflate counter-anions are clearly apparent in the X-ray crystal structure of **7b**.

Yeung and coworkers reported zwitterionic-type phosphonium salt-catalyzed intermolecular bromoesterification of olefins (Scheme 8.10) [27]. In the discussion

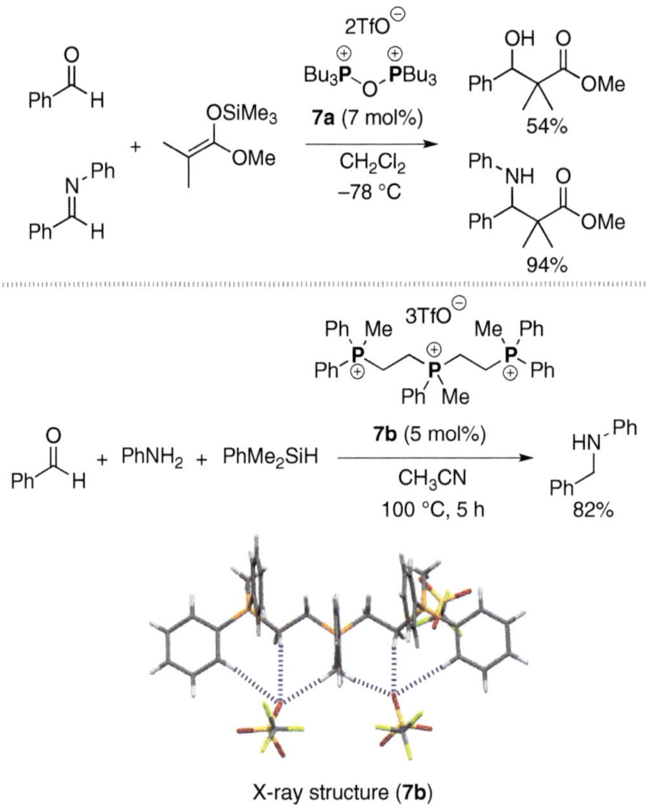

Scheme 8.9 Catalytic abilities of bis- and tris-quaternary phosphonium triflates. Source: Adapted from Mukaiyama *et al.* [25].

of plausible reaction mechanisms, the author suggested the importance of hydrogen-bonding interactions between acidic α-hydrogen atoms on the phosphonium salt moiety and carboxylate anion to provide a well-organized transition state that would promote the esterification.

8.5 Hydrogen-Bonding Catalysis of Tertiary Sulfonium Salts

Quaternary ammonium and phosphonium salts are known as an important class of catalysts, which are used to promote a wide variety of organic transformations, mainly as phase transfer and base catalysts. On the other hand, the catalytic abilities of tertiary sulfonium salts have yet to be appropriately demonstrated in organic synthesis, despite the fact that tertiary sulfonium salts are often utilized as useful reagents in organic reactions such as the Corey–Chaykovsky reactions [28]. Based on a discovery during phosphonium salt catalysis, as shown in Scheme 8.8, Hon

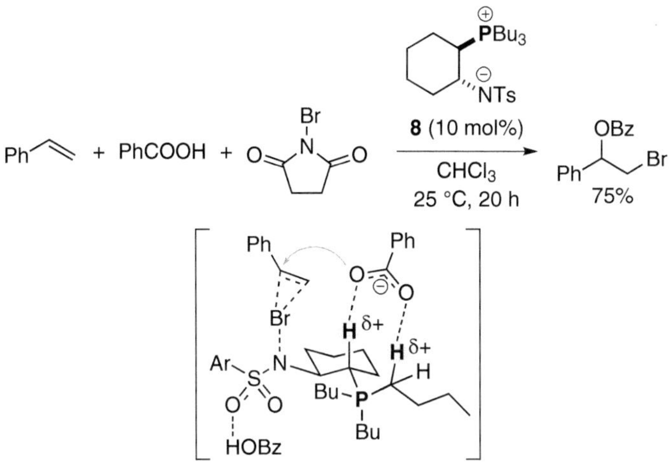

Scheme 8.10 Zwitterionic-type phosphonium salt catalyst **8** for bromoesterification. Source: Based on Ng *et al.* [27].

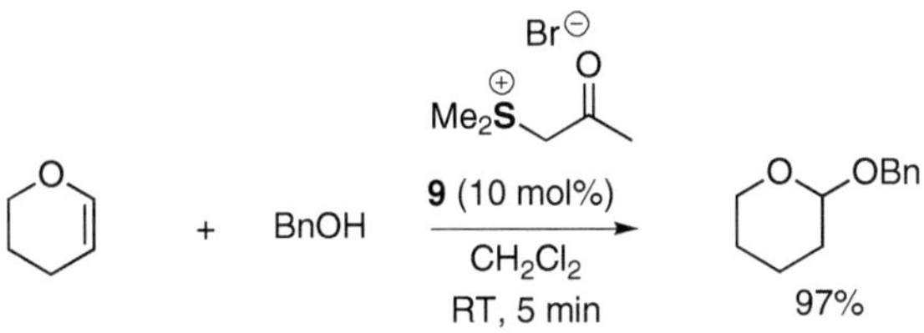

Scheme 8.11 Acetonyldimethylsulfonium bromide as a hydrogen-bonding catalyst. Source: Based on Hon *et al.* [23b].

et al. also reported that acetonyldimethylsulfonium bromide **9** is an efficient catalyst for acetalizations (Scheme 8.11) [23b].

Our research group also became interested in the hydrogen-bonding catalysis of cyclic tertiary sulfonium salts **10** when we determined that cyclic quaternary ammonium salts **1** worked as effective hydrogen-bonding catalysts on the basis of the characteristic properties of α-hydrogen atoms [29]. The structures of the α-hydrogen atoms that bound to the iodide anion compared favorably to the X-ray crystal structures of ammonium iodide **1a** and sulfonium iodide **10a** (Figure 8.7) [30]. Furthermore, based on the reported pK_a values, we expected the acidity of the α-hydrogen atoms of **10a** to approximate the acidity of **1a** [31].

With the important structural information of the cyclic tertiary sulfonium salts of type **10** in hand, the catalytic ability of **10** as a hydrogen-bonding catalyst was investigated in a Mannich-type reaction of *N*-acylisoquinoline as a benchmark reaction (Scheme 8.12). As mentioned in Scheme 8.3, ammonium iodide catalyst **1a** bearing carboxylate groups at the α-carbon atoms moderately promoted a Mannich-type reaction. A simple ammonium iodide catalyst **2a**, however, showed almost no acceleration of the reaction. On the other hand, a simple sulfonium iodide **10a** possessing higher acidic α-hydrogen atoms has moderately accelerated the reaction. These

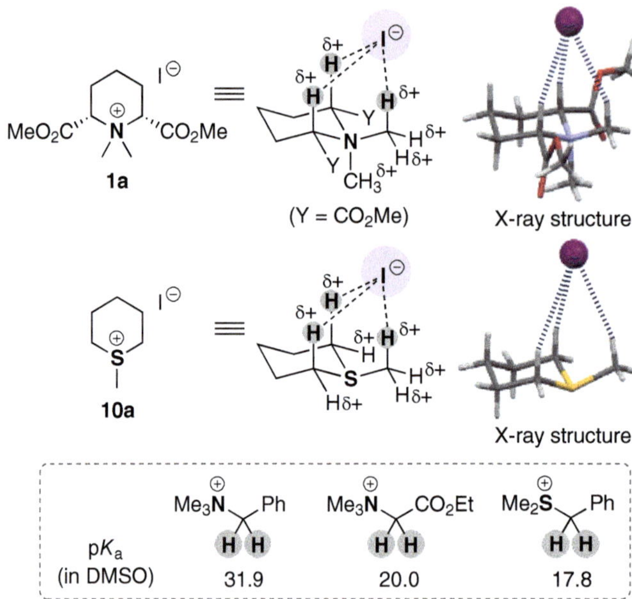

Figure 8.7 Structural similarity of ammonium iodide **1a** and sulfonium iodide **10a**. Source: Based on Gerdil [30].

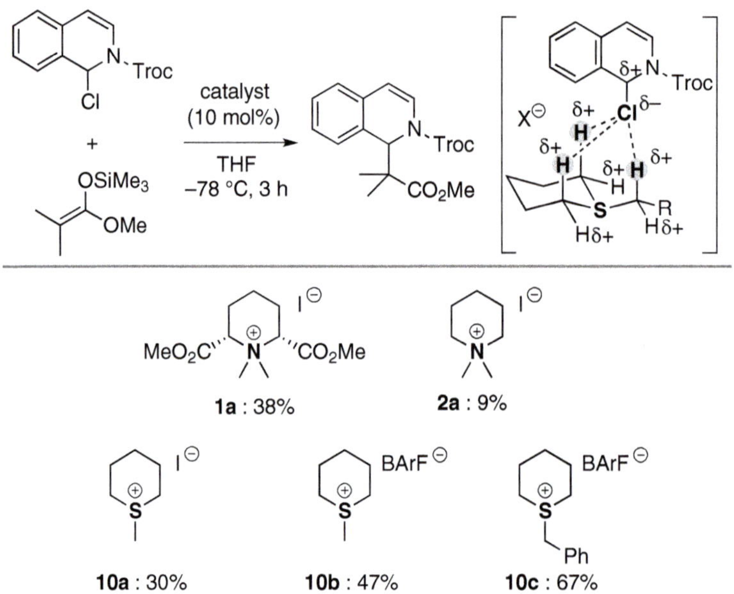

Scheme 8.12 Hydrogen-bonding catalysis of cyclic quaternary ammonium and tertiary sulfonium salts in a Mannich-type reaction.

results clearly suggested the importance of the acidity of the α-hydrogen atoms of cyclic tertiary sulfonium iodide **10a** as a hydrogen-bonding catalyst in promoting Mannich-type reactions. An exchange of the iodide counter-anion in sulfonium salt **10a** to $BArF^-$ as a non-coordinating counter-anion (**10b**) improved the results. The reaction with benzylsulfonium barfate **10c** further improved the results, and the target product was obtained in a 67% yield.

To expand the utility of cyclic tertiary sulfonium salts **10** as hydrogen-bonding catalysts, we next became interested in the activation of imines. An aza-Diels–Alder reaction of *N*-phenylbenzaldimine with a Danishefsky's diene was employed as a model reaction to evaluate the ability of sulfonium salts to activate imines (Scheme 8.13). Cyclic tertiary sulfonium iodide **10a** showed reactivity that was higher than that of piperidine-derived ammonium iodide **2a**, and the reaction with **10a** provided the target aza-Diels–Alder product in a good yield (81% yield). Sulfonium barfate **10b** further improved the catalytic activity to provide the product in a very high yield (98% yield). It should be noted that the reactions with sodium barfate (Na^+BArF^-) and tetraarylboric acid (H^+BArF^-) provided the aza-Diels–Alder product only in low yields (21% and 16% yields, respectively). These results clearly suggest the importance of a cyclic tertiary sulfonium salt moiety for the efficient promotion of aza-Diels–Alder reactions. Details of the activation modes of the imine by cyclic tertiary sulfonium salts **10** were investigated using DFT calculations [16, 17].

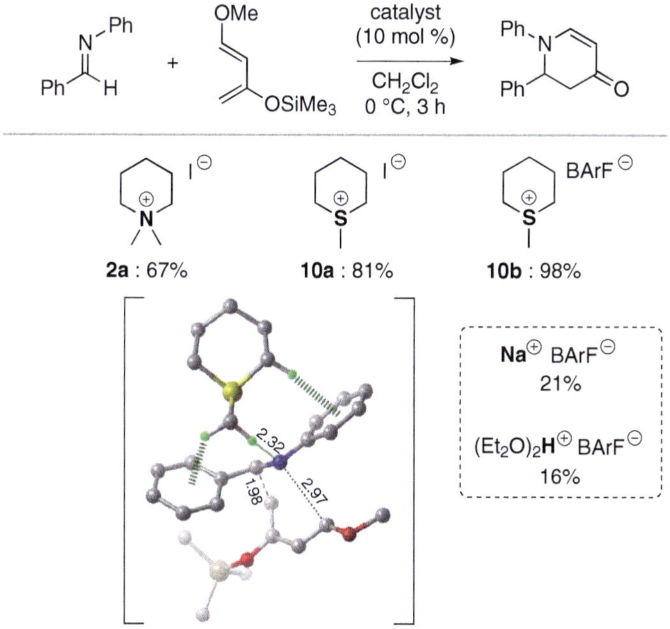

Scheme 8.13 Application of cyclic tertiary sulfonium salt catalysts to aza-Diels–Alder reactions.

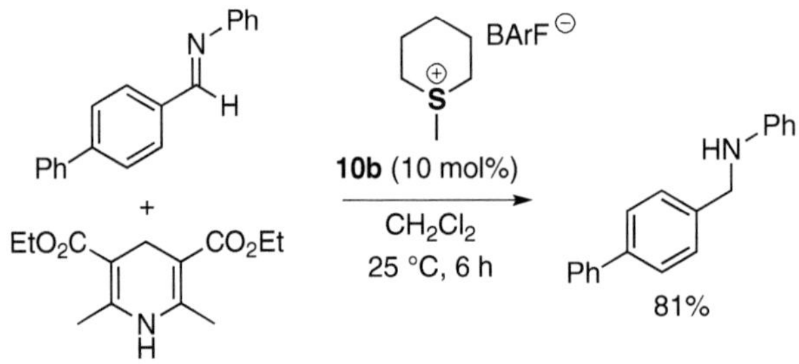

Scheme 8.14 Application of cyclic tertiary sulfonium barfate to the reduction of imines.

Furthermore, the cyclic tertiary sulfonium barfate catalyst **10b** could also be applied to the reduction of imines (Scheme 8.14). The reductions of imines with a Hantzsch ester were efficiently promoted by sulfonium barfate **10b** to provide the target secondary amine products in good yields.

8.6 Conclusion

Although the hydrogen-bonding donor abilities of quaternary ammonium salts were reported more than 25 years ago, until recently, hydrogen-bonding catalysis using quaternary ammonium salts has remained out of the spotlight. Now, many organic chemists have recognized the potential utility of not only quaternary ammonium salts but also quaternary phosphonium and tertiary sulfonium salts as hydrogen-bonding catalysts. The hydrogen-bonding donor ability of individual α-hydrogen atoms in these onium salts is not strong, but when arranged in appropriate positions via a rational design, the multiple hydrogen-bonding activation of electrophiles results in efficient hydrogen-bonding catalysis. Furthermore, X-ray structure analyses and DFT calculations of onium salt compounds have significantly promoted a deeper understanding of the hydrogen-bonding donor property for catalysis. This understanding of the alkyl-onium salts has introduced new possibilities into onium salt chemistry not only in the fields of catalytic and organic chemistries but also in the material and biological sciences.

References

1 (a) Olah, G.A. (1973). Carbocation and onium ion reagents. *Aldrichimica Acta* 6: 7–16.
(b) Vedejs, E. and West, F.G. (1986). Ylides by the desilylation of α-silyl onium salts. *Chem. Rev.* 86: 941–955.
(c) Prakash, G.K.S. (2006). Electrophilic intermediates and their reactions in superacids. *J. Org. Chem.* 71: 3661–3676.
(d) Laali, K.K. (ed.) (2007). *Recent Developments in Carbocation and Onium Ion Chemistry, ACS Symposium Series 965*. Washington, DC: American Chemical Society.

(e) Sweeney, J.B. (2009). Sigmatropic rearrangements of 'onium' ylids. *Chem. Soc. Rev.* 38: 1027–1038.
(f) Kwon, Y.-W., Choi, D.H., and Jin, J.-I. (2012). Optical, electro-optic and optoelectronic properties of natural and chemically modified DNAs. *Polym. J.* 44: 1191–1208.
(g) Guo, X. and Hu, W. (2013). Novel multicomponent reactions via trapping of protic onium ylides with electrophiles. *Acc. Chem. Res.* 46: 2427–2440.
(h) Pernak, J. (2013). Biological activity of the salts containing a quaternary nitrogen atom. *Przem. Chem.* 92: 1653–1656.
(i) Preshlock, S., Tredwell, M., and Gouverneur, V. (2016). ^{18}F-labeling of arenes and heteroarenes for applications in positron emission tomography. *Chem. Rev.* 116: 719–766.
(j) Haketa, Y. and Maeda, H. (2018). Dimension-controlled π-electronic ion-pairing assemblies. *Bull. Chem. Soc. Jpn.* 91: 420–436.
(k) Makvandi, P., Jamaledin, R., Jabbari, M. *et al.* (2018). Antibacterial quaternary ammonium compounds in dental materials: a systematic review. *Dent. Mater.* 34: 851–867.
(l) Roiser, L., Zielke, K., and Waser, M. (2018). Ammonium ylide mediated cyclization reactions. *Asian J. Org. Chem.* 7: 852–864.

2 (a) Kürti, L. and Czakó, B. (2005). *Strategic Applications of Named Reactions in Organic Synthesis*. San Diego: Elsevier Academic Press.
(b) Li, J.J. (2006). *Name Reactions*. Berlin: Springer.
(c) Li, J.J. (2007). *Name Reactions for Functional Group Transformations*. Hoboken: John Wiley & Sons.
(d) Wang, Z. (2009). *Comprehensive Organic Name Reactions and Reagents*. Hoboken: John Wiley & Sons.

3 (a) Dehmlow, E.V. and Dehmlow, S.S. (1993). *Phase Transfer Catalysis*, 3e. Weinheim: VCH.
(b) Starks, C.M., Liotta, C.L., and Halpern, M. (1994). *Phase-Transfer Catalysis*. New York: Chapman & Hall.
(c) Sasson, Y. and Neumann, R. (1997). *Handbook of Phase-Transfer Catalysis*. London: Blackie Academic & Professional.
(d) Halpern, M.E. (1997). *Phase-Transfer Catalysis, ACS Symposium Series 659*. Washington, DC: American Chemical Society.

4 For reviews on asymmetric phase-transfer catalysis, see: (a) O'Donnell, M.J. (2001). The preparation of optically active α-amino acids from the benzophenone imines of glycine derivatives. *Aldrichimica Acta* 34: 3–15.
(b) Ooi, T. and Maruoka, K. (2004). Asymmetric organocatalysis of structurally well-defined chiral quaternary ammonium fluorides. *Acc. Chem. Res.* 37: 526–533.
(c) Ooi, T. and Maruoka, K. (2007). Recent advances in asymmetric phase-transfer catalysis. *Angew. Chem. Int. Ed.* 46: 4222–4266.
(d) Maruoka, K. (2008). *Asymmetric Phase-Transfer Catalysis*. Weinheim: Wiley-VCH.

(e) Jew, S.-s. and Park, H.-g. (2009). Cinchona-based phase-transfer catalysts for asymmetric synthesis. *Chem. Commun.*: 7090–7103.
(f) Werner, T. (2009). Phosphonium salt organocatalysis. *Adv. Synth. Catal.* 351: 1469–1481.
(g) Enders, D. and Nguyen, T.V. (2012). Chiral quaternary phosphonium salts: a new class of organocatalysts. *Org. Biomol. Chem.* 10: 5327–5331.
(h) Novacek, J. and Waser, M. (2013). Bifunctional chiral quaternary ammonium salt catalysts: a rapidly emerging class of powerful asymmetric catalysts. *Eur. J. Org. Chem.*: 637–648. (i) Brak, K. and Jacobsen, E.N. (2013). Asymmetric ion-pairing catalysis. *Angew. Chem. Int. Ed.* 52: 534–561.
(j) Shirakawa, S. and Maruoka, K. (2013). Recent developments in asymmetric phase-transfer reactions. *Angew. Chem. Int. Ed.* 52: 4312–4348.
(k) Shirakawa, S. and Maruoka, K. (2014). Asymmetric phase-transfer reactions under base-free neutral conditions. *Tetrahedron Lett.* 55: 3833–3839.
(l) Kaneko, S., Kumatabara, Y., and Shirakawa, S. (2016). A new generation of chiral phase-transfer catalysts. *Org. Biomol. Chem.* 14: 5367–5376.
(m) Liu, S., Kumatabara, Y., and Shirakawa, S. (2016). Chiral quaternary phosphonium salts as phase-transfer catalysts for environmentally benign asymmetric transformations. *Green Chem.* 18: 331–341.
(n) O'Donnell, M.J. (2019). Benzophenone Schiff bases of glycine derivatives: versatile starting materials for the synthesis of amino acids and their derivatives. *Tetrahedron* 75: 3667–3696.
(o) Qian, D. and Sun, J. (2019). Recent progress in asymmetric ion-pairing catalysis with ammonium salts. *Chem. Eur. J.* 25: 3740–3751.

5 Liu, S., Maruoka, K., and Shirakawa, S. (2017). Chiral tertiary sulfonium salts as effective catalysts for asymmetric base-free neutral phase-transfer reactions. *Angew. Chem. Int. Ed.* 56: 4819–4823.

6 Nakamura, T., Okuno, K., Nishiyori, R., and Shirakawa, S. (2020). Hydrogen-bonding catalysis of alkyl-onium salts. *Chem. Asian J.* 15: 463–472.

7 For reviews on hydrogen-bonding catalysis, see: (a) Schreiner, P.R. (2003). Metal-free organocatalysis through explicit hydrogen bonding interactions. *Chem. Soc. Rev.* 32: 289–296.
(b) Pihko, P.M. (2004). Activation of carbonyl compounds by double hydrogen bonding: an emerging tool in asymmetric catalysis. *Angew. Chem. Int. Ed.* 43: 2062–2064.
(c) Doyle, A.G. and Jacobsen, E.N. (2007). Small-molecule H-bond donors in asymmetric catalysis. *Chem. Rev.* 107: 5713–5743.
(d) Auvil, T.J., Schafer, A.G., and Mattson, A.E. (2014). Design strategies for enhanced hydrogen-bond donor catalysts. *Eur. J. Org. Chem.*: 2633–2646.
(e) Nishikawa, Y. (2018). Recent topics in dual hydrogen bonding catalysis. *Tetrahedron Lett.* 59: 216–223.
(f) Reep, C., Sun, S., and Takenaka, N. (2019). C(sp^2)–H hydrogen-bond donor groups in chiral small-molecule organocatalysts. *Asian J. Org. Chem.* 8: 1306–1316.

8 (a) Reetz, M.T., Hütte, S., and Goddard, R. (1993). Tetrabutylammonium salts of CH-acidic carbonyl compounds: real carbanions or supramolecules? *J. Am. Chem. Soc.* 115: 9339–9340.
(b) Reetz, M.T., Hütte, S., and Goddard, R. (1995). Synthetic and mechanistic aspects of anionic polymerization of (METH)acrylates initiated by metal-free salts of CH-acidic compounds. *J. Phys. Org. Chem.* 8: 231–241.
(c) Reetz, M.T., Hütte, S., Goddard, R., and Robyr, C. (1996). Synthesis, structure, and stereoselective reaction of a chiral hydroxy-stabilized metal-free enolate. *Chem. Eur. J.* 2: 382–384.
(d) Reetz, M.T., Hütte, S., and Goddard, R. (1999). Supramolecular structure of the tetrabutylammonium salt of 2-phenylpropionitrile. *J. Prakt. Chem.* 341: 297–301.
(e) Goddard, R., Herzog, H.M., and Reetz, M.T. (2002). Cation–anion CH···O^- interactions in the metal-free phenolate, tetra-*n*-butylammonium phenol-phenolate. *Tetrahedron* 58: 7847–7850.

9 (a) Reetz, M.T. (1998). New methods for the anionic polymerization of α-activated olefins. *Angew. Chem. Int. Ed.* 27: 994–998.
(b) Cannizzaro, C.E. and Houk, K.N. (2002). Magnitudes and chemical consequences of R_3N^+–C–H···O=C hydrogen bonding. *J. Am. Chem. Soc.* 124: 7163–7169.
(c) Thomas, C., Milet, A., Peruch, F., and Bibal, B. (2013). Activation of carbonyl bonds by quaternary ammoniums and a (Na^+:crown-ether) complex: investigation of the ring-opening polymerization of cyclic esters. *Polym. Chem.* 4: 3491–3498.

10 (a) Ohshima, T., Shibuguchi, T., Fukuta, Y., and Shibasaki, M. (2004). Catalytic asymmetric phase-transfer reactions using tartrate-derived asymmetric two-center organocatalysts. *Tetrahedron* 60: 7743–7754.
(b) Cook, T.C., Andrus, M.B., and Ess, D.H. (2012). Quantum mechanical transition-state analysis reveals the precise origin of stereoselectivity in chiral quaternary cinchonidinium phase-transfer catalyzed enolate allylation. *Org. Lett.* 14: 5836–5839.
(c) Kamachi, T. and Yoshizawa, K. (2014). Enantioselective alkylation by binaphthyl chiral phase-transfer catalysts: a DFT-based conformational analysis. *Org. Lett.* 16: 472–475.
(d) Johnston, C.P., Kothari, A., Sergeieva, T. *et al.* (2015). Catalytic enantioselective synthesis of indanes by a cation-directed 5-*endo-trig* cyclization. *Nat. Chem.* 7: 171–177.
(e) Peng, Q. and Paton, R.S. (2016). Catalytic control in cyclizations: from computational mechanistic understanding to selectivity prediction. *Acc. Chem. Res.* 49: 1042–1051.
(f) Novacek, J., Izzo, J.A., Vetticatt, M.J., and Waser, M. (2016). Bifunctional ammonium salt catalyzed asymmetric α-hydroxylation of β-ketoesters by simultaneous resolution of oxaziridines. *Chem. Eur. J.* 22: 17339–17344.

(g) He, C.Q., Simon, A., Lam, Y.-h. *et al.* (2017). Model for the enantioselectivity of asymmetric intramolecular alkylations by bis-quaternized cinchona alkaloid-derived catalysts. *J. Org. Chem.* 82: 8645–8650.
(h) Yasui, M., Yamada, A., Tsukano, C. *et al.* (2020). Enantioselective acetalization by dynamic kinetic resolution for the synthesis of γ-alkoxybutenolides by thiourea/quaternary ammonium salt catalysts: application to strigolactones. *Angew. Chem. Int. Ed.* 59: 13479–13483.

11 (a) Couture, J.-F., Collazo, E., Ortiz-Tello, P.A. *et al.* (2007). Specificity and mechanism of JMJD2A, a trimethyllysine-specific histone demethylase. *Nat. Struct. Mol. Biol.* 14: 689–695.
(b) Berg, L., Mishra, B.K., Andersson, C.D. *et al.* (2016). The nature of activated non-classical hydrogen bonds: a case study on acetylcholinesterase–ligand complexes. *Chem. Eur. J.* 22: 2672–2681.
(c) Orman, M., Bodea, S., Funk, M.A. *et al.* (2019). Structure-guided identification of a small molecule that inhibits anaerobic choline metabolism by human gut bacteria. *J. Am. Chem. Soc.* 141: 33–37.
(d) Blom, A.E.M., Campello, H.R., Lester, H.A. *et al.* (2019). Probing binding interactions of cytisine derivatives to the α4β2 nicotinic acetylcholine receptor. *J. Am. Chem. Soc.* 141: 15840–15849.
(e) Itoh, Y., Nakashima, Y., Tsukamoto, S. *et al.* (2019). N^+–C–H···O Hydrogen bonds in protein-ligand complexes. *Sci. Rep.* 9: 767.

12 Thomas, C., Brut, S., and Bibal, B. (2014). Quaternary ammoniums and a cationic sodium complex as supramolecular catalysts in ring-opening of epoxides by amines. *Tetrahedron* 70: 1646–1650.

13 Shirakawa, S., Liu, S., Kaneko, S. *et al.* (2015). Tetraalkylammonium salts as hydrogen-bonding catalysts. *Angew. Chem. Int. Ed.* 54: 15767–15770.

14 Park, Y., Park, E., Jung, H. *et al.* (2011). Cinchona-derived ammonium salts-catalyzed aza Diels–Alder reaction of Danishefsky's diene with imines. *Tetrahedron* 67: 1166–1170.

15 Kumatabara, Y., Kaneko, S., Nakata, S. *et al.* (2016). Hydrogen-bonding catalysis of tetraalkylammonium salts in an aza-Diels–Alder reaction. *Chem. Asian J.* 11: 2126–2129.

16 Yamanaka, M., Mochizuki, A., Nakamura, T. *et al.* (2020). Trialkylsulfonium and tetraalkylammonium salts as hydrogen-bonding catalysts in an aza-Diels–Alder reaction: experimental and computational studies. *Heterocycles* 101: 580–592.

17 He, C.Q., Lam, C.C., Yu, P. *et al.* (2020). Catalytic effects of ammonium and sulfonium salts and external electric fields on aza-Diels–Alder reactions. *J. Org. Chem.* 85: 2618–2625.

18 Berkessel, A., Das, S., Pekel, D., and Neudörfl, J.-M. (2014). Anion-binding catalysis by electron-deficient pyridinium cations. *Angew. Chem. Int. Ed.* 53: 11660–11664.

19 Nagorny, P. and Sun, Z. (2016). New approaches to organocatalysis based on C—H and C—X bonding for electrophilic substrate activation. *Beilstein J. Org. Chem.* 12: 2834–2848.

20 Lam, Y.-P., Wang, X., Tan, F. *et al.* (2019). Amide/iminium zwitterionic catalysts for (trans)esterification: application in biodiesel synthesis. *ACS Catal.* 9: 8083–8092.

21 Ke, Z., Lam, Y.-P., Chan, K.-S., and Yeung, Y.-Y. (2020). Zwitterion-catalyzed deacylative dihalogenation of β-oxo amides. *Org. Lett.* 22: 7353–7357.

22 (a) Wittig, G. (1974). From diyls over ylides to my idyll. *Acc. Chem. Res.* 7: 6–14.
(b) Boutagy, J. and Thomas, R. (1974). Olefin synthesis with organic phosphonate carbanions. *Chem. Rev.* 74: 87–99.
(c) Maryanoff, B.E. and Reitz, A.B. (1989). The Wittig olefination reaction and modifications involving phosphoryl-stabilized carbanions. Stereochemistry, mechanism, and selected synthetic aspects. *Chem. Rev.* 89: 863–927.

23 (a) Hon, Y.-S. and Lee, C.-F. (1999). Acetonyltriphenylphosphonium bromide in organic synthesis: an extremely efficient catalyst for the protection and deprotection of alcohols as alkyl vinyl ethers. *Tetrahedron Lett.* 40: 2389–2392.
(b) Hon, Y.-S., Lee, C.-F., Chen, R.-J., and Szu, P.-H. (2001). Acetonyltriphenylphosphonium bromide and its polymer-supported analogues as catalysts in protection and deprotection of alcohols as alkyl vinyl ethers. *Tetrahedron* 57: 5991–6001.
(c) Hon, Y.-S. and Lee, C.-F. (2001). Acetonyltriphenylphosphonium bromide in organic synthesis: a useful catalyst in the cyclotrimerization of aldehydes. *Tetrahedron* 57: 6181–6188.

24 Chen, L., Xiao, B.-X., Du, W., and Chen, Y.-C. (2019). Quaternary phosphonium salts as active Brønsted acid catalysts for Friedel–Crafts reactions. *Org. Lett.* 21: 5733–5736.

25 (a) Mukaiyama, T., Matsui, S., and Kashiwagi, K. (1989). Effective activation of carbonyl and related compounds with phosphonium salts. The aldol and Michael reactions of carbonyl compounds with silyl nucleophiles and alkyl enol ethers. *Chem. Lett.*: 993–996.
(b) Mukaiyama, T., Kashiwagi, K., and Matsui, S. (1989). A convenient synthesis of β-aminoesters. The reaction of imines with ketene silyl acetals catalyzed by phosphonium salts. *Chem. Lett.*: 1397–1400.

26 Bayne, J.M., Fasano, V., Szkop, K.M. *et al.* (2018). Phosphorous(V) Lewis acids: water/base tolerant P_3-trimethylated trications. *Chem. Commun.* 54: 12467–12470.

27 Ng, W.-H., Hu, R.-B., Lam, Y.-P., and Yeung, Y.-Y. (2020). Zwitterion-catalyzed intermolecular bromoesterifications. *Org. Lett.* 22: 5572–5576.

28 (a) Corey, E.J. and Chaykovsky, M. (1965). Dimethyloxosulfonium methylide ($(CH_3)_2SOCH_2$) and dimethylsulfonium methylide ($(CH_3)_2SCH_2$). Formation and application to organic synthesis. *J. Am. Chem. Soc.* 87: 1353–1364.
(b) Li, A.-H., Dai, L.-X., and Aggarwal, V.K. (1997). Asymmetric ylide reactions: epoxidation, cyclopropanation, aziridination, olefination, and rearrangement. *Chem. Rev.* 97: 2341–2372.
(c) Aggarwal, V.K. and Winn, C.L. (2004). Catalytic, asymmetric sulfur ylide-mediated epoxidation of carbonyl compounds: scope, selectivity, and applications in synthesis. *Acc. Chem. Res.* 37: 611–620.

29 Kaneko, S., Kumatabara, Y., Shimizu, S. *et al.* (2017). Hydrogen-bonding catalysis of sulfonium salts. *Chem. Commun.* 53: 119–122.

30 Gerdil, R. (1974). The crystal and molecular structure of 1-methyl-1-thionia-cyclohexane iodide. *Helv. Chim. Acta* 57: 489–493.

31 Cheng, J.-P., Liu, B., and Zhang, X.-M. (1998). Radical substituent effects of α-sulfonium groups. *J. Org. Chem.* 63: 7574–7575.

9

Assisted and Dual Anion Binding in Amino and Nucleophilic Catalysis

Efraim Reyes[1], Qui-Nhi Duong[2], Liher Prieto[1], Olga García Mancheño[2], and José L. Vicario[1]

[1] *University of the Basque Country (UPV/EHU), Department of Organic Chemistry II, Faculty of Science and Technology, Barrio Sarriena s/n. Leioa C.P 48940, Bizkaia, Spain*
[2] *Institute of Organic Chemistry, Westfälische Wilhelms-Universität Münster, Corrensstraße 36, D-48149 Münster, Germany*

This chapter describes the application of the concept of anion-binding-assisted and dual catalysis in amino and nucleophilic catalyzed transformations. Hence, this chapter will first give an overview on dual amino/H-bond donor catalysis involving enamine and iminium activation approaches. Moreover, the advantages on the anion-binding/nucleophilic cocatalysis will be then discussed, focusing on the use of pyridine-type nucleophilic catalysts and excluding other examples that are integrated in other chapters (e.g. phosphine cocatalysis in Chapter 5).

9.1 Dual Amino/H-Bond Donor Catalysis

The renaissance of aminocatalysis has opened new venues for the development of synthetic methodology with enormous potential [1]. The ability of a primary or secondary amine to condense reversibly with an aldehyde or ketone to form a nucleophilic enamine species able to react with an electrophile and subsequently undergo hydrolysis that enables turnover of the catalyst is a concept that was established decades ago by two industrial research groups for the enantioselective intramolecular aldol reaction (see Scheme 9.1a) [2]. Further evolution of the concept, highlighted in the seminal L-proline-catalyzed aldol reaction reported in 2000 by Barbas, List, and Lerner [4], ended up in a blossoming field with a wide variety of highly effective reactions involving electrophiles of different nature that can be carried out with high enantio- and diastereoselectivity, provided that a chiral amine catalyst is employed [5]. Two important benefits aroused from this reaction manifold. On the one hand, the reaction proceeded in the absence of any metal reagent and could be performed in non-anhydrous solvents in open air. On the other hand, this opened the possibility of performing catalytic and enantioselective enolate-type chemistry involving aldehydes, that is, a challenging situation by

Anion-Binding Catalysis, First Edition. Edited by Olga García Mancheño.

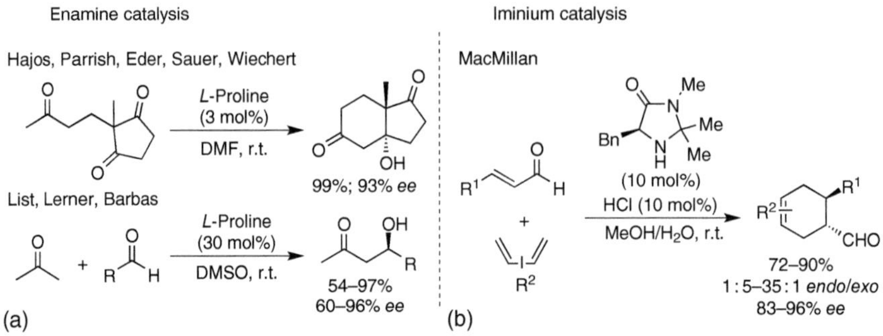

Scheme 9.1 Seminal examples on enantioselective aminocatalysis. Sources: (a) Adapted from Hajos and Parrish [2a, b], Eder *et al.* [2c, d], (b) Based on Jen *et al.* [3].

itself. Another breakthrough in this field came simultaneously with the report by MacMillan showing that chiral amines could also activate α,β-unsaturated aldehydes toward cycloaddition chemistry through the formation of an α,β-unsaturated iminium ion upon reversible condensation (Scheme 9.1b) [3]. This concept has found wide application in synthesis as the formation of the iminium ion leads to an activated chiral electrophile with the potential to undergo conjugate addition with a wide variety of nucleophiles [6] or even to engage in cascade reactions initiated by this initial 1,4-addition process [7].

Incorporating an H-bond donor entity either as a part of a bifunctional aminocatalyst [8] or as an external additive [9] has been widely used in many examples of successful reactions. The most commonly found situation involves the interaction of the H-bond donor with neutral reagents to provide activation of an electrophilic moiety or as a powerful stereodirecting element in bifunctional H-bond donor/aminocatalysts. As an alternative, the combination of aminocatalysis with anion-binding events opens a different scenario that will be discussed in Sections 9.1.1–9.1.4 that are classified according to the aminocatalytic activation manifold.

9.1.1 Enamine Activation

The seminal L-proline-catalyzed aldol reaction is a perfect example of the participation of a H-bond donor site at the structure of the aminocatalyst, which is subsequently engaged in the activation of the electrophile during the aldol reaction between the enamine intermediate and the external aldehyde reagent. Therefore, the reaction proceeds through a cyclic chair-like transition state with reduced conformational mobility, leading to outstanding levels of both diastereo- and enantioselectivity (Scheme 9.2). The key participation of this H-bond in the C—C bond formation has been supported through a series of both experimental and computational studies [10]. This dual activation manifold does not only account for the very high facial selectivity observed but also explains the excellent anti-selectivity observed for the intermolecular aldol reaction. In this case, it involves the exclusive participation of the *E* diastereoisomer of the enamine intermediate species, as the

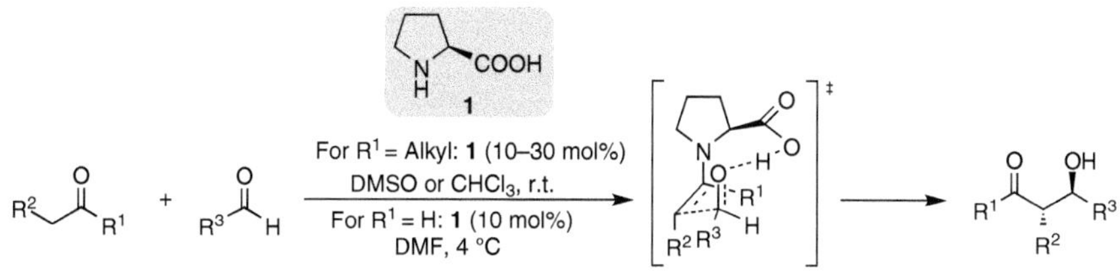

Scheme 9.2 The proline-catalyzed enantioselective cross-aldol reaction.

presence of the competitive *Z* diastereoisomer that would lead to opposite facial selectivity, which is disfavored both from the kinetic and thermodynamic point of view. In addition, this reactivity profile and stereochemical model can be applied to a variety of structurally related electrophiles such as imines, azodicarboxylates, nitrosobenzene, and some others [11].

Despite the excellent performance of L-proline in many of these reactions and the rather wide substrate scope, there are still several situations in which the reaction is limited either because of low conversion or poor enantiocontrol. This has been attributed in many cases to the poor solubility of L-proline in the typical organic solvents employed for the reaction. A possible solution has been found through the incorporation of achiral H-bond donor additives able to engage in H-bonding interactions with the carboxylate moiety that can increase the solubility of the catalyst. This has been demonstrated with the observation that a faster and more efficient aldol reaction between cyclohexanone and aromatic aldehydes takes place in a non-polar solvent when an equimolar amount (10 mol%) of Schreiner thiourea **2** together with L-proline (**1**) are used as a catalytic system (Scheme 9.3) [12]. Detailed experiments support the key role played by the H-bond donor additive in binding to the carboxylate anion, leading to a more soluble catalyst and also contributing

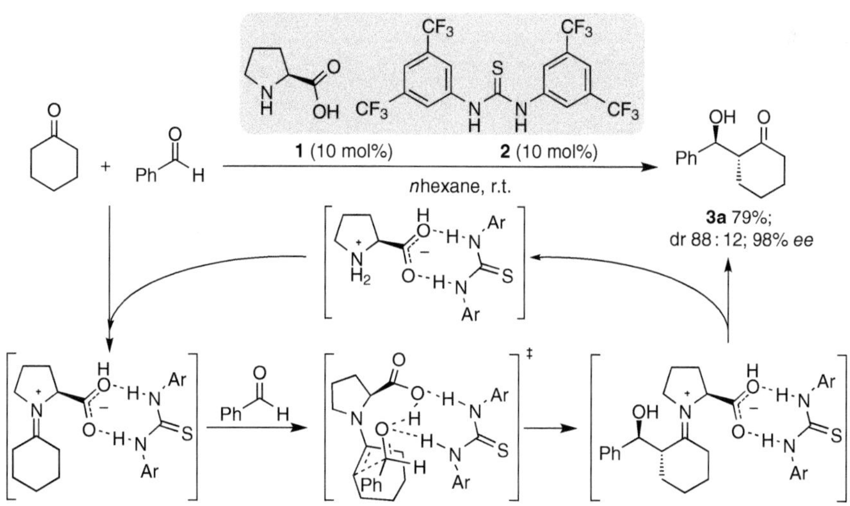

Scheme 9.3 The use of thiourea **2** as a H-bond donor additive in the proline-catalyzed aldol reaction. Source: Based on Reis *et al.* [12].

to stabilize the initial species formed upon condensation between the catalyst and the ketone donor [13]. This observation has been further extended to other mechanistically related reactions such as the Mannich reaction [14] or the nitroso-aldol reaction [15], and even to the use of other H-bond donor reagents able to bind to the carboxylate anion of L-proline in a similar way as thiourea **2** does [16].

In a similar way, a chiral H-bond donor additive can be incorporated in order to further enhance the enantioselectivity. This has provided a solution for the Michael reaction between ketones and nitroalkenes, in which L-proline by itself provides a very poor performance. In the example depicted in Scheme 9.4, the *matched* combination between L-proline and quinidine-derived thiourea **4** (5 mol% each) forms a quaternary ammonium salt as the catalyst assembly that performs well in the Michael reaction between a variety of both cyclic and acyclic ketones [17]. The anion-binding interaction between the thiourea moiety and the carboxylate group of L-proline provides a valuable stereodirecting element for the conjugate addition of the enamine intermediate to the nitroalkene moiety, which is also proposed to engage in additional H-bonding interactions with the protonated quinuclidine moiety. Interestingly, under these conditions, the reaction provides the opposite enantiomer of the Michael adduct that is obtained in the simple L-proline-catalyzed reaction, which also supports the bifunctional character of the catalyst and the key role played by the anion-binding event.

This type of anion-binding interactions through H-bond donor moieties has also played a key role for the success of the enantioselective α-alkylation of α,α-disubstituted aldehydes through S_N1-type reactions under enamine catalysis. In the example shown in Scheme 9.5, the hydrogen bond donor motif of primary amine/thiourea catalyst **6** is involved in the formation of this key carbocationic intermediate, establishing H-bonding interactions with the leaving group and therefore contributing to the stabilization of this species through the formation of an ion pair [18]. This intimate ion-paired intermediate also implies that the thiourea

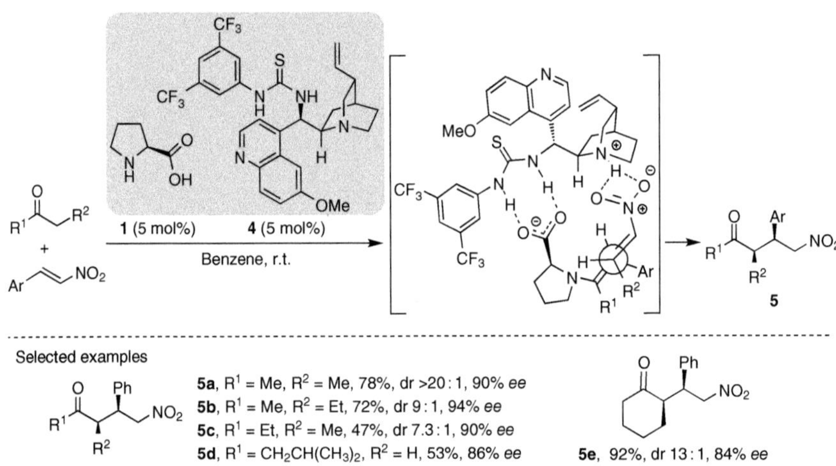

Scheme 9.4 The Michael reaction between aldehydes and nitroalkenes catalyzed by the combination of L-proline and quinidine-based thiourea **4**.

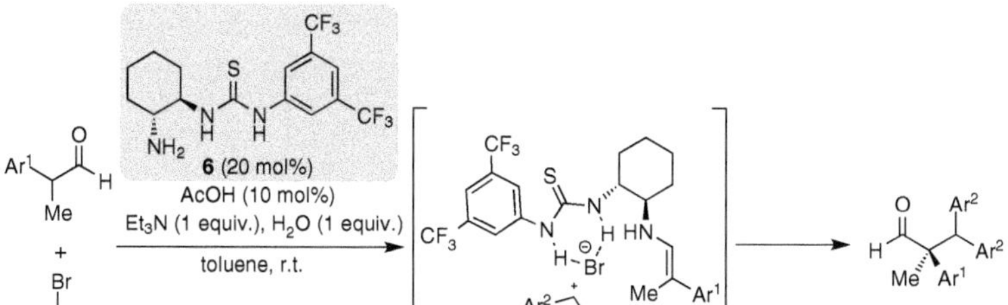

7a, Ar^1 = Ph, Ar^2 = Ph, 70%, 91% *ee*
7b, Ar^1 = Ph, Ar^2 = 4-FC_6H_4, 60%, 90% *ee*
7c, Ar^1 = 4-FC_6H_4, Ar^2 = Ph, 57%, 92% *ee*
7d, Ar^1 = 4-MeC_6H_4, Ar^2 = Ph, 59%, 85% *ee*

Scheme 9.5 α-Alkylation of α,α-disubstituted aldehydes catalyzed by **6**.

moiety is also involved in directing the trajectory of the electrophile toward one of the two diastereotopic faces of the enamine. The reaction is wide in scope with respect to the possible substitution patterns at the aldehyde but requires for the use of bromodiarylmethanes as proelectrophiles that form particularly stabilized benzhydryl cations.

As an alternative, the α-benzylation of aldehydes has also been described using benzyl bromides as the electrophilic reagent, in this case combining azabicyclic secondary amino acid **8** (30 mol%) as an aldehyde activator and guanidine **9** as a superstoichiometric additive (Scheme 9.6) [19]. This reaction does not involve the neat formation of a carbocation, and therefore, it does not involve the generation of a separated intermediate in which the anionic leaving group is captured through H-bonding interactions. However, the design behind the success of the reaction relies on the same ability of an H-bond donor to interact with a developing negative charge at the halide atom during the S_N2 pathway that is proposed for this reaction.

In particular, the guanidinium species would be involved in the activation of the leaving group by providing an electrostatic stabilization of the developing negative charge at the halide and also with some additional contribution of π–π interactions between the aryl substituent of the enamine intermediate and the benzyl bromide. The choice of the secondary amine catalyst is also crucial for the success of the reaction by providing an active secondary amine motif with high ability to form the desired enamine intermediate, but showing low tendency to undergo competitive

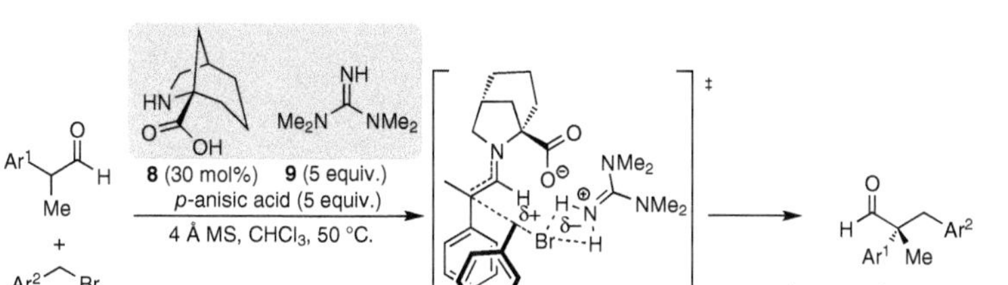

10a, Ar^1 = Ph, Ar^2 = Ph, 80%, 91% *ee*
10b, Ar^1 = 4-MeC_6H_4, Ar^2 = Ph, 82%, 90% *ee*
10c, Ar^1 = 4-BrC_6H_4, Ar^2 = Ph, 72%, 80% *ee*
10d, Ar^1 = Ph, Ar^2 = 4-BrC_6H_4, 80%, 88% *ee*

Scheme 9.6 α-Alkylation of α,α-disubstituted aldehydes catalyzed by **8** in the presence of guanidine **9**. Source: Based on List *et al.* [19].

deactivation via N-alkylation by the benzyl bromide reagent, which was the main observed pathway when other proline-type catalysts were tested in this reaction.

9.1.2 Dienamine Activation

The combination of the enamine activation manifold with the principle of vinylogy [20] enables the extension of the nucleophilic reactivity to remote positions of the starting pronucleophile [21]. After the seminal example of the use of dienamine intermediates in the catalytic enantioselective γ-amination of enals reported by Jørgensen [22], this approach has received a great deal of attention, also going further beyond the possibility of using simple dienamines [23] to more extended trienamine and tetraenamine intermediates [24]. However, because of the fact that the new stereocenters are being installed in positions that are located far away from where the chiral information of the amine catalyst is placed, together with the high degree of conformational freedom of dienamine, trienamine, or tetraenamine systems generated from acyclic molecules, the achievement of high stereocontrol can become a challenging situation for those catalysts that impart face selectivity exclusively through steric effects, as it is the case of the archetypical Jørgensen–Hayashi-type catalysts [25]. As a consequence, the use of bifunctional amine/H-bond donor catalysts able to interact with the incoming electrophile can provide a useful solution to this problem, leading to a more conformationally restricted organization of the reagents during the formation of new stereocenters. A good example of this strategy is shown with the use of bifunctional pyrrolidine/squaramide catalyst **11a** developed by Jørgensen, which was successfully applied to the formal enantioselective [2+2] cycloaddition between nitroalkenes and catalytically generated dienamines (Scheme 9.7) [26].

The ability of this type of catalysts to engage in anion binding through H-bonding with the squaramide moiety in order to provide an efficient stereodirecting effect that enhances the stereoselectivity of reactions through dienamine activation has been demonstrated in the [5+2] cycloaddition between *in situ*-generated oxidopyrilium ylides and enolizable enals (Scheme 9.8) [27]. This reaction consists of a

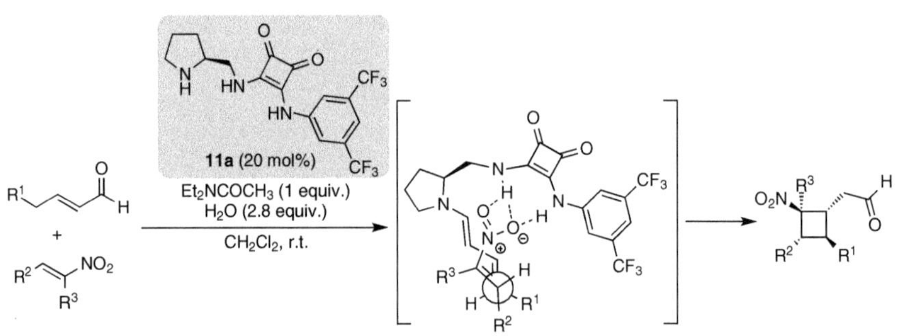

Scheme 9.7 Use of bifunctional catalyst **11a** in the formal [2+2] cycloaddition reaction between enals and nitroalkenes under dienamine activation. Source: Based on Albrecht *et al.* [26].

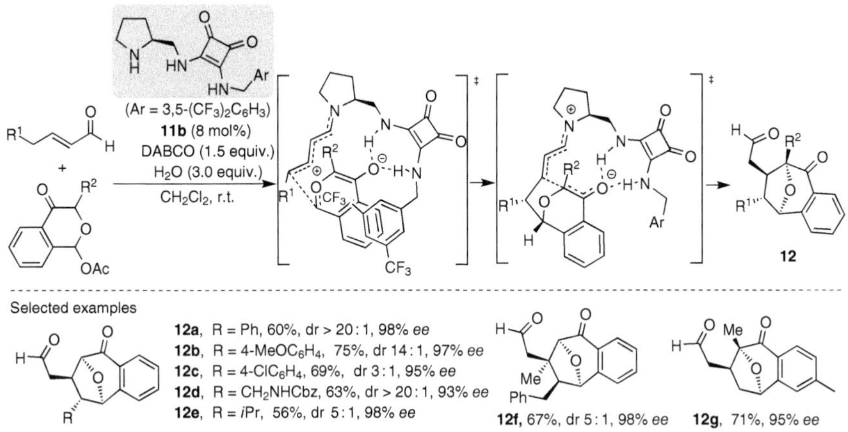

Scheme 9.8 Use of bifunctional catalyst **11b** in the formal [5+2] cycloaddition reaction between enals and oxidopyrilium ylides under dienamine activation. Source: Based on Orue *et al.* [27].

stepwise process in which, once the oxidopyrilium 1,3-dipolar reagent is formed by DABCO-promoted elimination of AcOH, the 1,2-addition of the dienamine to the oxonium moiety takes place, followed by the intramolecular Michael reaction. The initial step occurs through a rigid arrangement due to both H-bonding interaction between the oxidopyrilium reagent and the squaramide moiety, together with π–π interactions between the former and the (bis-trifluoromethyl)benzyl substituent located at one of the N-atoms of the squaramide [28]. The reaction was found to be rather wide in scope, providing a family of adducts with a broad variety of substitution patterns in the central 8-oxabicyclo[3.2.1]octane core, also including the possibility to form products with tetrasubstituted stereocenters with high yield and stereoselectivities.

Thioureas are also proficient catalysts for this type of chemistry. In particular, primary amine/thiourea **13** is an outstanding catalyst for the formal aza-Diels–Alder reaction between acyclic enones and dihydro-β-carbolines as the imine counterpart (Scheme 9.9) [29]. The reaction also requires for a Brønsted acid such as acetic acid as a cocatalyst that activates the cyclic imine through the formation of an acyliminium cation. In this case, the thiourea moiety participates through H-bonding interactions with the anionic counterion, directing the approach of the acyliminium salt to the 2-aminodiene intermediate through the formation of a close-contact ion pair. The reaction can be either via a concerted [4+2] cycloaddition or a stepwise Mannich/aza-Michael reaction, both possibilities leading to the observed final products. This reaction provides densely functionalized indoloquinolizidine derivatives with excellent yields, high diastereoselectivities, and excellent enantiocontrol, demonstrating the high proficiency of the transformation for a wide variety of substrates with different substitution patterns at both the enone and the cyclic imine reagent. The reaction could also be extended successfully to the use of 3,4-dihydroisoquinolines, providing a direct access to the benzoquinolizidine scaffold.

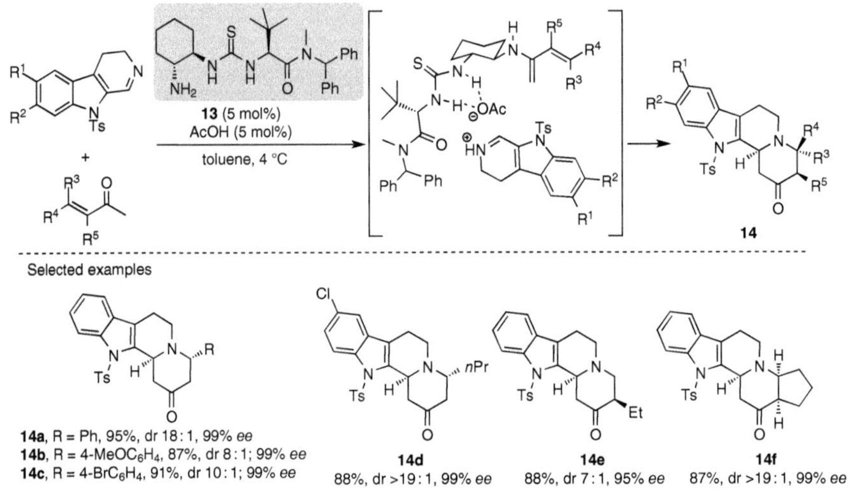

Scheme 9.9 The formal aza-Diels–Alder cycloaddition between enones and cyclic imines catalyzed by primary amine/thiourea **13**. Source: Based on Lalonde *et al.* [26].

9.1.3 Iminium Ion Activation

The combination of the iminium activation manifold with H-bond donor elements in catalysis has been widely explored in the literature, also including relevant examples in which the H-bond donor unit is engaged in anion-binding events. A particularly synergistic situation is found when the H-bond donor unit is placed as part of a bifunctional aminocatalyst, where the possibility of anion binding with the external negatively charged nucleophile that undergoes conjugate addition to the activated α,β-unsaturated iminium ion can provide a powerful stereodirecting element that contributes positively to the stereocontrol of the reaction. A good example of this behavior can be found in Scheme 9.10, in which primary amine/thiourea **15** was identified as an excellent catalyst for the conjugate addition of oxindoles to enals, generating adjacent stereocenters, one of them being an all-carbon quaternary one [30]. The authors support the formation of the iminium ion as the activated Michael acceptor species based on the necessity for a primary amine moiety for the reaction to take place efficiently. They also proposed that the thiourea is involved in the stabilization of the enol nucleophile through H-bonding interactions, which also determines the approach of the latter to the iminium ion.

As an alternative, the combined action of a chiral secondary or primary amine with an external achiral thiourea as the hydrogen bond donor has been successfully applied to several reactions in which this external hydrogen bond donor cocatalyst facilitates the reaction either by providing improved conversions or by enhancing the stereocontrol. An example of this behavior is shown in Scheme 9.11, in which the incorporation of Schreiner thiourea **2** led to a significant improvement of the yield for the cascade oxa-Michael/Michael reaction between 6-hydroxy-2*H*-pyran-3(6*H*)-one and enals catalyzed by diarylprolinol catalyst **17a** that induces the formation of an activated iminium ion upon condensation between

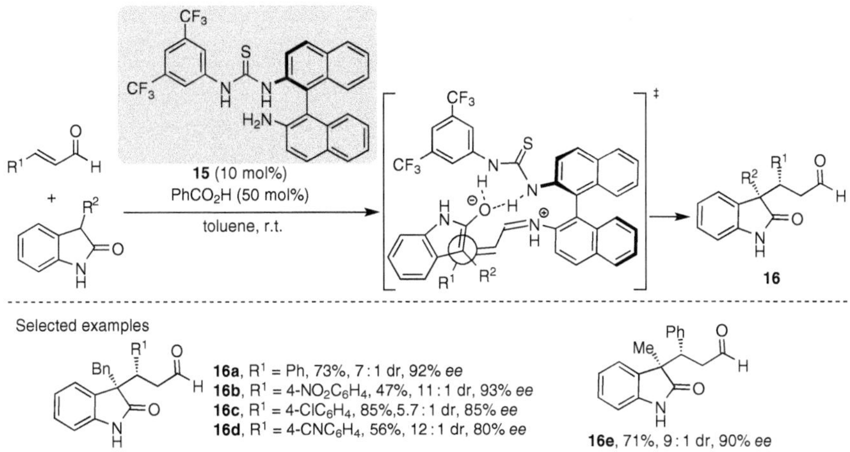

Scheme 9.10 Conjugate addition of oxindoles to enals catalyzed by primary amine/thiourea **15**.

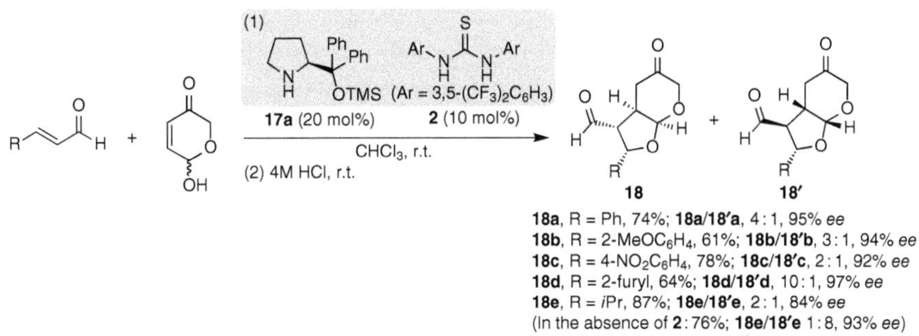

Scheme 9.11 Combined use of catalyst **17a** and thiourea **2** in the oxa-Michael–Michael cascade reaction between 6-hydroxy-2*H*-pyran-3(6*H*)-one and enals.

17a and the enal electrophile [31]. This reaction also proceeds together with the dynamic kinetic resolution of the pyranone starting material, being the H-bond donor cocatalyst involved in accelerating the racemization of this reagent in the reaction medium before the oxa-Michael reaction step takes place. In the absence of this H-bond donor cocatalyst, the reaction provided a different diastereoisomer that involved the preferential kinetic participation of the opposite enantiomer of the chiral *O*-pronucleophile compared to the same reaction in which **2** was incorporated as an additive.

Another illustrative and highly successful example on the combined use of this prolinol-type catalyst with an external H-bond donor cocatalyst is shown in the enantioselective vinylogous Michael reaction depicted in Scheme 9.12 [32]. In this case, bis-*N*-Tf-substituted ethylenediamine **19** is incorporated as the H-bond donor cocatalyst that engages in the activation of the Michael donor, generating a stabilized dienolate that is long-lived enough to undergo conjugate addition to the less-hindered diastereotopic face of the chiral α,β-unsaturated iminium ion

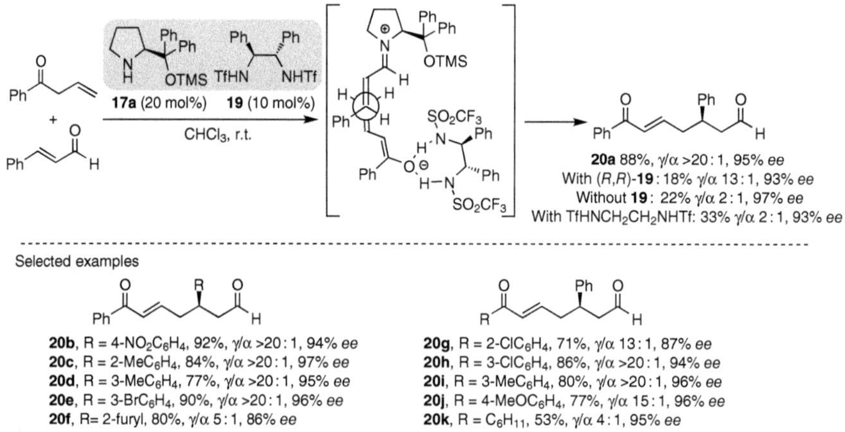

Scheme 9.12 Combined use of catalyst **17a** and diamine **19** in the vinylogous Michael reaction between β,γ-unsaturated ketones and enals. Source: Based on Gu *et al.* [32].

intermediate that is formed upon activation of the enal with the prolinol-derived catalyst. The reaction was found to proceed with a remarkable γ-selectivity with respect to the bis-nucleophilic nature of the dienolate donor reagent and also provided high yields and enantioselectivity for a rather wide range of substrates with different substitution patterns, although in all cases involving γ-unsubstituted β,γ-unsaturated ketones as Michael donors. The authors also demonstrated the synergistic nature of the catalyst combination between (*S*)-configured **17a** and (*R,R*)-diphenylethylenediamine cocatalyst **19**, observing that when the reaction was carried out in the presence of the enantiomer of **19** together with **17a**, the yield of the reaction was drastically reduced, although the enantioselectivity of the process remained the same. Using an achiral version of **19** such as bis-*N*-Tf-ethylenediamine provided a similar result, a decrease in the yield and high enantioselectivity. This type of combinations between chiral secondary amines and diamine **19** have been successfully employed in a variety of mechanistically related reactions [33].

In a different context, thiourea **2** has been employed to direct the reactivity of nitrones in the formal [3+2] cycloaddition with enals under iminium activation of the latter (Scheme 9.13) [34]. While these reagents typically participate as 1,3-C,O dipolar species when involved in cycloaddition chemistry, the incorporation of the thiourea cocatalyst results in the stabilization of a nitrone ylide reagent that underwent formal [3+2] cycloaddition as 1,3-C,N dipolar reagents, providing *N*-hydroxy pyrrolidine adducts instead of the isoxazoline adducts that are commonly obtained under standard conditions. As in the previous reaction, diarylprolinol-type catalyst **17a** was used as in the previous case for the activation of the enal and for achieving stereocontrol. Interestingly, mechanistic investigations also showed that thiourea **2** also played a key role in the reaction through the stabilization of the α,β-unsaturated iminium ion through H-bonding with the hydroxide counterion, accelerating the initial steps of the catalytic cycle, in which the staring enal is activated through condensation with the aminocatalyst and therefore increasing the concentration of this

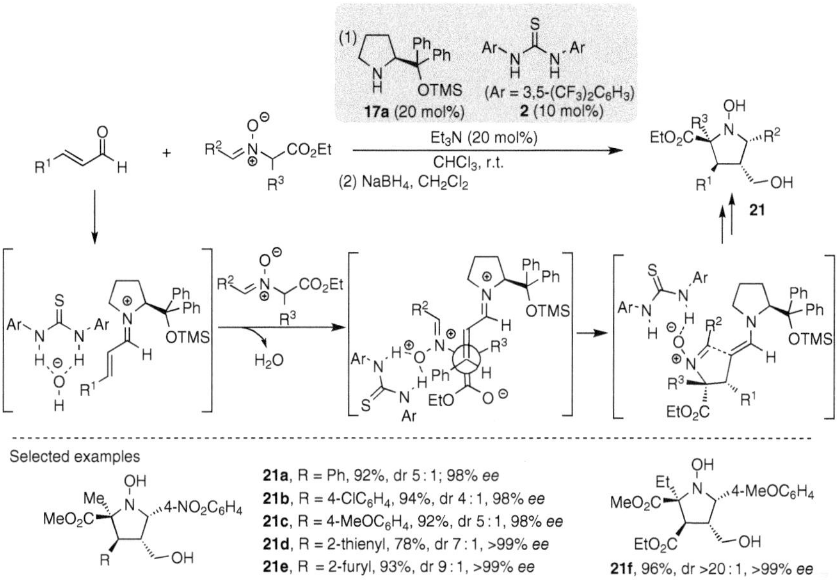

Scheme 9.13 Use of thiourea **2** to direct the regioselectivity in the [3+2] cycloaddition between nitrones and enals catalyzed by **17a**. Source: Based on Prieto *et al.* [34].

intermediate in solution. This H-bonded hydroxide anion subsequently participates as a Brønsted base in the formation of the nitrone ylide reagent involved in the key formal cycloaddition process with the activated iminium ion dipolarophile. The latter reaction was also demonstrated to proceed in a stepwise manner (rather than in a concerted manner). Therefore, it involves an initial Michael reaction between the nitrone ylide/thiourea complex and the α,β-unsaturated iminium ion, delivering an enamine intermediate that is followed by intramolecular Mannich reaction and final catalyst turnover.

This type of anion-binding interactions between an H-bond donor additive and the counterion of the iminion ion intermediate to enhance the proficiency of reactions under iminium activation have been explored with other systems [35], which also include more elaborated H-bond donor reagents. For example, the performance of MacMillan imidazolidinone **22**, which had been previously reported to be an outstanding catalyst for intermolecular Diels–Alder reactions [36], has been demonstrated to be significantly improved for the intramolecular version shown in Scheme 9.14 when bis-thiourea **23** is incorporated as a cocatalyst [37]. Remarkably, when the reaction was carried out in the presence of the combination of **22** and **23**, an outstanding rate enhancement was observed with respect to the reaction catalyzed by MacMillan imidazolidinone itself. This shows that this anion-binding effect to increase the population of the key reactive iminium ion intermediate can provide an efficient solution to those cases in which very slow reactions or high catalyst loadings are required. The authors also demonstrated that Schreiner thiourea **2** could be successfully used in the same reaction, also providing a similar rate enhancement effect.

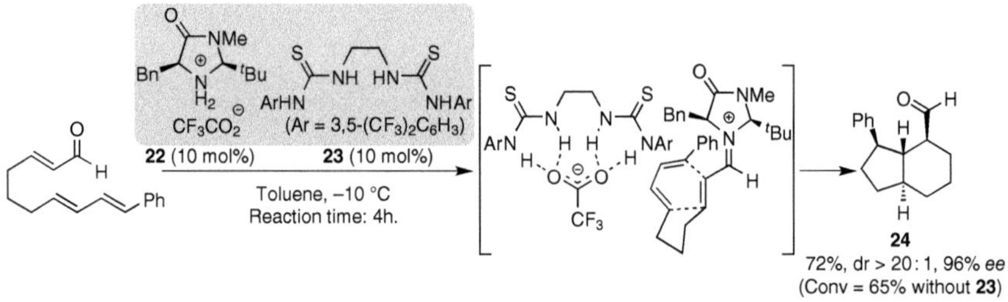

Scheme 9.14 Rate enhancement observed in the **22**-catalyzed intramolecular Diels–Alder reaction in the presence of thiourea **23**. Source: Based on Wang *et al.* [37].

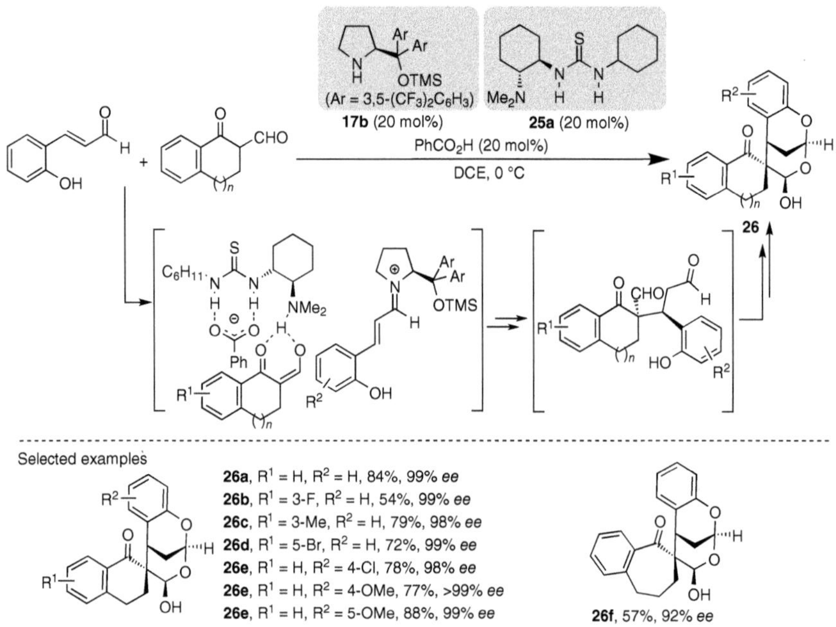

Scheme 9.15 Combined use of chiral thiourea **25a** and diarylprolinol **17b** as catalysts in the Michael reaction between α-formyl-substituted cyclic ketones and *o*-hydroxycinnamaldehydes. Source: Based on Wang *et al.* [38].

In another related example, the reaction between *o*-hydroxycinnamaldehydes and α-formyl substituted cyclic ketones has been found to proceed with improved efficiency using the *matched* combination between diarylprolinol-derived catalyst **17b** and tertiary amine/thiourea cocatalyst **25a** in the presence of an equimolar amount of a Brønsted acid additive (Scheme 9.15) [38]. This transformation consists on the initial Michael reaction between the active methylene pronucleophile and the activated α,β-unsaturated iminium ion, in which the Brønsted acid additive facilitates this initial condensation step between the aminocatalyst and the enal and remains bonded to the thiourea cocatalyst **25a** through anion-bonding interactions. This enables the communication between the two chiral catalysts in the key Michael reaction, being able to improve significantly the enantioselectivity with respect to

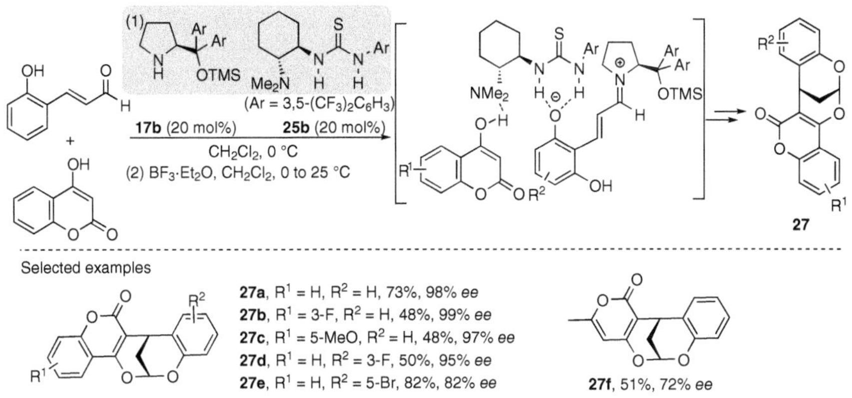

Scheme 9.16 Combined use of chiral thiourea **6b** and diarylprolinol **17b** as catalysts in the Michael reaction between 1,3-dicarbonyl compounds and *o*-hydroxycinnamaldehydes. Source: Based on Zhang *et al.* [39].

the reaction in which aminocatalyst **17b** is used individually. Once the Michael adduct is obtained after the catalyst has been released, the final product is obtained as a stable tricyclic product formed after two intramolecular hemiacetalization processes initiated by the addition of the phenol moiety to the remaining formyl substituent.

Interestingly, a different behavior has been observed when a related reaction involving *o*-hydroxycinnamaldehydes and related active methylene compounds such as cyclic 1,3-diketones as pronucleophiles is carried out using the same type of catalyst combination. In this case, the thiourea moiety is proposed to engage in anion-binding interactions with the phenoxide anion formed upon generation of the α,β-unsaturated iminium ion. This is then serving as an additional anchoring point for the formation of a reaction intermediate, in which all components of the reaction, the aminocatalyst, the Michael acceptor, the Michael donor, and the bifunctional tertiary amine/thiourea cocatalyst, are assembled together, providing maximum stereocontrol (Scheme 9.16) [39].

9.1.4 Vinylogous Iminium Ion Activation

The principle of vinylogy can also be combined with the iminium ion activation approach, enabling the δ-functionalization of α,β,γ,δ-diunsaturated aldehydes or ketones with an appropriate nucleophile. This approach has been significantly less developed than the corresponding vinylogous versions of the enamine activation manifold, and most examples involve the use of catalysts that exert stereocontrol exclusively based on steric effects [40]. Nevertheless, there is one illustrative example in the literature that shows the proficiency of bifunctional amine/H-bond donor catalyst **28** to promote the vinylogous Michael reaction of oxazolones with β-alkenyl-substituted cyclohexenones (Scheme 9.17) [41]. The reaction is fully regioselective toward the 1,6-addition product and provides excellent diastereo

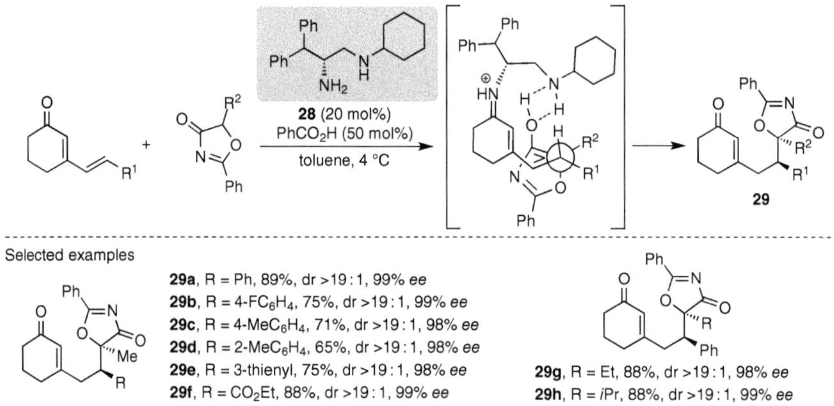

Scheme 9.17 Enantioselective 1,6-addition of oxazolones to α,β,δ,γ-unsaturated dienones through vinylogous iminium ion activation. Source: Based on Wei *et al.* [41].

and enantioselectivities for a wide variety of substrates tested. The reaction is also amenable to be applied to the use of oxindoles as Michael donors with a very similar performance.

9.2 Thiourea – Pyridine-Based Nucleophilic Dual Catalysis

Nucleophilic catalysis – also referred to as covalent catalysis – relies on the use of a Lewis base (nucleophile) that reacts with an electrophilic center to form a Lewis adduct as a key covalent intermediate [42]. Thus, nucleophiles can catalyze reactions by increasing the electrophilicity of the species they bind to and lowering the activation energy for a specific reaction. In particular, pyridine derivatives such as 4-(*N*,*N*-dimethylamino)pyridine (DMAP) have been widely used as nucleophilic catalysts, resulting in the identification of powerful synthetic methodologies [43–45]. In this section, the strategies based on the combination of pyridine-based catalysis and anion-binding catalysis will be discussed.

9.2.1 Kinetic Resolution and Desymmetrization of Amines

The kinetic resolution of racemic amines represent an important approach toward synthetically valuable chiral amines and is commonly realized enzymatically [46] or with the help of a chiral acylating reagent and a DMAP-type derivative to generate a chiral ion pair (**IP1**) as an active acylating agent (Scheme 9.18a) [44]. However, the later strategy is generally limited to the use of stoichiometric amounts of the chiral agent, while there is a scarce number of available chiral acylating reagents [45]. Moreover, the kinetic resolution of amines, especially primary, is rather challenging since their nucleophilicity competes with the reactivity of the nucleophilic promoters. To overcome these limitations, in 2009, a new dual catalysis concept by

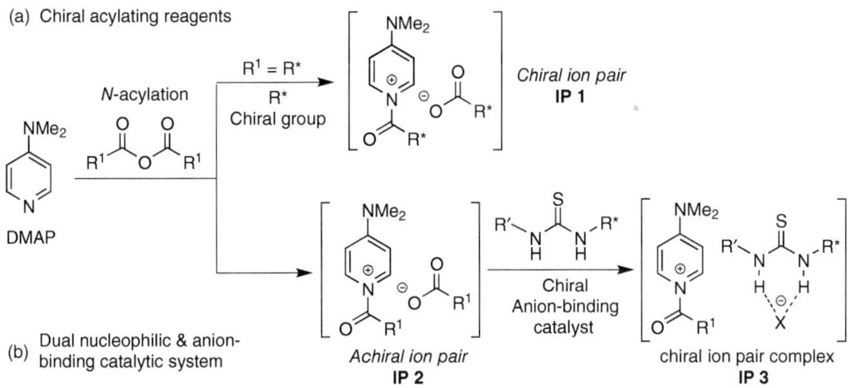

Scheme 9.18 Approaches for the generation of DMAP-type chiral ion pair acylating agents. Source: Adapted from Fu [44].

merging nucleophilic catalysis and chiral hydrogen bonding anion-binding catalysis was introduced by the group of Seidel [47]. The new methodology consists of an achiral ion pair (**IP2**), formed *in situ* by acylation of the achiral, readily available DMAP, which further interacts with a thiourea-containing chiral hydrogen-bonding catalyst to give the chiral ion pair (**IP3**) (Scheme 9.18b). The generated chiral ion pair **IP3** shows higher reactivity as an acylating agent than the achiral **IP2** and the acylating reagent itself, facilitating an asymmetric induction in the next step. This is possible as the binding of the chiral catalyst to the counter-anion in the ion pair lowers the interaction between the anion and the acyl pyridinium cation and makes the species more electrophilic and, thereby, a more reactive acylating agent.

Consequently, Seidel and coworkers enabled a successful kinetic resolution of simple racemic primary benzylic amines **30** by *N*-acylation to **31** with a high selective factor (*s*-factor) (Scheme 9.19) [47]. The reaction was performed with benzoic anhydride in the presence of 20 mol% DMAP and the chiral bis-thiourea catalyst **32** (20 mol%) in toluene at −78 °C, leading selectivity factors from 7 to 24 [48]. The scope of the reaction was broad, tolerating substitution in all positions of the aryl group **31b-d**. However, electron-poor benzylic amines **31e** were resolved with higher selectivity (up to 20 *s*-factor), while extending the aromatic phenyl moiety to naphthyl gave an enhanced selectivity ($s = 24$, **31f**).

Interestingly, the reaction does not work either in the absence of both DMAP and chiral thiourea catalyst **32** nor in the presence of only DMAP. In contrast, the addition of the thiourea catalyst led to a 40% conversion, but only with a low selectivity (*s*-factor 1.4). Considering the mechanism of the reaction, an activation of the anhydride with the thiourea catalyst via hydrogen bonding to form **A** can be envisioned [49]. However, it was proposed that in the presence of DMAP, this species **A** is switched to the chiral close ion pair intermediate **B** between the *N*-acylated DMAP and the thiourea-benzoate binding complex, which explains the upcoming higher selectivity (Scheme 9.20).

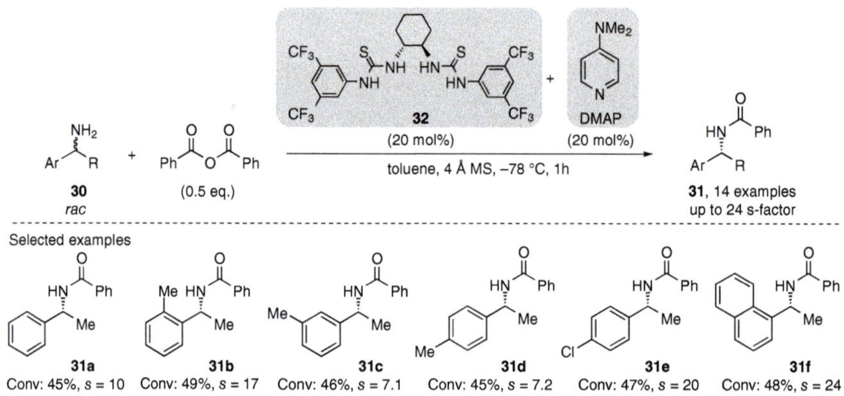

Scheme 9.19 Kinetic resolution of benzylic amines with the DMAP/chiral bis-thiourea catalytic system. Source: Based on De *et al.* [47].

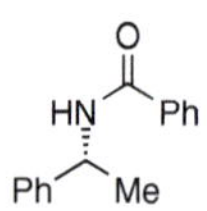

Catalyst	DMAP	Conv.	s-factor
✗	✗	<1	-
✗	✓	<2	-
✓	✗	40%	1.4
✓	✓	45%	10.0

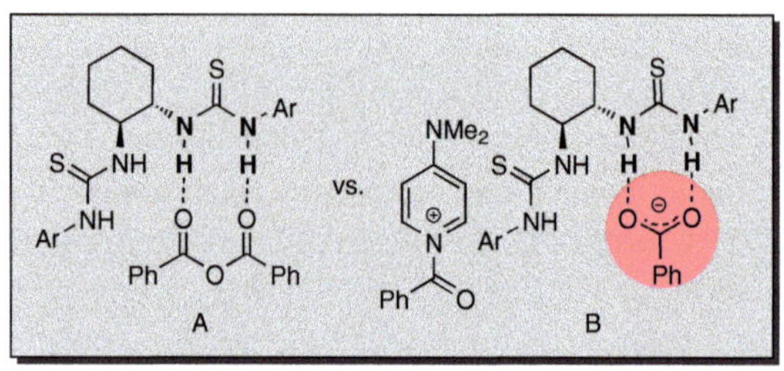

Scheme 9.20 Catalysts' effect and proposed activation modes.

Later on, the group of Seidel was able to extend this strategy for the kinetic resolution of other primary amines such as propargylic amines **33** (Scheme 9.21) [50]. After an intensive evaluation of different reaction parameters, including the modification of the thiourea–catalyst framework toward more potent catalysts, a significant

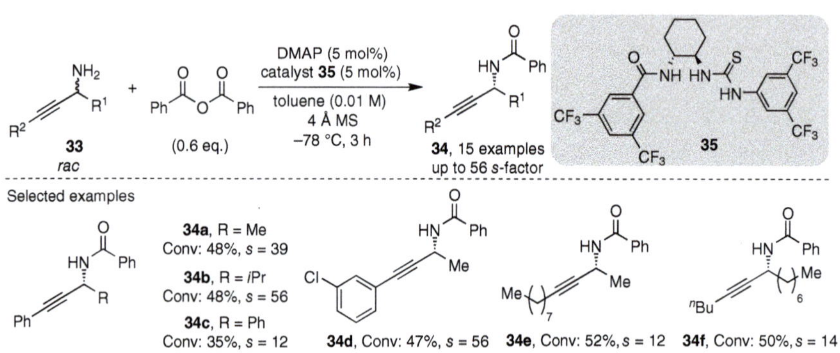

Scheme 9.21 Kinetic resolution of proparylamine derivatives. Source: Based on Klauber *et al.* [50].

improvement of the catalytic system was achieved. Hereby, the replacement of one thiourea subunit by a 3,5-bis(CF_3)benzamide moiety allowed the reduction of the catalyst loading of both catalysts DMAP and **35** from 20 to 5 mol%, while the selectivity was increased to a *s*-factor of up to 56. A variety of substituted arylpropargyl alkyl amines **33** were resolved successfully with high selectivity, while an aryl substitution (R = Ph, **34c**) and propargylic amines with two aliphatic residues led to moderate *s*-factors (s = 12–15, **34i-j**).

This catalytic system (DMAP/**35**) presenting a high activity was next employed for the kinetic resolution of the more nucleophilic and challenging allylic amines [51]. Hence, the first nonenzymatic kinetic resolution of primary allylic amines **36** was successfully achieved; however, only an *s*-factor of 12 was observed with 1-methyl-3-phenyl allylic amine as a model substrate. Therefore, a further optimization of the reaction conditions was conducted. Different achiral nucleophilic cocatalysts were explored, from which 4-pyrrolidinopyridine (PPY) provided the best results (Scheme 9.22). This cocatalyst led to slightly improved *s*-factors up to 20, especially with substrates bearing bulkier alpha substituents (R^3) such as isopropyl **37b** or *n*-butyl **37c**. Furthermore, the catalysts' loading could be decreased to 2 mol%, while maintaining the same levels of reactivity and selectivity. Additionally, allylic amines with extended π-systems and trisubstituted olefins are resolved with a satisfactory selectivity. However, a low *s*-factor of 3.5 was monitored with the related non-conjugated cyclohexenyl ethylamine **37j**.

In 2015, the groups of Seidel and Schreiner jointly analyzed the prototypical dual bis-thiourea **32** and amide–thiourea **35**/DMAP systems in order to gain insights into the mechanism of the kinetic resolution process via benzoylation [52]. Catalyst aggregation through multiple intra- and intermolecular H-bonding interactions within the thiourea unit was observed. This influences the anion-binding affinity and was argued to be at some extent responsible for obtaining improved selectivities upon reducing the catalyst loadings for the model reaction of racemic

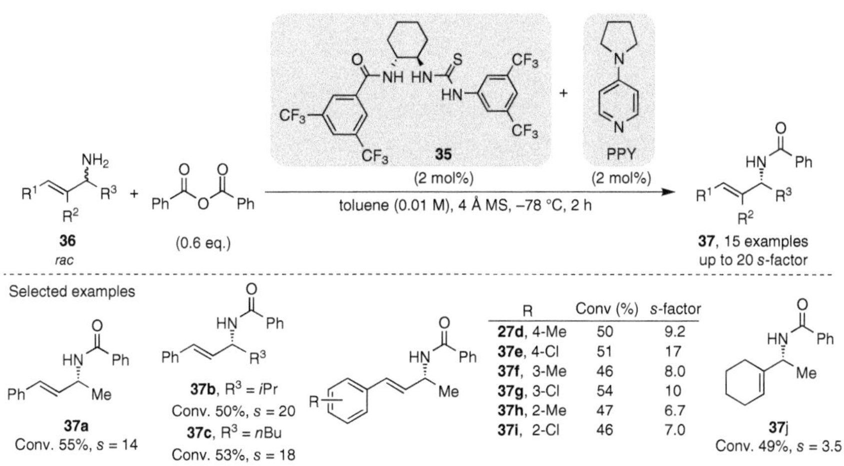

Scheme 9.22 Scope of the allylic amine resolution in the presence of benzoic anhydride.

1-phenylethanamine (e.g. catalyst **35** from 10 to 5 mol%; *s*-factors 8.3 and 13, respectively). Moreover, computational studies at the M06/6-31G(d,p) level of theory including solvent effects revealed additional secondary interactions within the chiral contact ion pair (Scheme 9.23), responsible for the efficient stereo-discrimination, and thus rationalizing the observed preferential benzylation of the *R*- over the *S*-enantiomer.

Based on the previous observations, the resolution of benzylic amines **30** was reevaluated (Scheme 9.24) [53]. Subsequently, different achiral nucleophilic cocatalysts were tested with the amide–thiourea catalyst **35**. Moreover, the effect of the cocatalyst solubility was investigated, arguing that a highly soluble nucleophilic catalyst should result in a more soluble ion pair and hence led to an improvement in the overall process. Hence, changing the methyl groups of DMAP by the bulker isopropyl **38a** or longer linear alkyl chains **38b-e** improved the selectivity, obtaining the best result with the *n*-propyl-substitution (4-(*N*,*N*-dipropylamino)pyridine, DPAP, **38b**) (from a *s*-factor of 13–27). Although in previous works with allylic amines PPY was superior than DMAP [51], in this case, no further improvement was observed with cyclic amines (either PPY nor 4-piperidinopyridine (PDPY)). Furthermore, the tested more nucleophilic tricyclic pyridine-type **38f-g** [54] cocatalysts, as well as *N*-methylimidazole (**38h**), 1,8-diazabicyclo[5.4.0] undec-7-ene (DBN, **38i**), or cyclic isothioureas **38j** [55], gave insufficient selectivity. Therefore, the best catalytic system comprises the thiourea–catalyst **35** and the DPAP using a catalyst loading of 5 mol%. A short scope evaluation was explored under the new developed systems and a dramatic improvement on the kinetic resolution of benzylic amines **30** was mastered in respect to the former results, achieving an *s*-factor up to 67.

The methodology could also be extended to the kinetic resolution of racemic 1,2-diarylethane-1,2-diamines **39** (Scheme 9.25) [56]. Once again, the DPAP/H-donor **35** catalytic system proved more efficient than when using DMAP, being able to increase the *s*-factor from 5.5 to 30. A variety of 1,2-diamines **39** were resolved, in which the selectivity was highly dependent on the substitution on aryl groups (*s*-factors ranking from 5 to 30). For practical reasons, a one-pot acylation of the remaining free amine with 2,2,2-trichloroethyl chloroformate (TrocCl) was performed, facilitating the separation of the enantiomeric enriched diamines **40**.

Following a similar approach, the desymmetrization of *meso*-diamines **39** was also reported by Seidel and coworkers (Scheme 9.26) [57]. The advantage of this process over the previous kinetic resolution relies on the possibility of obtaining quantitatively the desired highly enantioenriched monoacylated product. However, a base such a triethylamine is now required to reach full conversions. Hence, the H-donor **35** in combination with DMAP was successfully used for the desymmetrization of *meso*-1,2-diarylethane-1,2-diamines **39** with benzoic anhydride. High levels of enantioinduction were achieved (**12**, up to 95% *ee*), in particular for the *para*-halogen-substituted derivatives.

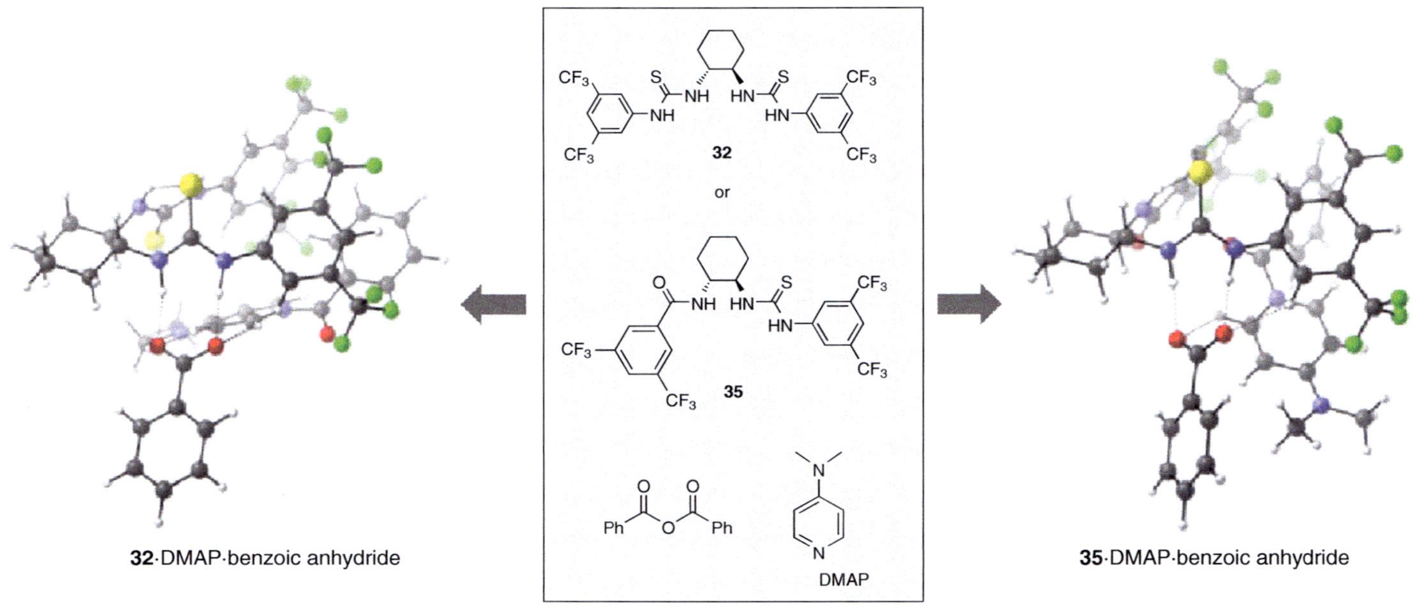

Scheme 9.23 Chiral ion pairs for acyl-transfer computed at M06/6-31G(d,p) level by Schreiner and Seidel. Source: Mittal *et al.* [52].

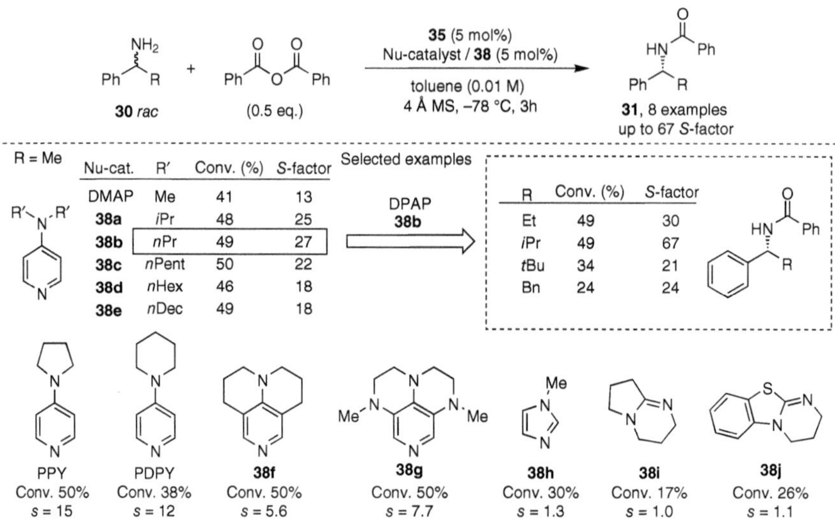

Scheme 9.24 Re-evaluation of the kinetic resolution of benzylic amines **30** with catalyst **35**. Source: Based on Mittal *et al.* [53].

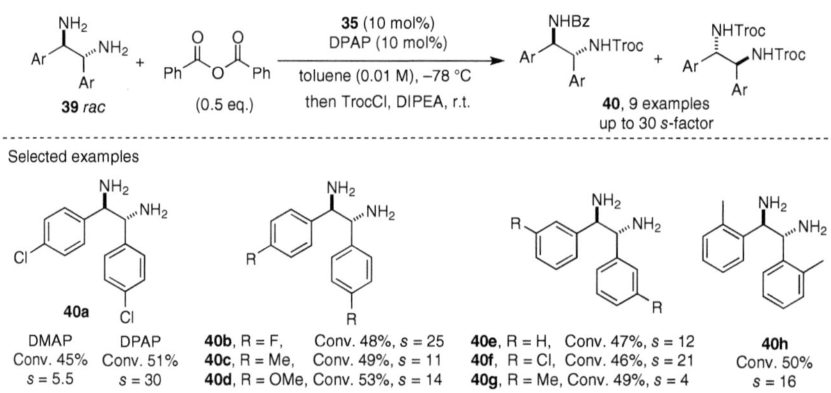

Scheme 9.25 Kinetic resolution of racemic 1,2-diamines **39**. Source: Based on Min *et al.* [56].

9.2.2 Asymmetric Steglich Rearrangement

The Steglich rearrangement of azlactones **42** [58] is a benchmark reaction used to test new nucleophilic organocatalyst, leading to valuable amino acid precursors [59]. Consequently, a cooperative catalysis approach embracing the combination of a chiral anion-binding catalyst and an achiral nucleophilic pyridine-type cocatalyst has also been explored toward the synthesis of chiral oxazolones **43** (Scheme 9.27) [60].

In 2011, Seidel and coworkers explored a large variety of chiral thioureas in combination with DMAP for the Steglich reaction of alkyl–aryl-disubstituted azlactones **42** into oxazolones **43** (Scheme 9.28) [60]. The best results were obtained with 20 mol% of DMAP-Jacobsen's thiourea **44** as a catalytic system in toluene

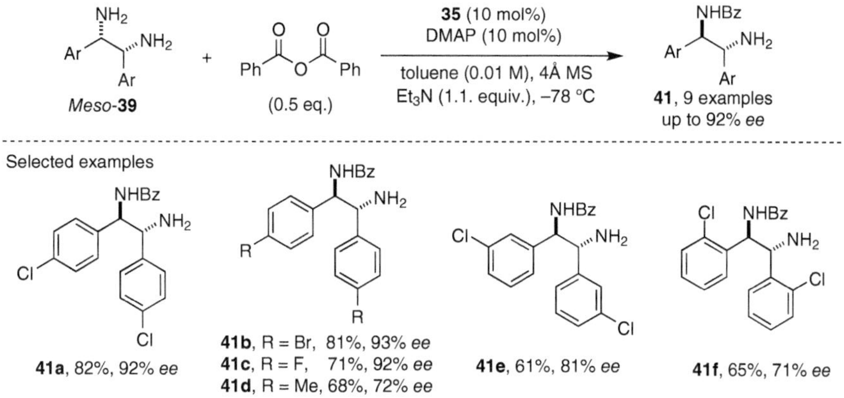

Scheme 9.26 Desymmetrization of *meso*-1,2-diamines. Source: Based on De *et al.* [57].

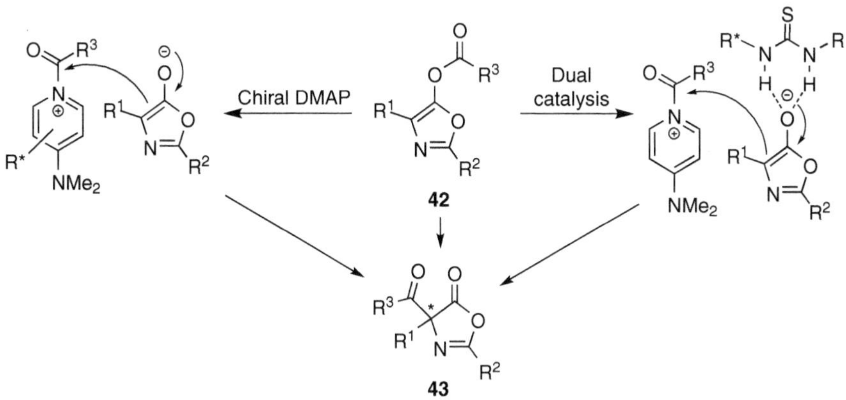

Scheme 9.27 Chiral nucleophile vs. cocatalytic approach for the asymmetric azlactone rearrangement. Source: Based on De *et al.* [60].

at −78 °C. Interestingly, the DMAP mono-catalyzed reaction proceeded sluggish under the standard conditions, showing the importance of the dual catalytic approach. Moreover, a notable effect of the aryl substitution of the azlactone on the enantioselectivity was observed. Thus, the electron-rich 3,5-dimethoxyphenyl group provided the best selectivities up to 91% *ee*.

Instead of using two catalysts in a cooperative manner as described above, the group of Suga reported in 2016 the efficient use of chiral DMAP-binapthyl-based bifunctional catalysts (Scheme 9.29) [61]. Thus, a large number of chiral DMAP derivatives that incorporate a binapthyl unit at C4 position of a pyridine ring were designed. With this strategy, several advantages were covered in one catalyst to fulfill a successful asymmetric induction. Firstly, the chiral 1,1′-bi-2-napthol (BINOL) is commercially available and inexpensive and serves as an enantioselective moiety of the catalyst. In addition, it is easy to modify, and therefore, different functional groups (FG) on the BINOL core are readily accessible [62, 63]. Two possible asymmetric acylated reaction types with these DMAP-binapthyl-based bifunctional

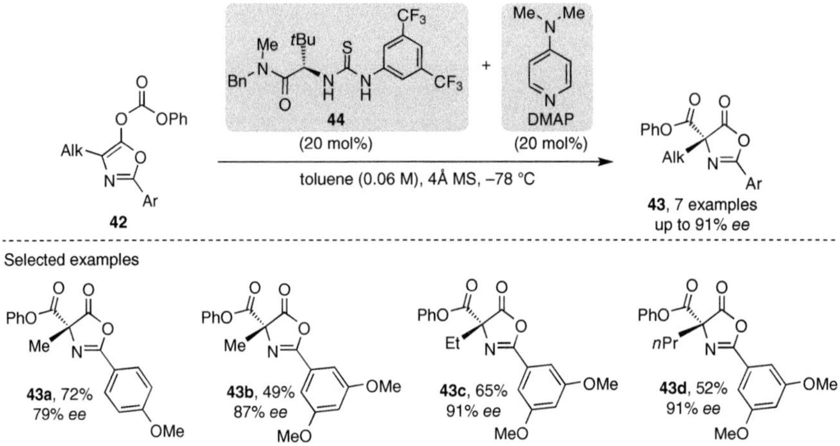

Scheme 9.28 DMAP-chiral thiourea cocatalyzed Steglich reaction. Source: Based on De *et al.* [60].

catalyst system can be envisioned: (i) a Steglich-type reaction involving attractive interactions between the functional groups of the catalyst structure and the corresponding counter-anion of the *N*-acylpyridinium salt (type 1, Scheme 9.29, left) or (ii) acylation reactions by interaction with a suitable nucleophile such as an alcohol (type 2, Scheme 9.29, right).

Various binapthyl-based chiral nucleophilic catalysts were applied in a Steglich rearrangement of oxindole derivatives **45** as model substrates (Scheme 9.30). It was revealed that the C_2-symmetric scaffold with tertiary alcohols unit is extremely essential for high enantioselective induction. Under the optimized conditions (THF, −20 °C), the desired products **46** were successfully afforded in quantitative yields and up to 98% *ee* using only 0.5 mol% of **47** as a catalyst. Therefore, different 3-alkyl- and allyl-substituted oxindoles were well tolerated, providing high enantioselectivities (**46a-c**, up to 98% *ee*). However, a 3-(hetero)aryl or a *N*-methyl instead of *N*-carbamoyl substitution led to a decreased on the *ee* (**46d-e**).

DFT calculations were carried out toward gaining a deeper insight into the influence of the tertiary hydroxyl groups at the chiral BINOL C3 and C3' positions. Thus, the energy lowest transition state (**TS-I**) showed an attractive hydrogen-bonding interaction between the substrate (enolate oxygen of the ion pair) and the *tert*-alcohol units of the catalyst. Furthermore, an additional electrostatic attraction between the substrate and an ortho hydrogen of the aryl units at the C3 site of the chiral catalyst takes place, further stabilizing the **TS** and all together enabling the observed high asymmetric transformation.

The versatile ability of this novel DMAP-binaptyl-based bifunctional catalyst **47** was further demonstrated in the kinetic resolution of different alcohols in the presence of an anhydride as an acylation agent (Scheme 9.31). *Rac*-benzylic carbinols **48** and *rac*-1,2-diols **50** were successfully resolved, the latter with an excellent selectivity up to 125 (**51**). Moreover, an efficiently desymmetrization of *meso*-1,2-diols such

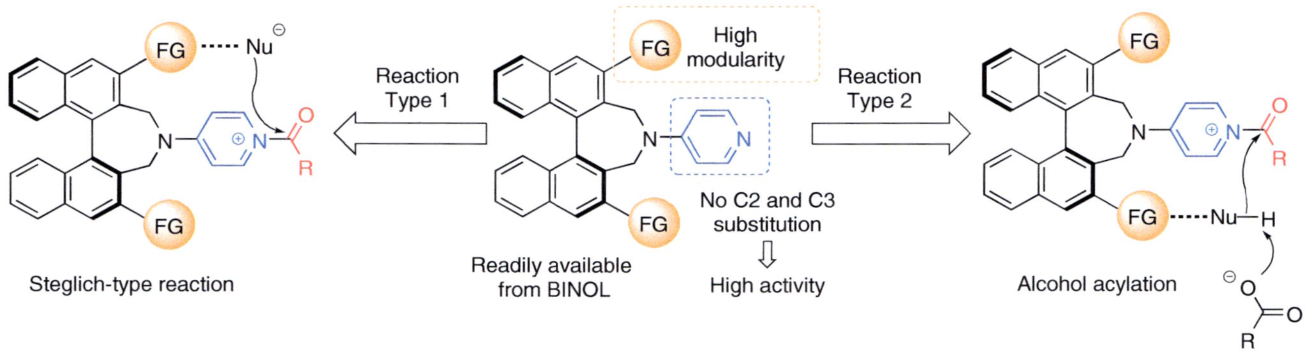

Scheme 9.29 General structure and reactivity of DMAP-binaptyl-based bifunctional catalysts. Source: Based on Mandai *et al.* [61].

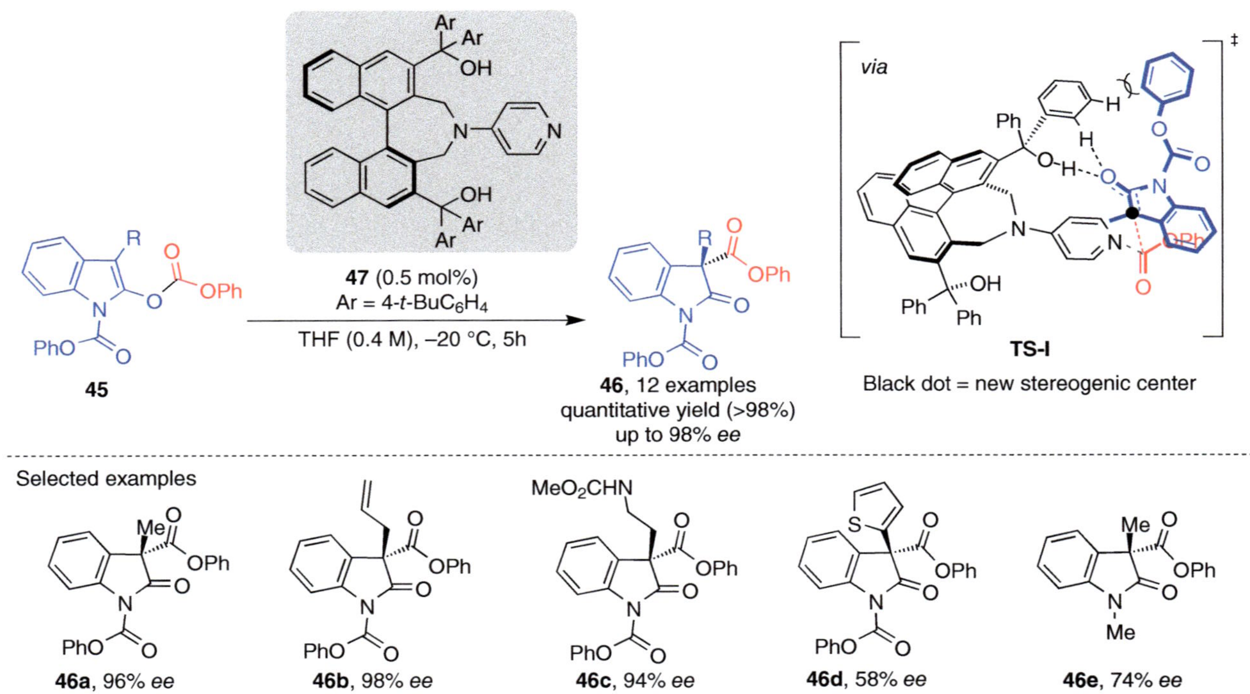

Scheme 9.30 Acylation by chiral DMAP-diol bifunctional catalysis.

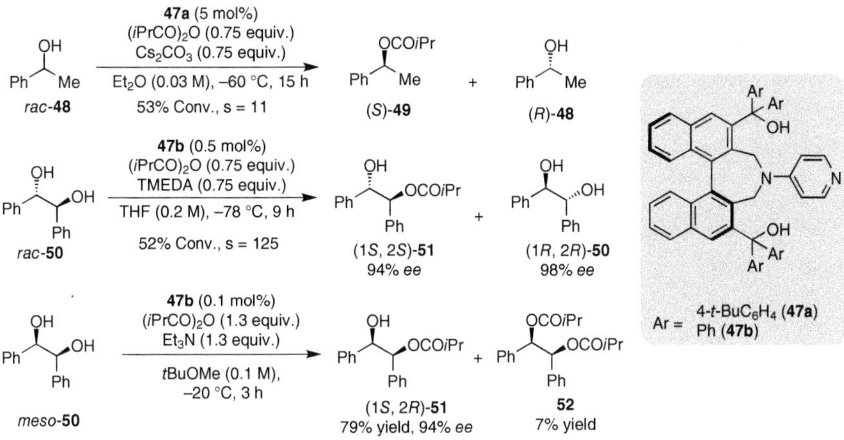

Scheme 9.31 Intermolecular acylation of different alcohols with catalyst **47**.

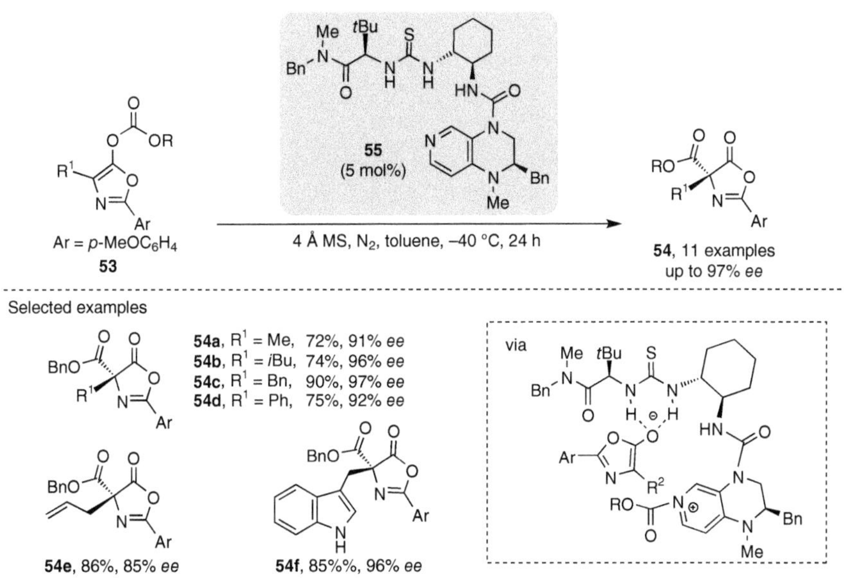

Scheme 9.32 Chiral DMAP-thiourea bifunctional catalyst for the Steglich rearrangement. Source: Based on Li *et al.* [64].

as *meso*-**50** furnished the corresponding desired mono-ester **51** in high yield (74%) and enantioselectivity (94% *ee*).

A similar approach was reported by Lin and coworkers in 2019, which developed the first chiral DMAP-thiourea-based bifunctional catalyst **55** for the asymmetric Steglich rearrangement of oxazolyl carbonates **53** into enatioenriched *C*-carboxyazlactones **54** (Scheme 9.32) [64].

Various oxazolyl–benzyl carbonates possessing alkyl, allyl, and (hetero)aryl functionalities were well tolerated, given the products in moderate to good yields and high enantioselectivities (up to 97% *ee*). The authors proposed the formation of a key chiral ion pair between the *N*-acyl-pyridinium-thiourea-based catalyst and the resulting enolate, which interacts with the thiourea moiety via a hydrogen bond. Finally, after the irreversible C-carboxylation of the enolate, the desired *C*-carboxyazlactone **54** is formed.

9.2.3 Other Acylation Reactions

Inspired by the work of Seidel on chiral thiourea-DMAP cocatalyzed acylations [47, 50, 51], in 2011, Jacobsen and coworkers reported a successful enantioselective acylation of silyl ketene acetals **56** toward valuable chiral α,α-disubstituted butyrolactones **58** (Scheme 9.33a) [65]. They made used of acyl fluorides **57** as acylating agents in the presence of a chiral arylpyrrolidine-based thiourea catalysts **30** and PPY as a cocatalyst. Different meta or para substitution at the benzoyl fluorides were well-tolerated **58f-j**, allowing the acylation of several *p*-substituted-aryl silyl ketene acetals **57a-d** into the corresponding butyrolactones **58** in good yields and high enantioselectivities (up to 95% yield and *ee*). Moreover, the preparative scale reaction was performed with only a 0.5 mol% catalyst loading, not requiring a chromatographic purification and leading to the essentially enantiopure lactone after crystallization.

Interestingly, the reaction did not proceed in the absence of the thiourea catalyst **59**. Therefore, it was assumed that the anion-binding catalyst is not only crucial for the enantioselective transformation but also plays an important role in the rate-determining step. In this regard, an activation of the benzoyl fluoride (**57**) was proposed to take place by the thiourea catalyst (**I**), facilitating the *N*-acylation of PPY and generation of a thiourea-bound acylpyridinium-fluoride ion pair **II** (Scheme 9.33b). This ion pair gets additional stabilization by the interaction between the thiourea–arene fragment and the cationic acylating unit. Subsequently, the rate-determining desilylation of the silyl ketene acetal proceeds to release TMSF and the free enolate, which interacts directly with the thiourea moiety (**III**). The final chiral product was obtained in the enantio-determining acylation step, with a concomitant release of both catalysts to close the catalytic cycle.

9.2.4 Anion-Binding-Assisted Polymerization Reactions

As shown in Sections 9.2.1–9.2.3, this type of anion-binding cocatalysis can be efficiently used in several acylation and Steglich reactions; however, this approach has been only recently explored in catalytic polymerization. In this regard, Tao and Wang developed a ring-opening polymerization (ROP) [66] of *O*-carboxyanhydrides **59** with a bifunctional catalyst **60** bearing a thiourea anion-binding unit and an anchored 4-(dimethylamino)pyridine group (Scheme 9.34) [67]. While other catalytic approaches for the ROP of carboxyanhydrides still suffer from low chain propagation or stereocontrol selectivity, this method allows the synthesis of highly isotactic polyesters **61** with high molecular weight (MW) up to 150.0 kDa.

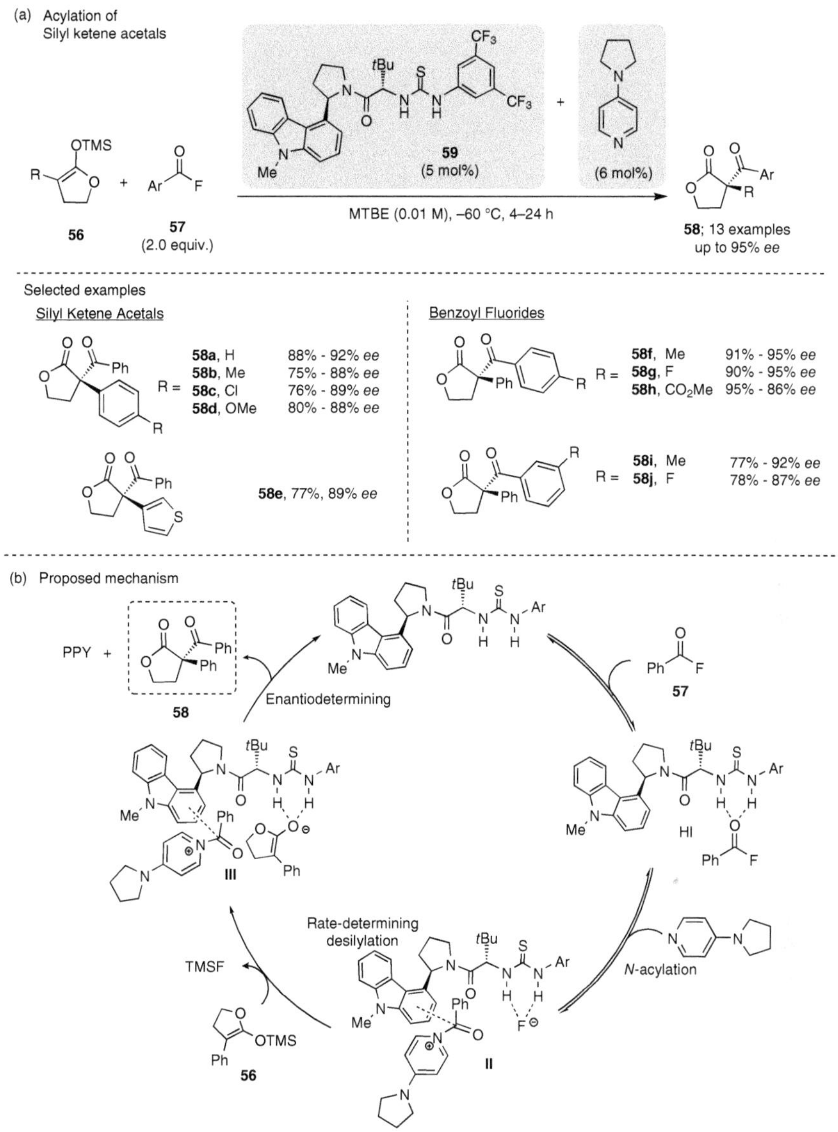

Scheme 9.33 PPY-chiral thiourea cocatalyzed acylation of cyclic silylketene acetals. Source: Based on Birrell *et al.* [65].

The anion binding of the thiourea unit to the carboxylate of the growing end of the polymer chain shows several positive effects such as an enhanced of the moisture tolerance and stability of the active polymerization species, as well as selective elongation process. Hence, after the initial nucleophilic opening of the cyclic carbonate by the DMAP unit, the thiourea facilitates the propagation step through a dynamic anion–carbonate binding and decarboxylation that ensures a high selectivity for the rind opening over possible side reactions. Moreover, DFT calculations

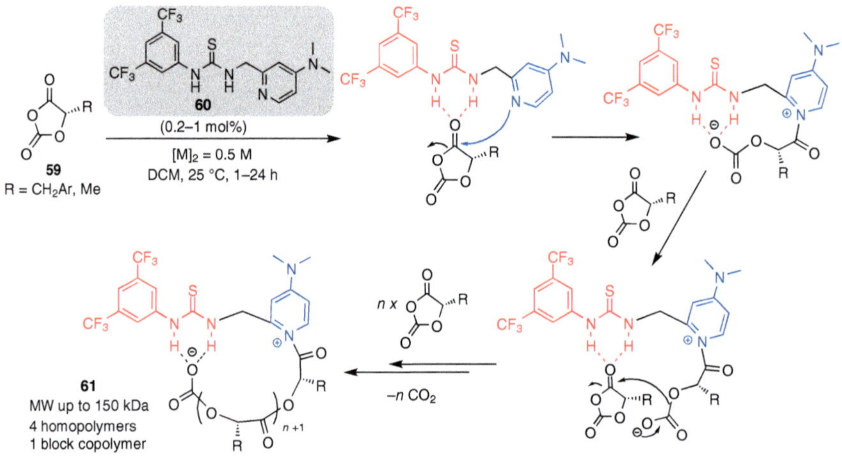

Scheme 9.34 DMAP-thiourea cocatalyzed ring-opening polymerization. Source: Based on Li *et al.* [67].

were conducted, reinforcing the experimental observations on the key role of the dynamic anion-binding interaction of the thiourea moiety, to open new possibilities in catalytic polymerization processes.

9.3 Conclusions

In this chapter, two different types of bifunctional or assisted co-organocatalytic approaches have been presented. On the one hand, the combination of anion binding with several activation modes within amino covalent catalysis such as enamine, dienamine, and imininum were discussed. Hence, it was shown that the anion-binding effects introduced through H-bonding of the reactants with the catalyst (mostly bearing thiourea units) can not only lead to an improvement on the enantioinduction in respect to the sole use of aminocatalysis but can lead to in some cases a different and structurally more challenging product outcome. On the other hand, the anion-binding–nucleophilic cocatalysis approach employing achiral pyridine derivatives to activate acylating agents for (dynamic) kinetic resolutions and desymmetrizations of racemic amines and meso-diamines in the presence of chiral HB-donor catalyst was next covered. This concept has been extended to other acylating reactions, such as the enantioselective acylation of silyl ketene acetals with acyl fluorides or the asymmetric Steglich rearrangement toward the formation of chiral azlactones. In the latter case, apart from the related cocatalytic systems, also bifunctional catalysts bearing both the pyridine and HB-donor groups proved highly effective.

Given that a large number of organic transformations proceed via ionic intermediates, the potential to modulate the reactivity and enhance enantiodifferentiation in a broader variety of reactions catalyzed by covalent amino or nucleophilic species could be envisioned. Moreover, it can also be anticipated that this type of HB-donor/covalent cocatalysis will provide new reactivities and synthetic approaches in the near future.

References

1 (a) List, B. and Maruoka, K. (2012). *Science of Synthesis, Asymmetric Organocatalysis*. Stuttgart, Germany: *Georg Thieme Verlag*. (b) Pellissier, H. (2011). *Recent Developments in Asymmetric Organocatalysis*. Cambridge: RSC Publishing. (c) Jacobsen, E.N. and MacMillan, D.W.C. (2010). Organocatalysis. *Proc. Nat. Acad. Sci. USA* 107: 20618–20619. (d) Bertelsen, S. and Jørgensen, K.A. (2009). Organocatalysis–after the gold rush. *Chem. Soc. Rev.* 38: 2178–2189. (e) MacMillan, D.W.C. (2008). The advent and development of organocatalysis. *Nature* 455: 304–308. (f) Holland, M.C. and Gilmour, R. (2015). Deconstructing covalent organocatalysis. *Angew. Chem. Int. Ed.* 54: 3862–3871.

2 (a) Hajos, Z. G. and Parrish, D. R. (1971). Asymmetric synthesis of optically active polycyclic organic compounds. German Patent DE 2,102,623, filled 20 January 1971. (b) Hajos, Z.G. and Parrish, D.R. (1974). Asymmetric synthesis of bicyclic intermediates of natural product chemistry. *J. Org. Chem.* 39: 1615–1621. (c) Eder, U., Sauer, G. and Wiechert, R. (1971). Optically active1,5-indanone and 1,6-naphthalenedione. German Patent DE 2,014,757. (d) Eder, U., Sauer, G., and Wiechert, R. (1971). New type of asymmetric cyclization to optically active steroid CD partial structures. *Angew. Chem. Int. Ed. Engl.* 10: 496–497.

3 Jen, W.S., Wiener, J.J.M., and MacMillan, D.W.C. (2000). New strategies for organic catalysis: the first enantioselective organocatalytic 1,3-dipolar cycloaddition. *J. Am. Chem. Soc.* 122: 9874–9875.

4 List, B., Lerner, R.A., and Barbas III, C.F. (2000). Proline-catalyzed direct asymmetric Aldol reactions. *J. Am. Chem. Soc.* 122: 2395–2396.

5 (a) Mukherjee, S., Yang, J.W., Hoffmann, S., and List, B. (2007). Asymmetric enamine catalysis. *Chem. Rev.* 107: 5471–5569. (b) Sulzer-Mossé, S. and Alexakis, A. (2007). Chiral amines as organocatalysts for asymmetric conjugate addition to nitroolefins and vinyl sulfones via enamine activation. *Chem. Commun.*: 3123–3135. (c) Kano, T. and Maruoka, K. (2008). Design of chiral bifunctional secondary amine catalysts for asymmetric enamine catalysis. *Chem. Commun.*: 5465–5473.

6 (a) Melchiorre, P. (2012). Cinchona-based primary amine catalysis in the asymmetric functionalization of carbonyl compounds. *Angew. Chem. Int. Ed.* 51: 9748–9770. (b) Erkkilae, A., Majander, I., and Pihko, P.M. (2007). Iminium catalysis. *Chem. Rev.* 107: 5416–5470. (c) Lelais, G. and MacMillan, D.W.C. (2006). Modern strategies in organic catalysis: the advent and development of iminium activation. *Aldrichim. Acta.* 39: 79–87.

7 (a) Nayak, S., Panda, P., Bhakta, S. *et al.* (2016). Current advances of organocatalytic Michael–Michael cascade reaction in the synthesis of highly functionalized cyclic molecules. *RSC Adv.* 6: 96154–96175. (b) Grondal, C., Jeanty, M., and

Enders, D. (2010). Organocatalytic cascade reactions as a new tool in total synthesis. *Nat. Chem.* 2: 167–178.

8 (a) Alemán, J., Parra, A., Jiang, H., and Jørgensen, K.A. (2011). Squaramides: bridging from molecular recognition to bifunctional organocatalysis. *Chem. Eur. J.* 17: 6890–6899. (b) Matos Paz, B., Jiang, H., and Jørgensen, K.A. (2015). Aminocatalysis: beyond steric shielding and hydrogen-bonding. *Chem. Eur. J.* 21: 1846–1853. (c) Doyle, A.G. and Jacobsen, E.N. (2007). Small-molecule H-bond donors in asymmetric catalysis. *Chem. Rev.* 107: 5713–5743. (d) Chauhan, P., Mahajan, S., Kaya, U. *et al.* (2015). Bifunctional amine-squaramides: powerful hydrogen-bonding organocatalysts for asymmetric domino/cascade reactions. *Adv. Synth. Catal.* 357: 253–281.

9 (a) Hong, L., Sun, W., Yang, D. *et al.* (2016). Additive effects on asymmetric catalysis. *Chem. Rev.* 116: 4006–4123. (b) Anebouselvy, K., Shruthi, K.S., and Ramachary, D.B. (2017). Asymmetric supramolecular organocatalysis: a complementary upgrade to organocatalysis. *Eur. J. Org. Chem.*: 5460–5483. (c) Meeuwissen, J. and Reek, J.N.H. (2010). Supramolecular catalysis beyond enzyme mimics. *Nat. Chem.* 2: 615–621.

10 (a) Allemann, C., Gordillo, R., Clemente, F.R. *et al.* (2004). Theory of asymmetric organocatalysis of Aldol and related reactions: rationalizations and predictions. *Acc. Chem. Res.* 37: 558–569. (b) Hoang, L., Bahmanyar, S., Houk, K.N., and List, B. (2003). Kinetic and stereochemical evidence for the involvement of only one proline molecule in the transition states of proline-catalyzed intra- and intermolecular Aldol reactions. *J. Am. Chem. Soc.* 125: 16–17. (c) List, B., Hoang, L., and Martin, H.J. (2004). New mechanistic studies on the proline-catalyzed aldol reaction. *Proc. Nat. Acad. Sci. USA* 101: 5839–5842. (d) Orlandi, M., Ceotto, M., and Benaglia, M. (2016). Kinetics versus thermodynamics in the proline catalyzed aldol reaction. *Chem. Sci.* 7: 5421–5427. (e) Sharma, A.K. and Sunoj, R.B. (2010). Enamine versus oxazolidinone: what controls stereoselectivity in proline-catalyzed asymmetric aldol reactions? *Angew. Chem. Int. Ed.* 49: 6373–6373. (f) Schmid, M.B., Zeitler, K., and Gschwind, R.M. (2010). The elusive enamine intermediate in proline-catalyzed aldol reactions: NMR detection, formation pathway, and stabilization trends. *Angew. Chem. Int. Ed.* 49: 4997–5003. (g) Clemente, F.R. and Houk, K.N. (2004). Computational evidence for the enamine mechanism of intramolecular Aldol reactions catalyzed by proline. *Angew. Chem. Int. Ed.* 43: 5766–5768. (h) Bahmanyar, S., Houk, K.N., Martin, H.J., and List, B. (2003). Quantum mechanical predictions of the stereoselectivities of proline-catalyzed asymmetric intermolecular Aldol reactions. *J. Am. Chem. Soc.* 125: 2475–2479. (i) Haindl, M.H., Hioe, J., and Gschwind, R.M. (2015). The proline enamine formation pathway revisited in dimethyl sulfoxide: rate constants determined via NMR. *J. Am. Chem. Soc.* 137: 12835–12842. (j) Renzi, P., Hioe, J., and Gschwind, R.M. (2017). Enamine/dienamine and brønsted acid catalysis: elusive intermediates, reaction mechanisms, and stereoinduction modes based on in situ nmr spectroscopy and computational studies. *Acc. Chem. Res.* 50: 2936–2948.

11 For some reviews or accounts: (a) Notz, W., Tanaka, F., and Barbas, C.F. III, (2004). Enamine-based organocatalysis with proline and diamines: the development of direct catalytic asymmetric Aldol, Mannich, Michael, and Diels–Alder reactions. *Acc. Chem. Res.* 37: 580–591. (b) List, B. (2004). Enamine catalysis is a powerful strategy for the catalytic generation and use of carbanion equivalents. *Acc. Chem. Res.* 37: 548–557. (c) Jensen, K.L., Dickmeiss, G., Jiang, H. *et al.* (2012). The diarylprolinol silyl ether system: a general organocatalyst. *Acc. Chem. Res.* 45: 248–264. (d) Duthaler, R.O. (2003). Proline-catalyzed asymmetric α-amination of aldehydes and ketones – an astonishingly simple access to optically active α-hydrazino carbonyl compounds. *Angew. Chem. Int. Ed.* 42: 975–978. (e) Merino, P. and Tejero, T. (2004). Organocatalyzed asymmetric α-aminoxylation of aldehydes and ketones – an efficient access to enantiomerically pure α-hydroxycarbonyl compounds, diols, and even amino alcohols. *Angew. Chem. Int. Ed.* 43: 2995–2997.

12 Reis, Ö., Eymur, S., Reis, B., and Demir, A.S. (2009). Direct enantioselective aldol reactions catalyzed by a proline–thiourea host–guest complex. *Chem. Commun.* 1088-1090.

13 (a) El-Hamdouni, N., Companyó, X., Rios, R., and Moyano, A. (2010). Substrate-dependent nonlinear effects in proline–thiourea-catalyzed Aldol reactions: unraveling the role of the thiourea co-catalyst. *Chem. Eur. J.* 16: 1142–1148. (b) Companyó, X., Valero, G., Crovetto, L. *et al.* (2009). Highly enantio- and diastereoselective organocatalytic desymmetrization of prochiral cyclohexanones by simple direct aldol reaction catalyzed by proline. *Chem. Eur. J.* 15: 6564–6568.

14 Demir, A.S. and Basceken, S. (2013). Study of asymmetric aldol and Mannich reactions catalyzed by proline–thiourea host–guest complexes in nonpolar solvents. *Tetrahedron: Asymmetry* 24: 515–525.

15 Poe, S.L., Bogdan, A.R., Mason, B.P. *et al.* (2009). Use of bifunctional ureas to increase the rate of proline-catalyzed α-aminoxylations. *J. Org. Chem.* 74: 1574–1580.

16 (a) Martínez-Castañeda, Á., Poladura, B., Rodríguez-Solla, H. *et al.* (2011). Direct Aldol reactions catalyzed by a heterogeneous guanidinium salt/proline system under solvent-free conditions. *Org. Lett.* 13: 3032–3035. (b) Zhou, Y. and Shan, Z. (2006). Chiral diols: new class of additives for direct Aldol reaction catalyzed by L-proline. *J. Org. Chem.* 71: 9510–9512.

17 (a) Mandal, T. and Zhao, C.-G. (2008). Modularly designed organocatalytic assemblies for direct nitro-Michael addition reactions. *Angew. Chem. Int. Ed.* 47: 7714–7717. See also: (b) Huang, H., Konda, S., and Zhao, J.C.-G. (2016). *Angew. Chem. Int. Ed.* 55: 2213–2216. (c) Rana, N.K., Huang, H., and Zhao, J.C.-G. (2014). Highly diastereodivergent synthesis of tetrasubstituted cyclohexanes catalyzed by modularly designed organocatalysts. *Angew. Chem. Int. Ed.* 53: 7619–7623. (d) Bhaskararao, B. and Sunoj, R.B. (2018). Two chiral catalysts in action: insights into cooperativity and stereoselectivity in proline and cinchona-thiourea dual organocatalysis. *Chem. Sci.* 9: 8738–8747.

18 (a) Brown, A.R., Kuo, W.-H., and Jacobsen, E.N. (2010). Enantioselective catalytic α-alkylation of aldehydes via an S_N1 pathway. *J. Am. Chem. Soc.* 132: 9286–9288.

For a related example involving indolylmethanols, see: (b) Guo, Z.-L., Xue, J.-H., Fu, L.-N. *et al.* (2014). The direct asymmetric alkylation of α-amino aldehydes with 3-indolylmethanols by enamine catalysis. *Org. Lett.* 16: 6472–6475.

19 List, B., Coric, I., Grygorenko, O.O. *et al.* (2014). The catalytic asymmetric α-benzylation of aldehydes. *Angew. Chem. Int. Ed.* 53: 282–285.

20 (a) Fuson, R.C. (1935). The principle of vinylogy. *Chem. Rev.* 16: 1–27. (b) Curti, C., Battistini, L., Sartori, A., and Zanardi, F. (2020). New developments of the principle of vinylogy as applied to π-extended enolate-type donor systems. *Chem. Rev.* 120: 2448–2612.

21 (a) Arceo, E. and Melchiorre, P. (2012). Extending the aminocatalytic HOMO-raising activation strategy: where is the limit? *Angew. Chem. Int. Ed.* 51: 5290–5292. (b) Jurberg, I.D., Chatterjee, I., Tannert, R., and Melchiorre, P. (2013). When asymmetric aminocatalysis meets the vinylogy principle. *Chem. Commun.* 49: 4869–4883. (c) Jiang, H., Albrecht, Ł., and Jørgensen, K.A. (2013). Aminocatalytic remote functionalization strategies. *Chem. Sci.* 4: 2287–2300.

22 Bertelsen, S., Marigo, M., Brandes, S. *et al.* (2006). Dienamine catalysis: organocatalytic asymmetric γ-amination of α,β-unsaturated aldehydes. *J. Am. Chem. Soc.* 128: 12973–12980.

23 (a) Ramachary, D.B. and Reddy, Y.V. (2012). Dienamine catalysis: an emerging technology in organic synthesis. *Eur. J. Org. Chem.*: 865–887. (b) Marcos, V. and Alemán, J. (2016). Old tricks, new dogs: organocatalytic dienamine activation of α,β-unsaturated aldehydes. *Chem. Soc. Rev.* 45: 6812–6832.

24 Kumar, I., Ramaraju, P., and Mir, N.A. (2013). Asymmetric trienamine catalysis: new opportunities in amine catalysis. *Org. Biomol. Chem.* 11: 709–716.

25 (a) Gotoh, H. and Hayashi, Y. (2013). Diarylprolinol silyl ethers: development and application as organocatalysts. In: *Sustainable Catalysis* (ed. P.J. Dunn), 287–316. Hoboken, NJ: John Wiley & Sons. (b) Reyes-Rodríguez, G.J., Rezayee, N.M., Vidal-Albalat, A., and Jørgensen, K.A. (2019). Prevalence of diarylprolinol silyl ethers as catalysts in total synthesis and patents. *Chem. Rev.* 119: 4221–4260. (c) Meninno, S., Volpe, C., and Lattanzi, A. (2019). Diaryl prolinols in stereoselective catalysis and synthesis: an update. *ChemCatChem* 11: 3716–3729. (d) Palomo, C. and Mielgo, A. (2006). Diarylprolinol ethers: expanding the potential of enamine/iminium-ion catalysis. *Angew. Chem. Int. Ed.* 45: 7876–7880. (e) Vega-Peñaloza, A., Paria, S., Bonchio, M. *et al.* (2019). Profiling the privileges of pyrrolidine-based catalysts in asymmetric synthesis: from polar to light-driven radical chemistry. *ACS Catal.* 9: 6058–6072.

26 Albrecht, Ł., Dickmeiss, G., Cruz Acosta, F. *et al.* (2012). Asymmetric organocatalytic formal [2+2]-cycloadditions via bifunctional H-bond directing dienamine catalysis. *J. Am. Chem. Soc.* 134: 2543–2546.

27 Orue, A., Uria, U., Reyes, E. *et al.* (2015). Catalytic enantioselective [5+2] cycloaddition between oxidopyrylium ylides and enals under dienamine activation. *Angew. Chem. Int. Ed.* 54: 3043–3046.

28 Roca Lopez, D., Uria, U., Reyes, E. *et al.* (2016). Mechanistic insights into the mode of action of bifunctional pyrrolidine-squaramide-derived organocatalysts. *Chem. Eur. J.* 22: 884–889.

29 Lalonde, M.P., McGowan, M.A., Rajapaksa, N.S., and Jacobsen, E.N. (2013). Enantioselective formal Aza-Diels–Alder reactions of enones with cyclic imines catalyzed by primary aminothioureas. *J. Am. Chem. Soc.* 135: 1891–1894.

30 Galzerano, P., Bencivenni, G., Pesciaioli, F. *et al.* (2009). Asymmetric iminium ion catalysis with a novel bifunctional primary amine thiourea: controlling adjacent quaternary and tertiary stereocenters. *Chem. Eur. J.* 15: 7846–7849.

31 Orue, A., Uria, U., Roca-Lopez, D. *et al.* (2017). Racemic hemiacetals as oxygen-centered pronucleophiles triggering cascade 1,4-addition/Michael reaction through dynamic kinetic resolution under iminium catalysis. *Development and mechanistic insights. Chem. Sci.* 8: 2904–2913.

32 Gu, Y., Wang, Y., Yu, T.-Y. *et al.* (2014). Rationally designed multifunctional supramolecular iminium catalysis: direct vinylogous Michael addition of unmodified linear dienol substrates. *Angew. Chem. Int. Ed.* 53: 14128–14131.

33 (a) Jia, Z.-L., Wang, Y., Zhao, C.-G. *et al.* (2017). Highly enantioselective construction of Hajos–Wiechert ketone skeletons via an organocatalytic vinylogous Michael/Stetter relay sequence. *Org. Lett.* 19: 2130–2133. (b) Jia, Z.-L., Wang, Y., Xu, G.-Q., and Xu, P.-F. (2017). Kinetic resolution via supramolecular iminium catalysis: multiactivation enables the asymmetric synthesis of β-aryl substituted aldehydes and densely functionalized cyclohexanes. *Chem. Commun.* 53: 4938–4941.

34 Prieto, L., Juste-Navarro, V., Uria, U. *et al.* (2017). Regioselectivity change in the organocatalytic enantioselective (3+2) cycloaddition with nitrones through cooperative hydrogen-bonding catalysis/iminium activation. *Chem. Eur. J.* 23: 2764–2768.

35 (a) Juste-Navarro, V., Prieto, L., Delso, I. *et al.* (2017). A case study of thiourea-assisted iminium formation by hydroxyl anion binding: kinetic, spectroscopic and computational evidences. *Adv. Synth. Catal.* 359: 4122–4128. For other examples on the combined use of Schreiner thiourea in combination with diarylprolinols in iminium catalysis, see: (b) Zhao, C.-G., Feng, Z.-T., Xu, G.-Q. *et al.* (2020). Highly enantioselective construction of strained spiro[2,3]hexanes through a Michael addition/ring expansion/cyclization cascade. *Angew. Chem. Int. Ed.* 59: 3058–3062.

36 Ahrendt, K.A., Borths, C.J., and MacMillan, D.W.C. (2000). New strategies for organic catalysis: the first highly enantioselective organocatalytic Diels–Alder reaction. *J. Am. Chem. Soc.* 122: 4243–4244.

37 Wang, Y., Yu, T.-Y., Zhang, H.-B. *et al.* (2012). Hydrogen-bond-mediated supramolecular iminium ion catalysis. *Angew. Chem. Int. Ed.* 51: 12339–12342.

38 Wang, C., Chen, Y.-H., Wu, H.-C. *et al.* (2019). The quinary catalyst–substrate complex induced construction of spiro-bridged or cagelike polyheterocyclic compounds via a substrate-controlled cascade process. *Org. Lett.* 21: 6750–6755.

39 Zhang, X.-Q., Lv, X.-J., Pei, J.-P. *et al.* (2020). An asymmetric multicatalytic reaction sequence of 2-hydroxycinnamaldehydes and enolic 1,3-dicarbonyl compounds to construct bridged bicyclic acetals. *Org. Chem. Front.* 7: 292–297.

40 Pioneering report: (a) Tian, X., Liu, Y., and Melchiorre, P. (2012). Aminocatalytic enantioselective 1,6 additions of alkyl thiols to cyclic dienones: vinylogous iminium ion activation. *Angew. Chem. Int. Ed.* 51: 6439–6442. For some reviews: (b) Lear, M.J. and Hayashi, Y. (2013). Remote 1,6-stereocontrol by iminium-mediated organocatalytic events. *ChemCatChem* 5: 3499–3501. (c) Chauhan, P., Kaya, U., and Enders, D. (2017). Advances in organocatalytic 1,6-addition reactions: enantioselective construction of remote stereogenic centers. *Adv. Synth. Catal.* 359: 888–912.

41 Wei, Y., Liu, Z., Wu, X. *et al.* (2015). Remote construction of chiral vicinal tertiary and quaternary centers by catalytic asymmetric 1,6-conjugate addition of prochiral carbon nucleophiles to cyclic dienones. *Chem. Eur. J.* 21: 18921–18924.

42 Punekar, N.S. (2018). Nucleophilic catalysis and covalent reaction intermediates. In: *ENZYMES: Catalysis, Kinetics and Mechanisms*. Singapore: Springer.

43 Spivey, A.C. and Arseniyadis, S. (2004). Nucleophilic catalysis by 4-(dialkylamino)pyridines revisited – the search for optimal reactivity and selectivity. *Angew. Chem. Int. Ed.* 43: 5436–5441.

44 (a) Fu, G.C. (2000). Enantioselective nucleophilic catalysis with "planar-chiral" heterocycles. *Acc. Chem. Res.* 33: 412–420. (b) Fu, G.C. (2004). Asymmetric catalysis with "planar-chiral" derivatives of 4-(dimethylamino)pyridine. *Acc. Chem. Res.* 37: 542–547.

45 Wurz, R.P. (2007). Chiral dialkylaminopyridine catalysts in asymmetric synthesis. *Chem. Rev.* 107: 5570–5595.

46 (a) Lee, J.H., Han, K., Kim, M.-J., and Park, J. (2010). Chemoenzymatic dynamic kinetic resolution of alcohols and amines. *Eur. J. Org. Chem.* 999-1015. (b) Turner, N.J. (2009). Directed evolution drives the next generation of biocatalysts. *Nat. Chem. Biol.* 5: 567–573. (c) Höhne, M. and Bornscheuer, U.T. (2009). Biocatalytic routes to optically active amines. *ChemCatChem* 1: 42–51.

47 De, C.K., Klauber, E.G., and Seidel, D. (2009). Merging nucleophilic and hydrogen bonding catalysis: an anion binding approach to the kinetic resolution of amines. *J. Am. Chem. Soc.* 131: 17060–17061.

48 Kagan, H.B. and Fiaud, J.C. (1988). *Kinetic Resolution In Topics in Stereochemistry*, vol. 18 (eds. E.L. Eliel and S.H. Wilen), 249–330. New York, Chichester: John Wiley & Sons Selectivity factor s is defined as s = [rate of fast-reacting enantiomer]/[rate of slow-reacting enantiomer].

49 (a) Oh, S.H., Rho, H.S., Lee, J.W. *et al.* (2008). A highly reactive and enantioselective bifunctional organocatalyst for the methanolytic desymmetrization of cyclic anhydrides: prevention of catalyst aggregation. *Angew. Chem. Int. Ed.* 47: 7872–7875. (b) Peschiulli, A., Gun'k, Y., and Connon, S.J. (2008). Highly enantioselective desymmetrization of meso anhydrides by a bifunctional thiourea-based organocatalyst at low catalyst loadings and room temperature. *J. Org. Chem.* 73: 2454–2457.

50 Klauber, E.G., De, C.K., Shah, T.K., and Seidel, D. (2010). Merging nucleophilic and hydrogen bonding catalysis: an anion binding approach to the kinetic resolution of propargylic amines. *J. Am. Chem. Soc.* 132: 13624–13626.

51 Klauber, E.G., Mittal, N., Shah, T.K., and Seidel, D. (2011). A dual-catalysis/anion-binding approach to the kinetic resolution of allylic amines. *Org. Lett.* 13: 2464–2467.

52 Mittal, N., Lippert, K.M., De, C.K. *et al.* (2015). A dual-catalysis anion-binding approach to the kinetic resolution of amines: insights into the mechanism via a combined experimental and computational study. *J. Am. Chem. Soc.* 137: 5748–5758.

53 Mittal, N., Sun, D.X., and Seidel, D. (2012). Kinetic resolution of amines via dual catalysis: remarkable dependence of selectivity on the achiral cocatalyst. *Org. Lett.* 14: 3084–3087.

54 (a) Singh, S., Das, G., Singh, O.V., and Han, H. (2007). Development of more potent 4-dimethylaminopyridine analogues. *Org. Lett.* 9: 401–404. (b) Rycke, N.D., Berionni, G., Couty, F. *et al.* (2011). Synthesis and reactivity of highly nucleophilic pyridines. *Org. Lett.* 13: 530–533.

55 (a) Birman, V.B., Li, X., and Han, Z. (2007). Nonaromatic amidine derivatives as acylation catalysts. *Org. Lett.* 9: 37–40. (b) Maji, B., Joannesse, C., Nigst, T.A. *et al.* (2011). Nucleophilicities and Lewis basicities of isothiourea derivatives. *J. Org. Chem.* 76: 5104–5112. (c) Kobayashi, M. and Okamoto, S. (2006). Unexpected reactivity of annulated 3*H*-benzothiazol-2-ylideneamines as an acyl transfer catalyst. *Tetrahedron Lett.* 47: 4347–4350.

56 Min, C., Mittal, N., De, C.K., and Seidel, D. (2012). A dual-catalysis approach to the kinetic resolution of 1,2-diaryl-1,2-diaminoethanes. *Chem. Commun.* 48: 10853–10855.

57 De, C.K. and Seidel, D. (2011). Catalytic enantioselective desymmetrization of meso-diamines: a dual small-molecule catalysis approach. *J. Am. Chem. Soc.* 133: 14538–14541.

58 Steglich, W. and Höfle, G. (1969). *N,N*-Dimethyl-4-pyridinamine, a very effective acylation catalyst. *Angew. Chem. Int. Ed.* 8: 981–981.

59 (a) Fisk, J.S., Mosey, R.A., and Tepe, J.J. (2007). The diverse chemistry of oxazol-5-(4*H*)-ones. *Chem. Soc. Rev.* 36: 1432–1440. (b) Mosey, R.A., Fisk, J.S., and Tepe, J.J. (2008). Stereoselective syntheses of quaternary substituted α-amino acids using oxazol-5-(4*H*)-ones. *Tetrahedron: Asymmetry* 19: 2755–2762. (c) Alba, A.-N.R. and Rios, R. (2011). Oxazolones in organocatalysis, new tricks for an old reagent. *Chem. Asian J.* 6: 720–734.

60 De, C.K., Mittal, N., and Seidel, D. (2011). A dual-catalysis approach to the asymmetric steglich rearrangement and catalytic enantioselective addition of O-acylated azlactones to isoquinolines. *J. Am. Chem. Soc.* 133: 16802–16805.

61 Mandai, H., Fujii, K., Yasuhara, H. *et al.* (2016). Enantioselective acyl transfer catalysis by a combination of common catalytic motifs and electrostatic interactions. *Nat. Commun.* 7: 11297.

62 Liang, D.-C., Luo, R.-S., Yin, L.-H. *et al.* (2012). A new synthetic route for axially chiral secondary amines with binaphthyl backbone and their applications in asymmetric Michael reaction of aldehydes to nitroalkenes. *Org. Biomol. Chem.* 10: 3071–3079.

63 Ooi, T., Kameda, M., and Maruoka, K. (2003). Design of N-spiro C_2-symmetric chiral quaternary ammonium bromides as novel chiral phase-transfer catalysts: synthesis and application to practical asymmetric synthesis of α-amino acids. *J. Am. Chem. Soc.* 125: 5139–5151.

64 Li, Q.-H., Zhang, G.-S., Wang, Y.-H. *et al.* (2019). A novel chiral DMAP–thiourea bifunctional catalyst catalyzed enantioselective Steglich and Black rearrangement reactions. *Org. Chem. Front.* 6: 2624–2629.

65 Birrell, J.A., Desrosiers, J.-N., and Jacobsen, E.N. (2011). Enantioselective acylation of silyl ketene acetals through fluoride anion-binding catalysis. *J. Am. Chem. Soc.* 133: 13872–13875.

66 (a) Dubois, P., Coulembier, O., and Raquez, J.M. (2009). *Handbook of Ring-Opening Polymerization.* Wiley-VCH: Weinheim. (b) Matyjaszewski, K. and Möller, M. (2012). *Ring-Opening Polymerization and Special Polymerization Processes in Polymer Science: A Comprehensive Reference*, vol. 4. Elsevier BV: Amsterdarm.

67 Li, M., Zhang, S., Zhang, X. *et al.* (2021). Unimolecular anion-binding catalysts for selective ring-opening polymerization of *O*-carboxyanhydrides. *Angew. Chem. Int. Ed.* 60: 6003–6012.

10

Anion-Binding Catalysis by Halogen, Chalcogen, Pnictogen, and Tetrel Bonding

Raffaella Papagna, Lukas Vogel, and Stefan M. Huber

Ruhr-Universität Bochum, Organic Chemistry I, Faculty for Chemistry and Biochemistry, Universitätsstraße 150, 44801 Bochum, Germany

This chapter describes applications of secondary interactions that are based on the elements of the seventh, sixth, fifth, and fourth main group of the periodic table: halogen bonding, chalcogen bonding, pnictogen bonding, and tetrel bonding. Herein, the term "secondary bonding" [1] will be used to summarize these forces. In the past decades, they have found wide applications in the field of crystal engineering, supramolecular chemistry, anion recognition, and, lately, organocatalysis. However, although these non-covalent interactions are closely related to the well-known and prevalent hydrogen bonding, their area of research is still at an early stage and continuously developing. Similar to the reviews published by Lim and Beer in 2018 [2], Huber and Sutar in 2019 [3], and Taylor in 2020 [4], this chapter will first introduce the nature as well as the history of secondary bonding and will then briefly discuss the design of appropriate receptors for anion binding (including differences to hydrogen-bonding-based systems), with an emphasis on more recent developments. The main focus, however, will be on the application of these interactions in organocatalysis.

10.1 History of Halogen Bonding

According to the definition by IUPAC published in 2013, halogen bonding (XB) is defined as the non-covalent interaction between an electrophilic halogenated compound RX (with R = electron-withdrawing group and X = Cl, Br, and I) and a Lewis base (LB) forming a halogen bond adduct [5]. In analogy with hydrogen bonds, the halogenated compound is called the XB donor and the LB is referred to as the XB acceptor. At first, this nomenclature may appear to be counterintuitive, as the LA usually accepts and the LB usually donates charge density.

Anion-Binding Catalysis, First Edition. Edited by Olga García Mancheño.

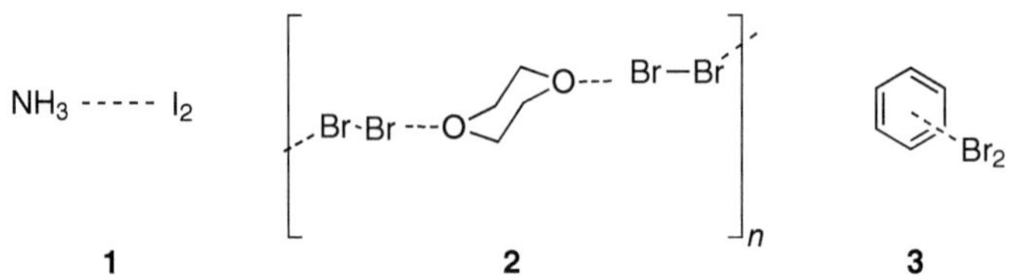

Figure 10.1 Halogen adducts described by Colin (**1**) and Hassel (**2**,**3**). Source: Based on Guthrie [7].

The first description of a halogen bond was presented by Colin in 1814, describing the interaction between elemental iodine and ammonia [6]. The right composition and stoichiometry of this structure, however, was established only 50 years later by Guthrie, who identified a 1 : 1 complexation (Figure 10.1, **1**) [7]. A theoretical explanation of how the halogen and the LB really interact with each other was first provided by Mulliken *et al.* in 1950, introducing the concept of charge transfer [8–11]. The first crystal structure, and thus the first visualization of the geometry of XB in the solid state, was published in 1954 by Hassel [12, 13]. He analyzed the interaction between elemental bromine and solvent molecules such as 1,4-dioxane (Figure 10.1, **2**) and benzene (Figure 10.1, **3**).

Simultaneously, Hildebrand and Benesi analyzed the interaction between elemental iodine and benzene in solution using UV–vis spectroscopy and determined the binding data of the XB interaction [14]. Unfortunately, the interaction between halogens and the LB did not generate much further interest and was virtually disregarded for over two decades. However, Sandorfy and Di Paolo investigated in 1974 the ability of various substituted fluorocarbons to break hydrogen bonds via IR spectroscopy [15]. This study showed both that halogen bonds are also formed between a LB and organohalogen compounds (**4**, Figure 10.2) and that the potency of breaking hydrogen bonds depends on the used halogen in the order F < Cl < Br < I. The term *halogen bond* itself was introduced by Dumas *et al.* in the mid-1970s, when they investigated the interaction between tetrahalomethanes (CX_4) and different bases [16]. Further examples of halogen bonding were published roughly 20 years later by Weiss *et al.* featuring crystal structures between iodide and iodomethylpiperidinium (**5**) [17] or *syn*-trinitroiodobenzene (**6**) [18]. The former likely constitutes the first case in which the adduct of a cationic halogen bond donor was structurally characterized. Subsequently, Legon *et al.* reported pioneering studies on gas-phase interactions between halogens and interhalogens with Lewis bases such as phosphine (**7**) [19–21]. In parallel, more detailed studies on halogen bonds in the solid state were performed by the groups of Desiraju [22, 23] and Allen *et al.* [24].

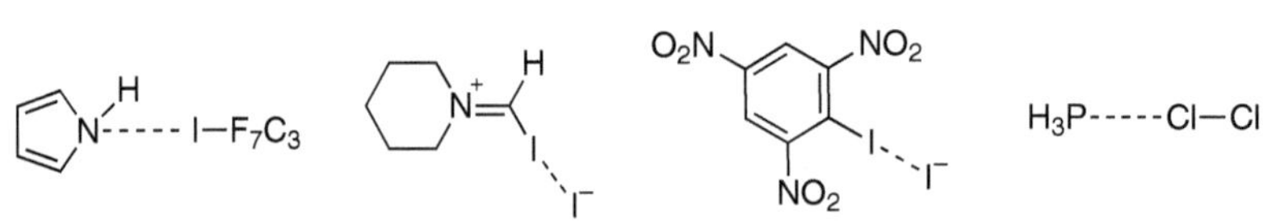

Figure 10.2 Various XB interactions published by Sandorfy (**4**), Weiss (**5**, **6**), and Legon (**7**).

In the following years, XB gained a lot of interest in crystal engineering, with seminal contributions by Metrangolo and Resnati [25–29]. Based on this preparatory work, the IUPAC finally established the definition of halogen bonds in 2013, as described above [5]. Apart from the general definition of halogen bonding, the high directionality of this interaction was also stressed, with the interaction angle between the halogen and the LB being typically close to 180° (Figure 10.3).

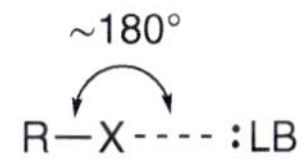

Figure 10.3 Schematic illustration of a halogen bond (also depicting the high directionality of the interaction).

Furthermore, this non-covalent interaction was described as follows:

> *The forces involved in the formation of the halogen bond are primarily electrostatic, but polarization, charge transfer, and dispersion contributions all play an important role. The relative roles of the different forces may vary from one case to the other* [5].

This description includes several contributions that certainly play an important role. Nonetheless, there is still a debate on the relative relevance of these individual terms. As previously mentioned, Mulliken first described halogen bonding as charge–transfer complexes [8–11] and his statement was later supported by numerous other researchers, for instance by Weiss and coworkers [30], Rosokha *et al.* [31, 32], Bickelhaupt and coworkers [33], and others [34–40]. From the molecular orbital point of view, the non-bonding (nb) orbital of the LB donates electron density into the empty antibonding orbital (σ^*) of the R—X bond (Figure 10.4). To enable a strong orbital overlap between the nb orbital and the σ^* orbital, the interaction angle has to be linear; otherwise, a repulsion between the p_y and p_z lone pair and the LB will occur. Hence, the closer the interaction angle is to 180°, the stronger the orbital overlap and thus the stronger the halogen bond will be.

According to the IUPAC definition, however, halogen bonds are primarily based on electrostatics. This concept is based on the so-called "σ-hole" established by Politzer *et al.*, who calculated the electrostatic potential surfaces of various methane substituted compounds [41, 42]. Figure 10.5 illustrates the calculated electrostatic potential surfaces of CF_4, CF_3Cl, CF_3Br, and CF_3I (from left to right). As indicated in the figure, a positive potential along the carbon halogen bond (indicated by the blue color directly in front of the halogen) is formed, which increases in the order $F \ll Cl < Br < I$ because of the increased polarizability of the respective halogen atom.

Furthermore, the magnitude of the σ-hole is also increased by electron-withdrawing substituents as used for this example. This anisotropic electron distribution is called

Figure 10.4 Molecular orbital interaction between the non-bonding (nb) orbital of the LB and the antibonding (σ^*) orbital of the R–X bond.

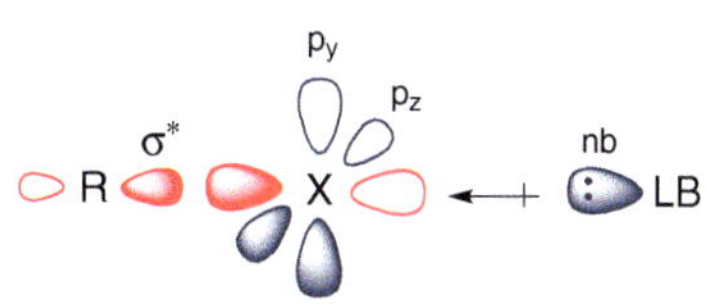

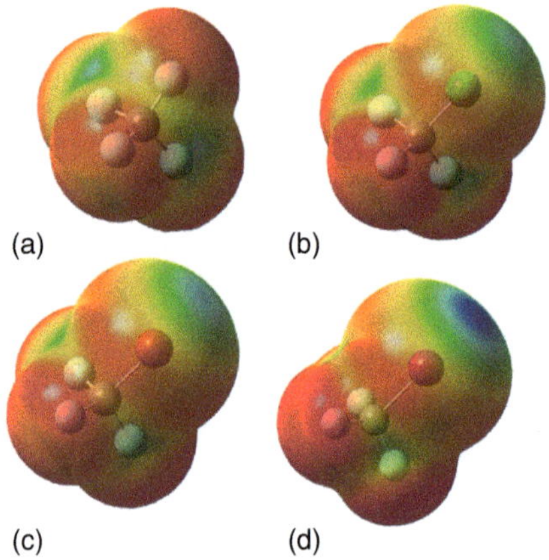

Figure 10.5 Calculated electrostatic potential surfaces of CF_4, CF_3Cl (a and b), CF_3Br, and CF_3I (c and d) at 0.001 electron/Bohr^3. Blue represents a positive and red a negative potential. Calculations were performed at the M06-2X/def2-TZVPP level. The figure is based on the work of Infante and coworkers in 2013 [43]. Source: Based on Huber *et al.* [43].

polar flattering, which can also be observed in X-ray diffraction [44]. It has to be mentioned that fluorine exhibits only in very few cases halogen bonding because of its high electronegativity [45–48], and although astatine is known to be radioactive, with a half-life time of 7.2 hours for its most stable isotope, there are recent studies showing halogen bonding involving also the heaviest halogen atom [49–51].

As already mentioned, halogen bonding features a bonding angle of approximately 180°. The orientation of the σ-hole supports this directionality because an attractive electrostatic interaction between the negatively polarized LB and the XB donor is only possible in this positive region. Marked deviations from this angle would result in a repulsion between the LB and the orthogonal lone pairs of the halogen (see orange belt around the halogen).

Although the IUPAC definition stresses the importance of electrostatics and even though the σ-hole model explains the majority of trends observed experimentally, the other factors should not be neglected, as they may become dominant for certain adducts. As an example, a recent study by Head-Gordon and coworkers has shown that the relative contribution of electrostatics and charge-transfer differs from case to case [52]. Moreover, Huber *et al.* published in 2020 an anionic organic compound able to form "anti-electrostatic" halogen bonds with iodide (Figure 10.6) [53]. This adduct between two anions demonstrates the limitations of predictions based on static σ-holes and underlines the potential importance of polarization and charge-transfer.

10.2 History of Chalcogen Bonding

Chalcogen bonding (ChB) [54, 55] (Figure 10.7) was defined in a similar manner as halogen bonding in a recently published IUPAC definition as an "attractive interaction between an electrophilic region associated with a chalcogen atom [...] and a nucleophilic region [...]" [56].

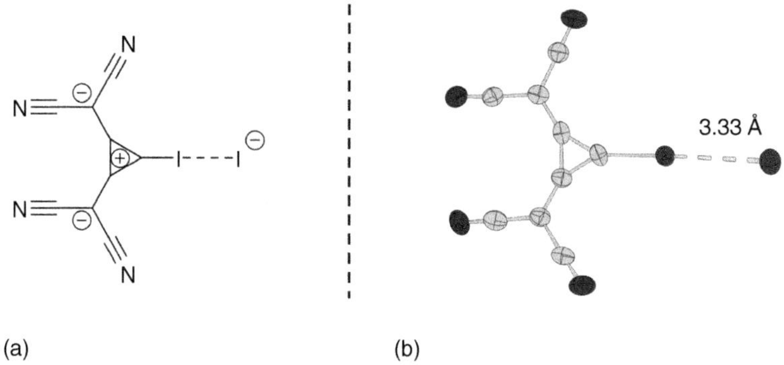

Figure 10.6 "Anti-electrostatic" XB between iodinated bis(dicyanomethylene) cyclopropanid and iodide (a) with the corresponding crystal structure (b) [53]. TDA counterion is omitted for clarity. Carbons are shown in gray, nitrogens in blue, and iodines in purple. Ellipsoids at 50% probability. Source: Based on Holthoff *et al.* [53].

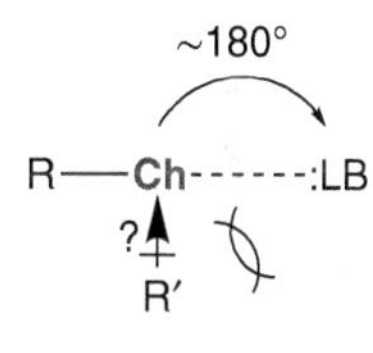

Figure 10.7 Schematic representation of a chalcogen bond highlighting the steric and electronic influences of the second substituent at the chalcogen atom (R′). Source: Adapted from Vogel *et al.* [54].

While it is often called a "σ-hole" interaction [1, 57, 58], the same individual attractive contributions as for halogen bonding (electrostatics, charge- transfer, and dispersion) also need to be considered for chalcogen bonding [59]. There is some evidence that the charge-transfer character is particularly important for strong chalcogen bonds [59], whereas dispersive forces dominate the interaction for weak ChB [60]. Electrostatics are a major contributor for all cases but seem to be especially relevant for medium-strength interactions. While the strength of chalcogen bonding is influenced in the same way as the one of halogen bonding, regarding the interacting atom and the electron-withdrawing nature of the substituent bound to the chalcogen [61, 62], an additional focus must be laid to the second substituent of the chalcogen atom (or the additional ones for chalcogen atoms in higher oxidation states). Not only does it have an electronic but also a steric effect that will impact the binding site significantly. While this is clearly an important aspect, detailed experimental studies regarding the influence of the second substituent are currently missing, limiting the knowledge to theoretical studies [63, 64], Another important factor regarding ChB donors is the redox chemistry of these compounds (especially for the heavier atoms), which needs to be considered in the catalyst design [65, 66].

The term "chalcogen bonding" has only been used in the most recent years [67]. Examples of chalcogen-containing molecules, where the chalcogen acts as a Lewis acid, have been known for several decades [68], and there are several examples where short contacts between chalcogen atoms and Lewis basic centers are essential to conductive behavior in organic materials [69–72]. However, there are barely any examples where anions are bound by the chalcogen center. Most early examples are known from crystal engineering and intramolecular applications [73, 74].

One noteworthy example of a chalcogen-bonded complex is the structure formed between elemental iodine and diphenyl diselenide, which also features strong halogen bonding [75]. This example can also be considered as a borderline case of chalcogen bonding, as the stronger chalcogen bond donor diphenyl ditelluride will cause the I—I and Te—Te bonds to break, forming 2 equiv of phenyltelluriumiodide (Scheme 10.1) [76].

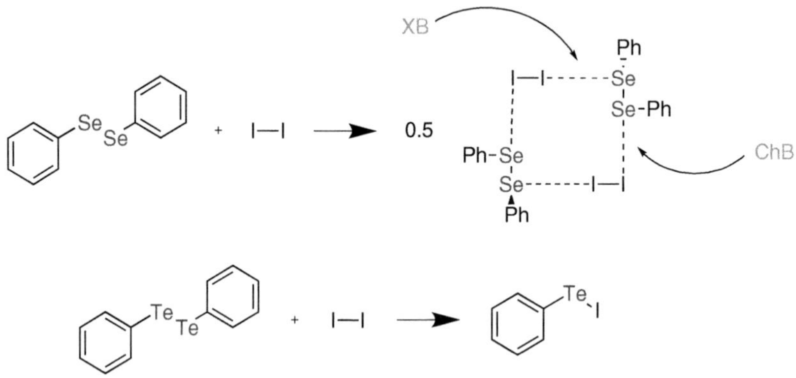

Scheme 10.1 Complex formation between diphenyl diselenide and elemental iodine (top) and reaction of diphenyl ditelluride and elemental iodine (bottom): stronger ChB contributes to a bond cleavage. Source: Based on Schulz and Klar [76].

Chalcogen-based systems have also been successfully used for chiral induction using rigidified ChB-complexed structures as intermediates. Pioneering work in this field was done by Tomoda and Wirth who used chiral selenides in addition reactions to styrene derivatives (Scheme 10.2) [77, 78].

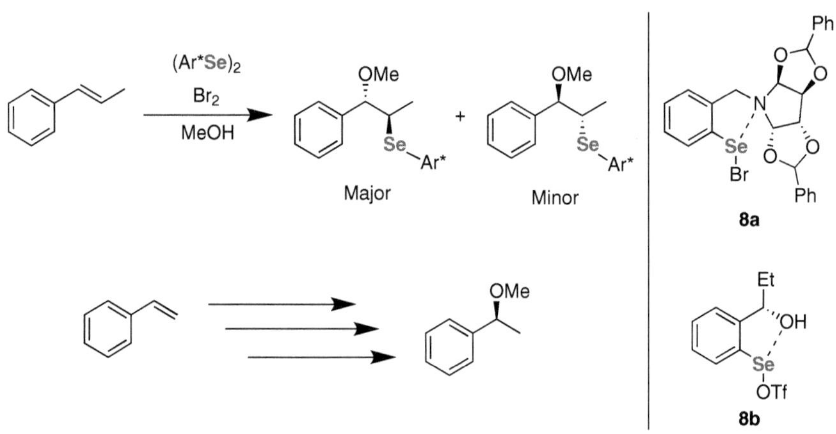

Scheme 10.2 Enantioselective addition of a methoxy group using chiral selenium agents **8a+b** for enantioinduction. Source: Adapted from Fujita *et al.* [77].

Based on these early examples and anion-binding studies (see Section 10.5), chalcogen bond donors were also applied as intermolecular Lewis acids in organocatalysis. While there are examples of them being used in anion-binding catalysis, which will be discussed in detail later, the first reaction that was published was the reduction of quinoline by Matile *et al.* (Scheme 10.3) [79]. For this reaction, a sulfur-based catalyst **9** was applied and, despite the usage of this comparably lowly polarizable chalcogen, its activity indicated that stronger performance could be expected when switching to selenium- or even tellurium-based compounds.

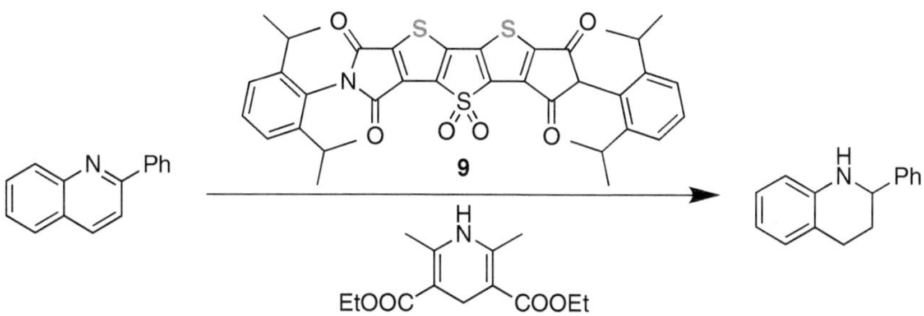

Scheme 10.3 Transfer hydrogenation by Benz *et al.* [79] catalyzed using a sulfur-based chalcogen bond donor. Source: Based on Benz *et al.* [79].

This could indeed be shown by using tellurium-based catalysts such as **10** (Scheme 10.4), which were applied in the activation of nitro and carbonyl groups in Michael addition reactions [80]. Especially, the activation of the nitro group is an interesting example, as the halogen bonding derivatives of the triazolium-based donors failed to catalyze this reaction.

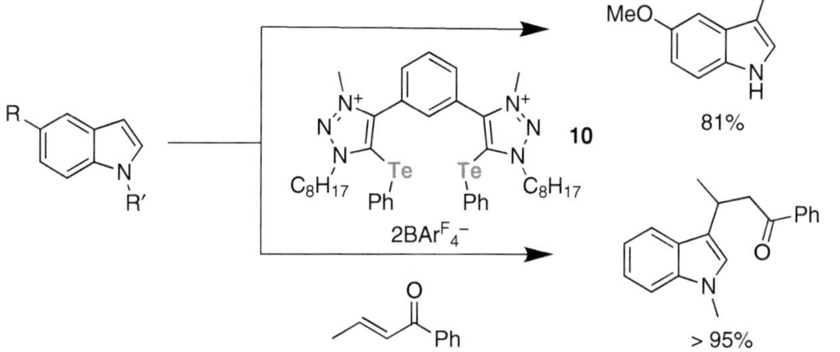

Scheme 10.4 Carbonyl and nitro activation in Michael-type reactions using a tellurium-based catalyst. Source: Based on Wonner *et al.* [80].

Another interesting class of compounds used as chalcogen bonding catalysts has been published by the group of Wang *et al.* (Scheme 10.5) [81, 82]. They were able to catalyze two different reactions, namely, a Rauhut–Currier-type reaction and a cascade reaction of indole with β-ketoaldehydes, using phosphonium-substituted selenides such as **11**.

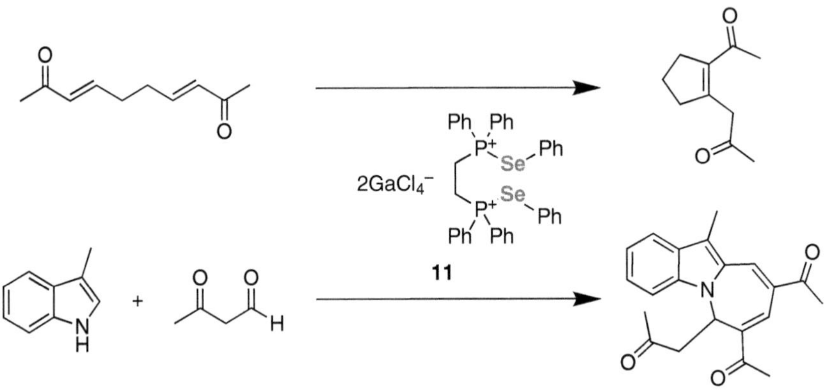

Scheme 10.5 Catalytic transformations by Wang *et al.* [81, 82] using a selenium-based chalcogen bond donor. Sources: Adapted from Wang *et al.* [81, 82].

10.3 History of Pnictogen and Tetrel Bonding

Compared to halogen and chalcogen bonding, pnictogen and especially tetrel bonding have been far less investigated. While there are numerous theoretical studies on the nature and properties of the latter two, experimental studies remain very scarce [83–85]. Overall, the same trends concerning the origins of the interactions and the influences on strength and directionality are observed as for XB and ChB. One important aspect to discuss especially for the heavier atoms for PnB (Sb and Bi) and TeB (Ge, Sn, and Pb), however, is a potential differentiation between classical Lewis acid/base adducts and the interactions considered herein [86]. Put differently, this addresses the question of where to draw the border between covalent and non-covalent bonding. While it is relatively straightforward to differentiate between these two alternatives for all cases on opposite ends of binding strength, this is likely not a black-and-white issue for intermediate cases but rather a continuum. Nevertheless, several applications have been published already [87, 88], some of which clearly point toward a pnictogen bonding (or tetrel bonding) mechanism [86, 89–91].

Pnictogen bonding was also successfully established in a handful of catalytic transformations, with two of them revolving around halide abstraction, which will be covered in detail later on [90, 91]. The first application of an antimony-based catalyst **12** in the activation of a neutral substrate was published by the group of

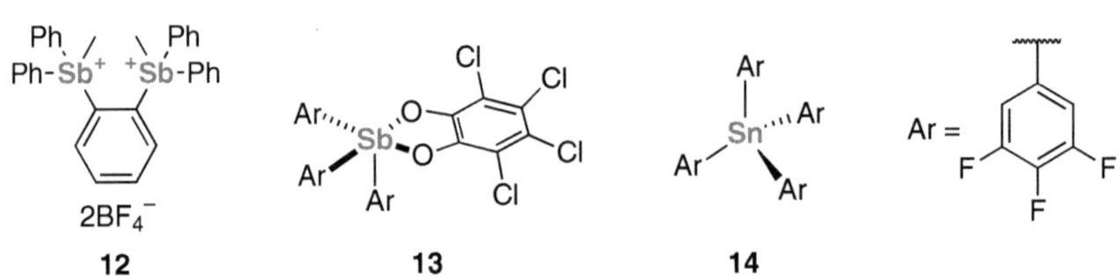

Figure 10.8 Pnictogen and tetrel bonding catalysts used by Gini *et al.* [86], Hirai *et al.* [87]. Source: Yang *et al.* [89].

Gabbaï (Figure 10.8) [89]. Using a dicationic bis-stibonium catalyst, the transfer hydrogenation of 2-phenylquinoline (see also Scheme 10.3) could be achieved successfully (with yields of up to 90% after just one hour). Matile and coworkers, on the other hand, used Sb(V)-compounds **13** and tetrel bonding catalyst **14** to activate brevetoxin-type polyether cyclizations [86]. In this study, it was also noted that the differentiation between pnictogen bonding and classical Lewis acid catalysis, which may be tricky for certain cases, is an important consideration to make.

10.4 Differences Between Hydrogen Bonding and Other Secondary Interactions

Hydrogen bonding has been known for many decades and has been studied intensively in different fields of chemistry, biochemistry, or biology. In contrast, although halogen bonds have also been known since the nineteenth century, interest in them has increased only over the last two decades and their research field is still quite young. Apart from this historical point of view, HB and XB are conceptually quite similar. Both non-covalent interactions feature a donor/acceptor interaction having the hydrogen/halogen as the acceptor moiety and the LB as the HB/XB acceptor. Furthermore, both interactions can be strengthened by increasing the electronegativity of the covalently bonded group. However, halogen bonds can also be enhanced by changing the halogen atom and thus by varying its polarizability, whereas hydrogen bonds are by definition restricted to one single interacting atom.

Another distinction to hydrogen bonds is the already mentioned directionality. Figure 10.9 shows the electrostatic potential surface of CF_3I and CF_3H. By having a closer look at the positive area (blue) of both compounds, it is evident that CF_3H has a much larger positive region than CF_3I. Hence, the XB angle needs to be $\approx 180°$, but the HB one can deviate markedly from such linearity [56, 92].

A further difference between hydrogen and halogen bonding is rooted in the electronic nature of the interacting atoms. While hydrogen is the smallest atom and thus less polarizable, halogens, especially bromine and iodine, are bigger in size and therefore more polarizable. Based on the HSAB concept, hydrogens are hence considered to be "hard" and halogens "soft" Lewis acids [93]. Furthermore, halogen bonds are much less sensitive towards solvent polarity than hydrogen bonds. For instance, Hunter and coworkers investigated the solvent dependency of complexes involving 4-(phenylazo)phenol, molecular iodine, tetramethylurea, and tetramethylthiourea in a series of different solvents by using UV/vis spectroscopy and proton NMR titration experiments [94]. They observed that while the strength of hydrogen bonds was sensitive to the solvent, decreasing with increasing polarity, halogen bonds (e.g. between iodine and tetramethylthiourea) were comparably little affected by the respective solvent.

These differences remain true for the other secondary interactions (ChB, PnB, and TeB); however, one further distinction has to be made. Whereas HB and halogen(I)-based XB both usually feature only one binding axis towards Lewis bases, other related interactions feature up to four such electrophilic axes, thus

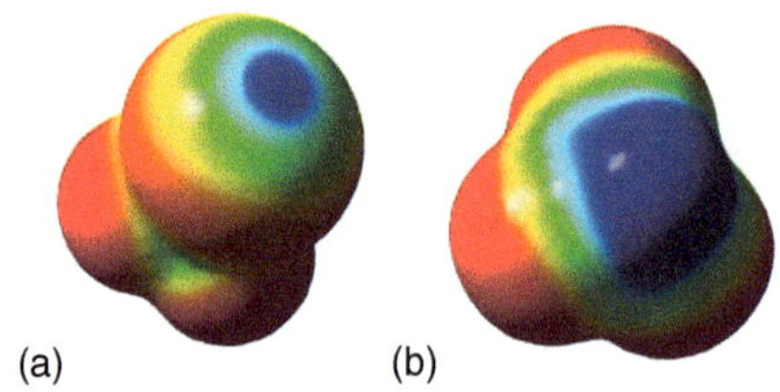

Figure 10.9 Electrostatic potential surface of CF_3I (a) and CF_3H (b). The calculations were performed with M06-2X/def2-TZVPP and the electrostatic potentials are plotted on the 0.001 a.u. density isosurface.

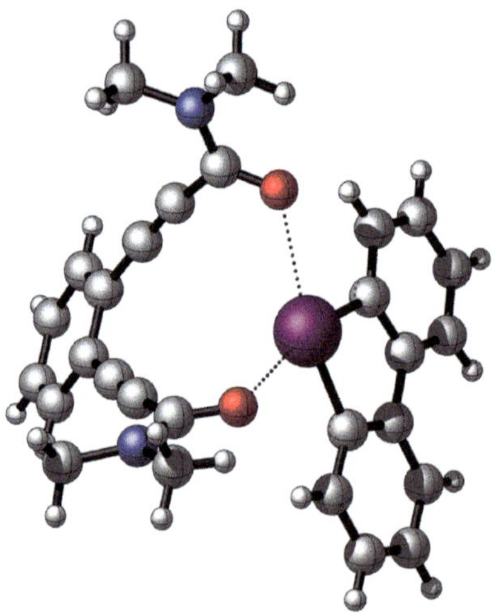

Figure 10.10 Crystal structure showing biaxial binding of an I(III)-based halogen bond donor with a diamide [95]. Structure illustrated with CYLview [96]. Source: Adapted from Heinen *et al.* [95].

enabling the simultaneous binding of up to four different substrates at once. This is even true for halogen(III)-based halogen bonding, as was recently demonstrated in the "biaxial" recognition of diesters and diamides (Figure 10.10), in which two different carbonyl groups are bound by the same halogen center [95]. From a topological point of view, this situation is very similar to chalcogen bonding [97].

10.5 Secondary Bonding in Anion Recognition

One type of halogen bond acceptors that can form very strong interactions both in the solid state and in solution are halides [18, 98]. Consequently, and due to the special properties of XB, such as their high directionality and tunability, halogen bonding has increasingly been applied in anion binding over the past decade.

In this regard, mono- and multidentate halogen bond donors were designed, which are able to bind to halides strongly and sometimes also selectively. However, the design of such XB donors can be very challenging since various factors have to be considered in order to achieve effective anion binding: (i) Multidentate halogen bond donors have to be large enough to "wrap around" the halide, which is typically larger than a neutral coordination center; (ii) the strict linearity of the interaction requires careful placement of XB interacting sites on XB donors, so that all individual interactions can satisfy this requirement; (iii) the substituents that are covalently bound to the interacting halogen have to be sufficiently electron-withdrawing or

Figure 10.11 Monodentate NaI receptor by Metrangoloe *et al.* [99].

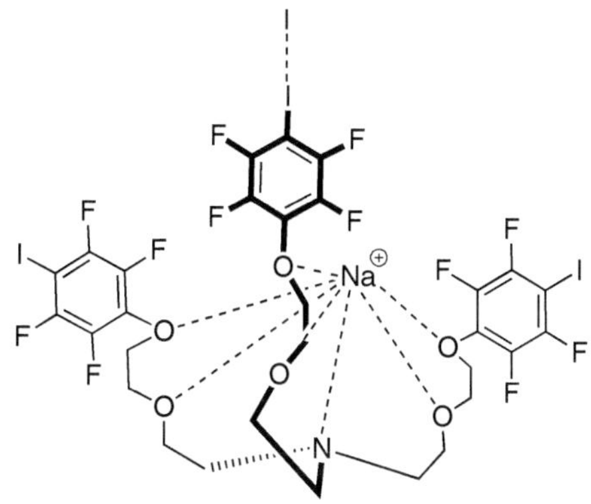

polarizing to enable strong XB, but they should also be easily accessible; and (iv) the resulting XB donors should be stable against air and moisture and should be soluble in various solvents in order to be able to analyze anion binding in different environments.

Presumably, the first XB receptor designed for anion binding in solution was synthesized by Metrangolo and Resnati in 2005 [99]. Their polyfluorinated, tripodal receptor was able to bind both the counter-cation (Na^+) and the halogen anion (I^-) due to two different recognition frameworks within the same molecule (Figure 10.11). Coordination of the halide was achieved via a 1 : 1 interaction with an iodopolyfluoroarene moiety.

Further monodentate receptors were studied by Taylor and coworkers, who determined binding constants between iodoperfluoroalkanes (C_6F_5I) and iodoperfluoroarenes ($C_8F_{17}I$) with different tetrabutylammonium salts (Bu_4NX) in acetone by ^{19}F NMR titrations [100]. Overall, both XB donors bind with the same anion affinity of $Cl^- > Br^- > I^- \gg TsO^- > NO_3^-$ (Table 10.1), but the K_a values of the iodoperfluoroalkane are higher than those of the iodoperfluoroarene. Since polyfluorinated scaffolds yield XB donors that bind anions relatively strongly because of their high electron deficiency, such core motifs have become popular for the design of anion receptors. Further enhancement of binding is usually achieved by multidentate binding of the receptor towards the anion, making use of cooperative effects. Accordingly, Taylor and coworkers developed both bi- and tridentate XB receptors based on *ortho*-iodo-substituted perfluoroarenes able to bind tetrabutylammonium salts in acetone as a solvent (Figure 10.12) [101–104]. When comparing the binding values of the bi- and tridentate receptors with those of the monodentate XB donors in Table 10.1, an enhancement of 1 order of magnitude can be observed relative to C_6F_5I, but roughly, the same results as for $C_8F_{17}I$ were detected. Moreover, the binding trend towards Lewis bases seems to be the same for mono- and bidentatedonors, decreasing in the order $Cl^- > Br^- > I^- \gg TsO^- > NO_3^-$.[1]

A further enhancement of the binding strength is possible by increased preorganization, as was demonstrated for the bidentate halogen bond donors **20a/b** (Figure 10.13), which contain four iodine substituents on a rigid polyfluorinated

1 For compound **19**, no binding constants were detected for I^- and NO_3^-, respectively.

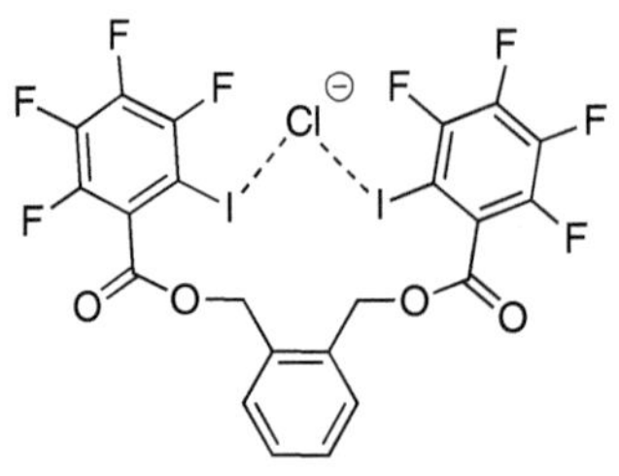

15
$K_a = 1.2 \times 10^3\ M^{-1}$

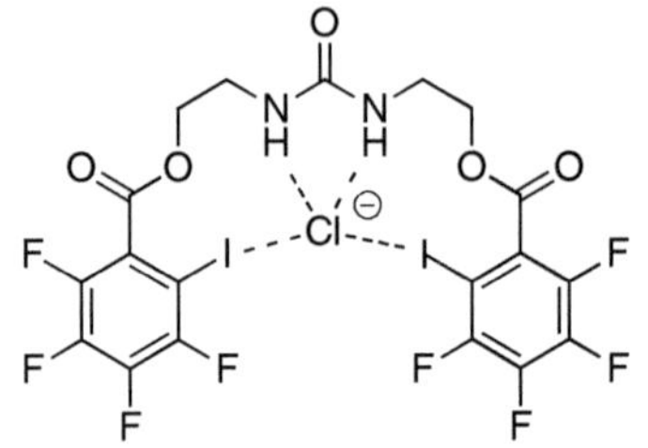

16
$K_a = 2.4 \times 10^3\ M^{-1}$

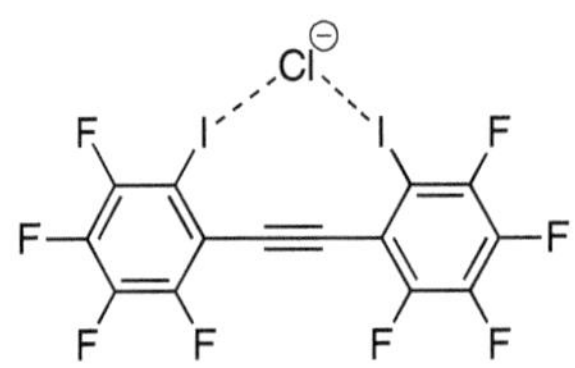

17
$K_a = 1.7 \times 10^3\ M^{-1}$

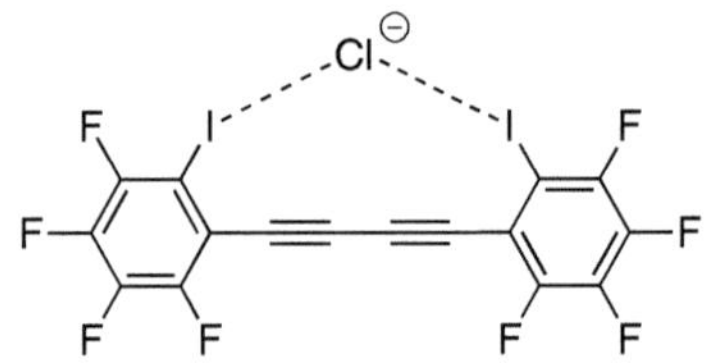

18
$K_a = 1.2 \times 10^3\ M^{-1}$

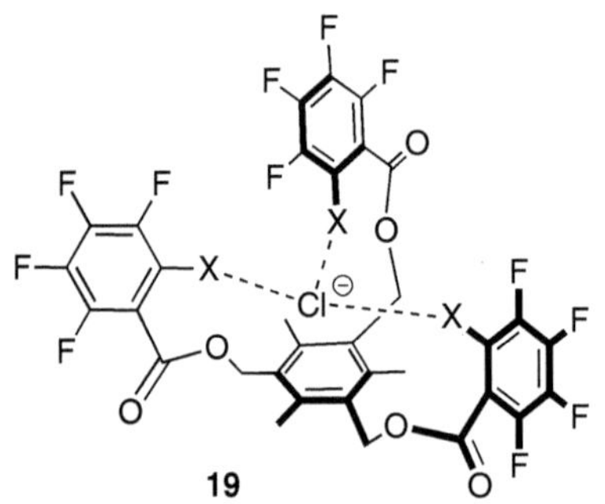

19

$K_a(H) = <5\ M^{-1}$
$K_a(I) = 1.8 \times 10^3\ M^{-1}$
$K_a(Br) = <5\ M^{-1}$
$K_a(F) = <10\ M^{-1}$

Figure 10.12 Bi- and tridentate XB receptors based on polyfluorinated scaffolds able to bind tetrabutylammonium salts in acetone-d_6 at 298 K and the corresponding binding constants K_a. Source: Adapted from Sarwar *et al.* [101].

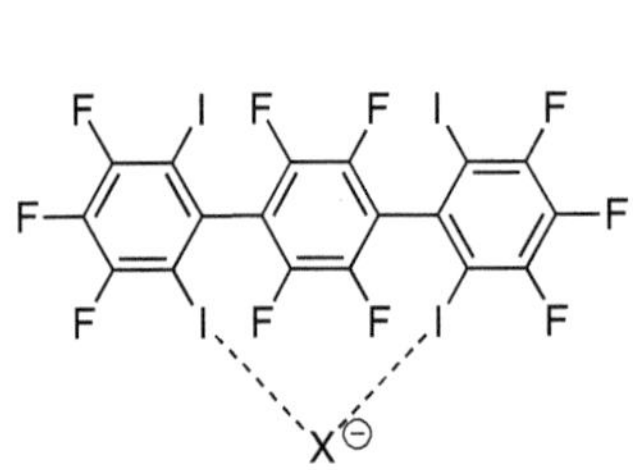

20a
$K_a\ (Bu_4N^+Cl^-) = 1.8 \times 10^4\ M^{-1}$
$K_a\ (Bu_4N^+Br^-) = 1.3 \times 10^4\ M^{-1}$

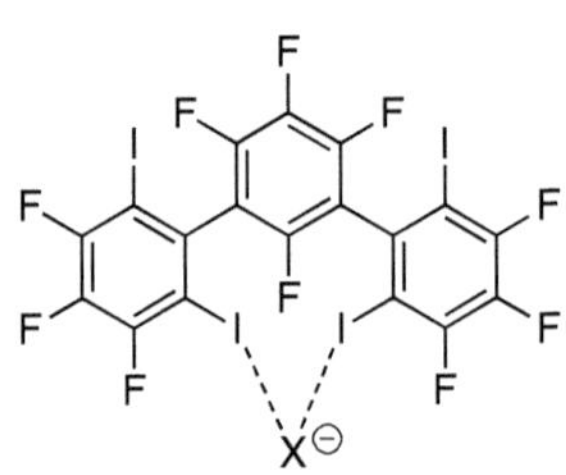

20b
$K_a\ (Bu_4N^+Cl^-) = 7.5 \times 10^4\ M^{-1}$
$K_a\ (Bu_4N^+Br^-) = 6.2 \times 10^4\ M^{-1}$

Figure 10.13 Association constants K_a of iodinated terphenyls as bidentate neutral anion receptors with tetrabutylammonium salts in THF at 303 K. Source: Based on Kniep *et al.* [105].

Table 10.1 Binding constants (K_a) of tetrabutylammonium halide salts with C_6F_5I and $C_8F_{17}I$ in acetone-d_6 at 295 K determined by ^{19}F NMR.

C_6F_5I		$C_8F_{17}I$	
X^-	K_a (M^{-1})	X^-	K_a (M^{-1})
Cl^-	1.5×10^2	Cl^-	2.2×10^3
Br^-	1.0×10^2	Br^-	1.0×10^3
I^-	4.4×10^1	I^-	3.3×10^2
TsO^-	$<0.5 \times 10^1$	TsO^-	—
NO_3^-	$<0.5 \times 10^1$	NO_3^-	—

terphenyl backbone [105]. For this study, isothermal titration calorimetry (ITC) was used to determine the association constants between XB donors **20a** and **20b** and tetrabutylammonium salts in THF at 303 K, showing again a higher binding affinity towards chloride vs. bromide.

Next to such polyfluorinated receptors, later-on halotriazoles (**22**, **23**) [106], halotriazolium (**21**, **24**) [39, 107, 108], haloimidazolium (**25**) [109], halobenzimidazolium (**26**) [110], and substituted pyridinium moieties (**27**) [39] also became very popular building blocks. All these compounds are easy to synthesize and to incorporate into various types of host frameworks. Figure 10.14 depicts some examples of the above-mentioned compounds and the corresponding association constants K_a with chloride anions. When comparing all the binding data, it immediately becomes obvious that cationic receptors bind much stronger than the neutral ones. This can be attributed to the additional Coulombic interaction due to the cationic unit within the molecule, leading to "charge-assisted" halogen bonding. Moreover, halotriazoliums were shown to strongly coordinate anions even in competitive solvents such as DMSO, as demonstrated, e.g. by the groups of Schubert [111] or Resnati [112].

Similar to halogen bonding, the study of chalcogen bonding in solution initially also focused on anion binding [113–116]. As a consequence, different systems have been developed for a range of applications, not only for anion recognition but also for anion transport and organocatalysis [4, 117]. Compared to XB, arguably a larger variety of structurally different ChB donors has been studied in anion recognition. Herein, the focus will be laid on the early examples, as these have laid the foundation towards catalytic applications.

These notable examples of anion-binding ChB donors have been reported by the groups of Taylor, Beer, and Gabbaï (Figure 10.15) [115, 116, 118]. Even though the latter was likely the first one to publish an anion receptor (**28**) that uses chalcogen bonding, the binding is not solely based on this interaction. Rather, the compound also features a Lewis-acidic boron center, which is used cooperatively with the chalcogen center in order to achieve strong fluoride binding [119]. The chalcogenated anion receptors by Beer [116] and Taylor [115], however, are considered to be based purely on chalcogen bonding. Their binding constants are as high as $1.3 \times 10^5\ M^{-1}$ (for the

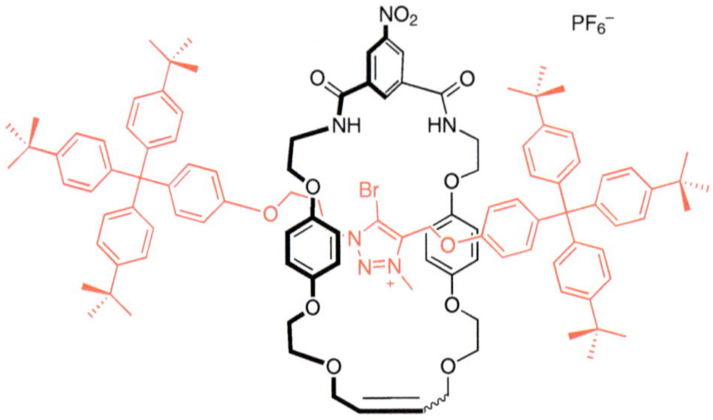

21
$K_a = 4.3 \times 10^2\ M^{-1}$

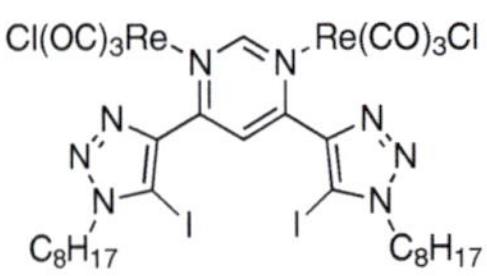

22
$K_a = 8.8 \times 10^2\ M^{-1}$

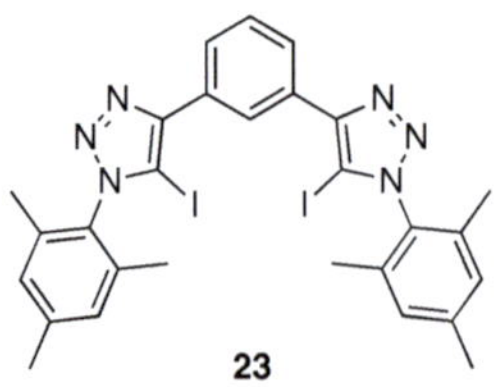

23
$K_a = 5\ M^{-1}$

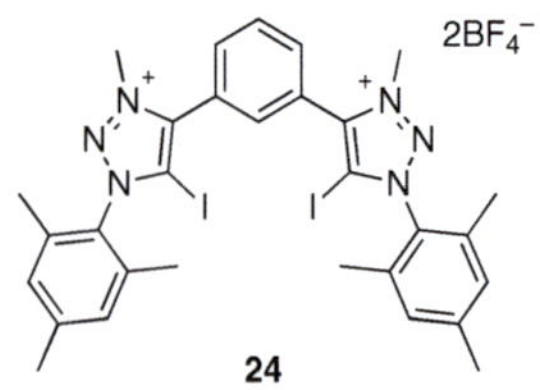

24
$K_a = 1.2 \times 10^3\ M^{-1}$

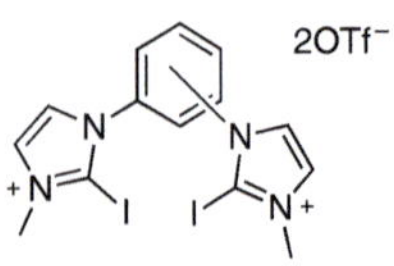

25
$K_a(m) = 5.2 \times 10^5\ M^{-1}$
$K_a(p) = 3.3 \times 10^4\ M^{-1}$

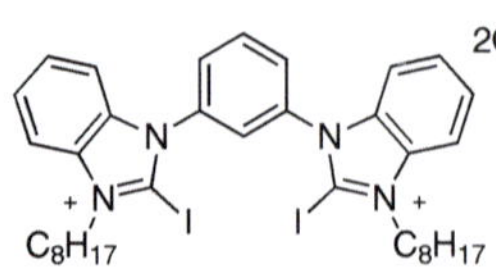

26
$K_a = 3.5 \times 10^6\ M^{-1}$

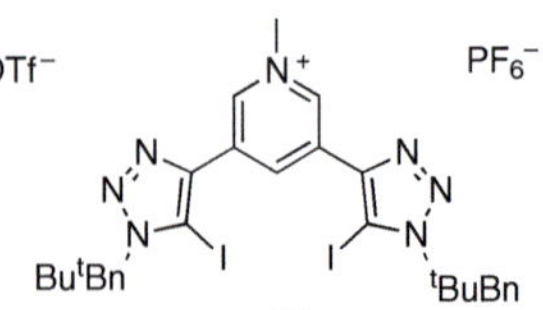

27
$K_a = 3.9 \times 10^2\ M^{-1}$

Figure 10.14 Examples of XB donors featuring halotriazoles (**22**, **23**), halotriazolium (**24**), haloimidazolium (**25**), halobenzimidazolium (**26**), and substituted pyridinium (**20**) as well as their corresponding binding constants K_a with tetrabutylammonium chloride.

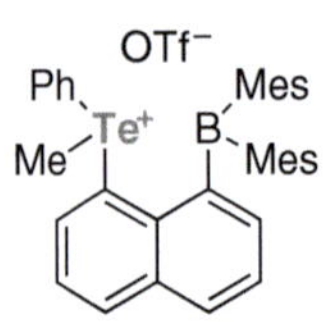

28

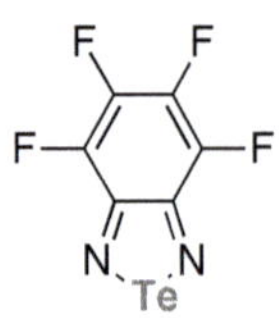

29

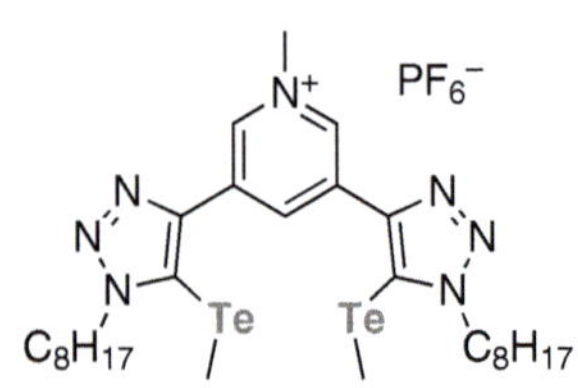

30

Figure 10.15 Selected chalcogen-bonding-based anion receptors. Sources: Garrett *et al.* [115], Lim *et al.* [116], Zhao and Gabbaï [118].

adduct of **29** with NBu_4Cl). Somewhat counterintuitively, the neutral ChB donors of Taylor *et al.* generally feature stronger anion binding than the overall cationic systems **30** from Beer *et al.* These results show that the structure–activity relationships are non-trivial and that not yet fully understood.

In contrast to XB and ChB, only little experimental evidence on the anion binding capabilities of pnictogen bond donors has been collected so far. Only two examples of Sb(III)-based donors [120, 121] had been published before the first application in organocatalysis by Matile and coworkers [90], which will be discussed in detail below. The first example, a monodentate anion receptor, is chloro-stibaindole **31** from Christianson and Gabbaï (Figure 10.16, left), which forms a complex upon addition of Et_4NCl with an association constant of $9.5 \times 10^4\ M^{-1}$ [121]. However due to the lability of the Sb—Cl bond in the compound, association towards other anions could not be measured as the chlorine was substituted resulting in a mixture of different complexes. The second example, a bidentate receptor by Cozzolino and coworkers is compound **32**, which was subjected to anion binding with halides in DMSO (Figure 10.16, right) [120]. A 3 : 1 binding model could be fitted towards chloride and bromide salts, which is proposed to be outcompeted by the Lewis-basic solvent for iodide binding.

Even though numerous examples for anion binding using tetrel-based systems have been published (for a collection of these cases see [4]), differences toward the other secondary interactions can be observed. Usually, the geometry around the tetrel center changes upon coordination of the anion, causing a shift from tetrahedral towards trigonal bipyramidal [119]. In line with the considerations made for the differentiation between pnictogen bonding and classical Lewis acid/base interactions mentioned above, this would mean that these cases are rather the latter and not tetrel bonding [86]. Nevertheless, some examples of strong anion-binding compounds will be given. One popular choice for these kinds of compounds are electron-deficient silicon compounds utilizing not only the Lewis acidity but also the fluorophilicity of the silicon atoms to form strong Si—F bonds (Figure 10.17). This way binding constants larger than $1 \times 10^9\ M^{-1}$ can be observed using the bidentate silicon compound **33** [122]. Another example with strong anion binding is the flexible bidentate bis-(organostannyl) receptor **34** reported by Jurkschat *et al.*, which gave association constants to chloride of up to $2.3 \times 10^3\ M^{-1}$ [123].

In conclusion, the design of strong anion-binding receptors has gained much interest over the past decade. With HB surely being the predominant and pioneering interaction for anion binding, XB and other secondary forces have now gradually emerged as competitors, often displaying even stronger binding constants in anion recognition in solution than their HB analogs.

Figure 10.16 Antimony-based anion receptors used for halide binding. Source: (left) Christianson and Gabbaï [121], (right) Qiu *et al.* [120].

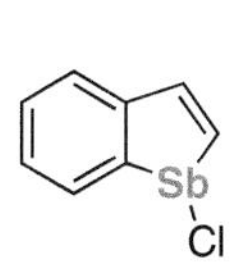

31

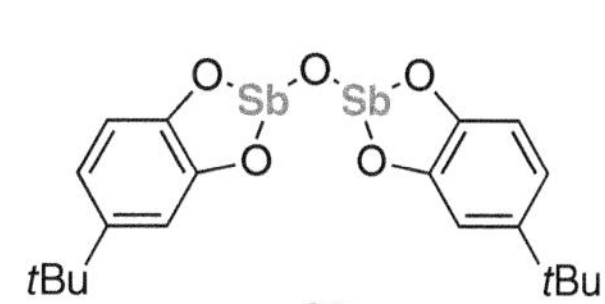

32

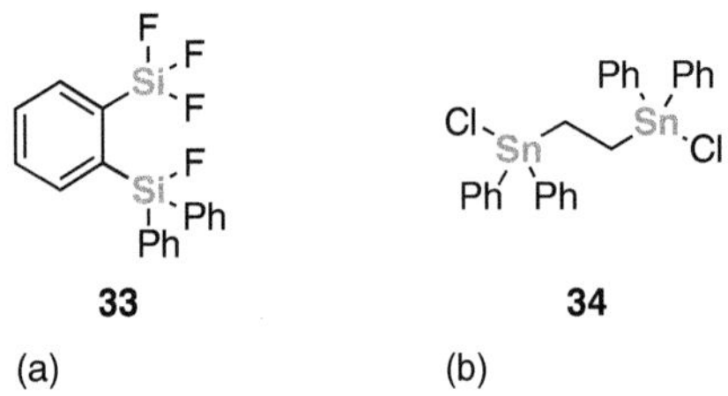

Figure 10.17 Tetrel-based anion receptors for fluoride binding (a) and chloride binding (b).

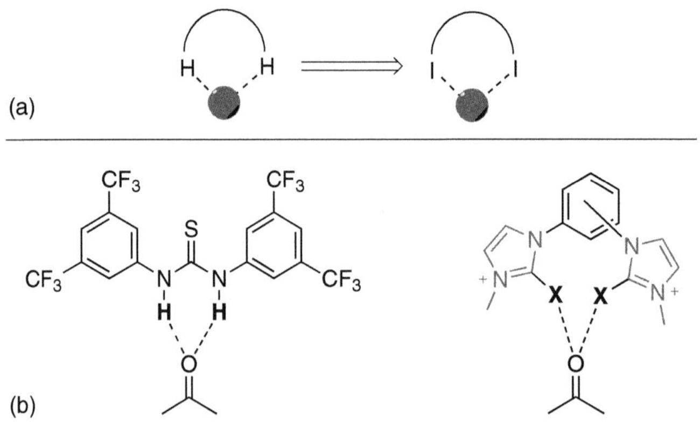

Figure 10.18 General idea of the concept transfer of hydrogen bond-assisted catalysis to halogen bond-assisted catalysis (a) and examples of a HB and XB carbonyl activation (b).

10.6 Halogen Bonding in Anion-Binding Catalysis

Inspired by the similarities to hydrogen bonding, chemist became interested to explore the ability of halogen bonding not only for anion recognition but also in organocatalysis. After anion binding had been established and quantified, this coordination was also sought to be used for the abstraction of halides from suitable substrates. With HB-based transformations as pioneering examples, which may also be performed enantioselectively [124], XB-based organocatalysis could benefit and build upon concepts already developed for these cases. Figure 10.18a, shows how the basic idea of HB-assisted catalysis by coordination of the corresponding hydrogen to the substrate (gray ball) is transferred to halogen bonding. Moreover, the XB catalyst is designed similarly to the HB catalyst (b), having both a bidentate structure and electron-withdrawing substituents for enhanced activation [125].

In general, two different modes of activation are discussed for XB organocatalysis. Since XB donors coordinate quite strongly to anions such as halides, this may shift S_N1-type equilibria in which these species act as leaving groups by their removal through complexation, thus accelerating the reaction (Scheme 10.6, right). This "anion-binding" catalysis is the focus of this chapter. One drawback of this approach is the fact that the resulting complex may be quite strongly bound, which inhibits the XB catalyst. Therefore, stoichiometric amounts of XB donor may be needed. The second activation mode is based on neutral substrates such as carbonyl or imine compounds. In this case, the XB donor coordinates to the LB, lowering its

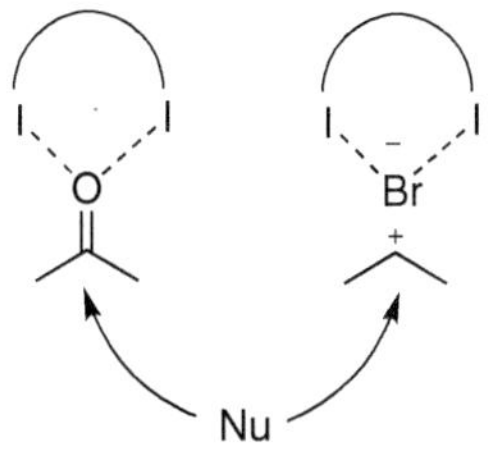

Scheme 10.6 Illustration of the two different activation modes of the XB donor depending on the substrate.

LUMO and thus activating it for nucleophilic attack (Scheme 10.6, left). This kind of XB organocatalysis is outside the scope of this chapter.

Moreover, a distinction could be made between transient and non-transient halogen bonding in organic synthesis. The former depicts a situation in which the bond between the halogen and the halogen-bearing group is broken in the progress of the reaction and is typically the case for halogenation reactions. Obviously, this would correspond to a decomposition of a potential organocatalyst and thus all examples in the following feature cases of non-transient halogen bonding, in which the XB donor remains intact during the course of the reaction. Readers interested in this distinction are referred to other articles for further elucidation [125, 126].

Presumably, the first rational use of XB donors in a transformation involving halide abstraction (and thus anion binding) is the Ritter-type reaction depicted in Scheme 10.7, which was activated by bidentate XB donors such as **25** [127–129]. After halide abstraction, the benzhydryl cation is formed, which is subsequently attacked by acetonitrile leading to *N*-benzhydryl acetamide **36**.

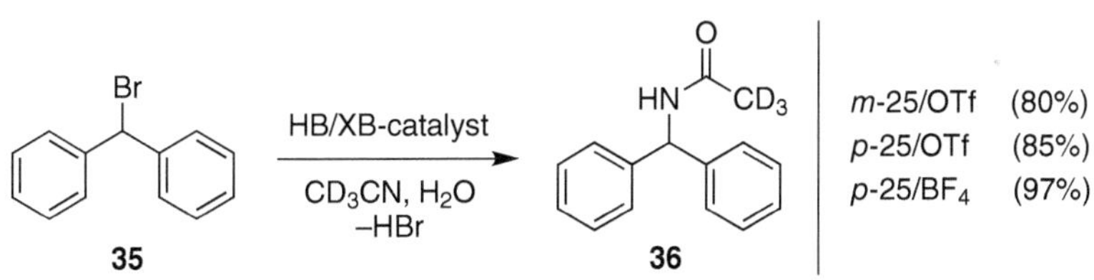

Scheme 10.7 Ritter-type halide abstraction reaction (left) and yields achieved with several XB donors (right). Sources: Adapted from Walter *et al.* [127], Kniep *et al.* [128, 129].

Furthermore, bidentate imidazolium-based halogen bond donors turned out to have a much higher catalytic activity – both for the meta (m) and the para (p) imidazolium-substituted derivatives – compared to monoiodinated derivatives, clearly demonstrating the enhancement by bidentate/cooperative binding. Additionally, the bromine-based analog of XB donor **25** was noted to be less active, in line with the more polarizable halogen forming stronger XB donors. Another important aspect to consider is the counter-anion used as it is crucial that it does not compete with the binding of the substrate. Therefore, less coordinating anions such as OTf^- or BF_4^- were chosen. Moreover, even though this Ritter-type reaction can also be catalyzed by other Lewis acids such as Ag^+ salts [130], it is not catalyzed by acid traces. Consequently, any hidden acid catalysis can be ruled out, indicating that the

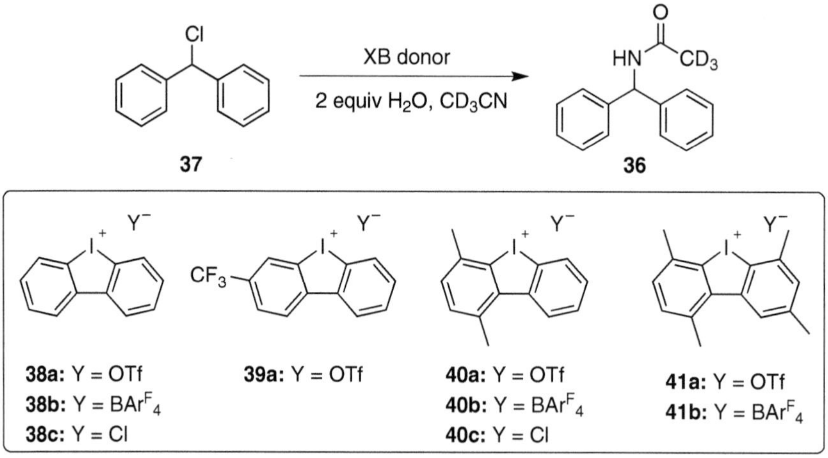

Scheme 10.8 Reaction of benzhydryl chloride **37** (top) with various iodolium compounds (bottom).

XB donor is indeed the activating agent. Aakeröy and Perera performed the same type of reaction using neutral bidentate XB donors, which afforded up to 90% yield after 96 hours [131].

Recently, Huber and coworkers extended the class of catalytically active organoiodine derivatives by introducing hypervalent iodolium compounds for halide abstraction reactions [132]. Although these compounds are only monodentate, preliminary calorimetric measurements showed that **38a** coordinates to tetrabutylammonium bromide in acetonitrile with a binding constant of roughly $10^6\ M^{-1}$, which is twice the one of bidentate XB donor ***47-m***. Consequently, these cyclic iodonium compounds were also considered as activators. Scheme 10.8 illustrates the reaction of benzhydryl chloride **37** – a more challenging substrate than bromide **35** mentioned above – with 1 equiv of an iodolium compound (**38–41**) in wet CD_3CN. Compounds **38a**, **38b**, and **39a** led to a product formation of approximately 80% after six days. Donors **40a** and **40b**, however, gave only 25–36% yield. This observation can be explained by the fact that one electrophilic axis of the iodine(III) is blocked by a methyl group, reducing its Lewis acidity. Following this trend, compounds **41a** and **41b** showed no conversion because these feature two methyl groups next to the iodine(III) center, blocking both axes. Thus, the crucial role of halogen bonding was confirmed.

In the following years, this type of activation was extended towards further substrates. Scheme 10.9 visualizes other halide-abstraction reactions based on XB: a Ritter-type reaction on 2,3,4,6-tetra-*O*-benzyl glycosyl chloride (a) [133], a Koenigs–Knorr-type glycosylation (b) [133], a semi-pinacol rearrangement (c) [134], as well as a Friedel–Crafts alkylation (d) [135].

However, as already mentioned, in all these reactions, stoichiometric amounts of activator are required because of the strong coordination between the XB donor and the abstracted halogen. The first actual anion-binding organocatalysis was realized in the form of the reaction between 1-chloroisochroman (**42**) and ketene silyl acetal

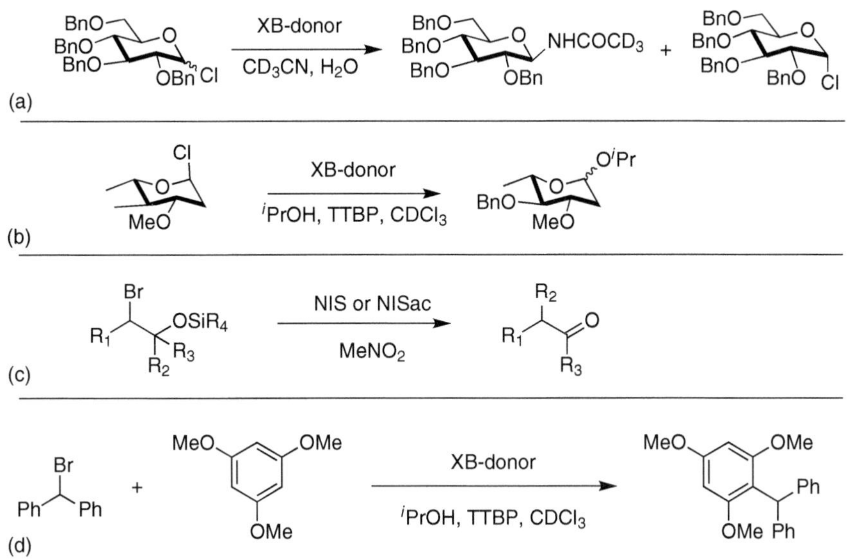

Scheme 10.9 Halide-abstraction reactions activated by halogen bonding. (a) Ritter-type reaction. (b) Koenigs–Knorr-type glycosylation. (c) Semipinacol rearrangement. (d) Friedel–Crafts alkylation.

(**43**), which had been already catalyzed by Jacobsen *et al.* via hydrogen bonding. The silyl protecting group seems to be a key element in this case, as the released chloride is intercepted by the cleaved TBS group, thus avoiding deactivation of the XB donor (Scheme 10.10) [110, 124]. A halogen-bond-catalyzed variant was first published by Huber and coworkers in 2013 using, e.g. a neutral polyfluorinated quarterphenyl (**45**), which achieved full conversion of ester **44** with 20 mol% catalyst loading (and up to 91% with 10 mol%) [105]. In comparison, a terphenyl analog of **45** was less active, with only 30% conversion (under the same conditions) – clearly indicating that the catalytic activity is affected by the number and orientation of iodine substituents. Having these results in hand, different cationic bidentate XB donors (**46-51**) were also tested using 10 mol% catalyst loading in THF as a solvent. The pyridinium-based catalyst **46** performed worst with only 14% yield, followed by the imidazolium **47** and benzimidazolium **50** with 25% and 49% yield, respectively. Higher yields were obtained when increasing the catalyst load to 20 mol% with 67% for **47** and 71% yield for XB donor **50.** The bidentate analogs **48** and **49**, on the other hand, showed higher catalytic activity with up to 67% yield (at 10 mol% catalyst load). For comparison, the brominated and chlorinated XB donors were also tested, confirming the trend of halogen bonding strength increasing in the order Cl < Br < I: no conversion was observed with chlorine as a halogen. To emphasize that the reaction is catalyzed by halogen bonding, the non-halogenated derivatives were tested as well, but provided much lower yield of product. A further dramatic increase in activity could be achieved, however, by employing a rigid preorganized catalyst (**51**), which gave 70% conversion after six hours using only 0.5 mol% catalyst loading.

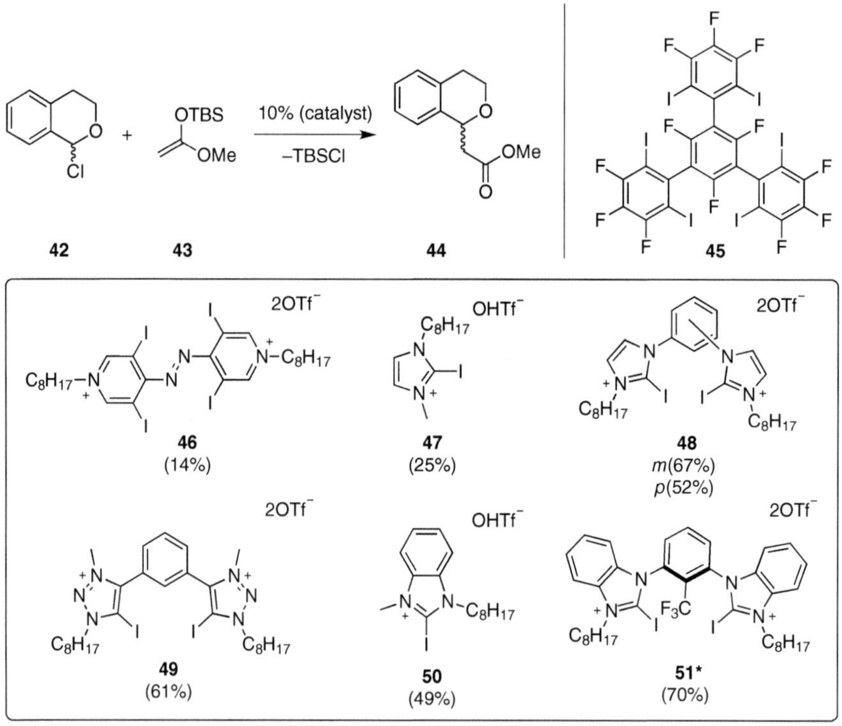

Scheme 10.10 Reaction between 1-chloroisochroman (**42**) and ketene silyl acetal (**43**) (top) and XB donors used (bottom). TBS = *tert*-butyldimethylsilyl. *Yield with 0.5 mol% catalyst load. Source: Adapted from Reisman *et al.* [124].

Other examples of halide abstraction reactions were studied by the group of Takemoto, achieving direct dehydroxylative coupling of activated alcohols with silylated nucleophiles (Scheme 10.11, top) [136]. Here, both the XB donor and iodine as halide source were used as catalysts. The *in situ* generated TMSI interacts with the XB donor forming a catalytic complex that coordinates to the alcohol. After cleavage of the latter, the resulting carbocation is then attacked by the used nucleophile. Recently, Yeung *et al.* published the addition of carbon-based nucleophiles to *N*-acyliminium ions via halogen bonding catalysis (Scheme 10.11, bottom) [137]. In this reaction, the used TMSCl cleaves the acetal moiety forming an inactive *N*-acyliminium ion, which undergoes an anion metathesis with the XB donor to give the reactive *N*-acyliminium triflate that is subsequently attacked by a nucleophile. Furthermore, an enantioselectivity of 44% *ee* could be achieved by using a chiral BINOL-based counterion. Surprisingly, this reaction was catalyzed best with a monodentate XB donor (**54**) instead of the usually stronger bidentate analogs.

Recognizing the potential of XB in organic synthesis, Takagi *et al.* used XBs for the polymerization of isobutyl vinyl ether (Scheme 10.12) [138]. For this cationic polymerization, both the XB donor and the initiator play an important role. The halogen substituent of the initiator is first cleaved by the XB donor and the resulting

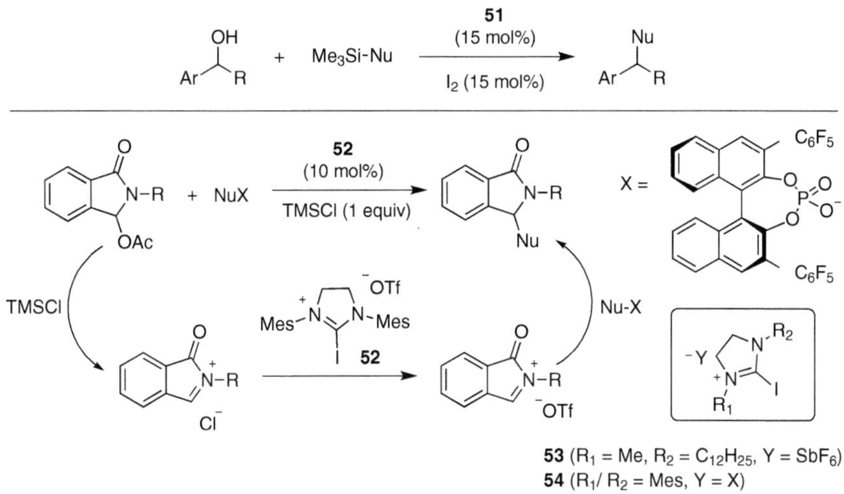

Scheme 10.11 Dehydroxylative coupling of activated alcohols with silylated nucleophiles (top) and functionalization of *N*,*O*-aminals (bottom). Source: (top) Based on Saito *et al.* [136], (bottom) Based on Chan and Yeung [137].

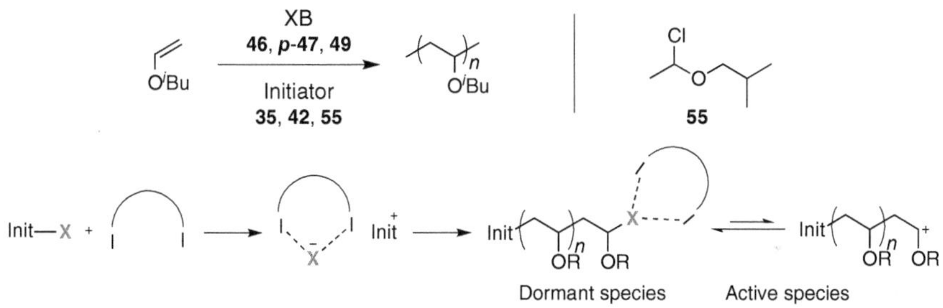

Scheme 10.12 Cationic polymerization via halogen bonding (top) and proposed mechanism (bottom). Source: Based on Takagi *et al.* [138].

carbocation (Int^+) then starts the chain reaction. The liberated halide recombines with the propagating end, resulting in a dormant species whereupon the halide is again cleaved by the XB donor, regenerating an active species that can then continue the propagation reaction.

Very recently, the group of Aoshima published another metal-free cationic polymerization of alkyl vinyl ethers and styrene derivatives using an iodine(III)-based XB donor (Scheme 10.13) [139]. Such "hypervalent" species had already been introduced as halogen bonding activators in a halide abstraction reaction before [132]. Similar to the previous polymerization example, both an initiator and the XB donor are crucial for the polymerization. In this case, tetrabutylammonium salts were used to generate the dormant species. The halogen substituent is then cleaved by the iodonium catalyst, forming the carbocation (active species) and thus the propagating species for the polymerization. An increase in the polymerization rate was observed

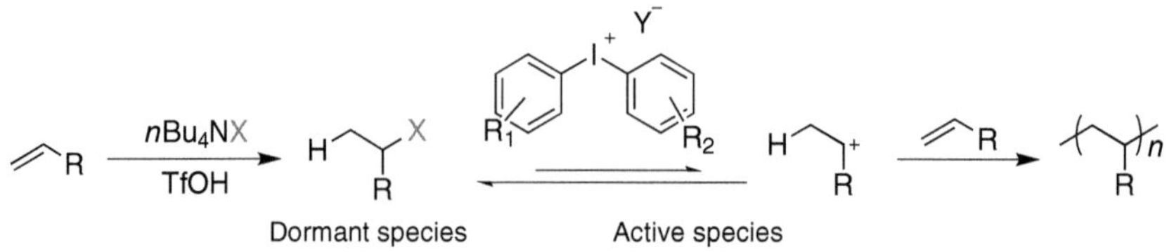

Scheme 10.13 Cationic polymerization of alkyl vinyl ethers with hypervalent iodine. Source: Based on Haraguchi *et al.* [139].

when electron-withdrawing substituents on the aromatic rings of the iodonium compound were used.

10.7 Chalcogen Bonding in Anion-Binding Catalysis

In contrast to the rapidly growing number of publications on XB-catalyzed reactions, studies on anion-binding catalysis or activation via ChB are much scarcer. This can be explained by the fact that halogen bonding has had some kind of head start over chalcogen bonding, as the corresponding research started several years earlier. Another potential issue with ChB catalysis is the fact that alkyl-substituted cationic ChB donors may decompose in the presence of halides via a nucleophilic substitution reaction [140]. This leads to dealkylation and thus deactivation of the catalyst (Scheme 10.14).

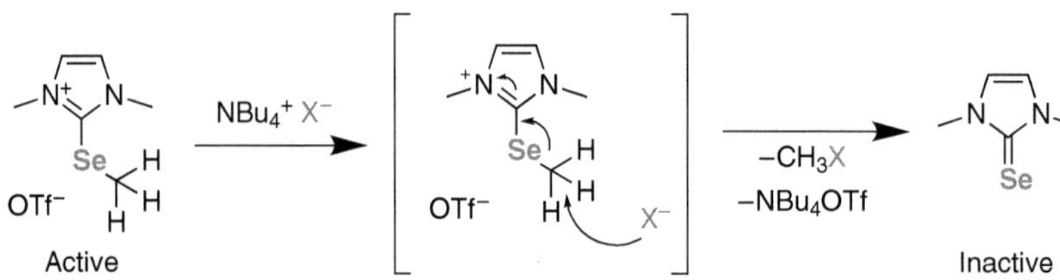

Scheme 10.14 Dealkylation of a chalcogen bond donor in the presence of a halide salt. Source: Based on Wonner *et al.* [140].

These considerations stem from the first example in which ChB donors were used in a carbon–halogen bond activation [140], which was published shortly after the pioneering work of Matile and coworkers on ChB organocatalysis [79]. While the latter dealt with the activation of neutral substrates, the former focused once again on the solvolysis of benzhydryl bromide as a benchmark reaction (compare Scheme 10.7). Similar to the XB variant, stoichiometric amounts of ChB donor needed to be used, and thus, this case does not qualify as catalysis, but it nevertheless highlights the strong potential of these Lewis acids for organic synthesis.

For this reaction, selenium-containing ChB donors based on a preorganized bidentate benzimidazolium motif (compare XB donor **51**) were synthesized and applied in the reaction. The first compound used was the methylated derivative;

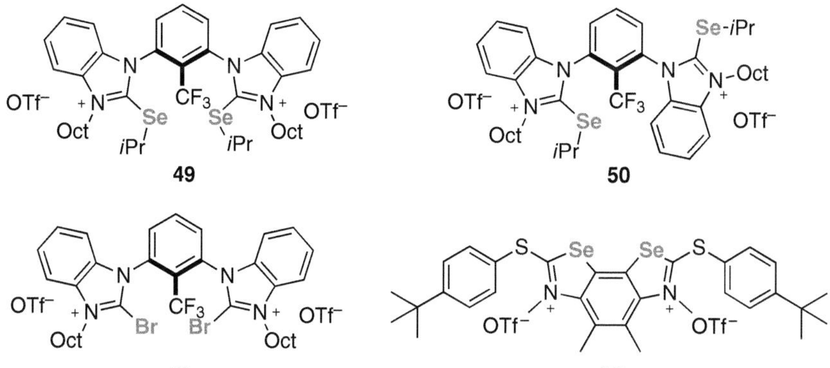

Figure 10.19 Activators used in the ChB-mediated solvolysis of benzhydryl bromide.

however, decomposition of the catalyst was observed during the reaction according to the mechanism described above. In order to increase stability, bulkier alkyl groups were attached to the selenium center. While octyl groups still resulted in a significant amount of decomposition, no such reaction was observed for the isopropyl-substituted ChB donor **49** (Figure 10.19). Yields of up to 64% after 140 hours could be observed, and through comparison experiments with different hydrogen bonding derivatives and possible decomposition products, the activation could clearly be assigned to the proposed chalcogen bonding mechanism. To gain a better insight into the activity, the brominated analog **51** of this compound was also tested, resulting in a yield of only 35% after the same time. Additionally, the initial rate constants of the various activators were determined, using the background reaction as a standard. The ChB donor was in this case roughly four times as active as the brominated derivative (k_{rel}(ChB) = 34 vs. k_{rel}(XB) = 9, Table 10.2).

This reaction was also investigated by the group of Matile using dicationic benzodiselenazole-based system **52**. While the initial rate constant (k_{rel} = 36) is similar to that of the bisbenzimidazolium-based compound, the ChB donor provided only a yield of 21% after 140 hours [141].

Still, in all these cases, the abstracted halide bound too strongly to the ChB donor to allow catalytic turnover, and thus, a recovery mechanism of the Lewis

Table 10.2 Yields and relative initial rate accelerations of selected activators used for the solvolysis of benzhydryl bromide.

Activator	Yield in % after 140 h	k_{rel}
—	10	1
49	64	34
50	45	23
51	35	9
52	21	36

acid would be necessary to allow organocatalysis. The first such example of actual anion-binding catalysis followed shortly after the studies described above and focused on the aforementioned reaction of 2-chloro-isochromane with ketene silyl acetals [142]. Here, as described above, the silyl group which is released from the acetal traps the liberated chloride and thus recovers the catalyst. In this study, the same ChB donors that were used for the benzhydryl bromide hydrolysis were applied; however, one major difference can be seen: no decomposition of the differently alkylated catalysts could be observed. This is due to the fact that a much lower temperature (−78 °C vs. r.t.) was used for the reaction that makes a nucleophilic substitution of the alkyl group less likely. Consequently, not only the isopropyl-substituted derivative but also the methylated and octylated ChB donors could be employed. Interestingly, the latter showed the highest activity of the selenylated species in this reaction, achieving a yield of 92% after 118 hours. The other derivatives provided the product in 66% and 74% yield, respectively. This demonstrates that the steric influence of the second selenium substituent has a strong influence on the performance of the catalyst, as the electronic parameters of the different alkyl groups deviate only slightly from one another. While no explanation to this phenomenon can be given thus far, it seems that sterically more demanding substituents increase the activity of these catalysts. This is even more striking when comparing the corresponding monodentate derivatives of these ChB donors. Due to the trifluoromethyl group on the bridging benzene ring, the rotation of the benzimidazolium substituents is hindered, allowing the isolation of *syn*- and *anti*-atropisomers. While the *syn*-compound is clearly binding in a bidentate manner because of the preorganized structure, the *anti*-isomer can only bind in a monodentate manner despite the presence of two binding sites. When comparing the activity of the differently substituted *anti*-compounds, the bulky isopropyl-substituted donor induces the highest conversion (42%), while the methyl-substituted one shows the lowest (19%). This could indicate that the higher activity observed for the octylated derivative may be due to a sweet spot in activity vs. steric hindrance, as two bulky groups close to the binding site might hinder the accessibility of the active center. The chalcogen-based catalysts were again compared to the corresponding brominated XB donor, and the latter was once again markedly less active.

Lastly, a further example using tellurium-based ChB donors was published by Matile and coworkers, which also featured a comparison with halogen and pnictogen bonding (which will be discussed in more detail below) [90]. In this study, bis-pentafluorophenyl-telluride and -selenide were used in the activation of the above described reaction of 2-chloro-isochromane with ketene silyl acetals and also in a Reissert-type substitution of isoquinoline (Scheme 10.15). Both reactions rely on an anion-binding mechanism to remove chloride from the active species. While only little activity of the selenium-based ChB donor was observed (6% yield for both products), the tellurium derivative provided a yield of 48% and 47%, respectively, clearly illustrating the increased chalcogen bonding potential of tellurium-based compounds.

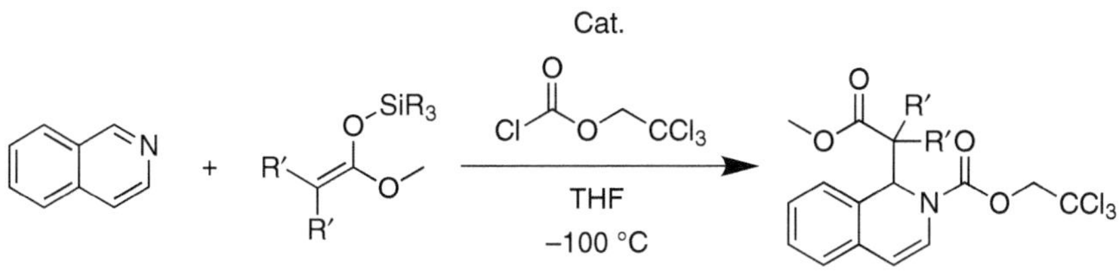

Scheme 10.15 Reissert-type substitution of isoquinoline catalyzed by Benz *et al.* [90]. For the list of catalysts used, see Table 10.3.

10.8 Pnictogen and Tetrel Bonding in Anion-Binding Catalysis

Compared to XB and even ChB, there is much less available literature on PnB or TeB catalysis. To the best of our knowledge, there is no example for the use of tetrel bonding in catalysis. While there are at least a handful of reactions that can be activated by pnictogen bonding, only one of those involves the activation of a carbon–halogen bond and thus an anion-binding mechanism: the above-mentioned publication by Matile and coworkers, which also included chalcogen and halogen bonding [90]. As catalyst candidates for the different interactions, pentafluorophenyl-substituted compounds were synthesized, i.e. C_6F_5X (X = I and Br), $(C_6F_5)_2Ch$ (Ch = Te and Se), and $(C_6F_5)_3Pn$ (Pn = P, As, and Sb). In order to obtain further information on the importance of these substituents, the pentafluorophenyl groups of the antimony compound were exchanged for phenyl groups once, twice, and thrice. First of all, the different catalysts were evaluated regarding their anion-binding strength toward chloride, using ^{19}F-NMR titrations against TBACl in THF. The respective dissociation constants are given in Table 10.3. Well in line with theoretical expectations, the compounds containing elements from the fourth period (Br, Se, and As) showed significantly higher dissociation constants than their respective analogs from the fifth period (I, Te, and Sb). In addition, an important trend could be observed when the different secondary interactions were compared: the binding strength increases in the order XB < ChB < PnB. This may be explained through the higher polarization of the higher substituted pnictogens compared to the chalcogens and halogens, as more electron-withdrawing substituents can be introduced into these molecules. Even though this should make tetrel bonding an even more promising candidate for these applications, the binding capabilities of TeB donors are diminished by the fact that the substituents sterically block any of the possible binding sites. Lastly, the dissociation constants of the differently substituted antimony-based donors also reveal that the additional electron-withdrawing substituents are likely the cause of the increased binding strength: the exchange of one pentafluorophenyl group toward a simple phenyl group increases the K_D value by 1 order of magnitude (from 19 to 570 μM), which interestingly is very close to the value of the tellurium derivative (470 μM).

Another observation that supports this reasoning comes from the catalytic application of the compounds in two different reactions, both revolving around

Table 10.3 Dissociation constants, reaction yields, and rate accelerations measured by Matile *et al.* using XB, ChB, and PnB donors.

Compound	K_D (μM)	Yield (%)[a]	Yield (%)[b]	k_{cat}/k_{uncat}[a]	k_{cat}/k_{uncat}[b]
C_6F_5Br	—	4	6	—	—
C_6F_5I	1 370 ± 30	47	15	50	5
$(C_6F_5)_2Se$	27 000 ± 4 000	6	6	—	—
$(C_6F_5)_2Te$	470 ± 70	47	48	52	39
$(C_6F_5)_3P$	—	3	—	—	—
$(C_6F_5)_3As$	13 300 ± 800	3	7	—	—
Ph_3Sb	—	3	8	—	—
$Ph_2(C_6F_5)Sb$	—	3	11	—	—
$Ph(C_6F_5)_2Sb$	570 ± 70	40	30	209	14
$(C_6F_5)_3Sb$	19 ± 7	53	91	4 090	99

Yield and k_{cat}/k_{uncat} are related to [a] = reaction of 2-chloroisochrome (Scheme 10.4) and [b] = Reissert-type reaction (Scheme 10.12).

the abstraction of chloride as the key step in the reaction mechanism. Here, once again the already discussed reaction of 2-chloro-isochromane with ketene silyl acetals as well as the Reissert-type substitution of isoquinoline was employed (see Schemes 10.4 and 10.12). In the latter case, no significant difference in the yields of the product were visible when XB, ChB, and PnB donors were compared; however, the rate acceleration of the reaction is significantly higher for the antimony compound relative to the iodine and tellurium derivatives. This is still true when one electron-withdrawing substituent on the antimony center is replaced, even though the yield does suffer in this case.

Looking at the reaction of 2-chloro-isochromane on the other hand, the same trend ($C_6F_5I < Ph(C_6F_5)_2Sb < (C_6F_5)_2Te < (C_6F_5)_3Sb$) as for the anion-binding studies can be observed. In this case, the slightly higher dissociation constant for $Ph(C_6F_5)_2Sb$ is also reflected in the yields and rate accelerations.

The group of Gabbaï also published the activation of benzhydryl bromide (see Scheme 10.7) – alongside transfer hydrogenation reactions of isoquinoline and *N*-benzylidene aniline with a Hantzsch Ester – using Sb(III)- and Sb(V)-based pnictogen bond donors (Figure 10.20 and Table 10.4) [91]. Interestingly, only catalytic amounts of the PnB donors had to be used in this case when the reaction temperature was increased to 40 °C. Expectedly, the phenyl-substituted derivatives **53a** and **54a** gave the lowest yield, while accumulating diphenylmethanol as a by-product. The more electron-deficient fluorine-substituted derivatives, however, showed much better performance, especially regarding the Sb(V)-compounds **54b** and **54c**, demonstrating that an increase in the oxidation state of the antimony is beneficial to the catalytic activity, which is similar to the trend observed for iodine(I)- and iodine(III)-based XB donors [132]. While these trends could also be

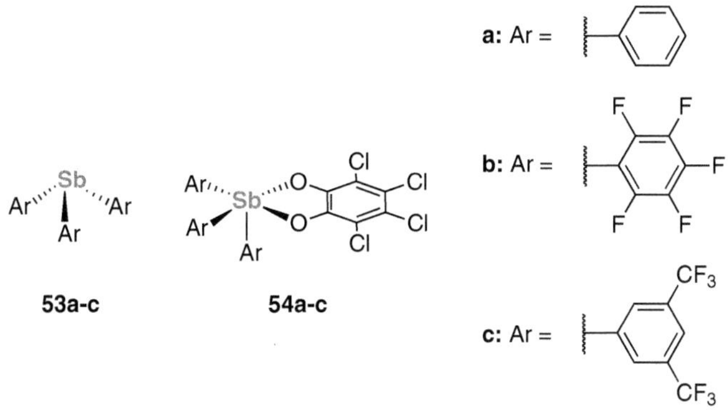

Figure 10.20 Antimony-based catalysts used by Yang *et al.* Source: Based on Yang *et al.* [91].

Table 10.4 Yields of the Ritter-like reaction of benzhydryl bromide (Scheme 10.7) using antimony-based catalysts (Figure 10.20).

Aryl group	Yield (%) using 53	Yield (%) using 54
a	10	12
b	44	55
c	21	77

Source: Based on Yang *et al.* [91].

supported by calculated fluorine affinity and LUMO calculations in the gas phase, perfluorinated donor **53b** showed an unexpectedly high conversion. This fact could be explained by a cooperative anion–π interaction formed toward the bromide by a pentafluorophenyl ring.

Overall, all this data stresses the high potential of pnictogen bonding compared to similar weak interactions and should raise further awareness for pnictogen bond donors as alternatives for Lewis acidic catalysts.

10.9 Conclusion

This chapter dealt with the application of weak interactions involving the group 14–17 elements in anion-binding organocatalysis. These secondary interactions have not nearly been studied as extensively as the much more well-known hydrogen bonding, but they are receiving increased interest in the past years, also due to several key differences that differentiate them from their classical analog.

Clearly, the most progress has been achieved with the use of halogen bonding, which could already be applied to several anion-binding reactions and has also successfully been employed in numerous other catalytic transformations. Different strategies important for catalyst design have been discussed (polarizability, cooperativity, and preorganization) and their impact has clearly been shown. So far, however, virtually all reported cases of catalysis related to anion-binding deal with halides, typically chloride and bromide. Thus, there seem to be many opportunities to push this concept further toward other leaving groups – even though some of them may be less suitable for halogen bonding than for hydrogen bonding (e.g. weaker coordination of oxoanions by halogen bond donors compared to hydrogen bond donors has been observed).

While halogen bonding catalysis has been studied for about 10 years now, the other "unorthodox" secondary interactions have only been in focus in this field for a relatively short amount of time, and thus understandably, there is less experience with chalcogen bonding and particularly pnictogen and tetrel bonding. In all cases in which halogen bonding catalysis was compared to chalcogen (and pnictogen) bonding, however, a clear trend toward better performance for catalysts based on elements of earlier groups was observed. It is well possible, then, that halogen bonding will have to share its currently prominent position with these other interactions and that all of them may become useful tools for the design of catalysts. Likely, halogen, chalcogen, and pnictogen bonding will at some point complement each other (and hydrogen bonding), with each one making use of its specific properties. Most surprising, maybe, is the fact that well into the twenty-first century, there is still so little we know about these long-ignored secondary interactions.

References

1 Alcock, N. (1972). Secondary bonding to nonmetallic elements. In: *Advances in Inorganic Chemistry and Radiochemistry* (eds. H.J. Emeléus and A.G. Sharpe), 1–58. Academic Press.
2 Lim, J. and Beer, P. (2018). Sigma-hole interactions in anion recognition. *Chemistry* 4: 731–783.
3 Sutar, R. and Huber, S. (2019). Catalysis of organic reactions through halogen bonding. *ACS Catal.* 9: 9622–9639.
4 Taylor, M. (2020). Anion recognition based on halogen, chalcogen, pnictogen and tetrel bonding. *Coord. Chem. Rev.* 413: 213270.
5 Desiraju, G., Ho, P., Kloo, L. *et al.* (2013). Definition of the halogen bond (IUPAC Recommendations 2013). *Pure Appl. Chem.* 85: 1711–1713.
6 Colin, M. (1814). Note sur quelques combinaisons de l'iode. *J. Phys. Chim. Hist. Nat. Arts* 10: 252–272.
7 Guthrie, F. (1863). XXVIII. – On the iodide of iodammonium. *J. Chem. Soc.* 16: 239–244.
8 Mulliken, R. (1950). Structures of complexes formed by halogen molecules with aromatic and with oxygenated solvents 1. *J. Am. Chem. Soc.* 72: 600–608.
9 Mulliken, R. (1952). Molecular compounds and their spectra. II. *J. Am. Chem. Soc.* 74: 811–824.

10 Mulliken, R. (1952). Molecular compounds and their spectra. III. The interaction of electron donors and acceptors. *J. Phys. Chem.* 56: 801–822.

11 Mulliken, R. and Person, W. (1969). *Molecular Complexes.* A lecture and reprint volume. New York: Wiley-Interscience.

12 Hassel, O., Hvoslef, J., Vihovde, E., and Sörensen, N. (1954). The structure of bromine 1,4-dioxanate. *Acta Chem. Scand.* 8: 873.

13 Hassel, O., Strømme, K., Haraldsen, H. *et al.* (1958). Structure of the crystalline compound benzene-bromine (1,1). *Acta Chem. Scand.* 12: 1146.

14 Benesi, H. and Hildebrand, J. (1949). A spectrophotometric investigation of the interaction of iodine with aromatic hydrocarbons. *J. Am. Chem. Soc.* 71: 2703–2707.

15 Di Paolo, T. and Sandorfy, C. (1974). On the hydrogen bond breaking ability of fluorocarbons containing higher halogens. *Can. J. Chem.* 52: 3612–3622.

16 Dumas, J., Peurichard, H., and Gomel, M. (1978). $CX_4\cdots$base interactions as models of weak charge-transfer interactions: comparison with strong charge-transfer and hydrogen-bond interactions. *J. Chem. Res., Synop* . 2: 54–55.

17 Weiss, R., Rechinger, M., and Hampel, F. (1994). Concerning the mechanism of α-elimination: hypervalent ion pairing in an iodocarbenium iodide. *Angew. Chem. Int. Ed.* 33: 893–895.

18 Weiss, R., Schwab, O., and Hampel, F. (1999). Ion-pair strain as the driving force for hypervalent adduct formation between iodide ions and substituted iodobenzenes: structural alternatives to Meisenheimer complexes. *Chem. Eur. J.* 5: 968–974.

19 Legon, A., Lister, D., and Thorn, J. (1994). Non-reactive interaction of ammonia and molecular chlorine: rotational spectrum of the 'charge-transfer' complex $H_3N\cdots Cl_2$. *J. Chem. Soc., Faraday Trans.* 90: 3205–3212.

20 Legon, A. and Warner, H. (1993). Isolation of stable intermediates in reactive gas mixtures: rotational spectrum of H_3P Cl_2 in a pulsed jet. *J. Chem. Phys.* 98: 3827–3832.

21 Bloemink, H., Legon, A., and Thorn, J. (1995). 'Charge-transfer' complexes of ammonia with halogens. Nature of the binding in H_3N BrCl from its rotational spectrum. *J. Chem. Soc., Faraday Trans.* 91: 781–787.

22 Desiraju, G. and Parthasarathy, R. (1989). The nature of halogen$\cdots$halogen interactions: are short halogen contacts due to specific attractive forces or due to close packing of nonspherical atoms? *J. Am. Chem. Soc.* 111: 8725–8726.

23 Pedireddi, V., Reddy, D., Goud, B. *et al.* (1994). The nature of halogen–halogen interactions and the crystal structure of 1,3,5,7-tetraiodoadamantane. *J. Chem. Soc., Perkin Trans.* 2: 2353–2360.

24 Allen, F., Kennard, O., Watson, D. *et al.* (1987). Tables of bond lengths determined by X-ray and neutron diffraction. Part 1. Bond lengths in organic compounds. *J. Chem. Soc., Perkin Trans.* 2: S1.

25 Metrangolo, P. and Resnati, G. (2001). Halogen bonding: a paradigm in supramolecular chemistry. *Chem. Eur. J.* 7: 2511–2519.

26 Metrangolo, P., Resnati, G., and Arman, H. (2008). *Halogen Bonding, Fundamentals and Applications.* Berlin, Heidelberg: Springer.

27 Metrangolo, P. and Resnati, G. (2015). *Halogen bonding, Fundamentals and applications*, vol. I and II. Berlin, Heidelberg: Springer.

28 Bertani, R., Sgarbossa, P., Venzo, A. *et al.* (2010). Halogen bonding in metal–organic–supramolecular networks. *Coord. Chem. Rev.* 254: 677–695.

29 Pennington, W., Resnati, G., and Taylor, M. (2013). Halogen bonding: from self-assembly to materials and biomolecules. *CrystEngComm* 15: 3057–3057.

30 Reed, A.E., Weinhold, F., Weiss, R., and Macheleid, J. (1985). Nature of the contact ion pair $CCl_3^+Cl^-$. a theoretical study. *Fortschr. Phys. Chem.*: 2688–2694.

31 Rosokha, S., Neretin, I., Rosokha, T. *et al.* (2006). Charge-transfer character of halogen bonding: molecular structures and electronic spectroscopy of carbon tetrabromide and bromoform complexes with organic σ- and π-donors. *Heteroat. Chem* 17: 449–459.

32 Rosokha, S., Stern, C., and Ritzert, J. (2013). Experimental and computational probes of the nature of halogen bonding: complexes of bromine-containing molecules with bromide anions. *Chem. Eur. J.* 19: 8774–8788.

33 Wolters, L., Schyman, P., Pavan, M. *et al.* (2014). The many faces of halogen bonding: a review of theoretical models and methods. *WIREs Comput. Mol. Sci.* 4: 523–540.

34 Hunter, C. (2004). Quantifying intermolecular interactions: guidelines for the molecular recognition toolbox. *Angew. Chem. Int. Ed.* 43: 5310–5324.

35 Cabot, R. and Hunter, C. (2009). Non-covalent interactions between iodo-perfluorocarbons and hydrogen bond acceptors. *Chem. Commun.* 2005–2007.

36 Kozuch, S. and Martin, J. (2013). Halogen bonds: benchmarks and theoretical analysis. *J. Chem. Theory Comput.* 9: 1918–1931.

37 Pinter, B., Nagels, N., Herrebout, W., and de Proft, F. (2013). Halogen bonding from a hard and soft acids and bases perspective: investigation by using density functional theory reactivity indices. *Chem. Eur. J.* 19: 519–530.

38 Wang, C., Danovich, D., Mo, Y., and Shaik, S. (2014). On the nature of the halogen bond. *J. Chem. Theory Comput.* 10: 3726–3737.

39 Robinson, S., Mustoe, C., White, N. *et al.* (2015). Evidence for halogen bond covalency in acyclic and interlocked halogen-bonding receptor anion recognition. *J. Am. Chem. Soc.* 137: 499–507.

40 Karpfen, A. (2008). Theoretical characterization of the trends in halogen bonding. In: *Halogen Bonding* (eds. P. Metrangolo and G. Resnati), 1–15. Berlin, Heidelberg: Springer.

41 Brinck, T., Murray, J., and Politzer, P. (1992). Surface electrostatic potentials of halogenated methanes as indicators of directional intermolecular interactions. *Int. J. Quantum Chem.* 44: 57–64.

42 Clark, T., Hennemann, M., Murray, J., and Politzer, P. (2007). Halogen bonding: the σ-hole. *J. Mol. Model.* 13: 291–296.

43 Huber, S., Scanlon, J., Jimenez-Izal, E. *et al.* (2013). On the directionality of halogen bonding. *Phys. Chem. Chem. Phys.* 15: 10350–10357.

44 Nyburg, S. (1979). "Polar flattening": non-spherical effective shapes of atoms in crystals. *Acta Cryst A* 35: 641–645.

45 Bloemink, H., Hinds, K., Holloway, J., and Legon, A. (1995). Characterisation of a pre-reactive intermediate in gas-phase mixtures of fluorine and ammonia: the rotational spectrum of the $H_3N...F_2$ complex. *Chem. Phys. Lett.* 245: 598–604.

46 Metrangolo, P., Murray, J., Pilati, T. *et al.* (2011). The fluorine atom as a halogen bond donor, viz. a positive site. *CrystEngComm* 13: 6593.

47 Metrangolo, P., Murray, J., Pilati, T. *et al.* (2011). Fluorine-centered halogen bonding: a factor in recognition phenomena and reactivity. *Cryst. Growth Des.* 11: 4238–4246.

48 Vent-Schmidt, T., Brosi, F., Metzger, J. *et al.* (2015). Fluorine-rich fluorides: new insights into the chemistry of polyfluoride anions. *Angew. Chem. Int. Ed.* 54: 8279–8283.

49 Guo, N., Maurice, R., Teze, D. *et al.* (2018). Experimental and computational evidence of halogen bonds involving astatine. *Nat. Chem.* 10: 428–434.

50 Hill, J. and Hu, X. (2013). Theoretical insights into the nature of halogen bonding in prereactive complexes. *Chem. Eur. J.* 19: 3620–3628.

51 Zhou, F., Liu, Y., Wang, Z. *et al.* (2019). A new type of halogen bond involving multivalent astatine: an ab initio study. *Phys. Chem. Chem. Phys.* 21: 15310–15318.

52 Thirman, J., Engelage, E., Huber, S., and Head-Gordon, M. (2018). Characterizing the interplay of Pauli repulsion, electrostatics, dispersion and charge transfer in halogen bonding with energy decomposition analysis. *Phys. Chem. Chem. Phys.* 20: 905–915.

53 Holthoff, J., Engelage, E., Weiss, R., and Huber, S. (2020). "Anti-electrostatic" halogen bonding. *Angew. Chem. Int. Ed.* 59: 11150–11157.

54 Vogel, L., Wonner, P., and Huber, S. (2019). Chalcogen bonding, an overview. *Angew. Chem. Int. Ed.* 58: 1880–1891.

55 Mahmudov, K., Kopylovich, M., Guedes da Silva, M., and Pombeiro, A. (2017). Chalcogen bonding in synthesis, catalysis and design of materials. *Dalton Trans.* 46: 10121–10138.

56 Aakeroy, C., Bryce, D., Desiraju, G. *et al.* (2019). Definition of the chalcogen bond (IUPAC Recommendations 2019). *Pure Appl. Chem.* 91: 1889–1892.

57 Murray, J., Lane, P., Clark, T., and Politzer, P. (2007). Sigma-hole bonding: molecules containing group VI atoms. *J. Mol. Model.* 13: 1033–1038.

58 Buergi, H. and Dunitz, J. (1987). Fractional bonds: relations among their lengths, strengths, and stretching force constants. *J. Am. Chem. Soc.* 109: 2924–2926.

59 Haberhauer, G. and Gleiter, R. (2020). The nature of strong chalcogen bonds involving chalcogen-containing heterocycles. *Angew. Chem. Int. Ed.* 59: 21236–21243.

60 Bleiholder, C., Werz, D., Köppel, H., and Gleiter, R. (2006). Theoretical investigations on chalcogen–chalcogen interactions: what makes these nonbonded interactions bonding? *J. Am. Chem. Soc.* 128: 2666–2674.

61 Iwaoka, M. and Tomoda, S. (1996). Nature of the intramolecular Se N nonbonded interaction of 2-selenobenzylamine derivatives. an experimental evaluation by ^{1}H, ^{77}Se, and ^{15}N NMR spectroscopy. *J. Am. Chem. Soc.* 118: 8077–8084.

62 Iwaoka, M. and Isozumi, N. (2012). Hypervalent nonbonded interactions of a divalent sulfur atom. Implications in protein architecture and the functions. *Molecules* 17: 7266–7283.

63 Lu, L., Lu, Y., Zhu, Z., and Liu, H. (2019). Pnictogen, chalcogen, and halogen bonds in catalytic systems: theoretical study and detailed comparison. *J. Mol. Model.* 26: 16.

64 Dong, W., Li, Q., and Scheiner, S. (2018). Comparative strengths of tetrel, pnicogen, chalcogen, and halogen bonds and contributing factors. *Molecules* 23: 1681.

65 Cao, H., Qian, R., and Yu, L. (2020). Selenium-catalyzed oxidation of alkenes: insight into the mechanisms and developing trend. *Catal. Sci. Technol.* 10: 3113–3121.

66 Breder, A. and Depken, C. (2019). Light-driven single-electron transfer processes as an enabling principle in sulfur and selenium multicatalysis. *Angew. Chem. Int. Ed.* 58: 17130–17147.

67 Kolb, S., Oliver, G., and Werz, D. (2020). Chemistry evolves, terms evolve, but phenomena do not evolve: from chalcogen-chalcogen interactions to chalcogen bonding. *Angew. Chem. Int. Ed.* 59: 22306–22310.

68 Gleiter, R., Haberhauer, G., Werz, D. *et al.* (2018). From noncovalent chalcogen-chalcogen interactions to supramolecular aggregates, experiments and calculations. *Chem. Rev.* 118: 2010–2041.

69 Vilela, F., Vobecka, Z., and Skabara, P. (2009). Organic conductors containing selenium and tellurium. In: *Patai's Chemistry of Functional Groups* (eds. S. Patai and Z. Rappoport), 1–30. Chichester: Wiley.

70 Würthner, F., Yao, S., Schilling, J. *et al.* (2001). ATOP dyes. Optimization of a multifunctional merocyanine chromophore for high refractive index modulation in photorefractive materials. *J. Am. Chem. Soc.* 123: 2810–2824.

71 Gsänger, M., Kirchner, E., Stolte, M. *et al.* (2014). High-performance organic thin-film transistors of J-stacked squaraine dyes. *J. Am. Chem. Soc.* 136: 2351–2362.

72 Bürckstümmer, H., Tulyakova, E., Deppisch, M. *et al.* (2011). Efficient solution-processed bulk heterojunction solar cells by antiparallel supramolecular arrangement of dipolar donor–acceptor dyes. *Angew. Chem. Int. Ed.* 50: 11628–11632.

73 Cozzolino, A., Elder, P., and Vargas-Baca, I. (2011). A survey of tellurium-centered secondary-bonding supramolecular synthons. *Coord. Chem. Rev.* 255: 1426–1438.

74 Bauzá, A., Quiñonero, D., Deyà, P., and Frontera, A. (2013). Halogen bonding versus chalcogen and pnicogen bonding: a combined Cambridge structural database and theoretical study. *CrystEngComm* 15: 3137–3144.

75 Kubiniok, S., Du Mont, W.-W., Pohl, S., and Saak, W. (1988). The reagent diphenyldiselane/iodine: no phenylselenenyl iodide but a charge transfer complex with cyclic moieties. *Angew. Chem. Int. Ed.* 27: 431–433.

76 Schulz, P. and Klar, G. (1975). Verbindungen mit Tellur in niedrigen Oxidationsstufen, I/compounds of Tellurium in low oxidation states, I. *Z. Naturforsch., B* 30: 40–42.

77 Fujita, K.-i., Iwaoka, M. and Tomoda, S. (1994). Synthesis of diaryl diselenides having chiral pyrrolidine rings with C2 symmetry. Their application to the asymmetric methoxyselenenylation of trans-β-methylstyrenes. *Chem. Lett.* 23: 923–926.

78 Wirth, T. (1995). Asymmetric reaction of arylalkenes with diselenides. *Angew. Chem. Int. Ed.* 34: 1726–1728.

79 Benz, S., López-Andarias, J., Mareda, J. *et al.* (2017). Catalysis with chalcogen bonds. *Angew. Chem. Int. Ed.* 56: 812–815.

80 Wonner, P., Dreger, A., Vogel, L. *et al.* (2019). Chalcogen bonding catalysis of a nitro-Michael reaction. *Angew. Chem. Int. Ed.* 58: 16923–16927.

81 Wang, W., Zhu, H., Liu, S. *et al.* (2019). Chalcogen–chalcogen bonding catalysis enables assembly of discrete molecules. *J. Am. Chem. Soc.* 141: 9175–9179.

82 Wang, W., Zhu, H., Feng, L. *et al.* (2020). Dual chalcogen–chalcogen bonding catalysis. *J. Am. Chem. Soc.* 142: 3117–3124.

83 Brammer, L. (2017). Halogen bonding, chalcogen bonding, pnictogen bonding, tetrel bonding: origins, current status and discussion. *Faraday Discuss.* 203: 485–507.

84 Bauzá, A., Mooibroek, T., and Frontera, A. (2013). Tetrel-bonding interaction: rediscovered supramolecular force? *Angew. Chem. Int. Ed.* 52: 12317–12321.

85 Zahn, S., Frank, R., Hey-Hawkins, E., and Kirchner, B. (2011). Pnicogen bonds: a new molecular linker? *Chem. Eur. J.* 17: 6034–6038.

86 Gini, A., Paraja, M., Galmés, B. *et al.* (2020). Pnictogen-bonding catalysis: brevetoxin-type polyether cyclizations. *Chem. Sci.* 11: 7086–7091.

87 Hirai, M., Cho, J., and Gabbaï, F. (2016). Promoting the hydrosilylation of benzaldehyde by using a dicationic antimony-based lewis acid: evidence for the double electrophilic activation of the carbonyl substrate. *Chem. Eur. J.* 22: 6537–6541.

88 Yang, M. and Gabbaï, F. (2017). Synthesis and properties of triarylhalostibonium cations. *Inorg. Chem.* 56: 8644–8650.

89 Yang, M., Hirai, M., and Gabbaï, F. (2019). Phosphonium-stibonium and bis-stibonium cations as pnictogen-bonding catalysts for the transfer hydrogenation of quinolines. *Dalton Trans.* 48: 6685–6689.

90 Benz, S., Poblador-Bahamonde, A., Low-Ders, N., and Matile, S. (2018). Catalysis with pnictogen, chalcogen, and halogen bonds. *Angew. Chem. Int. Ed.* 57: 5408–5412.

91 Yang, M., Tofan, D., Chen, C.-H. *et al.* (2018). Digging the σ-hole of organoantimony lewis acids by oxidation. *Angew. Chem. Int. Ed.* 57: 13868–13872.

92 Steiner, T. (2002). The hydrogen bond in the solid state. *Angew. Chem. Int. Ed.* 41: 48–76.

93 Pearson, R. (1963). Hard and soft acids and bases. *J. Am. Chem. Soc.* 85: 3533–3539.

94 Robertson, C., Perutz, R., Brammer, L., and Hunter, C. (2014). A solvent-resistant halogen bond. *Chem. Sci.* 5: 4179–4183.

95 Heinen, F., Engelage, E., Cramer, C., and Huber, S. (2020). Hypervalent iodine(III) compounds as biaxial halogen bond donors. *J. Am. Chem. Soc.* 142: 8633–8640.

96 C. Y. Legault. (2009). CYLview 1.0b. Université de Sherbrooke. http://www.cylview.org (accessed 11. November 2020).

97 Scilabra, P., Terraneo, G., and Resnati, G. (2019). The chalcogen bond in crystalline solids: a world parallel to halogen bond. *Acc. Chem. Res.* 52: 1313–1324.

98 Metrangolo, P., Neukirch, H., Pilati, T., and Resnati, G. (2005). Halogen bonding based recognition processes: a world parallel to hydrogen bonding. *Acc. Chem. Res.* 38: 386–395.

99 Mele, A., Metrangolo, P., Neukirch, H. *et al.* (2005). A halogen-bonding-based heteroditopic receptor for alkali metal halides. *J. Am. Chem. Soc.* 127: 14972–14973.

100 Sarwar, M., Dragisic, B., Salsberg, L. *et al.* (2010). Thermodynamics of halogen bonding in solution, substituent, structural, and solvent effects. *J. Am. Chem. Soc.* 132: 1646–1653.

101 Sarwar, M., Dragisic, B., Sagoo, S., and Taylor, M. (2010). A tridentate halogen-bonding receptor for tight binding of halide anions. *Angew. Chem. Int. Ed.* 49: 1674–1677.

102 Sarwar, M., Dragisić, B., Dimitrijević, E., and Taylor, M. (2013). Halogen bonding between anions and iodoperfluoroorganics: solution-phase thermodynamics and multidentate-receptor design. *Chem. Eur. J.* 19: 2050–2058.

103 Gillis, E., Demireva, M., Sarwar, M. *et al.* (2013). Structure and energetics of gas phase halogen-bonding in mono-, bi-, and tri-dentate anion receptors as studied by BIRD. *Phys. Chem. Chem. Phys.* 15: 7638–7647.

104 Chudzinski, M., McClary, C., and Taylor, M. (2011). Anion receptors composed of hydrogen- and halogen-bond donor groups, modulating selectivity with combinations of distinct noncovalent interactions. *J. Am. Chem. Soc.* 133: 10559–10567.

105 Kniep, F., Jungbauer, S., Zhang, Q. *et al.* (2013). Organocatalysis by neutral multidentate halogen-bond donors. *Angew. Chem. Int. Ed.* 52: 7028–7032.

106 Mole, T., Arter, W., Marques, I. *et al.* (2015). Neutral bimetallic rhenium(I)-containing halogen and hydrogen bonding acyclic receptors for anion recognition. *J. Organomet. Chem.* 792: 206–210.

107 Mercurio, J., Knighton, R., Cookson, J., and Beer, P. (2014). Halotriazolium axle functionalised [2]rotaxanes for anion recognition: investigating the effects of halogen-bond donor and preorganisation. *Chem. Eur. J.* 20: 11740–11749.

108 Tepper, R. and Schubert, U. (2018). Halogen bonding in solution, anion recognition, templated self-assembly, and organocatalysis. *Angew. Chem. Int. Ed.* 57: 6004–6016.

109 Walter, S., Kniep, F., Rout, L. *et al.* (2012). Isothermal calorimetric titrations on charge-assisted halogen bonds: role of entropy, counterions, solvent, and temperature. *J. Am. Chem. Soc.* 134: 8507–8512.

110 Jungbauer, S. and Huber, S. (2015). Cationic multidentate halogen-bond donors in halide abstraction organocatalysis: catalyst optimization by preorganization. *J. Am. Chem. Soc.* 137: 12110–12120.

111 Tepper, R., Schulze, B., Jäger, M. *et al.* (2015). Anion receptors based on halogen bonding with halo-1,2,3-triazoliums. *J. Org. Chem.* 80: 3139–3150.

112 Cametti, M., Raatikainen, K., Metrangolo, P. *et al.* (2012). 2-Iodo-imidazolium receptor binds oxoanions via charge-assisted halogen bonding. *Org. Biomol. Chem.* 10: 1329–1333.

113 Benz, S., Macchione, M., Verolet, Q. *et al.* (2016). Anion transport with chalcogen bonds. *J. Am. Chem. Soc.* 138: 9093–9096.

114 Garrett, G., Carrera, E., Seferos, D., and Taylor, M. (2016). Anion recognition by a bidentate chalcogen bond donor. *Chem. Commun.* 52: 9881–9884.

115 Garrett, G., Gibson, G., Straus, R. *et al.* (2015). Chalcogen bonding in solution, interactions of benzotelluradiazoles with anionic and uncharged Lewis bases. *J. Am. Chem. Soc.* 137: 4126–4133.

116 Lim, J., Marques, I., Thompson, A. *et al.* (2017). Chalcogen bonding macrocycles and [2]rotaxanes for anion recognition. *J. Am. Chem. Soc.* 139: 3122–3133.

117 Lim, J., Liew, J., and Beer, P. (2018). Thermodynamics of anion binding by chalcogen bonding receptors. *Chem. Eur. J.* 24: 14560–14566.

118 Zhao, H. and Gabbaï, F. (2010). A bidentate Lewis acid with a telluronium ion as an anion-binding site. *Nat. Chem.* 2: 984.

119 Farnham, W. and Harlow, R. (1981). Stereomutation at pentacoordinate silicon by intramolecular ligand exchange. *J. Am. Chem. Soc.* 103: 4608–4610.

120 Qiu, J., Song, B., Li, X., and Cozzolino, A. (2017). Solution and gas phase evidence of anion binding through the secondary bonding interactions of a bidentate bis-antimony(iii) anion receptor. *Phys. Chem. Chem. Phys.* 20: 46–50.

121 Christianson, A. and Gabbaï, F. (2017). A Lewis acidic, π-conjugated stibaindole with a colorimetric response to anion binding at Sb(III). *Organometallics* 36: 3013–3015.

122 Tamao, K., Hayashi, T., Ito, Y., and Shiro, M. (1992). Pentacoordinate anionic bis(siliconates) containing a fluorine bridge between two silicon atoms. Synthesis, solid-state structures, and dynamic behavior in solution. *Organometallics* 11: 2099–2114.

123 Jurkschat, K., Hesselbarth, F., Dargatz, M. *et al.* (1990). 1,2-bis(Organostannyl)ethanes as powerful bidentate Lewis acids. Crystal structures of $(Ph_2ClSnCH_2)_2{\cdot}(Me_2N)_2PO$ and $[Ph_3P\ N\ PPh_3][(Ph_2ClSnCH_2)_2Cl]$. *J. Organomet. Chem.* 388: 259–271.

124 Reisman, S., Doyle, A., and Jacobsen, E.N. (2008). Enantioselective thiourea-catalyzed additions to oxocarbenium ions. *J. Am. Chem. Soc.* 130: 7198–7199.

125 Bulfield, D. and Huber, S. (2016). Halogen bonding in organic synthesis and organocatalysis. *Chem. Eur. J.* 22: 14434–14450.

126 Guha, S., Kazi, I., Nandy, A., and Sekar, G. (2017). Role of Lewis-base-coordinated halogen(i) intermediates in organic synthesis: the journey from unstable intermediates to versatile reagents. *Eur. J. Org. Chem.* 37: 5497–5518.

127 Walter, S., Kniep, F., Herdtweck, E., and Huber, S. (2011). Halogen-bond-induced activation of a carbon-heteroatom bond. *Angew. Chem. Int. Ed.* 50: 7187–7191.

128 Kniep, F., Rout, L., Walter, S. *et al.* (2012). 5-Iodo-1,2,3-triazolium-based multidentate halogen-bond donors as activating reagents. *Chem. Commun.* 48: 9299–9301.

129 Kniep, F., Walter, S., Herdtweck, E., and Huber, S. (2012). 4,4′-Azobis(halopyridinium) derivatives: strong multidentate halogen-bond donors with a redox-active core. *Chem. Eur. J.* 18: 1306–1310.

130 Cast, J. and Stevens, T.S. (1953). *N*-alkylation of nitriles in presence of silver sulphate. *J. Chem. Soc.* 12: 4180–4181.

131 Perera, M. and Aakeröy, C. (2019). Organocatalysis by a multidentate halogen-bond donor: an alternative to hydrogen-bond based catalysis. *New J. Chem.* 43: 8311–8314.

132 Heinen, F., Engelage, E., Dreger, A. *et al.* (2018). Iodine(III) derivatives as halogen bonding organocatalysts. *Angew. Chem. Int. Ed.* 57: 3830–3833.

133 Castelli, R., Schindler, S., Walter, S. *et al.* (2014). Activation of glycosyl halides by halogen bonding. *Chem. Asian J.* 9: 2095–2098.

134 Tsuji, N., Kobayashi, Y., and Takemoto, Y. (2014). Electrophilic iodine(I) compounds induced semipinacol rearrangement via C—X bond cleavage. *Chem. Commun.* 50: 13691–13694.

135 Dreger, A., Engelage, E., Mallick, B. *et al.* (2018). The role of charge in 1,2,3-triazol(ium)-based halogen bonding activators. *Chem. Commun.* 54: 4013–4016.

136 Saito, M., Tsuji, N., Kobayashi, Y., and Takemoto, Y. (2015). Direct dehydroxylative coupling reaction of alcohols with organosilanes through Si—X bond activation by halogen bonding. *Org. Lett.* 17: 3000–3003.

137 Chan, Y.-C. and Yeung, Y.-Y. (2019). Halogen-bond-catalyzed addition of carbon-based nucleophiles to *N*-acylimminium ions. *Org. Lett.* 21: 5665–5669.

138 Takagi, K., Yamauchi, K., and Murakata, H. (2017). Halogen-bonding-mediated and controlled cationic polymerization of isobutyl vinyl ether: expanding the catalytic scope of 2-iodoimidazolium salts. *Chem. Eur. J.* 23: 9495–9500.

139 Haraguchi, R., Nishikawa, T., Kanazawa, A., and Aoshima, S. (2020). Metal-free living cationic polymerization using diaryliodonium salts as organic Lewis acid catalysts. *Macromolecules* 53: 4185–4192.

140 Wonner, P., Vogel, L., Düser, M. *et al.* (2017). Carbon–halogen bond activation by selenium-based chalcogen bonding. *Angew. Chem. Int. Ed.* 56: 12009–12012.

141 Benz, S., Besnard, C., and Matile, S. (2018). Chalcogen-bonding catalysis: from neutral to cationic benzodiselenazole scaffolds. *Helv. Chim. Acta* 101: e1800075.

142 Wonner, P., Vogel, L., Kniep, F., and Huber, S. (2017). Catalytic carbon–chlorine bond activation by selenium-based chalcogen bond donors. *Chem. Eur. J.* 23: 16972–16975.

11

New Trends and Supramolecular Approaches in Anion-Binding Catalysis

María C. Pérez-Aguilar, Melania Gómez-Martínez, Jan Kuhlmann, and Olga García Mancheño

Westfälische Wilhelms-Universität Münster, Institute of Organic Chemistry, Corrensstraße 36, D-48149 Münster, Germany

11.1 General Introduction

Anion-binding catalysis is gaining increasing attention in the past few years because of its modularity and capability to be combined with other types of catalysis, which offer new possibilities for discovery. Taking this into account, this chapter aims at giving an overview of some emerging directions of anion-binding catalysis that are still underrepresented within the existing broad portfolio already described in previous chapters. Hence, this chapter will first present the combination of photoredox and anion-binding catalysis as a new trend in this field for attaining novel or enhanced reactivities, followed by anion-binding assisted metal catalysis with bifunctional ligands that can coordinate both the active metal center and the counteranion of the substrate, which expands the type of reactions that can be targeted. The latest advances in supramolecular anion-binding receptors with catalytic activity will be next discussed according to the type of systems such as knots, rotaxanes, molecular motors, and macrocycles. Finally, the concept and recent applications of anion–π catalysis are introduced. Although it does not always imply an anionic species, this relatively novel approach in catalysis exploits the stabilization on π–acidic aromatic surfaces of partial negative charges in polar transition states, which are ubiquitous in organic transformations, to promote reactions and allow unprecedented reactivities.

11.2 Dual Photoredox and Anion-Binding Catalysis

Photocatalysis utilizes the energy of light to drive a chemical reaction in the presence of a catalyst that can absorb the applied light and get directly involved in the transformation in its excited state [1]. In this way, the use of photocatalysts provides active intermediates that can alter the rate of the reaction or even allow unconventional

Anion-Binding Catalysis, First Edition. Edited by Olga García Mancheño.

reactivities that are difficult or forbidden under standard classical thermal settings. For that reason, an increasing interest as well as a large of variety catalytic systems has been achieved in the past decades in this research area.

Although the photocatalysis term dates back to 1911 [2], its practical application has been based on the breakthrough realization of water electrolysis using titanium dioxide (TiO_2) by Fujishima and Honda in 1972 [3]. Since then, many important applications have been described, initially by means of powerful UV light as the energy source [4]. However, the use of visible light in photocatalytic organic synthesis has developed rapidly over the past years and emerged as a powerful synthetic tool because it allows milder, cheaper, controlled, and more selective transformations [5].

Particularly, the merging between visible light photocatalysis and organocatalysis has become a field of great interest in the past decade [6]. The cooperation of both catalytic systems provides new pathways and synthetic possibilities as well as solves some challenging drawbacks, such as the stereocontrol during the transformation (Figure 11.1) [15]. Pioneering work by the group of MacMillan on the combination of amino and photocatalysis already showed the benefits of such dual strategies in photochemical synthesis (Figure 11.1a) [7]. As a result, this approach was effectively employed with chiral amino catalysts to promote different kinds of reactions via enamine [16] or iminium [17] catalysis such as different α-functionalizations of aldehydes or Michael additions with excellent enantioselectivities. In contrast, Melchiorre developed a new conceptual idea that relies on the generation of a chiral donor–acceptor (EDA, electron donor–acceptor) complex in the media between the *in situ* formed enamine species and the substrate to induce chirality during photochemical transformations (Figure 11.1b) [8]. A further covalent approach for the asymmetric C–C functionalization of aldehydes was reported by Rovis and DiRocco who utilized chiral *N*-heterocyclic carbenes (NHC) as organocatalysts for the umpolung of aldehydes, while the photoredox catalyst (PC) oxidizes the amino coupling substrate (Figure 11.1c) [9]. Another concept toward asymmetric methods was introduced by the group of Bach in 2005, who designed different single bifunctional photocatalysts that affect complementary hydrogen bonding (HB) with the substrate to achieve a better fixation, enhancing the enantioinduction (Figure 11.1d) [10]. After this work, other examples of hydrogen bonding and photoredox cocatalysis have been reported using HB donors such as squaramides to activate carbonyl substrates (Figure 11.1e) [11]. Besides that, both Lewis or Brønsted bases and Brønsted acids have also been employed, providing key intermediates by deprotonation and protonation of the substrates before or after the photochemical process, respectively (Figure 11.1f,g) [12]. Moreover, several approaches in ion-pairing catalysis have also recently evoked significant attention for reactions implying ionic intermediates in photoredox catalysis. Those include both cation-directed and anion-directed photocatalytic processes. In this regard, the group of Ooi reported a cation-directed Ir-photocatalyzed coupling of *in situ* generated α-amino radical addition to imines using a chiral tetraaminophosphonium salt (Figure 11.1h) [13]. Alternatively, Nicewicz and coworkers modified oxopyrilium photocatalysts by

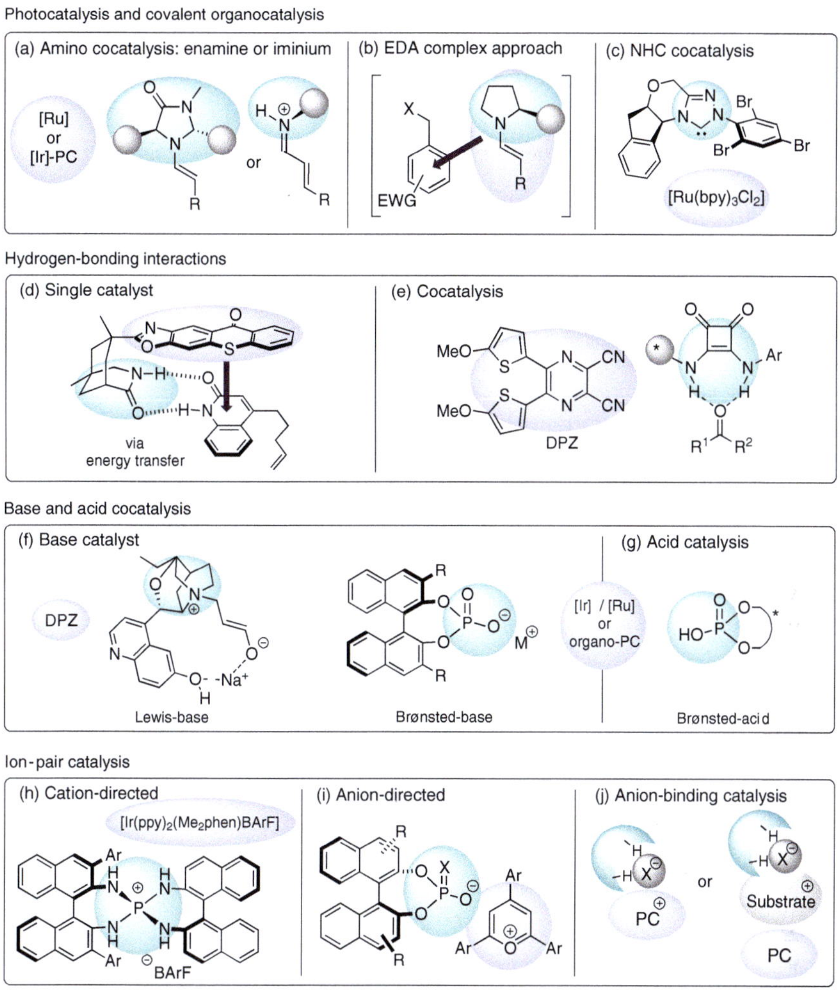

Figure 11.1 Dual photoredox – organocatalytic approaches (a–j). Photocatalysts (PC) or photoactive species are highlighted in lila and organocatalysts in green. Source: (a) Based on Nicewicz and MacMillan [7]. (b) Adapted from Ref. [8]. (c) Based on DiRocco and Rovis [9]. (d) Adapted from Refs [10]. (e) Source: Based on Liu *et al.* [11]. (f, g) Adapted from Ref. [12]. (h) Source: Uraguchi *et al.* [13]. (i) Source: Based on Morse *et al.* [14].

exchange of their non-chiral counteranions to a chiral phosphate ion, allowing moderate enantioinductions in a Diels–Alder reaction (Figure 11.1i) [14].

Surprisingly, within the growing field of ion-pairing catalysis, only a few reports concerning the promising combination of photoredox catalysis and non-covalent anion-binding catalysis have been reported so far (Figure 11.1j). This can be attributed to several challenges accounted when merging a photoredox catalyst with an anion-binding catalyst, whose activity might be reduced or quenched by undesired complexations or blocking of the active sites. Moreover, the PC or the substrate must contain an anionic motif that allows the suitable interaction with the

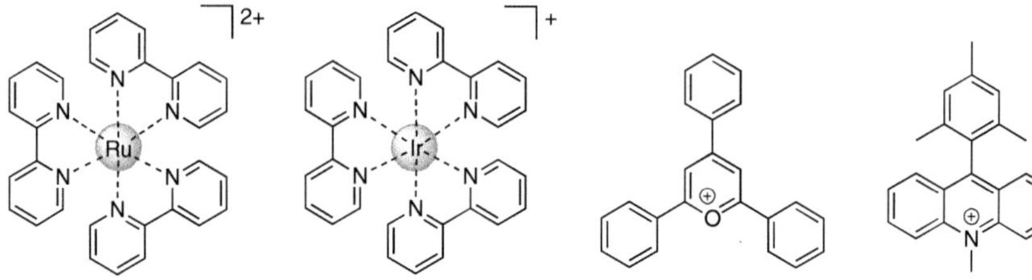

[Ru(bpy)$_3$]$^{+2}$ [Ir(ppy)$_2$(dtbbpy)]$^{+}$ TPP^{+} (2,4,6-Triphenylpyrylium) MesAcr^{+} (Fukuzumi)

Figure 11.2 Representative commonly employed ionic photoredox catalysts.

anion-binding catalyst. Fortunately, there are different well-known ionic photoredox catalysts based on both transition metal and organic photoredox-active species (Figure 11.2). Importantly, the HB donor catalyst–counteranion complex must form a contact ion pair with the cationic substrate or intermediate for achieving effective chiral inductions when aiming at enantioselective transformations. In addition, very different solvent media are ideal for each type of process. Mostly, polar solvents are used in photoredox catalysis, while asymmetric anion binding is mostly carried out in non-polar solvents to guarantee the formation of the chiral contact ion pair responsible for the chirality transfer.

In this field, pioneering work was done by Jacobsen and Stephenson, who presented the first example for the combination of visible light photoredox and enantioselective anion-binding catalysis (Scheme 11.1) [18]. Consequently, they embraced the α-oxidation and asymmetric nucleophilic addition of silyl ketene acetals **3** to *N*-aryl tetrahydroisoquinolines (THIQs) **1** toward the synthesis of β-amino acid derivatives **4**. However, the reaction had to be performed sequentially because of incompatibility issues between the two catalytic systems. In the first step, THIQ **1** is oxidized by Ru(bpy)$_3$$^{+2}$ and CCl_4 as an oxidant in MeCN into the corresponding intermediate **2**. After switching the solvent to the less polar methyl-*tert*-butylether (MTBE), a chiral thiourea-based anion-binding catalyst **5** is added to abstract the halogen atom of **2** and generate an iminium salt, which forms

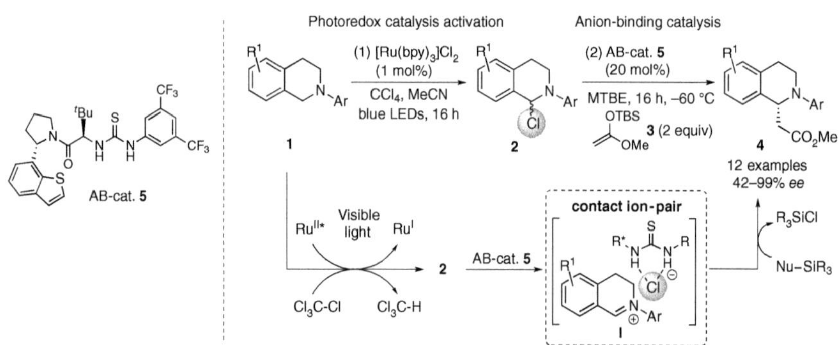

Scheme 11.1 Photoredox activation and asymmetric anion-binding catalysis sequential approach. Source: Based on Bergonzini *et al.* [18].

a chiral contact ion pair **I** with **5** upon binding of its counteranion and is responsible for inducing the enantioselectivity in the following addition of the nucleophile. The use of Cl^- as a counteranion for both the Ru–PC and the iminium intermediate was found as the optimal combination, providing moderate to good yields and high enantioselectivities (up to 99% ee).

As depicted with the previous work, conducting photoredox and anion-binding catalysis in a one-pot approach without requiring sequential addition of reagents, catalysts, and/or change of the solvent system represents a particular challenge. In this regard, in 2019, Yoon and coworkers investigated the influence of the counteranion on the reactivity of ruthenium(II)-based photocatalysts [19] (Scheme 11.2). Therefore, various Ru(II) catalysts were synthesized with different organic and inorganic counteranions (**X**) and tested in a radical cationic Diels–Alder reaction of anethole (**6**) as model reaction.

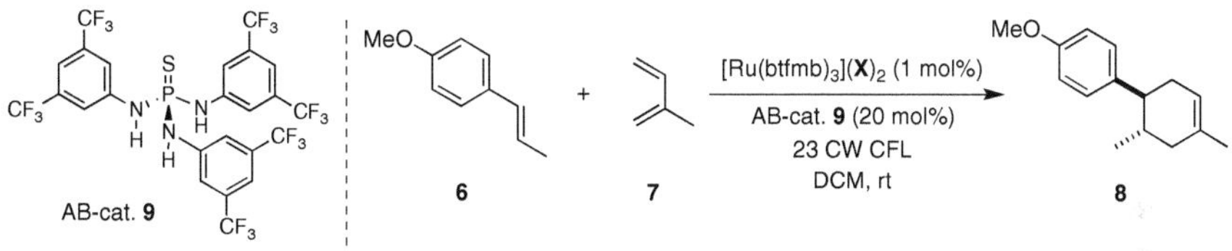

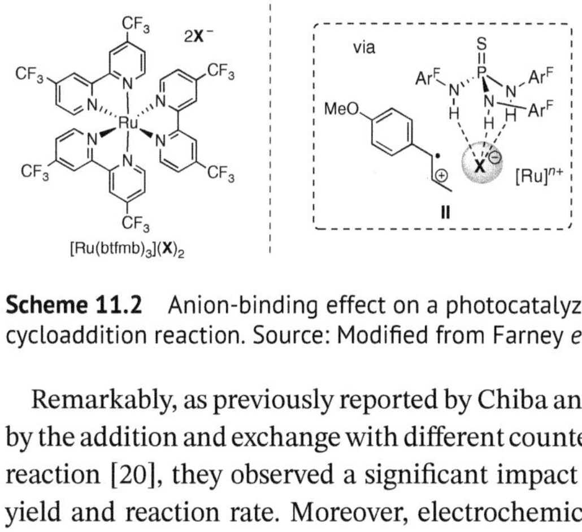

Entry	X	9	Time	Yield (%)
1	$BAr^F{}_4{}^-$	—	20 min	98
2	TfO^-	—	24 h	55
3	$Ar^FSO_3{}^-$	—	24 h	14
4	$Ar^FSO_3{}^-$	✓	2 h	99
5	—	✓	2 h	0

Scheme 11.2 Anion-binding effect on a photocatalyzed radical-cation Diels–Alder cycloaddition reaction. Source: Modified from Farney *et al.* [19].

Remarkably, as previously reported by Chiba and coworkers using $Ru(pbz)_3(PF_6)_2$ by the addition and exchange with different counteranions in this model Diels–Alder reaction [20], they observed a significant impact of the nature of the anion on the yield and reaction rate. Moreover, electrochemical investigations showed that ion pairing affects the triplet-state energy, as well as the ground-state electrochemical potential. To investigate and determine this influence, a H-bonding anion-binding cocatalyst was employed. The aim was to disrupt the Coulombic interactions between the counteranion and both the Ru catalyst and the reactive radical–cation transitional species of the reaction (**II**). Therefore, a thiophosphotriamide **9** was employed as an anion binder to abstract the strongly coordinating $Ar^FSO_3{}^-$ counteranion and form a less tight ion pair **II**. The influence of the anion-binding cocatalyst **9** was remarkable, leading to a notable increased yield and a reduced reaction time. It was proposed that the less coordinative anion effect by weakening the ion pair upon anion binding might result in the formation of a more potent triplet excited state PC oxidant and longer radical–cation chain length. This approach shows impressively that anion-binding catalysis can remarkably modulate the activity of the PC and reaction efficiency, providing new entries in the broad field of photoredox catalysis.

Alternatively, the use of organophotocatalyst instead of metal-based species has also been proposed by Jiang and coworkers, who reported on the assisted anion-binding protonation upon photocatalytic reduction of ketones and imines using dicyanopyrazine (DPZ) and a chiral non-covalent catalyst as a diguanidinium salt **10** or a urea-based catalyst **11** as a cocatalytic system, respectively [21] (Scheme 11.3).

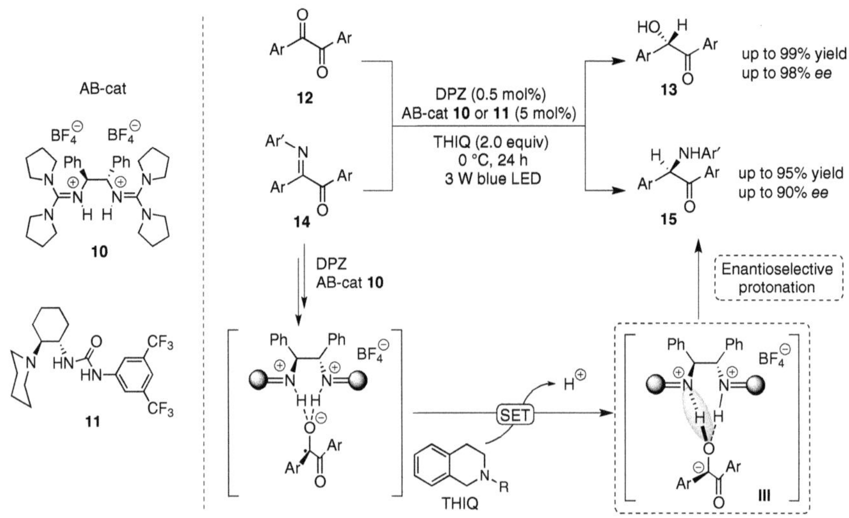

Scheme 11.3 Asymmetric reduction of 1,2-dicarbonyls **12** and α-ketoimines **14** employing photoredox catalysis and chiral non-covalent catalysis. Source: Based on Lin *et al.* [21].

The enantioselective protonation of 1,2-diketones **12** and 1,2-ketoketimines **14** proceeded smoothly through the controlled H^+ interchange between the protonated HB catalyst and the substrate (**III**), providing the corresponding chiral ketoalcohols **13** and ketoamines **15** with good to excellent enantioselectivities (up to 98% *ee*). It was found out that this reduction takes place through an EDA mechanism, thus not requiring the photocatalyst DPZ. However, the enantiomeric inductions were mostly significantly lower without implying the photoredox catalytic system.

Recently, an elegant example was developed by Ooi and coworkers [22], in which a highly enantioselective [3+2]-cycloaddition of *N*-cyclopropylurea **16** with α-alkylstyrenes **17** was described (Scheme 11.4). In this case, they used a directing group for anchoring the anion of the PC through non-covalent interactions and control both the reactivity and selectivity of the transformation. Therefore, an urea motif was selected because of its anion-binding ability and an iridium-based photocatalyst (**Ir-complex A**) with a chiral anion was employed. The cyclopropylurea can coordinate the chiral anion and stabilizes the radical–ion pair (**IV**) to induce chirality during radical addition with an α-alkylarene. This approach achieves a series of aminocyclopentanes **18** with very good diastereoselectivities and excellent enantioselectivities (up to 97% ee).

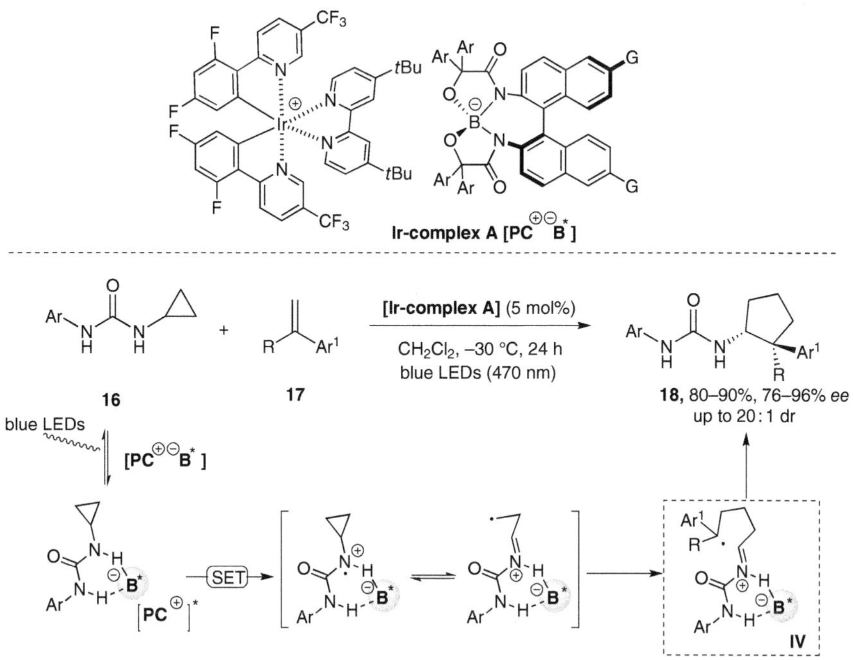

Scheme 11.4 Directing group strategy for the enantioselective [3+2]-cycloaddition of *N*-cyclopropylurea **16** with α-alkylstyrenes **17** mediated by a chiral iridium-borate photocatalyst.

11.3 Combination of Metal and Anion-Binding Catalysis

Besides the above-described photoredox cocatalyzed processes involving metal species, the idea of combining anion binding and metal catalysis is largely appealing as it might offer new possibilities and solutions to the often-low reactivity and/or stereocontrol of processes proceeding only through non-covalent interactions. Therefore, the introduction of metals might enhance the reactivity significantly. However, most transition metals are sensible to the media and can easily suffer ligand exchange, which could lead to the inactivation of the system by coordination at the anion-binding catalytic site. Therefore, the anion-binding and the metal-active sites need to be clearly separated and defined, which represents a great challenge. Although this field is still rather unexplored, the efficient implementation of the concept of anion-binding/metal cocatalysis for a few types of reactions, including hydrogenations, hydroformylations, or cross-coupling processes, has been recently achieved, and it will be described in the next sections.

11.3.1 Anion-Binding Assisted Hydrogenation Reactions

One of the most studied transformations utilizing the anion-binding/metal dual catalysis approach is the hydrogenation reaction. Based on the well-reported methods using chiral transition metal catalysts, the beneficial influence of anion-binding species has been introduced to the portfolio of asymmetric hydrogenations. In particular, the developed catalytic systems rely so far on the use of Rh and Ir species, which are very well known as highly efficient metals for hydrogenations, in combination with a bifunctional HB donor–bisphosphine ligand. Hence, in 2013, Zhang and coworkers designed a novel bifunctional catalyst, later called ZhaoPhos [23], which presents the two distinct functional areas that is required for the complexation of the metal center (ferrocenyl–bisphosphine) and the anion recognition (thiourea) (Scheme 11.5). Moreover, the thiourea hydrogen bonding moiety is sterically shielded by a meta-substituted difluoro methyl–phenyl group, hence hindering the metal ligation to this site. The ZhaoPhos ligand was first applied for the asymmetric hydrogenation of β-disubstituted nitroalkenes **19** in the presence of $[Rh(COD)Cl]_2$ as a metal source and 5 atm of H_2 in a protic solvent such as isopropanol [23a]. In this case, they clearly showed the need of the HB donor to bind the nitro group of the substrate and fix the conformation, being then able to reach high yields and enantiomeric excesses up to 99%, whereas the use of simple and classical chiral ferrocenyl–bisphosphine failed to induce chirality.

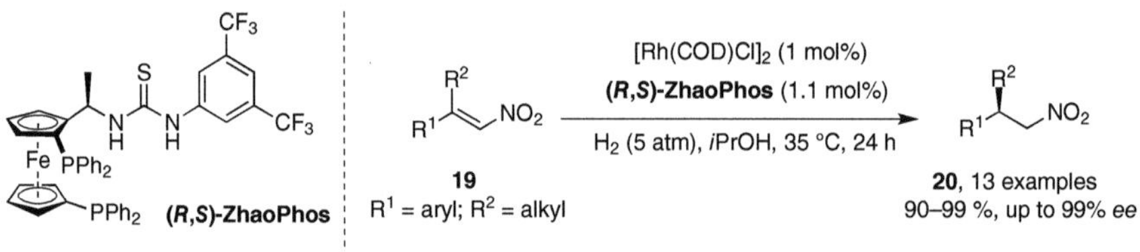

Scheme 11.5 Development of ZhaoPhos ligand and first application in asymmetric hydrogenation.

A year later, the same group reported on the first application of ZhaoPhos implying its anion-binding properties for the synthesis of chiral amines **22** by hydrogenation of the corresponding iminium hydrochloric acid salts **21** (Scheme 11.6) [24]. They found that the nature of the counterion of the iminium salt was crucial and reached the best results with chloride. Moreover, ^{1}H-NMR measurements indicated that the anion binding of chloride was directly involved in the reaction. Under these conditions, low catalytic loadings of 1–2 mol% of the Rh catalyst and 10 or 50 atm of H_2 were used, leading to the products in good yields (34–97%) and high enantiomeric excesses.

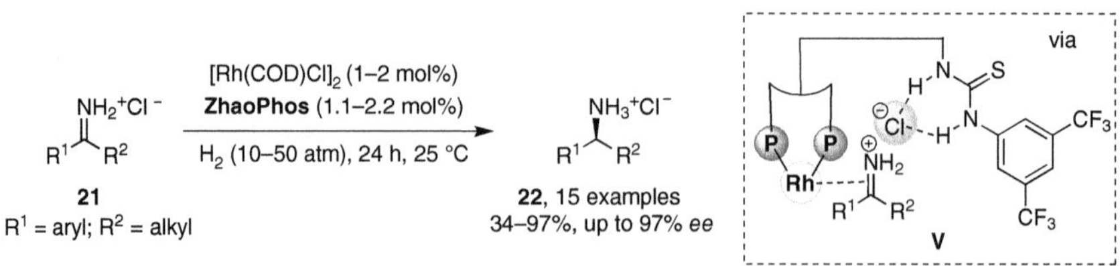

Scheme 11.6 Anion-binding assisted Rh-catalyzed hydrogenation of iminium salts **21**. Source: Based on Zhao *et al.* [24].

In 2016, the extension of this chemistry to the asymmetric hydrogenation of isoquinolines and quinolines was described (Scheme 11.7a) [25]. In this work, the *N*-heteroarene **23** was *in situ* protonated with HCl as Brønsted acid to form the corresponding ionic substrate, which was then activated by binding the anion by the thiourea group at the ZhaoPhos-Rh system. Under the optimized reaction conditions, $[Rh(COD)Cl]_2$ (1 mol%) and ZhaoPhos (2 mol%) in DCM/*i*PrOH at room temperature, high conversions, and enantioselectivities up to 99% were reached. Similarly, in 2018, the same group made use of the anion-binding properties of ZhaoPhos to bind the chloride anion of *in situ* formed indole hydrochloric acid salts as activated substrates for hydrogenation (Scheme 11.7b) [26]. The scope of this reaction includes a large variety of substituted indoles **28**, showing a great selectivity for the (*S*)-product **29** (up to 99% *ee*). Density functional theory (DFT) calculations on the mechanism of the reaction were carried out, revealing an inner-sphere mechanism for the hydride transfer. Consequently, a first oxidative addition of H_2 to the Rh(I) species forms an active Rh(III)–hydride intermediate **A**, which

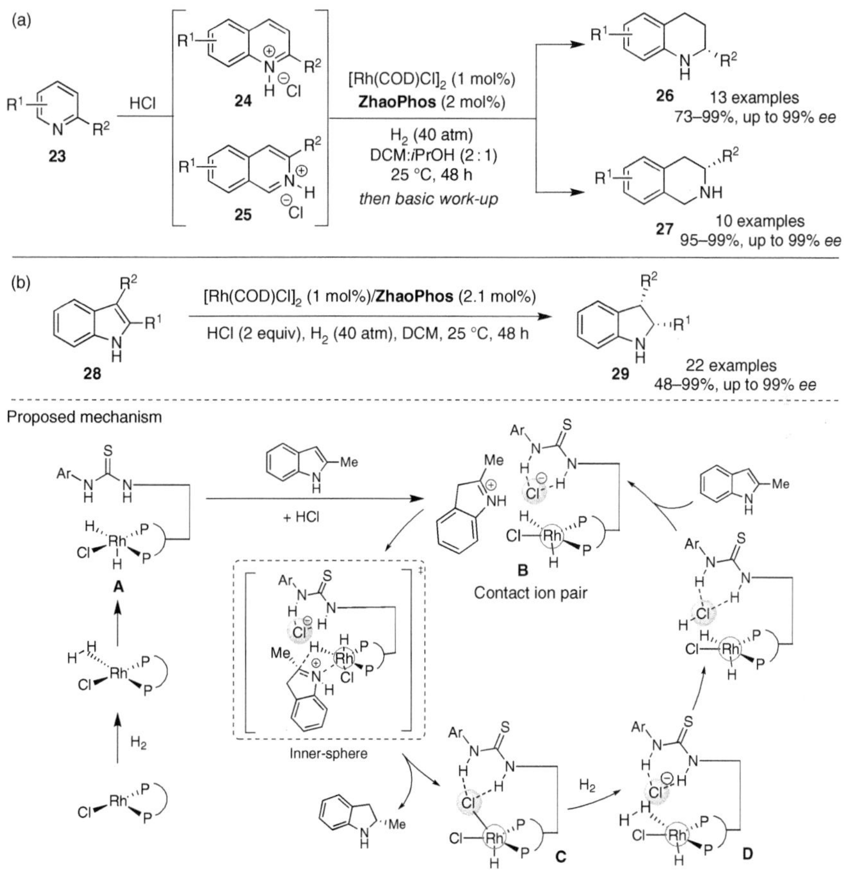

Scheme 11.7 ZhaoPhos-Rh-catalyzed asymmetric hydrogenation of *in situ* protonated N-heteroarenes. Source: (a) Based on Wen *et al.* [25]. (b) Based on Wen *et al.* [26].

forms an ion pair complex **B** with the protonated substrate. Then, a *Re*-face hydride transfer at the C2 position of the indole occurs. After dissociation of the product **29**, a Rh–chloride–thiourea-bridged complex **C** is formed, which coordinates another H_2 molecule upon dissociation of a chloride ligand, regenerating the active Rh–dihydride **D** with concomitant formation of HCl.

Later on, the combination of ZhaoPhos and an iridium catalyst was used to perform a ring closure reaction to form tetrahydroquinolines and tetrahydroisoquinolines by asymmetric reductive amination (Scheme 11.8) [27]. Thus, after the intramolecular reaction between the carbonyl group and the amino group in **30** or **31** in the presence of HCl, the corresponding iminium salt is formed. This intermediate is next activated by binding of the dual catalyst to its chloride counteranion, forming a chiral contact ion pair **VI** from which the stereodetermining hydrogenation step takes place. A large variety of substituted substrates were efficiently cyclized in this way, leading to the targeted chiral heterocycles **32** and **33** in up to 98% *ee*. Moreover, ZhaoPhos modifications were also carried out, as the mono N-methylation of the thiourea unit, showing the importance of the bidentate HB donor character of this group to provide high enantioselectivities.

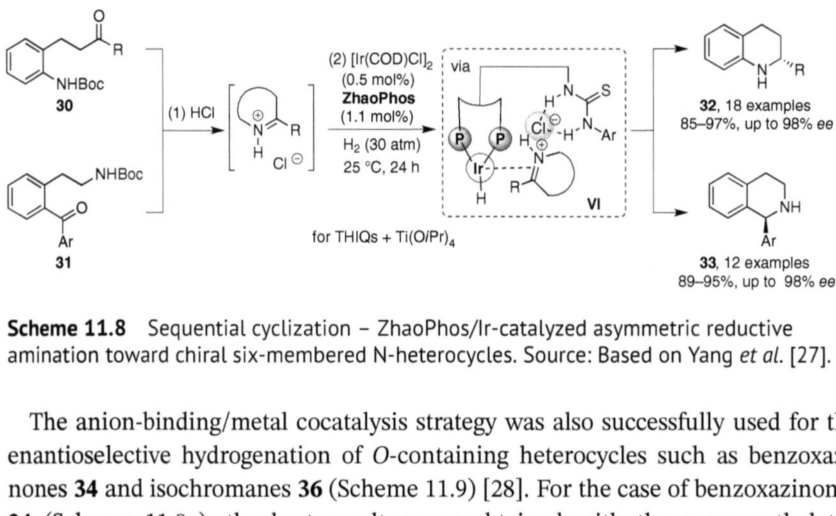

Scheme 11.8 Sequential cyclization – ZhaoPhos/Ir-catalyzed asymmetric reductive amination toward chiral six-membered N-heterocycles. Source: Based on Yang *et al.* [27].

The anion-binding/metal cocatalysis strategy was also successfully used for the enantioselective hydrogenation of *O*-containing heterocycles such as benzoxazinones **34** and isochromanes **36** (Scheme 11.9) [28]. For the case of benzoxazinones **34** (Scheme 11.9a), the best results were obtained with the monomethylated ZhaoPhos ligand (Me-ZhaoPhos) in combination with $[Ir(COD)Cl]_2$ (0.1 mol%) and 10 atm of H_2, reaching excellent yields and enantioselectivities (up to 99% *ee*) [28a]. Once again, the use of a strong acid, such as hydrochloric or triflic acid, resulted in the highest conversion and selectivity, suggesting a similar anion-binding assistant mechanism as shown before. The reaction scope is broad and could be extended to other heterocycles such as quinoxazolinones or their related seven-membered diazaheterocycles.

More recently, the standard ZhaoPhos–iridium system was employed to hydrogenate oxocarbenium ions generated *in situ* upon acid-mediated leaving group cleavage in 1-methoxy isochromane derivatives **36** (Scheme 11.9b) [28b]. The scope of this reaction includes different substituents at the aryl-motif of the isochromane, as well

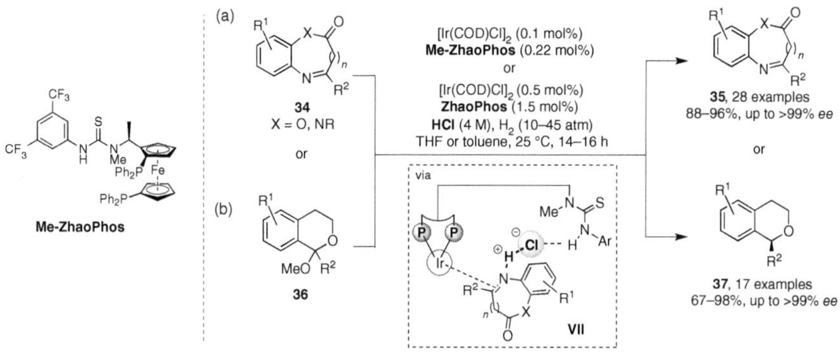

Scheme 11.9 Iridium-catalyzed hydrogenation of benzoxazinones and isochromanes. Source: (a) Han *et al.* [28a]. (b) Based on Yang *et al.* [28b]

as alkyl groups in the C1-position, leading to the chiral 1-substituted isochromanes **37** in high yields and good to excellent enantioselectivities (81 to ≥99% ee).

11.3.2 Hydroformylation Reactions

Hydroformylation is an important process for the production of highly valuable aldehyde compounds from alkenes, which requires the action of a transition metal catalysts and pressure of a CO/H_2 mixture. The choice of the catalytic system can modulate and control the regioselectivity toward the linear or branched aldehyde, as well as chemoselectivity between the desired aldehyde and the further reduced alcohol derivative. In the field of anion-assisted catalysis, several catalytic systems have been developed that rely on the recognition of carboxylates present in the substrates, hence pre-orienting them for a more effective selectivity.

In 2008, Šmejkal and Breit reported the use of a bifunctional ligand **40** for a Rh-catalyzed decarboxylative hydroformylation of α,β-unsaturated carboxylic acids **38** (Scheme 11.10) [29]. The most effective ligand comprises a monodentate phosphine motif for the metal complexation and a guanidine as the anion-binding domain. In the proposed catalytic cycle, the guanidine unit binds the carboxylate, orienting the substrate for coordination to the preformed Rh–carbonyl hydride complex **E**. Next, the insertion of the formyl group at the α-position and a hydrogen atom at the contiguous carbon takes place, generating the intermediate **F**. Finally, a decarboxylation occurs, delivering the product **39** and liberating the active catalytic species **G** for the complexation to another substrate molecule. Impressively, the reaction requires as low as 0.5 mol% of the Rh catalysts and tolerates a variety of functional groups such as thioethers, alcohols, ketones, carbamides, and even other non-conjugated olefins. However, as a recognition of the substrate plays an important role, the yields of the hydroformylated products depend heavily on the substitution at α,β-unsaturated carboxylic acid (47–97%).

Few years later, the same group extended the use of the Rh/bifunctional ligand as catalytic system for the hydroformylation of alkene-containing carboxylic acids **41**

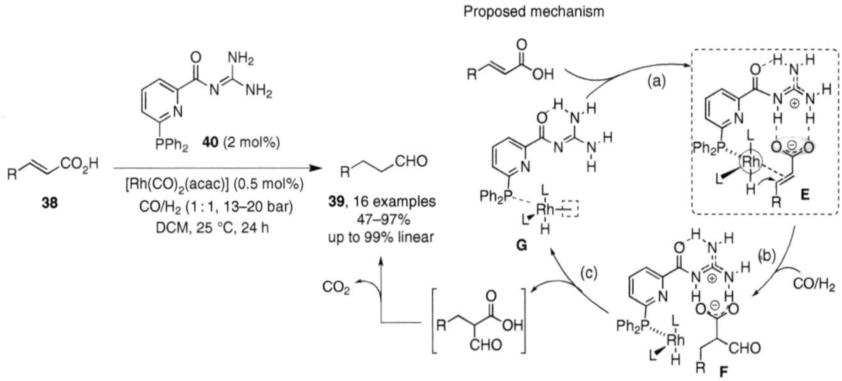

Scheme 11.10 Decarboxylative hydroformylation of α,β-unsaturated carboxylic acids by Šmejkal and Breit [29]. Source: Based on Šmejkal and Breit [29].

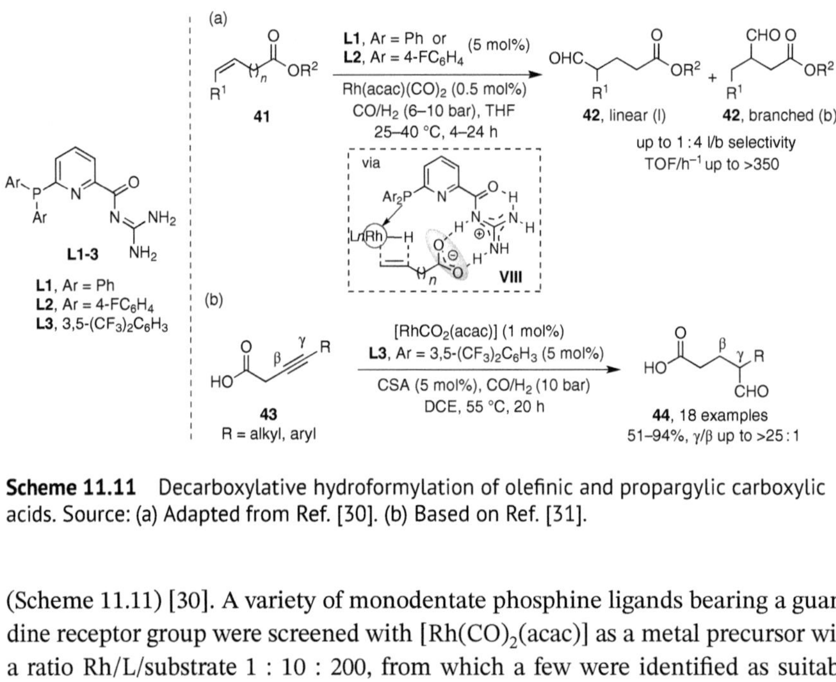

Scheme 11.11 Decarboxylative hydroformylation of olefinic and propargylic carboxylic acids. Source: (a) Adapted from Ref. [30]. (b) Based on Ref. [31].

(Scheme 11.11) [30]. A variety of monodentate phosphine ligands bearing a guanidine receptor group were screened with $[Rh(CO)_2(acac)]$ as a metal precursor with a ratio Rh/L/substrate 1 : 10 : 200, from which a few were identified as suitable as catalyst for high regioselective hydroformylations toward the linear aldehyde products **42**. For example, in the case of allyl substrates (Scheme 11.11a) [30], besides the original ligand **L1** (selectivity l/b > 20), the bis(4-flurophenyl)phosphine substituted ligand **L2** provided enhanced results in terms of both turnover frequency (TOF) and selectivity (up to 350 TOF/h^{-1}, and l/b ≤ 41). Moreover, competitive experiments with the corresponding allyl ester and a terminal alkene confirmed the important role of the bidentate ligand for the recognition of the carboxylic acid derivative, which shows a significant higher reactivity and selectivity. Propargyl carboxylic acids **43** were also efficiently hydroformylated and further reduced to provide γ-formyl aliphatic carboxylic acids **44** as the major products with a γ/β selectivity up to >25 : 1 (Scheme 11.11b) [31]. The best results were obtained with a 3,5-bis(trifluoromethyl)phenyl substitution at the phosphine ligand and the use of camphor sulfonic acid (CSA) in 5 mol% as an additive.

Alternatively, the group of Reek introduced a new bidentate ligand DIM-Phos (**45**) for hydroformylation reactions, which was designed by combining 7,7′-diamido-2,2′diindolylmethane (DIM), as the receptor unit for anions, with two terminal triphenylphosphine groups to coordinate the Rh metal center (Scheme 11.12a) [33]. Under the hydroformylation conditions (5 bar CO/H_2 (1 : 1), r.t.), a mixture of the ligand **45**, [Rh(acac)CO_2], the substrate (**47** or **49**), and DIPEA as a base to ensure the *in situ* formation of the carboxylate in a 1 : 3 : 100 : 150 ratio provided the best results. Under these conditions, the carboxylate group binds into the DIM cavity, and good to high conversions and linear selectivities were obtained for various substrates. Conversely, neutral derivatives such as esters or the free carboxylic acids (no addition of a base) did not readily react. The best match was found for the β-unsaturated acid (>50 l/b ratio); however, the scope could be extended to ω-unsaturated derivatives allowing a wide-ranking remote chemoselectivity control. Similarly, the reaction was also possible with phosphates **47**, although they were less reactive and the temperature had to be increased to 40 °C to achieve a high conversion and selectivity (l/b up to >40). Furthermore, the selective hydroformylation of vinyl 2- and 3-carboxyarenes **51** was also achieved chemo-

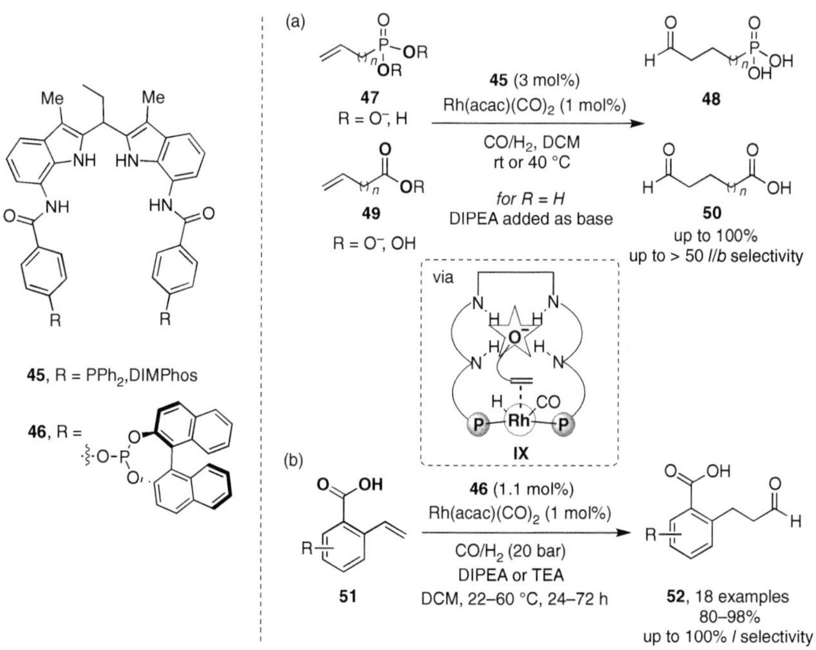

Scheme 11.12 Regioselective Rh-catalyzed hydroformylation reported by Reek and coworkers [32]. Source: (a) Adapted from Ref. [33]. (b) Based on Ref. [32].

and regioselectively (up to 100% linear product), when using the bifunctional chiral phosphite catalyst **46** presenting a similar bidentate ligand motif and integral anion-binding cavity (Scheme 11.12b) [32].

11.3.3 Anion-Binding – Metal-Catalyzed C–C Forming Reactions

Other transformations that were found applicable for the anion-binding/metal cocatalysis strategy include more classical C—C bond forming reactions such as alkynylations [34] or allylic substitutions [35], as well as C–O couplings such as Cu-catalyzed *O*-arylations already discussed in Chapter 6 [36].

Concerning C–C coupling reactions, Zhao and Seidel reported in 2015 on a Cu(I)-catalyzed multicomponent reaction to form chiral propargylic amines **56** from the corresponding aldehydes **53**, secondary amines **54**, and acetylenes **55** using a bifunctional chiral thiourea–carboxylic acid cocatalyst **57** (Scheme 11.13) [34]. Upon condensation of the aldehyde and amine, the corresponding iminium ion is generated *in situ*. This species can then form a chiral contact ion pair **X** with the thiourea catalyst, which was designed as an anion receptor to internally stabilize the catalyst's conjugate base that is used as a counteranion. Following this mechanism, a great scope in this A^3-reaction was shown, providing the desired chiral amines in good yields and enantiomeric excesses up to 96% in CH_2Cl_2 at 0 °C.

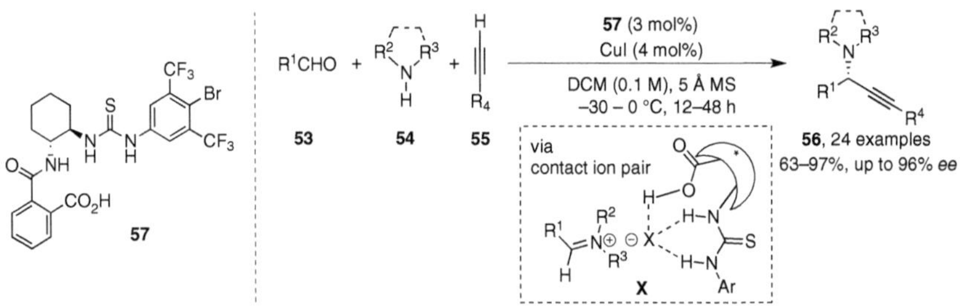

Scheme 11.13 A^3-reaction toward chiral propargylic amines using dual catalysis by Zhao and Seidel [34]. Source: Based on Dydio *et al.* [34].

Lastly, in 2016, Reek, Moberg, and coworkers used a bifunctional bis(indolylamide) methane–bisphosphite ligand **60** [35], fairly similar to the one initially used by Reek and coworkers in 2011 [33a], for palladium-catalyzed asymmetric allylic substitutions (Scheme 11.14). Allylic carbonates **58** were used as substrates, which were reacted with methylmalonate or benzylamine as a nucleophile. In this work, they aimed to induce the chirality by including a chiral anionic cofactor (**Co1** or **Co2**) to form a chiral contact ion pair assembly within the reactive cationic Pd–allyl complex intermediate. Accordingly, simple (*S*)-2-hydroxy-3-methylbutyric acid (**Co2**) as cofactor gave rise to the products **59** in excellent yields and *ee*-values of up to 57%.

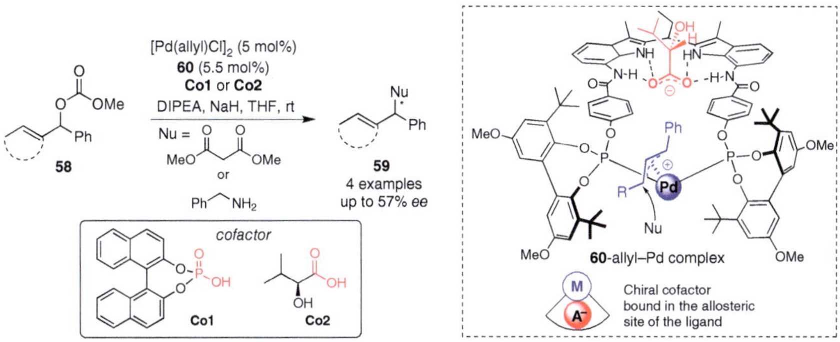

Scheme 11.14 Palladium-catalyzed allylic substitution reported by Reek, Moberg, and coworkers [35].

11.4 Supramolecular Approaches Involving Anion-Binding Catalysis

The significant importance of anions in biological systems, as well as in our environment, has been widely studied. In fact, the anions have turned to be essential for life because their presence or their perfect regulation can be either favorable or harmful to living organisms [37]. For example, chloride is linked with diseases such as cystic fibrosis, the iodide is involved in the biosynthesis of hormones by the thyroid gland, or bicarbonate maintains the pH value in the body. For that reason, the progress in the design and construction of supramolecular anion receptors that can be used in the selective binding and sensing of anions has been appealing [37b]. Moreover, in comparison with cation receptors, the selective anion coordination represents a challenge because of their intrinsic properties such as different geometries, low charge densities, pH dependency, and high free hydration energies. In the past decades, a large number of works based on the synthesis of the anion receptor and its binding properties, as well as the use of anions in sensing, ion extraction, anion-induced supramolecular assemblies, catalysis, and transmembrane transport, have been reported [38]. Particularly, in the context of homogeneous catalysis, supramolecular organocatalysis has emerged as a novel research field and has advanced notably. However, their application in anion-based catalysis is still significantly underdeveloped. In this section, a concise overview and discussion based on the structures of different supramolecular receptors, including knots, rotaxanes, molecular motors, and macrocycles (Figure 11.3), showing their anion-binding catalytic efficiency and applications will be provided. Although highly interesting examples can be found based on helical foldamer multidentate structures in the literature, they will not be discussed in this section as they are presented in detail in Chapter 7.

11.4.1 Mechanically Interlocked Molecules in Anion-Binding Catalysis

The mechanically interlocked molecules (MIMs) can be defined as structures that present an entanglement between different molecules or molecular components.

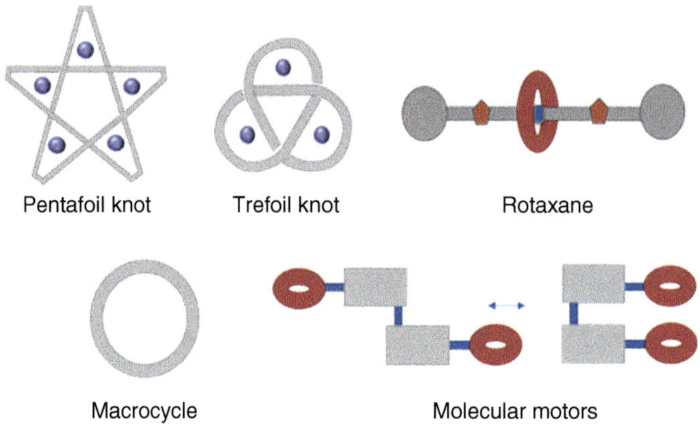

Figure 11.3 Schematic representation of the different structures.

When two or more molecular parts are linked via a mechanical bond, it gives rise to [*n*]rotaxanes or [*n*]catenanes. This kind of structures can also be afforded via topological bonds, meaning that it is not possible to represent the molecule in a plane without crossings, such as the pentafoil knot [39]. In recent years, the interest in MIMs has been increasingly grown because of the unique dynamic properties, switchability, and non-trivial architectures that they present [40]. Moreover, it is possible to control the synthesis and the design of the specific motifs required in the cavities of the MIMs for the selective binding of the guest species. These features have been exploited in a large range of applications, especially in host–guest recognition or catalysis [41]. Particularly, their role as catalysts is lately one of their most prominent implementations because of their notable enhanced selectivity, presenting in most of the cases better outcomes compared with the non-interlocked motifs [42]. However, the application and utility of these structures in anion-binding catalysis are still severely underdeveloped and remains limited.

11.4.1.1 Molecular Knots as Anion-Binding Catalysts

Molecular knots are present in a huge number of organic molecules such as DNA, RNA, or proteins. The benefits of these strands are still not clear but impose restrictions that can improve some properties such as intrinsic chirality, strong and selective ion binding, and catalytic activity [43]. Indeed, various small-molecule knots than can act like host for different guests as metal ions (facilitating their synthesis in some cases), organic molecules, or anion have been developed. Despite the latter one is particularly interesting for the anion-binding catalysis, only a couple of applications have been published in the past five years.

The first example of the employment of a metal–organic pentafoil knot in the allosteric regulation of Lewis-acid-catalyzed reactions such as a Diels–Alder cycloaddition or a Michael addition was described by Leigh and coworkers in 2016 (Figure 11.4) [44].

This important goal was achieved by the formation of a catalytically active carbocation through a cleavage of carbon–halogen bond employing a synthetic

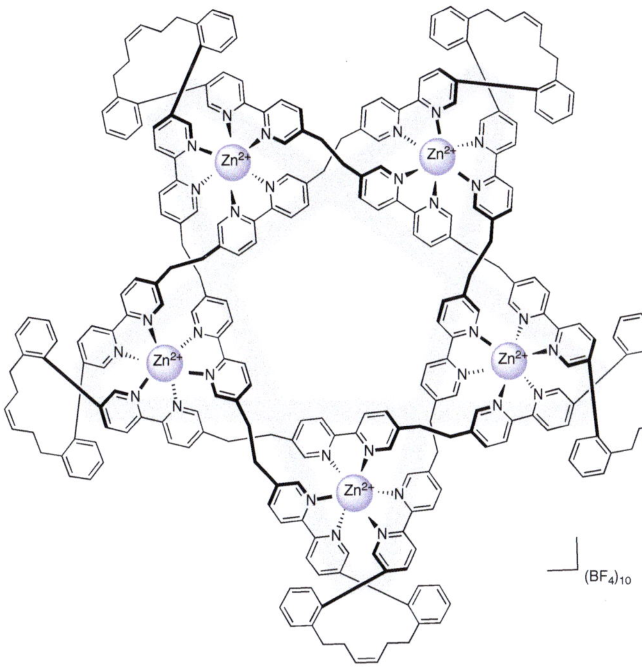

Figure 11.4 Structure of Leigh's Zn(II)-pentafoil knot organocatalyst $[Zn_5\mathbf{61}](BF_4)_{10}$.

Zn(II)-based pentafoil knot catalyst $[Zn_5\mathbf{61}](BF_4)_{10}$. This halogen abstraction was possible because of the stabilization of the anion through a combination of CH$\cdots$X$^-$ hydrogen bonding and long-range M$\cdots$X$^-$ electrostatic interactions. In consequence, the central cavity of these pentafoil knots presents strong binding constants for the ions Cl^- and Br^-. Furthermore, the reaction did not take place in the absence of the knot and neither in the presence of the unknotted and demetallated analogous, showing evidence of the knot-promoted catalysis (Scheme 11.15a). In this way, the fast and reversible coordination kinetics of Zn(II) permitted the initiation of the allosteric catalysis under mild conditions (a), abstracting the bromide from the bromodiphenylmethane to give the benzhydryl carbocation **67**, which subsequently promoted the Lewis acid carbocation catalysis (b). The catalytic activity of the pentafoil knot could be recovered removing the bromide ion from the cavity employing Na_4EDTA, giving rise to demetallated inactive knot (c). Finally, treating the metal-free knot with $Zn(BF_4)_2$ regenerated the active Zn(II)-pentafoil knot and allowed the catalyst to turn over (d). These kinds of catalysts have also been employed in the hydrolysis and the Ritter reaction of the bromodiphenylmethane **69** (Scheme 11.15b).

The next contribution in this field was published three years later by Mukherjee, Trabolsi, and coworkers who described the hydrolysis of different bromo Baylis–Hillman adducts **72** catalyzed by a Cu(II)-trefoil knot $[Cu_3\mathbf{71}](TFA)_6$ (Scheme 11.16) [45]. In this work, they studied the capability of binding bromide anions within the empty cavity of three different M(Cu, Cd, and Zn)-based trefoil

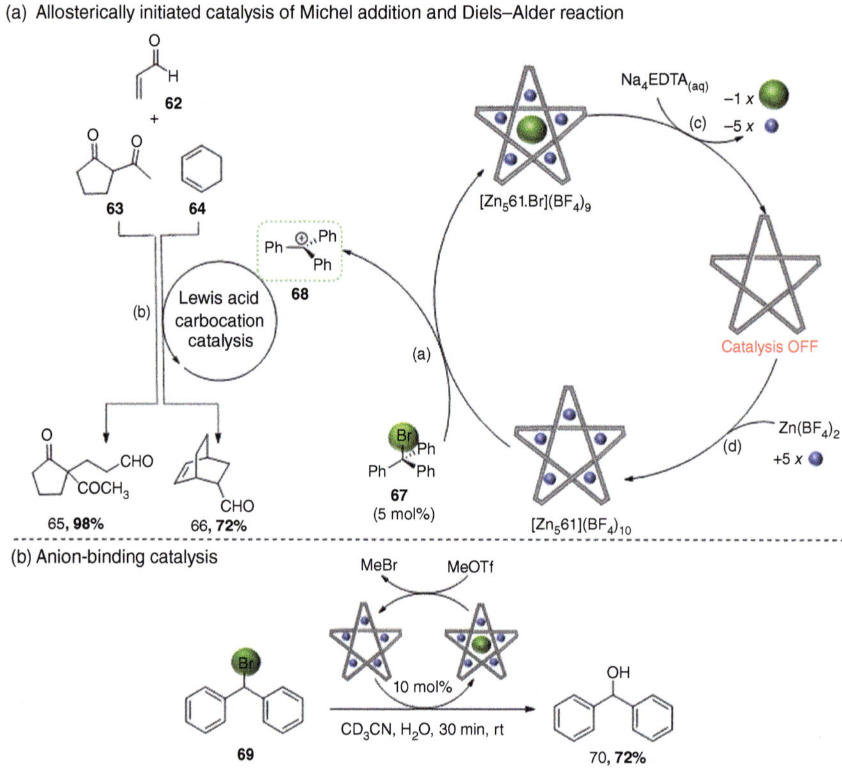

Scheme 11.15 (a) Zn(II)-pentafoil knot catalyzed Michael addition and Diels–Alder reactions by anion abstraction toward the catalytic trityl cation **68**. (b) $[Zn_5$**61**$](BF_4)_{10}$-catalyzed hydrolysis of Ph_2CHBr **69**.

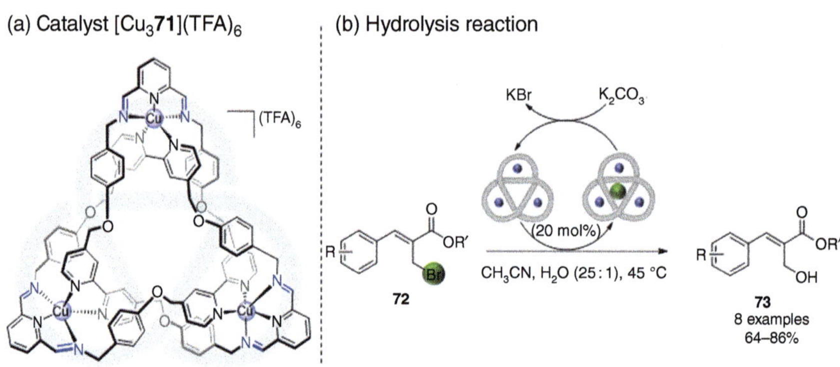

Scheme 11.16 (a) Structure of the Mukherjee and Trabolski's Cu(II)-based trefoil knot organocatalyst $[Cu_3$**71**$](TFA)_6$. (b) The Cu(II)-trefoil knot as a catalyst in the hydrolysis of bromo Baylis–Hillman adducts **72**. Source: Based on Prakasam *et al.* [45].

knots. The obtained results revealed the suitability of the Cu(II)-trefoil knots as a catalyst in this concerted S_N2 reaction because of the balance between the attractive host–guest electrostatic interactions and the shape and size of the cavity. Finally, the efficiency of the process could be enhanced by the regeneration of the catalyst through an anion exchange employing K_2CO_3.

11.4.1.2 Rotaxanes as Anion-Binding Catalysts

Rotaxanes are probably the most employed MIM structures. Among they were discovered half century ago and present numerous applications such as nanoscale devices, functional molecular systems, or molecular machines; their role as catalysts has emerged as a promising area in the past decades [46]. Particularly, [2]rotaxanes (formed by a macrocycle and a thread) based catalysts have been mostly used for this purpose because of the fact that the movement along the thread of the macrocycle can be controlled by one or various recognition sites [47]. Especially interesting are the rotaxanes that contain two or more stations for their macrocycles, in which their binding can be modulated in a reversible way by external stimuli.

Based on this concept, Leigh and coworkers developed in 2017 a switchable [2]rotaxane catalyst that shows a dual functionality and is able to response to the acidity of the media [48]. This rotaxane is formed by a dibenzo-24-crown-8 macrocycle that can be coordinated with the central ammonium moiety in a protonated media or, on contrary, to one of the triazolium rings with the addition of a base (Scheme 11.17). In this way, the protonated organocatalyst form (**74-H**$^+$) showed activity as anion-binding catalyst with the triazolium moieties exposed and accessible in a bent conformation that can act as the binding site.

The effectiveness of this catalyst was demonstrated in the Ritter reaction of bromodiarylmethanes **69** and the reaction of 1-chloroisochroman **76** with different ketene silyl acetal **77** (Scheme 11.18a). Both transformations did not proceed in the absence of the catalyst, the non-interlocked or in the presence of the deprotonated rotaxane (**74**), demonstrating that the macrocycle is blocking the binding ability of the triazolium rings in this form of the rotaxane.

Moreover, a tandem reaction was developed to prove the *in situ* dual function "on/off" of this switchable catalyst (Scheme 11.18b). In this process, the catalyst

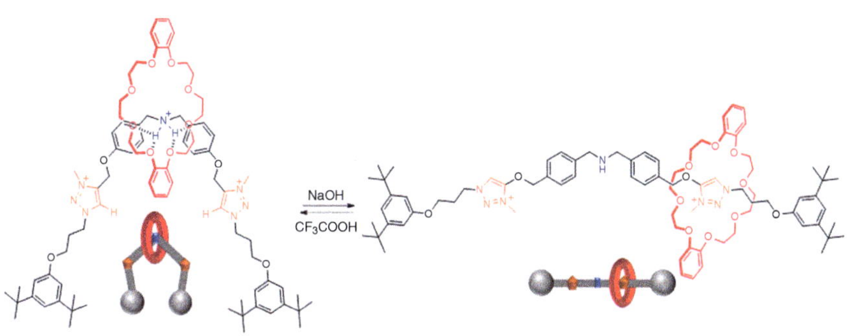

74-H$^+$ **74**

Scheme 11.17 Acid–base switchable Leigh's rotaxane organocatalyst.

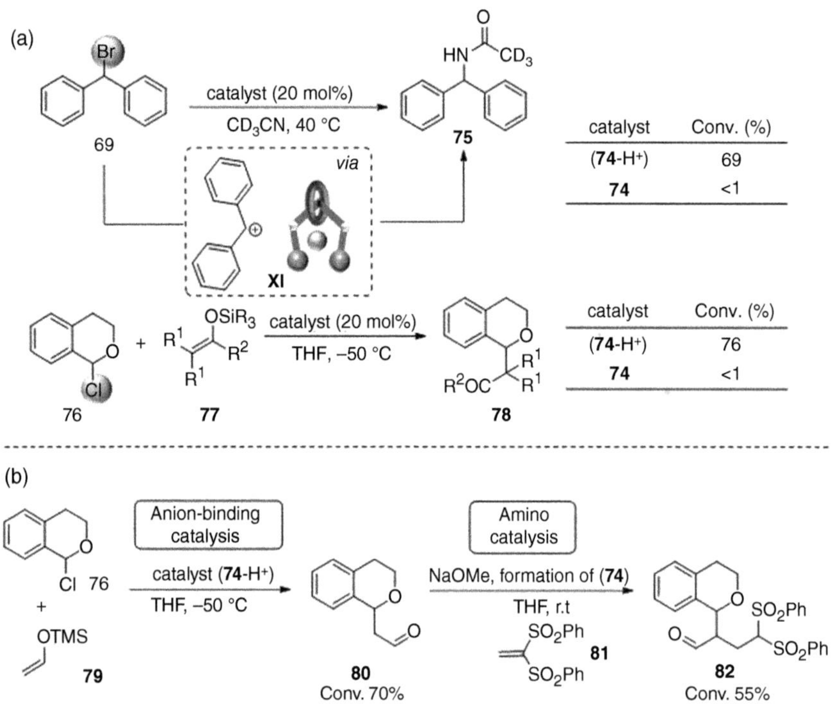

Scheme 11.18 (a) The study of the acid–base switchable catalyst in anion-binding catalysis. (b) Tandem reaction controlling the position of the macrocycle in the rotaxane-based organocatalyst.

(**74-H**$^+$) first promoted an anion-binding catalyzed alkylation reaction between the **76** and **79**, followed by a nucleophilic addition of the vinyl bis-sulfone **81** via an enamine activation catalyzed by the deprotonated form (**74**) to give rise to α,α-disubstituted aldehyde product **82**. According to the molecular machine definition, this molecular rotaxane catalyst could also be framed in the next section. However, taking into account its structure, it has been described separately.

11.4.2 Molecular Motors in Anion-Binding Catalysis

Artificial molecular machines that can achieve a responsive allosteric regulation of the catalytic activity depending on an external stimulus is one of the most popular field with a remarkable progress [49], awarded with the 2016 Novel Prize in Chemistry [50].

The design of a molecular motor-based photoresponsive chiral anion receptor and its applicability in anion-binding catalysis has been recently described in 2020 by the pioneer group in this area (Scheme 11.19a) [51]. In this work, Feringa coworkers studied a molecular motor core bearing two oligotriazole-based branches and its response to an external applied stimulus in the stereodivergent addition of the silyl ketene acetal **84** to chloroisochroman **76** (Scheme 11.19b). As is shown in Scheme 11.19, because of the double bond in the motor function,

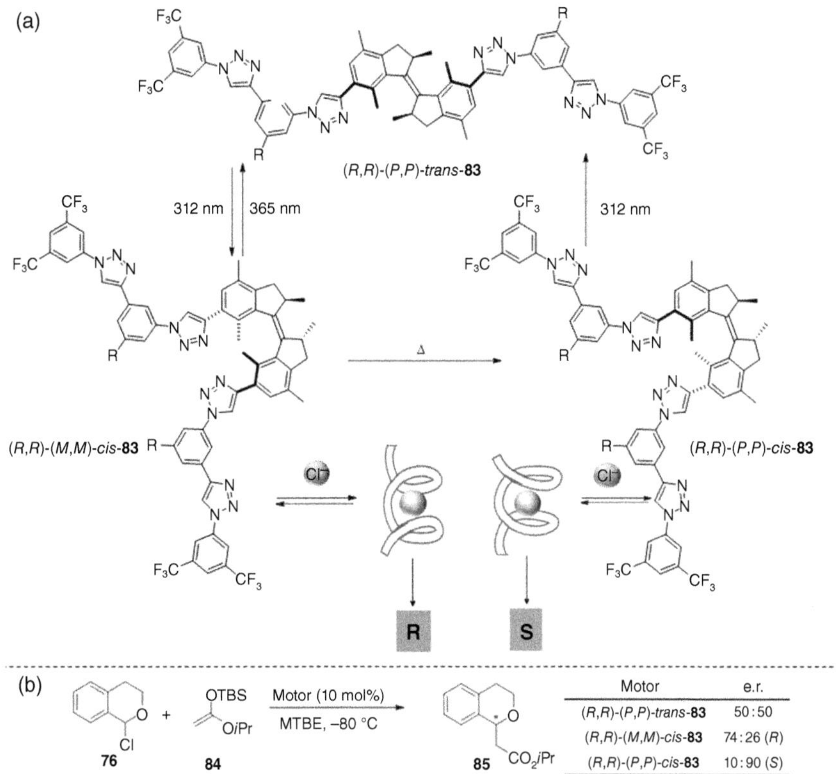

Scheme 11.19 (a) Feringa's oligotriazole-based molecular motors for stereodivergent anion-binding catalysis. (b) Molecular motor-catalyzed addition of the silyl ketene acetal **84** to the chloroisochroman **76**. Source: (a) Based on Dorel and Feringa [51].

the (*R*,*R*)-(*P*,*P*)-*trans*-**83** can be interconverted in a reversible way by irradiation or heating into the pseudoenantiomers (*R*,*R*)-(*M*,*M*)-*cis*-**83** and (*R*,*R*)-(*P*,*P*)-*cis*-**83**, both prompted to form a helical-based structure in the presence of an anion. Depending on the switchable tetratriazole catalyst present during the nucleophilic addition reaction, the opposite enantiomers of the final product can be obtained. Therefore, the presence of the (*R*,*R*)-(*M*,*M*)-*cis*-**83** yielded preferentially the **85**-*R*-enantiomer, whereas the (*R*,*R*)-(*P*,*P*)-*cis*-**83** generated the **85**-*S*-enantiomer, both with a high enantioinduction.

11.4.3 Macrocycles in Anion-Binding Catalysis

Functionalized macrocycles are the most employed supramolecular structures because of their synthetic accessibility providing defined binding pockets with catalytically active moieties [52]. Therefore, they can offer multivalent coordination sites as well as the activation of the substrates, giving rise to transformations otherwise unfavored. Moreover, their high binding affinity allows the trapping of "indirect" catalytic species as counteranions. Despite the wide use of macrocyclic architectures such as crown ethers for binding cations, the application and role of macrocycles in anion receptor catalysis is still underdeveloped. Herein, the recent progress in this field will be covered.

The first example for the use of these macrocycles in anion-binding catalysis was published by Jacobsen and coworkers in 2017 [53], in which a catalytic method for β-selective glycosylation was described (Scheme 11.20). Based on nucleophilic substitution reactions via abstraction of the anionic leaving group, they studied the reaction between α-pyranosyl chlorides **87** and neutral alcohols **88** in the presence of different thiourea derivatives. The employment of chiral monomeric thiourea gave rise to low yields and poor selectivity (~50 : 50 α:β); meanwhile, the α-product was predominantly afforded in the absence of the catalyst. However, the β-products were mainly achieved in excellent yields and good enantioselectivities (up to 8 : 92, α:β) when either the macrocyclic bis-thiourea (*S*,*S*)-**86** or the (*R*,*R*)-**86** enantiomer was employed. These results suggested that the reaction in the presence of the macrocycle underwent with an inversion of the configuration, taking place in a concerted stereospecific fashion. For that reason, the stereochemistry of the final β-glycosides **89** depends on the configuration of the electrophiles and not of the configuration of the catalyst.

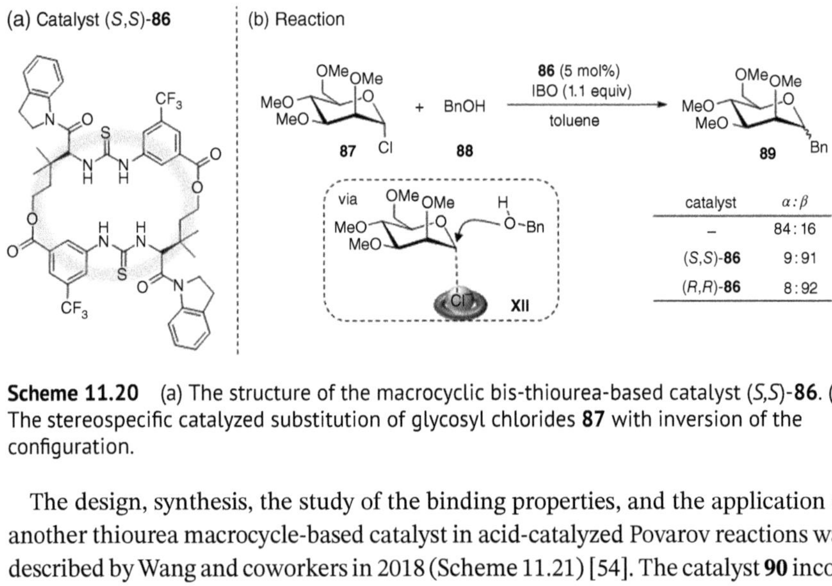

catalyst	α:β
–	84:16
(S,S)-**86**	9:91
(R,R)-**86**	8:92

Scheme 11.20 (a) The structure of the macrocyclic bis-thiourea-based catalyst (*S*,*S*)-**86**. (b) The stereospecific catalyzed substitution of glycosyl chlorides **87** with inversion of the configuration.

The design, synthesis, the study of the binding properties, and the application of another thiourea macrocycle-based catalyst in acid-catalyzed Povarov reactions was described by Wang and coworkers in 2018 (Scheme 11.21) [54]. The catalyst **90** incorporates two thioureas units and CF_3 groups at their aryl substituents that increase the N–H acidity facilitate the H-bonding formation of the C–H of resorcinol ring and the anion as well as improve the structure rigidity. Hence, the strong binding generated in the cavity made possible the improvement of the catalytic effectiveness by trapping the counteranion of the acid catalyst and, in consequence, inducing an acidity enhancement on the substrate. This macrocyclic catalyst **90** was applied in a Povarov transformation between different imino substrates **91** and 2,3-dihydrofuran **93** in the presence of ethane disulfonic acid (EDSA) to give rise to tetrahydroquinolines **94**. The reaction did not work in the presence of the acyclic version of the

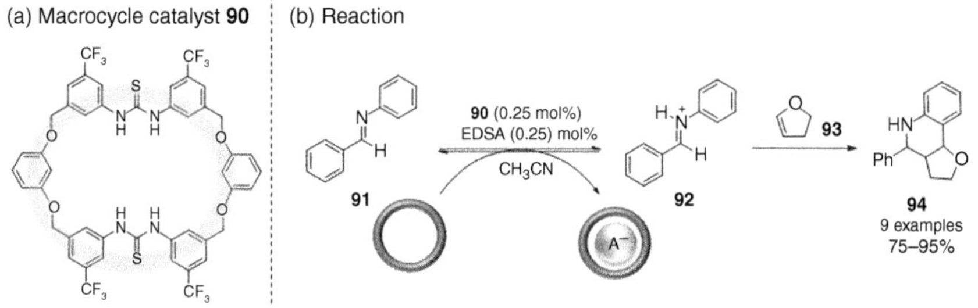

Scheme 11.21 (a) Structure of the macrocycle **90**. (b) EDSA-catalyzed Povarov reaction. The formation of the protonated imide is promoted by the macrocycle-anion trapping. $A^- = (^-O_3SCH_2CH_2SO_3^-)$. Source: Based on Ning *et al.* [54].

catalyst or when only the macrocycle or the EDSA were used, demonstrating in this way the cooperative effect of the macrocycle and the acid in the catalytic activity improvement. Finally, it was possible to employ only 0.25 mol% of the catalyst, maintaining a high reactivity.

The same group has recently described the chiral version of these bis-diarylthiourea macrocycle catalysts **95** by introducing two BINOL moieties for the generation of the chiral environment within the cavity (Scheme 11.22) [55]. Based on the same principle of acid co-catalysis, they carried out an enantioselective Friedel–Crafts reaction of indoles **98** with imines **96** employing a combination of the chiral macrocycle **95** and EDSA. The final products **99** were obtained in very good yield and excellent enantioselectivity proving the synergistic effect of the acid and the macrocycle to achieve a very high catalytic efficiency and effective chiral induction.

Until now, the design of the macrocycle employed in anion-binding catalysis was basically based on thiourea moieties in order to binding anionic guests within the cavity. Another supramolecular structure that can be used in order to trap counteranion species is a ditopic receptor [56]. This type of receptors bears two different recognition sites that can bind in a cooperative way an anion and a cation. As a result of this cooperation, electrostatic and hydrogen bonding interactions can be generated and, consequently, they present an enhanced anion-binding affinity and selectivity. The use of the ditopic aromatic amide macrocycles in anion-binding catalysis was

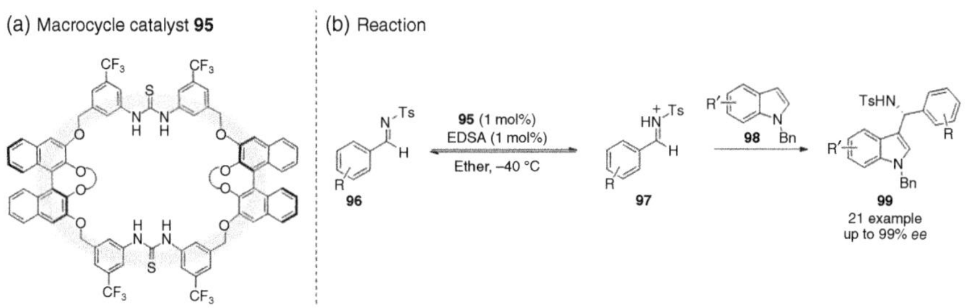

Scheme 11.22 (a) Structure of the chiral bis-thiourea-based macrocycle **95**. (b) The chiral macrocycle as a catalyst in the enantioselective Friedel–Crafts reaction between different indoles **98** and aldimines **96**. Source: Based on Ning *et al.* [55].

first described by Yuan and coworkers in 2019 (Scheme 11.23) [57]. In this work, the catalytic potential of the receptor **100**, which presents a prebound alkali metal improving the coordination of the anion, was tested in the benchmark Ritter reaction of bromodiphenylmethane **69** and acetonitrile. Moreover, they proved that the catalytic activity depends on the cation binding and, therefore, it can be switch on and off in a reversible way with the employment of HSO_4^-.

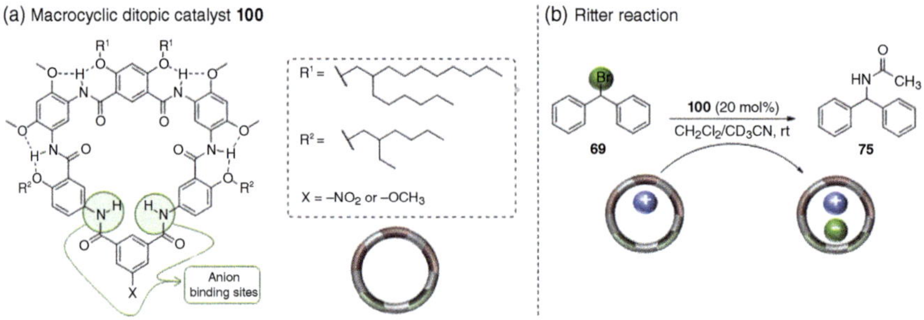

Scheme 11.23 (a) Structure of the macrocyclic ditopic receptor **100**. (b) The convergent ditopic amide macrocycle as a catalyst in the Ritter reaction between Ph_2CHBr **69** and acetonitrile. Source: Kang *et al.* [57].

11.5 Anion–π Catalysis

Aromatic rings possess the ability to interact with ionic compounds because of their delocalized clouds of π-electrons above and below their plane that can lead to π-basic or π-acid surfaces [58]. Thus, unsubstituted or electron-rich aromatic ring systems provide negatively charged, π-basic surfaces that predominately interact with cations, while aromatic units with withdrawing groups lower the π-basicity resulting in a complementary π-acidic surface that can then interact with anions [59]. This counterintuitive affinity of arenes to anionic species has been exploited in different research areas such as material sciences, complexation chemistry, synthesis, and more recently catalysis [58, 60]. To characterize and quantify the π-basic or acid properties of the aromatic systems, the quadrupole moment (Q_{ZZ}) is used, which can be determined by theoretical DFT calculations [61]. Hence, while a π-basic aromatic presents a negative quadrupole moment ($Q_{ZZ} < 0$), an acid π-molecule shows a positive quadrupole ($Q_{ZZ} > 0$) (Figure 11.5, left). Following this line, unsubstituted aromatic systems as benzene with a negative quadrupole ($Q_{ZZ} = -9$ B (B, Buckingham units) will predominately interact with cations, whereas strongly withdrawing substituted aromatic systems such as hexafluorobenzene (**101**) ($Q_{ZZ} = +10$ B) show a notable π-acidic surface and attract anions (Figure 11.5a) [61]. Other interesting structures presenting enhanced anion–π interactions are extended aromatic systems such as naphthalenediimides **102** (NDIs) [62] or perylenediimides **103** (PDIs) [63] (Figure 11.5b,c). This type of π-surfaces can be easily functionalized to give rise to strongly electron-poor aromatic systems with high π-acidity.

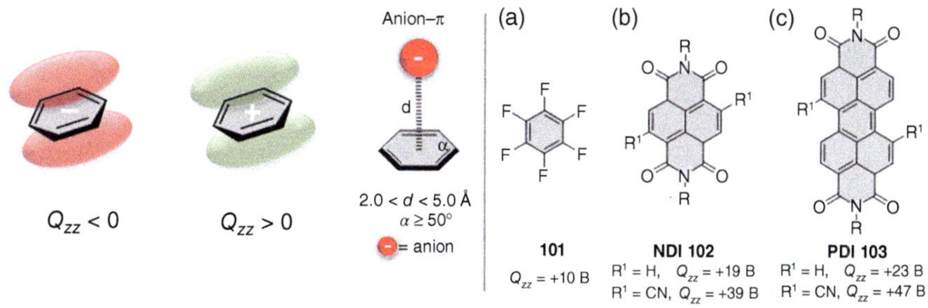

Figure 11.5 Coordination ability of aromatic surfaces to ions: aromatic rings with electron-rich (red) and electron-poor (green) surfaces. (a) Hexafluorobenzene, (b) naphthalenediimides (NDIs), and (c) perylenediimides (PDI) as representative π-acid surfaces. Source: Adapted from Refs. [61–63].

As mentioned above, in the past few years, the ability of π-acidic aromatic systems to interact with anions has been envisioned to be useful in catalysis. In this regard, this feature has been introduced as a secondary interaction in the design of novel catalyst structures to enhance the catalytic activity or enantiomeric inductions by a more efficient binding of the anionic substrates, intermediates, or transition states [64]. Following this idea, the so-called anion–π catalysis has lately emerged [58b, 60, 65], which implies as one of the principal interactions for the catalytic activity the stabilization or binding of key anionic species on the aromatic π-acidic surfaces of the catalyst. As a consequence, several types of catalytic transformations assisted or promoted by anion–π interactions on catalysts π-surfaces have been developed and will be presented in the following sections.

11.5.1 Anion–π-Catalyzed Kemp Elimination Reaction

The pioneering studies on anion–π catalysis were carried out by Matile and coworkers in 2013, who initially considered the possibility of applying a π-acid surface in a catalyst to stabilize anionic intermediates and transition states (TS). To prove this concept, this group investigated the ring opening of benzisoxazoles by N–O cleavage to form the corresponding benzonitrile (*Kemp* elimination) as a benchmark reaction with a modified NDI-carboxylate as π-acid catalyst **104** [66] (Scheme 11.24). Deprotonation of a benzisoxazole substrate **105** through the carboxylate functional group of the catalyst led to a partially negatively charged **TS** that is mainly stabilized by anion–π interactions. Upon ring opening, the anionic phenolate intermediate **107** – also stabilized by anion–π interactions – is formed and further gets protonated by the generated NDI–carboxylic acid species **108** to deliver the final phenolic nitrile derivative **110**. Control experiments with different NDIs showed that the increase of the π-acidity of the surface also enhanced the stabilization of the transition state. This group further synthesized different anion–π catalysts based on NDIs with cyano, sulfoxide, and sulfone substituents, which increased remarkably the π-acidity of the catalysts and performance in catalysis [66b].

Shortly after, Wheeler and Lu reported a computational study concerning the activation barrier of the previous reported *Kemp* elimination on π-surfaces [67]. They

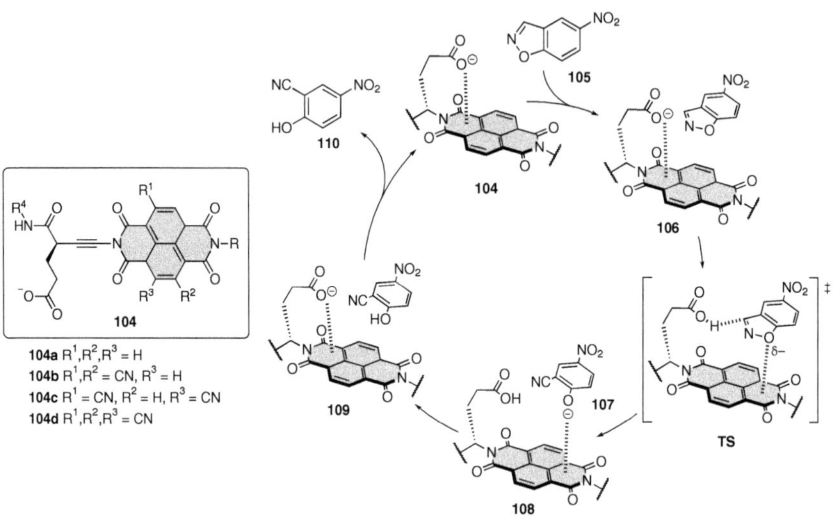

Scheme 11.24 Catalytic Kemp elimination. TS stabilized by anion–π interactions with the NDI surface. Source: Adapted from Ref. [66].

noted that anion–π interactions of NDIs decrease in the presence of a competing linked carboxylate, which could be attributed to an energy penalization caused by breaking the intern carboxylate–π surface within the catalyst. Thus, taking also into account the interaction between the attached carboxylate of the catalyst **104** with its π-acidic surface, slightly higher activation energies of the **TS** were determined. As a result, they propose new catalyst structures bearing additional nitrile groups (**104c** and **104d**) to change the electrostatic potential (ESP) and lower the energy of the TS by anion–π interactions (Scheme 11.24, left). In these structures, the increase of key anion–π stabilization responsible for catalysis was predicted significantly greater for the TS than the catalyst–substrate complex (CS), allowing for effective anion–π catalysis.

11.5.2 Anion–π Interactions in Enolate Chemistry

Some of the most important types of transformations in chemistry and biology dealing with anionic intermediates are based on enolates, which makes them a logical choice for developing further anion–π catalysis. Consequently, in 2014, the group of Matile designed a new NDI system **111** bearing two hydroxy moieties for the generation of an anchored malonate **112** by esterification with malonyl chloride to affect catalysis via enolate chemistry [68] (Scheme 11.25).

Hence, upon deprotonation with 1,1,3,3-tetramethylguanidine (TMG), the formed enolate intermediate **113** is then stabilized by the anion–π interactions with the NDI surface, accelerating the addition to Michael acceptors such as a cyclic α,β-unsaturated ketone or a nitroalkene (Scheme 11.25a) [68a]. Moreover, this chemistry could be extended to an anionic cascade reaction for the formation of coumarins **118** from salicylic aldehydes **116** by a Knoevenagel condensation – intramolecular transesterification sequence on the π-surface [68a, b]

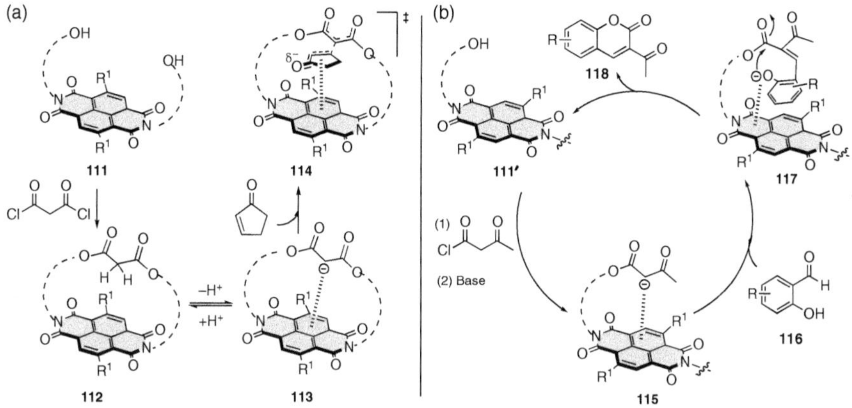

Scheme 11.25 Enolate stabilization on a π-acidic surface for accelerating nucleophilic addition reactions. Source: Adapted from Ref. [68].

(Scheme 11.25b). Several experiments, including ^{1}H-NMR titration and monitoring of the reaction, showed that the malonate on the π-acidic NDI (**115**) is more stable than the free enolate and the anion–π interactions are responsible for the stabilization of the reactive intermediate.

Besides the above studies with the malonate unit anchored to the catalyst structure, non-covalent systems have been next evolved. Hence, the addition of a malonic acid half thioester **119** to nitroolefins **120** has emerged as one of the most prominent benchmark reactions to investigate the anion–π catalytic properties of novel π-acid surface systems (Scheme 11.26) [69–72]. Both experimental and computational studies have confirmed that the binding of the charge-delocalized "enol-type" intermediate on π-acidic surfaces ($\mathbf{I_A}$) induces the formation of the valuable addition product **121**, while weaker anion–π interactions with the charge-localized "carboxylate" species ($\mathbf{I_B}$) provides mainly the undesirable decarboxylation product **122** [69] (Scheme 11.26a). Several π-acid systems ranking from mono [69a], dimer [69b], and π-stacked foldamer [69c] NDI structures **123** with a tertiary amine pedant substituent, to peptide stapling macrocyclic anion–π catalysts [69d], were efficiently implemented in this reaction. Moreover, it was proven the beneficiary effect of applying electric fields for assisting the transformation by immobilized anion–π catalysts on conductive In/TiO_2-surfaces [70], as it enhanced the polarization on the π-surface and increases its anionic intermediate and TS affinity, catalytic activity (>100-fold rate enhancement), and selectivity toward **121** (Scheme 11.26b).

The combination of anion–π catalysis and supramolecular chemistry have also been developed and tested in this model enolate addition reaction. Initially, catalytic active C_{60} fullerenes with an amine base group or a triad based on C_{60}–amine–polarizability enhancer such as another fullerene or an NDI unit were designed and investigated in catalysis [71] (Scheme 11.26c). Because of the induced π-acidity in fullerene–fullerene–amine triads **124**, these systems showed the highest selectivity, whereas NDI–fullerene–amine triads presented a similar activity as the simple fullerene-amine used as control. Later on, multi-walled carbon nanotubes

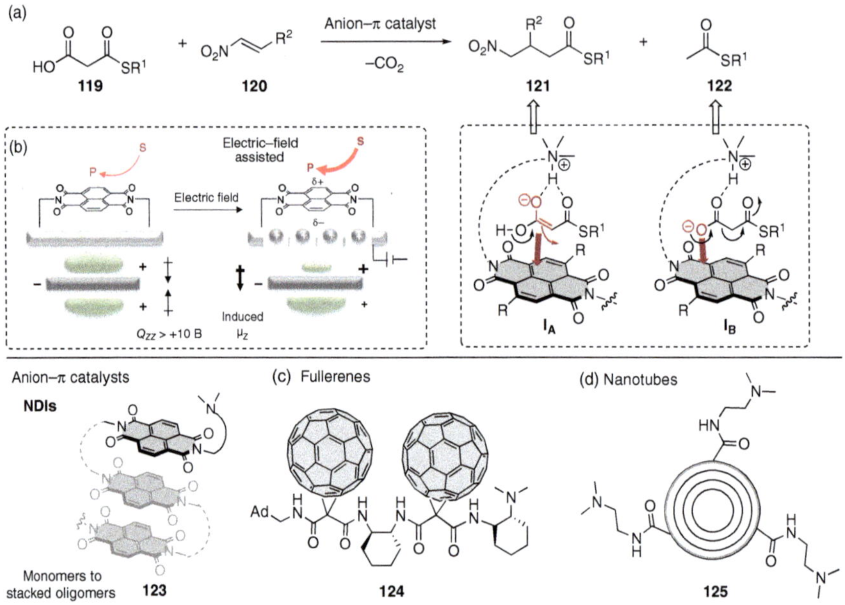

Scheme 11.26 Different approaches for the model anion–π-catalyzed nitro-Michael reaction with malonic half thioesters. Source: Adapted from Ref. [69].

(MWCNTs) **125** also modified with tertiary amine linkers were demonstrated as potent anion–π catalysts (Scheme 11.26d) [72]. Single-walled carbon nanotubes (SWCNTs) displayed lower activity than MWCNTs, indicating that the anion–π stabilization takes place not only through the polarizability along the tubes but also between the layers by anion–π interactions, which opens new opportunities for anion–π catalysis with carbon-based nanostructures.

11.5.3 Epoxide-Opening – Ether Cyclization Reactions

Epoxide ring-opening and subsequent ether cyclization sequential reactions are well-established processes in chemistry and biology, as for example, cascade cyclizations toward cyclic polyether complex building blocks and metabolites [73]. Following the Baldwin rules, five (tetrahydrofurane-type, THF) and six-membered rings (tetrahydropirane-type, THP) can then be selectively formed preferentially over an *exo-tet* TS. Contrary to classical cationic cyclizations, Matile and coworkers recently applied anion–π catalysis to stabilize the polarized TS of such transformations [74] (Scheme 11.27). Within this approach, they observed that the ring-opening cyclization proceeded on π-acidic aromatic surfaces such as of the simple hexafluorobenzene (C_6F_6, **101**) or the more reactive NDIs **129** though primary anion–π interactions, not requiring additional activating groups in the catalyst (Scheme 11.27a) [74a]. Moreover, this process showed an autocatalytic behavior, leading to the same products **127** or **128** derived from more classical procedures such as Brønsted acid catalysis.

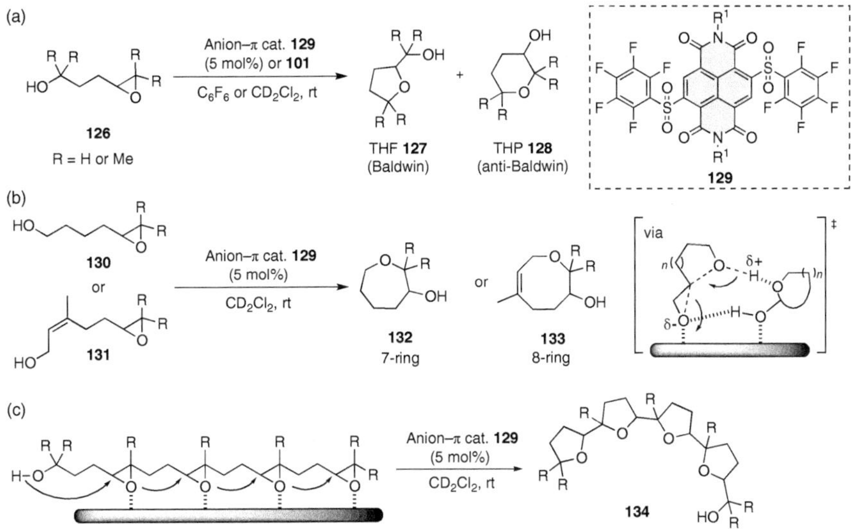

Scheme 11.27 Epoxide-opening – ether cyclization cascades realized by primary anion–π catalysis. Source: Adapted from Ref. [74].

Considering the potential of this new type of activation for entropically unfavored processes upon anion–π stabilization of the TS, the same group expanded this chemistry to the formation of larger seven- (**132**) and eight-membered rings (**133**), not accessible by common acid catalysis [74b]. This was successfully achieved with a strong π-acidic NDI **129** bearing electron-withdrawing sulfone groups with pentafluorophenyl substituents (Scheme 11.27b). Followed by this observation, this approach was recently used to mimic biosynthetic routes toward cascade cyclization reactions of polyethers [74c] (Scheme 11.27c). Hence, an anionic version of the well-known cation–π-catalyzed steroid-type cyclization was realized on π-acidic surfaces, giving access to complex, value-added synthetic Baldwin oligo-oxolane motifs **134**.

11.5.4 Enantioselective Anion–π Catalysis

Anion–π catalysis has also proven efficient to promote enantioselective transformations by introducing chiral information onto the π-acid system. In this regard, a series of multi-functional NDIs with a proline unit and a glutamic acid residue on the imide rests and sulfur groups on the aromatic core were designed for the nitro-Michael reaction with aldehydes through enamine catalysis [75] (Scheme 11.28). Although no anion was involved in the reaction, the anion–π interactions between the NDI surface and the polar nitro-group in the **TS-I** allowed for good diastereo- and enantioselectivities (up to 26 : 1 *d.r.* and 91% *ee*) with only 5 mol% of the thioether-substituted catalyst **137** in toluene. Moreover, the group of Matile also introduced new perylenediimides as a platform for chiral bifunctional anion–π catalysts. In contrast to the well-established NDIs, the surface and quadrupole moment of the PDIs are remarkably greater with respect to the pattern

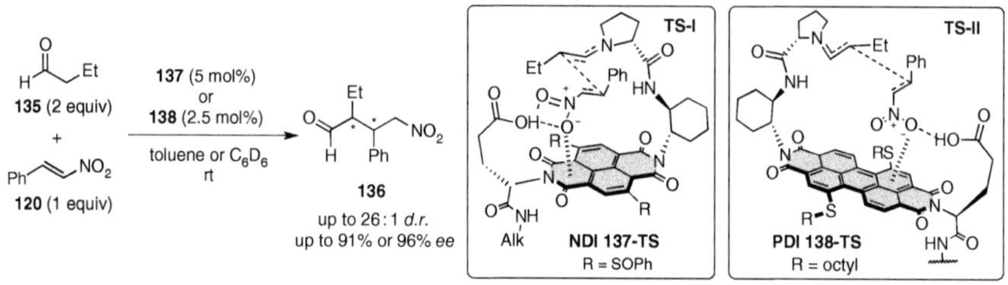

Scheme 11.28 Enantioselective enamine-type catalysis by TS stabilization on a NDI or PDI π-acidic surface. Source: (a) Based on Akamatsu and Matile [75]. (b) Based on Wang *et al.* [76]

NDI with similar substitution (**138**), leading to a more efficient fixation of the substrate and stabilization of the transition state (**TS-II**) (Scheme 11.28, right) [76]. Thus, the large surface of PDIs enabled decreasing the catalyst loading to 2.5 mol%, while giving rise the Michael adducts in an excellent high enantioselectivity up to 96% ee.

Alternatively, bifunctional NDI–amine catalysts promoted other asymmetric Michael reactions for the formation of two non-adjacent stereocenters [77] or the transamination of CF_3-substituted ketimines [78]; however, only a low or moderate stereocontrol was achieved (up to 22% *ee* and 37% *ee*, respectively). Further efforts toward enantioselective anion–π catalysis have then been underwent. In this regard, a series of tertiary alkylamine-modified NDIs and fullerenes were investigated in asymmetric Diels–Alder reactions (Scheme 11.29) [71a, 79]. While the fullerene π-systems **139** catalyzed the reaction of hydroxypyrone **140** with *N*-methyl maleimide with moderate both endo- and enantioselectivity (up to 1.8 : 1

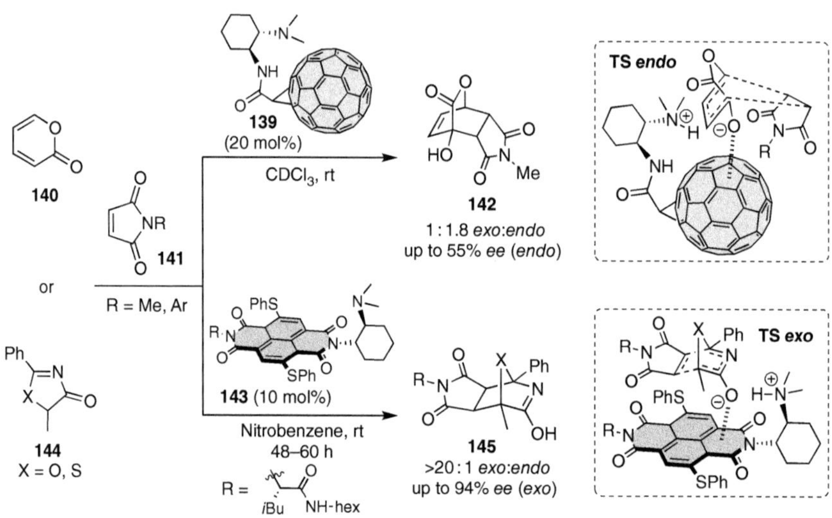

Scheme 11.29 Asymmetric anion–π-catalyzed Diels–Alder reactions. Source: Adapted from Refs. [71a, 79]

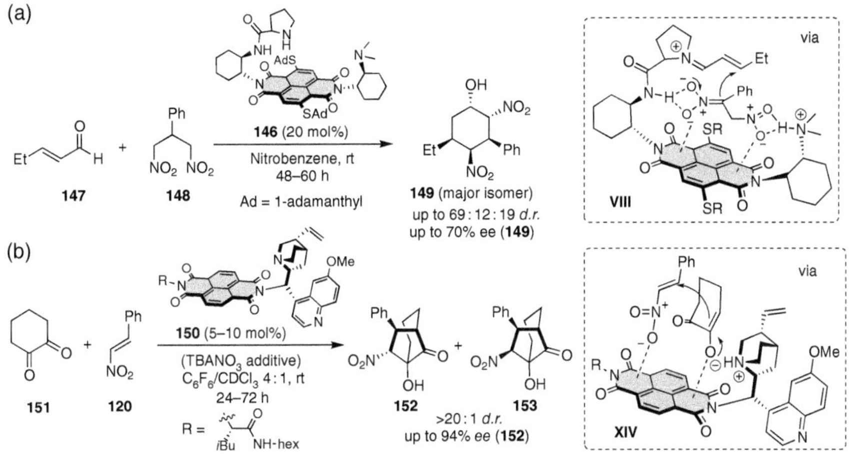

Scheme 11.30 Anion–π-catalyzed enantioselective cascade reactions toward multiple stereocenters. Source: (a) Based on Liu *et al.* [80a].

endo:*exo* and 55% *ee*) [71a], the NDI **143** promoted the cycloaddition of oxa- and thiazolones **144** with maleimides **141** in excellent *exo* selectivity (>20 : 1 *exo*:*endo*) and enantioinduction (up to 94% *ee*) [79].

More interestingly, the group of Matile also aimed at exploited anion–π interactions to stabilized charged dislocations over distance in cascade reactions involving anionic intermediates that can lead to complex molecules with multiple stereocenters in a single reaction setup [80]. In 2016, the first anion–π-catalyzed cascade reaction to form cyclohexane derivatives with five stereogenic centers was reported using a trifunctional NDI system **146** bearing a proline and a tertiary amine groups one at each *N*-imide end [80a] (Scheme 11.30a). This π-system affected an iminium/Henry sequence with the α,β-unsaturated aldehyde **147** and the dinitroalkane **148**, leading to the targeted cyclic products in moderate to good diastereo- and enantioselectivity (up to 70% *ee* for the major isomer **149**). To avoid or minimize charge repulsion with the cationic iminium species, the proline unit of the catalyst was placed far from the π-acidic surface, while favoring the π-acid stabilization of the active nitronate intermediates involved in the conjugated addition and subsequent Henry reaction.

Shortly after, a highly challenging synthesis of bicyclic products with four contiguous stereogenic centers using anion–π catalysis was also developed using a similar Michael–Henry cascade approach [80b]. In this case, the cascade reaction is initiated by the Michael addition of the conjugated enolate of cyclohexanedione **151** to a nitroolefin **120**, for which both species are coordinated with the π-acidic surface of **150** (Scheme 11.30b). The generated nitronate undergoes an intramolecular Henry reaction to form the second five-membered cycle. This Michael–Henry cascade reaction is an outstanding example for the application of anion–π catalysis, leading to good diastereoselectivities and remarkable high enantiomeric excesses (up to >20 : 1 *d.r.* and up to 94% *ee* of **152**).

11.5.5 Miscellaneous

The combination of photochemistry and anion–π interactions is a novel approach in this research field, which opens new opportunities and unique reaction pathways. In 2019, Frontera, Pericàs, and coworkers developed a visible-light-mediated method for the amidation of (hetero)arenes **155** with dinitrophenyl hydroxylamines **154** based on the stabilization of the transition state by primary anion–π interactions [81] (Scheme 11.31). Hence, this process is initiated by the coordination of carbonate anion over the π-acidic surface of the arylhydroxylamine (**157**), which weakens its N—O bond, and can then be easily cleaved to an amidyl radical **160** upon irradiation to the anion–π complex (to the excited species **157***) and single electron transfer to from **158**. The radical **160** then reacts with a (hetero)aromatic system, leading to the final product **156** after a SET with the *in situ* formed radical anion $CO_3^{\cdot-}$ and subsequent rearomatization by deprotonation. Although not catalytic on the π-acidic promoter employed, this example already shows the high potential of the combination of photo- and anion–π chemistry.

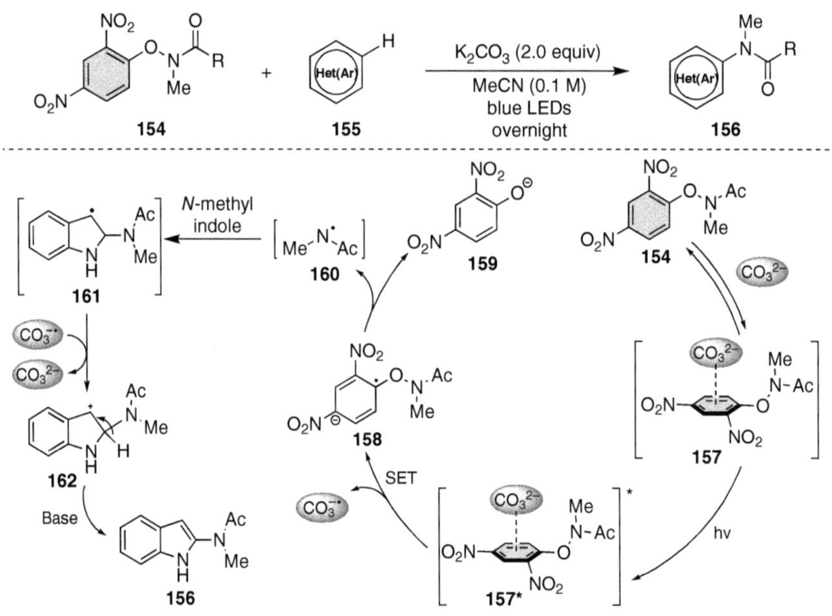

Scheme 11.31 Light-induced amidation of aromatic systems through anion–π interactions. Source: Based on Buglioni *et al.* [81].

11.6 Conclusion and Outlook

The design of new anion-binding receptors for their application in catalysis remains an emerging area that is attracting a considerable attention because it offers new possibilities in synthetic chemistry. In a relatively short period of time, inspiring novel applications have been reported, as it is illustrated in this chapter. In this regard, we have shown the first examples on the combination of anion-binding catalysis with photocatalysis, which provide great prospects in the achievement of enantioenriched products. Moreover, particular progress has been made in the use of anion-binding interactions in metal catalysis through the design of highly efficient bifunctional systems. As it has been detailed in this section, it is a promising area because of the excellent selectivity and reactivity achieved for a broad number of substrates beyond the more explored methods relied on solely organic catalysis. Concerning supramolecular anion-binding approaches, significant advances have been made by the design of some sophisticated structures and their elegant application in catalysis. Particularly interesting is the employment of switchable and responsive structures that can adapt their properties and catalytic activity upon applying different external stimuli. Finally, anion–π interactions have been lately introduced as the primary activation mode in the tool box of catalysis, allowing the stabilization of transition states and negatively charged intermediates through the catalytic cycle. Thus, anion–π catalysis on π-acidic surfaces of aromatic systems such as simple electron-deficient benzenes, naphthalenediimides, perylenediimides, or fullerenes has recently emerged, providing access to unconventional reactivities.

However, despite their great potential for future applications and innovative discoveries, the fields of photo-, metal-, and supramolecular anion-binding catalysis still present several challenges. For example, more accessible synthesis of the chiral supramolecular catalysts constitutes a continuous demand, while the control of the stereoselectivity by merging anion-binding catalysis with photocatalysis or homogeneous metal catalysis remains underdeveloped. Moreover, to solve some of the current catalytic activity issues, the forthcoming extension of hydrogen-bond to halogen- or chalcogen-bond-based receptor toward enhanced anion affinity can be envisioned. Hence, further efforts in these directions need to be undertaken, which will certainly lead to further important developments in the near future.

References

1 (a) Buzzetti, L., Crisenza, G.E.M., and Melchiorre, P. (2019). Mechanistic studies in photocatalysis. *Angew. Chem. Int. Ed.* 58: 3730–3747. (b) Romero, N.A. and Nicewicz, D.A. (2016). Organic photoredox catalysis. *Chem. Rev.* 116: 10075–10166. (c) Oelgemöller, M. (2016). Solar photochemical synthesis: from the beginnings of organic photochemistry to the solar manufacturing of commodity chemicals. *Chem. Rev.* 116: 9664–9682. (d) Bach, T. and Hehn, J.P. (2011). Photochemical reactions as key steps in natural product synthesis. *Angew. Chem. Int. Ed.* 50: 1000–1045. (e) Ravelli, D., Dondi, D., Fagnoni, M., and Albini, A. (2009). Photocatalysis. A multi-faceted concept for green chemistry. *Chem. Soc. Rev.* 38: 1999–2011. (f) Hoffmann, N. (2008). Photochemical reactions as key steps in organic synthesis. *Chem. Rev.* 108: 1052–1103. (g) Fagnoni, M., Dondi, D., Ravelli, D., and Albini, A. (2007). Photocatalysis for the formation of the C—C bond. *Chem. Rev.* 107: 2725–2756. (h) Serpone, N. and Pelizzetti, E. (1989). *Photocatalysis: Fundamentals and Applications*. New York: *Wiley*.

2 Eibner, A. (1911). Action of light on pigments I. *Chem-ZTG.* 35: 753–755.

3 Fujishima, A. and Honda, K. (1972). Electrochemical photolysis of water at a semiconductor electrode. *Nature* 238: 37–38.

4 Kärkäs, M.D., Porco, J.A., and Stephenson, C.R.J. (2016). Photochemical approaches to complex chemotypes: applications in natural product synthesis. *Chem. Rev.* 116: 9683–9747.

5 (a) Rigotti, T. and Alemán, J. (2020). Visible light photocatalysis – from racemic to asymmetric activation strategies. *Chem. Commun.* 56: 11169–11190. (b) Stephenson, C.R.J., Yoon, T.P., and MacMillan, D.W.C. (2018). *Visible Light Photocatalysis in Organic Chemistry*. Wiley-VCH Verlag GmbH&Co: Weinheim. (c) Marzo, L., Pagire, S.K., Reiser, O., and König, B. (2018). Visible-light photocatalysis: does it make a difference in organic synthesis? *Angew. Chem. Int. Ed.* 57: 10034–10072. (d) Chen, J.-R., Hu, X.-Q., Lu, L.-Q., and Xiao, W.-J. (2016). Exploration of visible-light photocatalysis in heterocycle synthesis and functionalization: reaction design and beyond. *Acc. Chem. Res.* 49: 1911–1923. (e) Prier, C., Rankic, D.A., and MacMillan, D.W.C. (2013). Visible light photoredox catalysis with transition metal complexes: applications in organic synthesis. *Chem. Rev.* 113: 5322–5363.

6 Skubi, K.L., Blum, T.R., and Yoon, T.P. (2016). Dual catalysis strategies in photochemical synthesis. *Chem. Rev.* 116: 10035–10074.

7 Nicewicz, D.A. and MacMillan, D.W.C. (2008). Merging photoredox catalysis with organocatalysis: the direct asymmetric alkylation of aldehydes. *Science* 322: 77–81.

8 (a) Silvi, M., Verrier, C., Rey, Y.P. *et al.* (2017). Visible-light excitation of iminium ions enables the enantioselective catalytic β-alkylation of enals. *Nat. Chem.* 9: 868. (b) Arceo, E., Jurberg, I.D., Álvarez-Fernández, A., and Melchiorre, P. (2013). Photochemical activity of a key donor–acceptor complex can drive stereoselective catalytic α-alkylation of aldehydes. *Nat. Chem.* 5: 750.

9 DiRocco, D.A. and Rovis, T. (2012). Catalytic asymmetric acylation of tertiary amines mediated by a dual catalysis mode: *N*-heterocyclic carbene and photoredox catalysis. *J. Am. Chem. Soc.* 134: 8094–8097.

10 (a) Alonso, R. and Bach, T. (2014). A chiral thioxanthone as an organocatalyst for enantioselective [2+2] photocycloaddition reactions induced by visible light. *Angew. Chem. Int. Ed.* 53: 4368–4371. (b) Bauer, A., Westkamper, F., Grimme, S., and Bach, T. (2005). Catalytic enantioselective reactions driven by photoinduced electron transfer. *Nature* 436: 1139–1140.

11 Liu, Y., Li, J., Ye, X., and Jiang, Z. (2016). Organocatalytic asymmetric formal arylation of benzofuran-2(3*H*)-ones with cooperative visible light photocatalysis. *Chem. Commun.* 52: 13955–13958.

12 (a) Wei, G., Zhang, C., Bures, F. *et al.* (2016). Enantioselective aerobic oxidative C(sp^3)–H olefination of amines via cooperative photoredox and asymmetric catalysis. *ACS Catal.* 6: 3708–3712. (b) Rono, L.J., Yaila, H.G., Wang, D.Y. *et al.* (2013). Enantioselective photoredox catalysis enabled by proton-coupled electron transfer: development of an asymmetric aza-pinacol cyclization. *J. Am. Chem. Soc.* 135: 17735–17738. (c) Shin, N.Y., Ryss, J.M., Zhang, X. *et al.* (2019).

Light-driven deracemization enabled by excited-state electron transfer. *Science* 366: 364–369.

13 Uraguchi, D., Kinoshita, N., Kizu, T., and Ooi, T. (2015). Synergistic catalysis of ionic Brønsted acid and photosensitizer for a redox neutral asymmetric α-coupling of *N*-arylaminomethanes with aldimines. *J. Am. Chem. Soc.* 137: 13768–13771.

14 Morse, P.D., Nguyen, T.M., Cruz, C.L., and Nicewicz, D.A. (2018). Enantioselective counter-anions in photoredox catalysis: the asymmetric cation radical Diels–Alder reaction. *Tetrahedron* 26: 3266–3272.

15 (a) Jiang, C., Chen, W., Zheng, W.-H., and Lu, H. (2019). Advances in asymmetric visible-light photocatalysis, 2015–2019. *Org. Biomol. Chem.* 17: 8673–8689. (b) Wang, C. and Lu, Z. (2015). Catalytic enantioselective organic transformations via visible light photocatalysis. *Org. Chem. Front.* 2: 179–190. (c) Meggers, E. (2015). Asymmetric catalysis activated by visible light. *Chem. Commun.* 51: 3290–3301. (d) Brimioulle, R., Lenhart, D., Maturi, M.M., and Bach, T. (2015). Enantioselective catalysis of photochemical reactions. *Angew. Chem. Int. Ed.* 54: 3872–3890. (e) Müller, C. and Bach, T. (2008). Chirality control in photochemical reactions: enantioselective formation of complex photoproducts in solution. *Aust. J. Chem.* 61: 557. (f) Wessig, P. (2006). Organocatalytic enantioselective photoreactions. *Angew. Chem. Int. Ed.* 45: 2168–2171.

16 (a) Córdova, A., Sudén, H., Engqvist, M. *et al.* (2004). The direct amino acid-catalyzed asymmetric incorporation of molecular oxygen to organic compounds. *J. Am. Chem. Soc.* 126: 8914–8915. (b) Nagib, D.A., Scott, M.E., and MacMillan, D.W.C. (2009). Enantioselective α-trifluoromethylation of aldehydes via photoredox organocatalysis. *J. Am. Chem. Soc.* 131: 10875–10877. (c) Melchiorre, P. (2009). Light in aminocatalysis: the asymmetric intermolecular α-alkylation of aldehydes. *Angew. Chem. Int. Ed.* 48: 1360–1363. (d) Shih, H., Vander Wal, M.N., Grange, R.L., and MacMillan, D.W.C. (2010). Enantioselective α-benzylation of aldehydes via photoredox organocatalysis. *J. Am. Chem. Soc.* 132: 13600–13603. (e) Allen, A.E. and MacMillan, D.W.C. (2010). The productive merger of iodonium salts and organocatalysis: a non-photolytic approach to the enantioselective α-trifluoromethylation of aldehydes. *J. Am. Chem. Soc.* 132: 4986–4987. (f) Cecere, G., König, C.M., Alleva, J.L., and MacMillan, D.W.C. (2013). Enantioselective direct α-amination of aldehydes via a photoredox mechanism: a strategy for asymmetric amine fragment coupling. *J. Am. Chem. Soc.* 135: 11521–11524. (g) Welin, E.R., Warkentin, A.A., Conrad, J.C., and MacMillan, D.W.C. (2015). Enantioselective α-alkylation of aldehydes by photoredox organocatalysis: rapid access to pharmacophore fragments from β-cyanoaldehydes. *Angew. Chem. Int. Ed.* 54: 9668–9672.

17 Murphy, J.J., Bastida, D., Paria, S. *et al.* (2016). Asymmetric catalytic formation of quaternary carbons by iminium trapping of radicals. *Nature* 532: 218–222.

18 Bergonzini, G., Schindler, C.S., Wallentin, C.-J. *et al.* (2014). Photoredox activation and anion binding catalysis in the dual catalytic enantioselective synthesis of β-amino esters. *Chem. Sci.* 5: 112–116.

19 Farney, E.P., Chapman, S.J., Swords, W.B. *et al.* (2019). Discovery and elucidation of counteranion dependence in photoredox catalysis. *J. Am. Chem. Soc.* 141: 6385–6391.

20 Shida, N., Imada, Y., Nagahara, S. *et al.* (2019). Interplay of arene radical cations with anions and fluorinated alcohols in hole catalysis. *Commun. Chem.* 2: 24.

21 Lin, L., Bai, X., Ye, X. *et al.* (2017). Organocatalytic enantioselective protonation for photoreduction of activated ketones and ketimines induced by visible light. *Angew. Chem. Int. Ed.* 56: 13842–13846.

22 Uraguchi, D., Kimura, Y., Ueoka, F., and Ooi, T. (2020). Urea as a redox-active directing group under asymmetric photocatalysis of iridium-chiral borate ion pairs. *J. Am. Chem. Soc.* 46: 19462–19467.

23 (a) Zhao, Q., Li, S., Huang, K. *et al.* (2013). A novel chiral bisphosphine-thiourea ligand for asymmetric hydrogenation of β,β-disubstituted nitroalkenes. *Org. Lett.* 15: 4014–4017. (b) Zhao, Q., Chen, C., Wen, J. *et al.* (2020). Noncovalent interaction-assisted ferrocenyl phosphine ligands in asymmetric catalysis. *Acc. Chem. Res.* 53: 1905–1921.

24 Zhao, Q., Wen, J., Tan, R. *et al.* (2014). Rhodium-catalyzed asymmetric hydrogenation of unprotected NH imines assisted by a thiourea. *Angew. Chem. Int. Ed.* 53: 8467–8470.

25 Wen, J., Tan, R., Liu, S. *et al.* (2016). Strong Brønsted acid promoted asymmetric hydrogenation of isoquinolines and quinolines catalyzed by a Rh–thiourea chiral phosphine complex via anion binding. *Chem. Sci.* 7: 3047–3051.

26 Wen, J., Fan, X., Tan, R. *et al.* (2018). Brønsted-acid-promoted Rh-catalyzed asymmetric hydrogenation of *N*-unprotected indoles: a cocatalysis of transition metal and anion binding. *Org. Lett.* 20: 2143–2147.

27 Yang, T., Yin, Q., Gu, G., and Zhang, X. (2018). A one-pot process for the enantioselective synthesis of tetrahydroquinolines and tetrahydroisoquinolines via asymmetric reductive amination (ARA). *Chem. Commun.* 54: 7247–7250.

28 (a) Han, Z., Liu, G., Wang, R. *et al.* (2019). Highly efficient Ir-catalyzed asymmetric hydrogenation of benzoxazinones and derivatives with a Brønsted acid cocatalyst. *Chem. Sci.* 10: 4328–4333. (b) Yang, T., Sun, Y., Wang, H. *et al.* (2020). Iridium-catalyzed enantioselective hydrogenation of oxocarbenium ions: a case of ionic hydrogenation. *Angew. Chem. Int. Ed.* 59: 6108–6114.

29 Šmejkal, T. and Breit, B. (2008). A supramolecular catalyst for the decarboxylative hydroformylation of α,β-unsaturated carboxylic acids. *Angew. Chem. Int. Ed.* 47: 3946–3949.

30 (a) Šmejkal, T. and Breit, B. (2008). A supramolecular catalyst for regioselective hydroformylation of unsaturated carboxylic acids. *Angew. Chem. Int. Ed.* 47: 311–315. (b) Šmejkal, T., Gribkov, D., Geier, J. *et al.* (2010). Transition-state stabilization by a secondary substrate–ligand interaction:a new design principle for highly efficient transition-metal catalysis. *Chem. Eur. J.* 16: 2470–2478.

31 Fang, W. and Breit, B. (2018). Tandem regioselective hydroformylation–hydrogenation of internal alkynes using a supramolecular catalyst. *Angew. Chem. Int. Ed.* 57: 14817–14821.

32 Dydio, P., Detz, R.J., de Bruin, B., and Reek, J.N.H. (2014). Beyond classical reactivity patterns: hydroformylation of vinyl and allyl arenes to valuable β- and γ-aldehyde intermediates using supramolecular catalysis. *J. Am. Chem. Soc.* 136: 8418–8429.

33 (a) Dydio, P., Dzik, W.I., Lutz, M. *et al.* (2011). Remote supramolecular control of catalyst selectivity in the hydroformylation of alkenes. *Angew. Chem. Int. Ed.* 50: 396–400. (b) Dydio, P., Detz, R.J., and Reek, J.N.H. (2013). Precise supramolecular control of selectivity in the Rh-catalyzed hydroformylation of terminal and internal alkenes. *J. Am. Chem. Soc.* 135: 10817–10828.

34 Zhao, C. and Seidel, D. (2015). Enantioselective A^3 reactions of secondary amines with a Cu(I)/acid–thiourea catalyst combination. *J. Am. Chem. Soc.* 137: 4650–4653.

35 Théveau, L., Bellini, R., Dydio, P. *et al.* (2016). Cofactor-controlled chirality of tropoisomeric ligand. *Organometallics* 35: 1956–1963.

36 Bhattarai, B., Tay, J.H., and Nagorny, P. (2015). Thiophosphoramides as cooperative catalysts for copper-catalyzed arylation of carboxylates with diaryliodonium salts. *Chem. Commun.* 51: 5398–5401.

37 (a) Gale, P.A., Davis, J.T., and Quesada, R. (2017). Anion transport and supramolecular medicinal chemistry. *Chem. Soc. Rev.* 46: 2497–2519. (b) Beer, P.D. and Gale, P.A. (2001). Anion recognition and sensing: the state of the art and future perspectives. *Angew. Chem. Int. Ed.* 40: 486–516.

38 (a) Chen, L., Berry, S.N., Wu, X. *et al.* (2020). Advances in anion receptor chemistry. *Chem.* 6: 61–141. (b) Gale, P.A., Howe, E.N.W., and Lu, X. (2016). Anion receptor chemistry. *Chem.* 1: 351–422. (c) Jia, C., Zuo, W., Zhang, D. *et al.* (2016). Anion recognition by oligo-(thio)urea-based receptors. *Chem. Commun.* 52: 9614–9627. (d) Busschaert, N., Caltagirone, C., Van Rossom, W., and Gale, P.A. (2015). Applications of supramolecular anion recognition. *Chem. Rev.* 115: 8038–8155. (e) Evans, N.H. and Beer, P.D. (2014). Advances in anion supramolecular chemistry: from recognition to chemical applications. *Angew. Chem. Int. Ed.* 53: 11716–11754. (f) Custelcean, R. (2010). Anions in Crystal Engineering. *Chem. Soc. Rev.* 39: 3675–3685. (g) Amendola, V., Fabbrizzi, L., and Mosca, L. (2010). Anion recognition by hydrogen bonding: urea-based receptors. *Chem. Soc. Rev.* 39: 3889–3915. (h) Lankshear, M.D. and Beer, P.D. (2007). Interweaving anion templation. *Acc. Chem. Res.* 40: 657–668. (i) Gale, P.A. (2001). Anion receptor chemistry: highlights from 1999. *Coord. Chem. Rev.* 213: 79–128.

39 (a) Bruns, C.J. and Stoddart, J.F. (2016). *The Nature of the Mechanical Bond: From Molecules to Machines*. Hoboken, NJ: Wiley and Sons. (b) Dietrich-Buchecker, C. and Sauvage, J.P. (1999). *Molecular Catenanes, Rotaxanes and Knots: A Journey Through the World of Molecular Topology*. Weinheim: Wiley-VCH.

40 (a) Denis, M. and Goldup, S.M. (2017). The active template approach to interlocked molecules. *Nat. Rev. Chem.* 1: 0061. (b) Van Dongen, S.F.M., Cantekin, S., Elemans, J.A.A.W. *et al.* (2014). Functional interlocked systems. *Chem. Soc. Rew.* 43: 99–122. (c) Spence, G.T. and Beer, P.D. (2013). Expanding the scope of the anion templated synthesis of interlocked structures. *Acc. Chem. Res.* 46: 571–586.

41 (a) Bak, K.M., Porfyrakis, K., Davis, J.J., and Beer, P.D. (2020). Exploiting the mechanical bond for molecular recognition and sensing of charged species. *Mater. Chem. Front.* 4: 1052–1073. (b) Pairault, N. and Niemeyer, J. (2018). Chiral mechanically interlocked molecules – applications of rotaxanes, catenanes and molecular knots in stereoselective chemosensing and catalysis. *Synlett* 29: 689–698.

42 Martinez-Cuevas, A., Saura-Sanmartin, A., Alajarin, M., and Berna, J. (2020). Mechanically interlocked catalysts for asymmetric synthesis. *ACS Catal.* 10: 7719–7733.

43 Fielden, S.D.P., Leigh, A.D., and Woltering, S.L. (2017). Molecular knots. *Angew. Chem. Int. Ed.* 56: 11166–11194.

44 Marcos, V., Stephens, A.J., Jaramillo-Garcia, J. *et al.* (2016). Allosteric initiation and regulation of catalysis with a molecular knot. *Science* 352: 1555–1559.

45 Prakasam, T., Devaraj, A., Saha, R. *et al.* (2019). Metal-organic self-assembled trefoil knots for C—Br bond activation. *ACS Catal.* 9: 1907–1914.

46 Evans, N.H. (2017). Chiral catenanes and rotaxanes: fundamentals and emerging applications. *Chem. Eur. J.* 24: 3101–3112.

47 Kwamen, C. and Niemeyer, J. (2020). Functional rotaxanes in catalysis. *Chem. Eur. J.* 27: 175–186.

48 Eichstaedt, K., Jaramillo-Garcia, J., Leigh, D.A. *et al.* (2017). Switching between anion-binding catalysis and aminocatalysis with a rotaxane dual-function catalyst. *J. Am. Chem. Soc.* 139: 9376–9381.

49 (a) Dorel, R. and Feringa, B.L. (2019). Photoswitchable catalysis based on the isomerisation of double bonds. *Chem. Commun.* 55: 6477–6486. (b) Kassem, S., van Leeuwen, T., Lubbe, A.S. *et al.* (2017). Artificial molecular motors. *Chem. Soc. Rev.* 46: 2592–2621.

50 (a) Sauvage, J.P. (2017). From chemical topology to molecular machines. *Angew. Chem. Int. Ed.* 56: 11080–11093. (b) Stoddart, J.F. (2017). Mechanically interlocked molecules (MIMs) molecular shuttles, switches, and machines. *Angew. Chem. Int. Ed.* 56: 11094–11125. (c) Feringa, B.L. (2017). The art of building small: from molecular switches to motors. *Angew. Chem. Int. Ed.* 56: 11060–11078.

51 Dorel, R. and Feringa, B.L. (2020). Stereodivergent anion binding catalysis with molecular motors. *Angew. Chem. Int. Ed.* 59: 785–789.

52 Kauerhof, F. and Niemeyer, J. (2020). Functionalized macrocycles in supramolecular organocatalysis. *ChemPlusChem.* 85: 889–899.

53 Park, Y., Harper, K.C., Kuhl, N. *et al.* (2017). Macrocyclic bis-thioureas catalyze stereospecific glycosylation reactions. *Science* 355: 162–166.

54 Ning, R., Ao, Y.-F., Wang, D.-X., and Wang, Q.-Q. (2018). Macrocycle-enabled counternanion trapping for improved catalytic efficiency. *Chem. Eur. J.* 24: 4268–4272.

55 Ning, R., Zhou, H., Nie, S.-X. *et al.* (2018). Chiral macrocycle-enabled counteranion trapping for boosting highly efficient and enantioselective catalysis. *Angew. Chem. Int. Ed.* 59: 10894–10898.

56 Kim, S.K. and Sessler, J.L. (2010). Ion pair receptors. *Chem. Soc. Rev.* 39: 3784–3809.

57 Kang, K., Lohrman, J.A., Nagarajan, S. *et al.* (2019). Convergent ditopic receptors enhance anion binding upon alkali metal complexation for catalyzing the ritter reaction. *Org. Lett.* 21: 652–655.

58 (a) Mahadevi, A.S. and Sastry, G.N. (2013). Cation–π interaction: its role and relevance in chemistry, biology, and material science. *Chem. Rev.* 113: 2100–2138. (b) Schottel, B.L., Chifotides, H.T., and Dunbar, K.R. (2008). Anion-π interactions. *Chem. Soc. Rev.* 37: 68–83.

59 Garau, C., Quiñonero, D., Frontera, A. *et al.* (2003). Anion–π interactions: must the aromatic ring be electron deficient? *New J. Chem.* 27: 211–214.

60 (a) Frontera, A., Gamez, P., Mascal, M. *et al.* (2011). Putting anion–π interactions into perspective. *Angew. Chem. Int. Ed.* 50: 9564–9583. (b) Rather, I.A., Wagay, S.A., and Ali, R. (2020). Emergence of anion–π interactions: the land of opportunity in supramolecular chemistry and beyond. *Coord. Chem. Rev.* 415: 213327.

61 (a) Williams, J.H. (1993). The molecular electric quadrupole moment and solid-state architecture. *Acc. Chem. Res.* 26: 593–598. (b) Giese, M., Albrecht, M., and Rissanen, K. (2015). Anion–π interactions with fluoroarenes. *Chem. Rev.* 115: 8867–8895.

62 Al Kobaisi, M., Bhosale, S.V., Latham, K. *et al.* (2016). Functional naphthalene diimides: synthesis, properties, and applications. *Chem. Rev.* 116: 11685–11796.

63 Würthner, F., Saha-Möller, C.R., Fimmel, B. *et al.* (2016). Perylene bisimide dye assemblies as archetype functional supramolecular materials. *Chem. Rev.* 116: 962–1052.

64 Neel, A.J., Hilton, M.J., Sigman, M.S., and Toste, F.D. (2017). Exploiting non-covalent π-interactions for catalyst design. *Nature* 543: 637–646.

65 (a) Bauzá, A., Deyà, P., and Frontera, A. (2015). Anion–π interactions in supramolecular chemistry and catalysis. In: *Noncovalent Forces, Challenges and Advances in Computational Chemistry and Physics*, vol. 19 (ed. S. Scheiner), 471–500. Cham: Springer. (b) Frontera, A. and Ballester, P. (2017). Anion–π interactions: theoretical studies, supramolecular chemistry and catalysis. In: *Aromatic Interactions: Frontiers in Knowledge and Application*, vol. 3 (eds. D.W. Johnson and F. Hof), 39–37. Royal Society of Chemistry. (c) Kan, X., Liu, H., Pan, Q. *et al.* (2018). Anion–π interactions: from concept to application. *Chin. Chem. Lett.* 29: 261–266. (d) Bornhof, A., Akamatsu, M., Zhao, Y. *et al.* (2018). The emergence of anion–π catalysis. *Acc. Chem. Res.* 51: 2255–2263.

66 (a) Zhao, Y., Domoto, Y., Orentas, E. *et al.* (2013). Catalysis with anion–π interactions. *Angew. Chem. Int. Ed.* 52: 9940–9943. (b) Domoto, Y., Gajewy, J., Wilson, A. *et al.* (2014). Anion–π catalysis. *J. Am. Chem. Soc.* 136: 2101–2111.

67 Lu, T. and Wheeler, S.E. (2014). Quantifying the role of anion–π interactions in anion–π catalysis. *Org. Lett.* 16: 3268–3271.

68 (a) Zhao, Y., Sakai, N., and Matile, S. (2014). Enolate chemistry with anion–π interactions. *Nat. Commun.* 5: 3911. (b) Miros, F.N., Huang, G., Zhao, Y. *et al.* (2015). Coumarin synthesis on π-acidic surfaces. *Supramol. Chem.* 27: 303–309.

(c) Miros, F.N., Zhao, Y., Sargsyan, G. *et al.* (2016). Enolate stabilization by anion–π interactions: deuterium exchange in malonate dilactones on π-acid surfaces. *Chem. Eur. J.* 22: 2648–2657.

69 (a) Cotelle, Y., Benz, S., Avestro, A.-J. *et al.* (2016). Anion–π catalysis of enolate chemistry: rigidified leonard turns as a general motif to run reactions on aromatic surfaces. *Angew. Chem. Int. Ed.* 55: 4275–4279. (b) Zhao, Y., Benz, S., Sakai, N., and Matile, S. (2015). Selective acceleration of disfavored enolate addition reactions by anion–π interactions. *Chem. Sci.* 6: 6219–6223. (c) Bornhof, A.-B., Bauzá, A., Aster, A. *et al.* (2018). Synergistic anion–$(\pi)_n$–π catalysis on π-stacked foldamers. *J. Am. Chem. Soc.* 140: 4884–4892. (d) Pham, A.-T. and Matile, S. (2020). Peptide stapling with anion-π catalysts. *Chem. Asian J.* 15: 1562–1566.

70 Akamatsu, M., Sakai, N., and Matile, S. (2017). Electric-field-assisted anion–π catalysis. *J. Am. Chem. Soc.* 139: 6558–6561.

71 (a) Loṕx01E55;ez-Andarias, J., Frontera, A., and Matile, S. (2017). Anion–π catalysis on fullerenes. *J. Am. Chem. Soc.* 139: 13296–13299. (b) López-Andarias, J., Bauzá, A., Sakai, N. *et al.* (2018). Remote control of anion–π catalysis on fullerene-centered catalytic triads. *Angew. Chem. Int. Ed.* 57: 10883–10887.

72 Bornhof, A.-B., Vázquez-Nakagawa, M., Rodríguez-Pérez, L. *et al.* (2019). Anion-π catalysis on carbon nanotubes. *Angew. Chem. Int. Ed.* 58: 16097–16100.

73 Vilotijevic, I. and Jamison, T.F. (2009). Epoxide-opening cascades in the synthesis of polycyclic polyether natural products. *Angew. Chem. Int. Ed.* 48: 5250–5281.

74 (a) Zhang, X., Hao, X., Liu, L. *et al.* (2018). Primary anion–π catalysis and autocatalysis. *J. Am. Chem. Soc.* 140: 17867–17871. (b) Paraja, M. and Matile, S. (2020). Primary anion–π catalysis of epoxide-opening ether cyclization into rings of different sizes: access to new reactivity. *Angew. Chem. Int. Ed.* 59: 6273–6277. (c) Paraja, M., Hao, X., and Matile, S. (2020). Polyether natural product inspired cascade cyclizations: autocatalysis on π-acidic aromatic surfaces. *Angew. Chem. Int. Ed.* 59: 15093–15097.

75 Akamatsu, M. and Matile, S. (2016). Expanded chiral surfaces for asymmetric anion–π catalysis. *Synlett* 27: 1041–1046.

76 Wang, C., Miros, F.N., Mareda, J. *et al.* (2016). Asymmetric anion–π catalysis on perylenediimides. *Angew. Chem. Int. Ed.* 55: 14422–14426.

77 Zhang, X., Liu, L., López-Andarias, J. *et al.* (2018). Anion-π catalysis: focus on nonadjacent stereocenters. *Helv. Chim. Acta* 101: e1700288.

78 Liu, L. and Matile, S. (2017). Anion-π transaminase mimics. *Supramol. Chem.* 29: 702–706.

79 Liu, L., Cotelle, Y., Bornhof, A.-B. *et al.* (2017). Anion–π catalysis of Diels–Alder reactions. *Angew. Chem. Int. Ed.* 56: 13066–13069.

80 (a) Liu, L., Cotelle, Y., Avestro, A.-J. *et al.* (2016). Asymmetric anion–π catalysis of iminium/nitroaldol cascades to form cyclohexane rings with five stereogenic centers directly on π-acidic surfaces. *J. Am. Chem. Soc.* 138: 7876–7879. (b) Liu, L., Cotelle, Y., Klehr, J. *et al.* (2017). Anion–π catalysis: bicyclic products with four contiguous stereogenic centers from otherwise elusive diastereospecific domino reactions on π-acidic surfaces. *Chem. Sci.* 8: 3770–3774.

81 Buglioni, L., Mastandrea, M.M., Frontera, A., and Pericàs, M.A. (2019). Anion–π interactions in light-induced reactions: role in the amidation of (hetero)aromatic systems with activated *N*-aryloxyamides. *Chem. Eur. J.* 25: 11785–11790.

Index

Anion-Binding Catalysis, First Edition. Edited by Olga García Mancheño.

b

C

u